D1702804

BIOENERGETICS

New Comprehensive Biochemistry

Volume 9

General Editors

A. NEUBERGER
London

L.L.M. van DEENEN
Utrecht

ELSEVIER
AMSTERDAM · NEW YORK · OXFORD

BIOENERGETICS

Editor

L. ERNSTER

Stockholm

1984

ELSEVIER
AMSTERDAM · NEW YORK · OXFORD

© 1984 Elsevier Science Publishers B.V.

All rights reserved. No part of this publication may be reproduced, stored in a retrieval system or transmitted in any form or by any means, electronic, mechanical, photocopying, recording or otherwise without the prior written permission of the publisher, Elsevier Science Publishers B.V./Biomedical Division, P.O. Box 211, 1000 AE Amsterdam, The Netherlands.
Special regulations for readers in the USA: this publication has been registered with the Copyright Clearance Center Inc. (CCC), Salem, Massachusetts. Information can be obtained from the CCC about conditions under which photocopying of parts of this publication may be made in the USA. All other copyright questions, including photocopying outside of the USA, should be referred to the publisher.

ISBN for the series: 0-444-80303-3
ISBN for the volume: 0-444-80579-6

Published by:
Elsevier Science Publishers B.V.
P.O. Box 211
1000 AE Amsterdam
The Netherlands

Sole distributors for the USA and Canada:
Elsevier Science Publishing Company, Inc.
52 Vanderbilt Avenue
New York, NY 10017
USA

Library of Congress Cataloging in Publication Data
Main entry under title:

Bioenergetics.

(New comprehensive biochemistry; v. 9)
Includes bibliographies and index.
1. Bioenergetics. 2. Energy metabolism.
I. Ernster, L. II. Series.
QD415.N48 vol. 9 574.19'2 s [574.19'283] 84-21273
[QH510]
ISBN 0-444-80579-6

Printed in The Netherlands

Introduction

"Research is to see what everybody has seen and think what nobody has thought"

Albert Szent-Györgyi: Bioenergetics
(Academic Press, New York, 1957)

Bioenergetics is the study of energy transformations in living matter. It is now well established that the cell is the smallest biological entity capable of handling energy. Every living cell has the ability, by means of suitable catalysts, to derive energy from its environment, to convert it into a biologically useful form, and to utilize it for driving life processes that require energy. In recent years, research in bioenergetics has increasingly been focused on the first two of these three aspects, i.e., the reactions involved in the capture and conversion of energy by living cells, in particular those taking place in the energy-transducing membranes of mitochondria, chloroplasts and bacteria. This area, often referred to as *membrane bioenergetics*, is also the main topic of the present volume. This trend is, however, relatively new; for example, it was not reflected in the contents of the previous volume on Bioenergetics in this series that appeared in 1967. As an introduction to the chapters that follow it appears appropriate, therefore, to give a brief historical background of these new developments. For details, the reader is referred to the large number of historical reviews on bioenergetics that have appeared over the past years, a selection of which is listed after this introduction.

Bioenergetics as a scientific discipline began a little over 200 years ago, with the discovery of oxygen. Priestley's classical observation that green plants produce and animals consume oxygen, and Lavoisier's demonstration that oxygen consumption by animals leads to heat production, are generally regarded as the first scientific experiments in bioenergetics. At about the same time Scheele, who discovered oxygen independently of Priestley, isolated the first organic compounds from living organisms. These developments, together with the subsequent discovery by Ingen-Housz, Senebier and de Saussure that green plants under the influence of sunlight take up carbon dioxide from the atmosphere in exchange for oxygen and convert it into organic material, played an important role in the development of concepts leading to the enunciation of the First Law of Thermodynamics by Mayer in 1842.

A recurrent theme in the history of bioenergetics is *vitalism*, i.e., the reference to 'vital forces', beyond the reach of physics and chemistry, to explain the mechanism of life processes. For about half a century following Scheele's first isolation of organic material from animals and plants it was believed that these compounds, which all contained carbon, could only be formed by living organisms — hence the name *organic* — a view which, however, was not shared by some chemists, e.g., Liebig and Wöhler. Indeed, in 1828 Wöhler succeeded for the first time in synthesizing an organic compound, urea, in the laboratory. This breakthrough was soon followed by other organic syntheses. Thus, the concept that only living organisms can produce organic compounds could not be maintained.

At the same time, however, it became increasingly evident that living organisms could produce these compounds better, more rapidly and with greater specificity, than could the chemist in his test tube. The idea, first proposed by Berzelius in 1835, that living organisms contained catalysts for carrying out their reactions, received increasing experimental support. Especially the work of Pasteur in the 1860s on fermentation by brewer's yeast provided firm experimental basis for the concept of biocatalysis. Pasteur's work was also fundamental in showing that fermentation was regulated by the accessibility of oxygen — the 'Pasteur effect' — which was the first demonstration of the regulation of energy metabolism in a living organism. In attempting to explain this phenomenon Pasteur was strongly influenced by the cell theory developed in the 1830s by Schleiden and Schwann, according to which the cell is the common unit of life in plants and animals. Pasteur postulated that fermentation by yeast required, in addition to a complement of active catalysts — 'ferments' — also a *force vitale* that was provided by, and dependent on, an intact cell structure. This 'vitalistic' view was again strongly opposed by Liebig, who maintained that it should be possible to obtain fermentation in a cell-free system. This indeed was achieved in 1897 by Büchner, using a press-juice of yeast cells.

In the early 1900s important progress was made toward the understanding of the role of phosphate in cellular energy metabolism. Following Büchner's demonstration of cell-free fermentation, Harden and Young discovered that this process required the presence of inorganic phosphate and a soluble, heat-stable cofactor which they called cozymase (later identified as the coenzyme nicotinamide adenine dinucleotide). These discoveries opened the way to the elucidation of the individual enzyme reactions and intermediates of glycolysis. The identification of various sugar phosphates by Harden and Young, Robison, Neuberg, Embden, Meyerhof and others, and the clarification of the role of cozymase in the oxidation of 3-phosphoglyceraldehyde by Warburg are the most important landmarks of this development.

A milestone in the history of bioenergetics was the discovery of ATP and creatine phosphate by Lohmann and by Fiske and Subbarow in 1929. Their pioneering findings that working muscle splits creatine phosphate and that the creatine so formed can be rephosphorylated by ATP, were followed in the late 1930s by Engelhardt's and Szent-Györgyi's fundamental discoveries concerning the role of ATP in muscle contraction. At about the same time Warburg demonstrated that the oxidation of 3-phosphoglyceraldehyde is coupled to ATP synthesis and Lipmann identified acetyl phosphate as the product of pyruvate oxidation in bacteria. In 1941, Lipmann developed the concept of 'phosphate-bond energy' as a general principle for energy transfer between energy-generating and energy-utilizing cellular processes.

It seemed that it was only a question of time until most of these processes could be reproduced and investigated using isolated enzymes.

Parallel to these developments, however, vitalism re-entered the stage in connection with studies of cell respiration. In 1912 Warburg reported that the respiratory activity of tissue extracts was associated with insoluble cellular structures. He called these structures 'grana' and suggested that their rôle is to enhance the activity of the iron-containing respiratory enzyme, *Atmungsferment*. Shortly thereafter Wieland, extending earlier observations by Battelli and Stern, reached a similar conclusion regarding cellular dehydrogenases. Despite diverging views concerning the nature of cell respiration — involving an activation of oxygen according to Warburg and an activation of hydrogen according to Wieland — they both agreed that the role of the cellular structure may be to enlarge the catalytic surface. Warburg referred to the 'charcoal model' and Wieland to the 'platinum model' in attempting to explain how this may be achieved.

In 1925 Keilin described the cytochromes, a discovery that led the way to the definition of the respiratory chain as a sequence of redox catalysts comprising the dehydrogenases at one end and *Atmungsferment* at the other, thereby bridging the gap in opinion between Warburg and Wieland. Using a particulate preparation from mammalian heart muscle, Keilin and Hartree subsequently showed Warburg's *Atmungsferment* to be identical to Keilin's cytochrome a_3. They recognized the need for a cellular structure for cytochrome activity, but visualized that this structure may not be necessary for the activity of the individual catalysts, but rather for facilitating their mutual accessibility and thereby the rates of interaction between the different components of the respiratory chain. Such a function, according to Keilin and Hartree, could be achieved by 'unspecific colloidal surfaces'. Interestingly, the possible role of phospholipids was not considered in these early studies and it was not until the 1950s that the membranous nature of the Keilin-Hartree heart-muscle preparation and its mitochondrial origin were recognized.

During the second half of the 1930s important progress was made in elucidating the reaction pathways and energetics of aerobic metabolism. In 1937 Krebs formulated the citric acid cycle and the same year Kalckar presented his first observations leading to the demonstration of aerobic phosphorylation, using a particulate system derived from kidney homogenates. Earlier, Engelhardt had obtained similar indications with intact pigeon erythrocytes.

Extending these observations, Belitser and Tsybakova concluded from experiments with minced muscle in 1939 that at least two molecules of ATP are formed per atom of oxygen consumed. These results suggested that phosphorylation probably occurs coupled to the respiratory chain. That this was the case was further suggested by measurements reported in 1943 by Ochoa, who deduced a P/O ratio of 3 for the aerobic oxidation of pyruvate in heart and brain homogenates. In 1945 Lehninger demonstrated that a particulate fraction from rat liver catalyzed the oxidation of fatty acids, and in 1948–1949 Friedkin and Lehninger provided conclusive evidence for the occurrence of respiratory chain-linked phosphorylation in this system using β-hydroxybutyrate or reduced nicotinamide adenine dinucleotide as substrate.

Although mitochondria had been observed by cytologists since the 1840s, the elucidation of their function had to await the availability of a method for their isolation. Such a method, based on fractionation of tissue homogenates by differential centrifugation, was developed by Claude in the early 1940s. Using this method, Claude, Hogeboom and Hotchkiss concluded in 1946 that the mitochondrion is the exclusive site of cell respiration. Two years later this conclusion was further substantiated by Hogeboom, Schneider and Palade with well-preserved mitochondria isolated in a sucrose medium and identified by Janus Green staining. In 1949 Kennedy and Lehninger demonstrated that mitochondria are the site of the citric acid cycle, fatty acid oxidation and oxidative phosphorylation.

In 1952–1953 Palade and Sjöstrand presented the first high-resolution electron micrographs of mitochondria. These micrographs served as the basis for the now generally accepted notion that mitochondria are surrounded by two membranes, a smooth outer membrane and a folded inner membrane giving rise to the *cristae*. In the early 1950s evidence also began to accumulate indicating that the inner membrane is the site of the respiratory-chain catalysts and the ATP-synthesizing system. In the following years research in many laboratories was focused on the mechanism of electron transport and oxidative phosphorylation, using both intact mitochondria and 'submitochondrial particles' consisting of vesiculated inner-membrane fragments.

Studies with intact mitochondria, performed in the laboratories of Boyer, Chance, Cohn, Green, Hunter, Kielley, Klingenberg, Lardy, Lehninger, Lindberg, Lipmann, Racker, Slater and others, provided information on problems such as the composition, kinetics and the localization of energy-coupling sites of the respiratory chain, the control of respiration by ATP synthesis and its abolition by uncouplers, and various partial reactions of oxidative phosphorylation. Most of the results could be explained in terms of the occurrence of non-phosphorylated high-energy compounds as intermediates between electron transport and ATP synthesis, a chemical coupling mechanism envisaged by several laboratories and first formulated in general terms by Slater. However, intensive efforts to demonstrate the existence of such intermediates proved unsuccessful.

Studies with beef-heart submitochondrial particles initiated in Green's laboratory in the mid-1950s resulted in the demonstration of ubiquinone and of non-heme iron proteins as components of the electron-transport system, and the separation, characterisation and reconstitution of the four oxidoreductase complexes of the respiratory chain. In 1960 Racker and his associates succeeded in isolating an ATPase from submitochondrial particles and demonstrated that this ATPase, called F_1, could serve as a coupling factor capable of restoring oxidative phosphorylation to F_1-depleted particles. These preparations subsequently played an important role in elucidating the role of the membrane in energy transduction between electron transport and ATP synthesis.

A somewhat similar development took place concerning studies of the mechanism of photosynthesis. Although the existence of chloroplasts and their association with chlorophyll had been known since the 1830s and their identity as the site of carbon dioxide assimilation was established in 1881 by Engelmann using isolated chloroplasts, it was not until the 1930s that the mechanism of photosynthesis began to be clarified. In 1938 Hill demonstrated that isolated chloroplasts evolve oxygen upon illumination and beginning in 1945 Calvin and his associates elucidated the pathways of the dark-reactions of photosynthesis leading to the conversion of carbon dioxide to carbohydrate.

The latter process was shown to require ATP, but the source of this ATP was unclear and a matter of considerable dispute. The breakthrough came in 1954 when Arnon and his colleagues demonstrated light-induced ATP synthesis in isolated chloroplasts. The same year Frenkel described photophosphorylation in cell-free preparations from bacteria. Photophosphorylation in both chloroplasts and bacteria was found to be associated with membranes, in the former case with the thylakoid membrane and in the latter with structures derived from the plasma membrane, called chromatophores. In the following years work in a number of laboratories, including those of Arnon, Avron, Chance, Duysens, Hill, Jagendorf, Kamen, Kok, San Pietro, Trebst, Witt and others, resulted in the identification and characterization of various catalytic components of photosynthetic electron transport. Chloroplasts and bacteria were also shown to contain ATPases similar to the F_1-ATPase of mitochondria.

By the beginning of the 1960s it was evident that both oxidative and photosynthetic phosphorylation were dependent on an intect membrane structure, and that this requirement probably was related to the interaction of the electron-transport and ATP-synthesizing systems rather than the activity of the individual catalysts. However, contemporary thinking concerning the mechanism of ATP synthesis was dominated by the chemical coupling hypothesis and did not readily envision a role for the membrane. This impasse was broken in 1961 when Mitchell first presented his chemiosmotic hypothesis, according to which energy transfer between electron transport and ATP synthesis takes place by way of a transmembrane proton gradient.

Mitchell's hypothesis was first received with skepticism, but in the mid-1960s evidence began to accumulate in favour of the chemiosmotic coupling mechanism. It was shown that electron-transport complexes and ATPases, when present in either native or artificial membranes, are capable of generating a transmembrane proton gradient and that this gradient can serve as the driving force for electron transport-linked ATP synthesis. Agents that abolished the proton gradient uncoupled electron transport from phosphorylation. Proton gradients were also shown to be involved in various other membrane-associated energy-transfer reactions, such as the energy-linked nicotinamide nucleotide transhydrogenase, the synthesis of inorganic pyrophosphate, the active transport of ions and metabolites, mitochondrial thermogenesis in brown adipose tissue and light-driven ATP synthesis and ion transport in *Halobacteria*. The chapters of this volume give an overview of our present state of knowledge concerning these processes.

The central problem in this field at present is to clarify the mechanisms involved in membrane-associated energy transduction at the molecular level. What are the molecular mechanisms by which energy-transducing catalysts translocate protons across the membrane? Is the generation of a proton gradient the primary event in energy conservation or is it preceded by chemical, e.g., conformational changes in the catalysts involved? Is communication by way of a transmembrane proton gradient the only means by which energy-transducing catalysts interact or are there mechanisms for more direct, 'localized' interactions between them within the membrane? How is the biosynthesis of energy-transducing catalysts regulated, e.g., in relation to subunit stoichiometry or, in the case of eukaryotes, to the coordination of nuclear and organellar gene expression? How are the subunits of energy-transducing catalysts processed and assembled in the membrane, and what is the relationship of these processes to the energy-state of the cell?

These are some of the current problems that are discussed in various chapters of this volume. Progress regarding these problems has long been dependent on knowledge of the structures, reaction mechanisms and biogenesis of the individual energy-transducing catalysts and their relationship to, and interactions within, the membranes in which they reside. Such information has been forthcoming during the last few years at an accelerating rate and further rapid progress can be foreseen. Looking at the field as a whole, one is left with the impression that, perhaps for the first time, bioenergeticists are taking full advantage of the powerful arsenal of methods and concepts of molecular biology and, vice versa, molecular biologists are becoming genuinely engaged in the fundamental problems of bioenergetics. What we may be witnessing is the fall of the last bastion of vitalism, the transition of *membrane bioenergetics* into *molecular bioenergetics*.

In terminating this introduction it is a true pleasure to express my thanks to the authors of the various chapters for having accepted the invitation to contribute to this volume and, in particular, for their efforts to submit their manuscripts in time, which made it possible to publish this volume while its contents are still reasonably up-to-date. I also wish to thank my colleague Kerstin Nordenbrand at the Arrhenius Laboratory for her valuable help with the editorial work, and Jim Orr, Desk Editor of the Biomedical Division, Elsevier Science Publishers B.V., for friendly and efficient cooperation.

Lars Ernster

Department of Biochemistry
Arrhenius Laboratory
University of Stockholm
S-106 91 Stockholm
Sweden

Some reviews on topics related to the history of bioenergetics

(In chronological order of appearance)

Rabinowitch, E.I. (1945) Photosynthesis and Related Processes. Interscience, New York.

Lindberg, O. and Ernster, L. (1954) Chemistry and Physiology of Mitochondria and Microsomes. Springer-Verlag, Vienna.

Krebs, H.A. and Kornberg, H.L. (1957) A survey of the energy transformations in living matter. Ergeb. Physiol. 49, 212–298.

Novikoff, A.B. (1961) Mitochondria (Chondriosomes). In The Cell (Brachet, J. and Mirsky, A.E., eds.) Academic Press, New York, Vol. II, pp. 299–421.

Lehninger, A.L. (1964) The Mitochondrion. Benjamin, New York.

Keilin, D. (1966) The History of Cell Respiration and Cytochrome. Cambridge University Press, Cambridge.

Slater, E.C. (1966) Oxidative Phosphorylation. Comprehensive Biochemistry, Vol. 14, pp. 327–396. Elsevier, Amsterdam

Kalckar, H.M. (1969) Biological Phosphorylations. Development of Concepts. Prentice-Hall, Englewood, NJ.

Krebs, H.A. (1970) The history of the tricarboxylic acid cycle. Perspect. Biol. Med. 14, 154–170.

Lipmann, F. (1971) Wonderings of a Biochemist. Wiley-Interscience, New York.

Fruton, J.S. (1972) Molecules and Life. Wiley-Interscience, New York.

Arnon, D.I. (1977) Photosynthesis 1950–1975. Changing concepts and perspectives. In Photosynthesis I (Trebst, A. and Avron, M., eds.) Encyclopedia of Plant Physiology, New Series, Springer-Verlag, Heidelberg, Vol. 5, pp. 7–56.

Boyer, P.D., Chance, B., Ernster, L., Mitchell, P., Racker, E. and Slater, E.C. (1977) Oxidative phosphorylation and photophosphorylation. Annu. Rev. Biochem. 46, 955–1026.

Racker, E. (1980) From Pasteur to Mitchell: A hundred years of bioenergetics. Fed. Proc. 39, 210–215.

Bogorad, L. (1981) Chloroplasts. J. Cell Biol. 91, 256s–270s.

Ernster, L. and Schatz, G. (1981) Mitochondria: A historical review. J. Cell Biol. 91, 227s–255s.

Skulachev, V.P. (1981) The proton cycle: History and problems of the membrane-linked energy transduction, transmission, and buffering. In Chemiosmotic Proton Circuits in Biological Membranes. (Skulachev, V.P. and Hinkle, P.C., eds.) pp. 3–46. Addison-Wesley, Reading, MA.

Slater, E.C. (1981) A short history of the biochemistry of mitochondria. In Mitochondria and Microsomes. (Lee, C.P., Schatz, G. and Dallner, G., eds.) pp. 15–43. Addison-Wesley, Reading, MA.

Tzagoloff, A. (1982) Mitochondria. Plenum Press, New York.

Hoober, J.K. (1984) Chloroplasts. Plenum Press, New York.

Non-conventional abbreviations

AcPyAD	acetyl pyridine adenine dinucleotide
AMP-PNP	adenyl imidodiphosphate
ATPS (β and/or γ)	thiophosphate analogs of ATP
BAL	2,3-dimercaptopropanol (British Anti Lewisite)
CAT	carboxyatractyloside (carboxyatractylate)
CCCP (ClCCP)	carbonylcyanide *p*-chloromethoxyphenylhydrazone
DBMIB	2,5-dibromo-3-methyl-6-isopropylbenzoquinone
DCCD	*N*,*N′*-dicyclohexylcarbodiimide
DCMU	3-(3,4-dichlorophenyl)-1-dimethylurea
DMPC	dimyristoyl phosphatidylcholine
DMSO	dimethylsulfoxide
DNP	2,4-dinitrophenol
DNP-INT	2-iodo-6-isopropyl-3-methyl-2,4,4′-trinitro-diphenylether
DTNB	5,5′-dithiobis(2-nitro-benzoate)
EDTA	ethylenediamine tetraacetate
EGTA	ethyleneglycol tetraacetate
ELISA	enzyme-linked immunosorbent assay
ETF	electron-transferring flavoprotein
FCCP	carbonylcyanide *p*-fluoromethoxyphenylhydrazone
FPLC	fast protein liquid chromatography
FSBA	*p*-fluorosulfonylbenzoyl-5-adenosine
HMHQQ	7-(*n*-heptadecyl)mercapto-6-hydroxy-5,8-quinolinequinone
HOQNO	2-*n*-heptyl-4-hydroxyquinoline *N*-oxide
NBDCl	4-chloro-7-nitro-2-oxal-1,3-diazole
NEM	*N*-ethylmaleimide
OSCP	oligomycin sensitivity conferring protein
PMS	phenazine methosulfate
PP_i	inorganic pyrophosphate
PPase	pyrophosphatase
Q	ubiquinone
RuBP	ribulose-1,5-biphosphate
SRP	signal recognition particle
TPP^+	triphenylphosphonium ion
UHDBT	5-*n*-undecyl-6-hydroxy-4,7-dioxobenzothiazol
1799	2,6-dihydroxy-1,1,1,7,7,7-hexafluoro-2,6-bis-(trifluoromethyl)-heptanone-4-[bis(hexafluoroacetyl)acetone]

Contents

Chapter 8. Metabolite transport in mammalian mitochondria, by K.F. LaNoue and A.C. Schoolwerth 221

Chapter 9. The uptake and release of calcium by mitochondria, by E. Carafoli and G. Sottocasa 269

Chapter 10. Thermogenic mitochondria, by J. Nedergaard and B. Cannon 291

Chapter 11. Bacteriorhodopsin and related light-energy converters, by J.K. Lanyi 315

Chapter 12. Biogenesis of energy-transducing systems, by N. Nelson and H. Riezman 351

Ernster (ed.) Bioenergetics

© *1984 Elsevier Science Publishers B.V.*

CHAPTER 1

Thermodynamic aspects of bioenergetics

KAREL VAN DAM and HANS V. WESTERHOFF *

Laboratory of Biochemistry, B.C.P. Jansen Institute, University of Amsterdam, Plantage Muidergracht 12, Amsterdam, The Netherlands

1. Introduction

By definition, bioenergetics is concerned with the transformations of energy in biological systems. Thermodynamics, though originally focussing on how heat could be used to do useful work in steam engines, was soon extended to analyze the interconversion of many more forms of energy.

Yet, statements of classical thermodynamics are quantitative only when the system studied is in equilibrium. Biology is interested in living systems. Such systems are always out of equilibrium so that classical thermodynamics has a limited potential for biology. The work of Onsager [1] suggested that thermodynamics could be extended to the description of non-equilibrium systems. The theory behind the application of this 'near-equilibrium' [2] non-equilibrium thermodynamics (NET) to biological systems has been elaborated in great detail [3–5]. It provided insight into the thermodynamic implication of the coupling of (in terms of Gibbs free energy) 'uphill' to 'downhill' processes and into the resulting thermodynamic efficiency of energy coupling. Yet, this near-equilibrium NET never became as generally used, and even accepted, as for instance the enzyme kinetics developed by Michaelis and Menten. Reasons for this were the following.

(*i*) The validity of the near-equilibrium non-equilibrium thermodynamics could only be guaranteed if all processes were 'close' to equilibrium, where 'close' should be interpreted as $\Delta G \ll 1.5$ kJ/mol. Most of the interesting processes in living systems are much farther from equilibrium.

(*ii*) This near-equilibrium non-equilibrium thermodynamics described biological systems as black boxes (some exceptions are found in Refs. 3, 7 and 8). It lacked the ambition of relating in- and output characteristics of the biological system to the mechanisms of operation of the metabolism within that black box. It is precisely those mechanisms that are of great interest to most biological chemists and physicists.

* Present address: National Institutes of Health, Building 2, Room 310, Bethesda, MD 20205, U.S.A.

(*iii*) On many points, near-equilibrium non-equilibrium thermodynamics seemed to be in conflict with that was already known from enzyme kinetics: it predicted that reaction rates would go to infinity when the substrate concentration would do so, whereas enzyme catalyzed reactions exhibit a maximum rate.

After the extension of thermodynamics to near-equilibrium systems had thus turned out to be of limited use in biological systems, a number of authors contributed to yet another extension of thermodynamics, i.e., to (biological) systems in which most reactions are enzyme-catalyzed. The latter extension also allows one to quantitatively relate the metabolic behaviour of biological systems to the characteristics of the enzymes within them. Meanwhile, this extension has been used to extract mechanistic information from experimental data obtained in a number of free-energy transducing systems.

Although it will burden the student of bioenergetics with some mathematical gymnastics, we feel that further progress in the understanding of a number of, still unsolved, elementary problems in bioenergetics, is impossible without quantitative analysis of the functioning of biological free-energy transducing systems. Examples of such problems are the extent of localization of the energy transducing protons; the occurrence of 'slip' (see below) in the proton pumps; the stoicheiometries at which protons are pumped; the extent to which specific enzymes control free-energy transduction.

Therefore, we shall present here a taste of the modern thermodynamic approaches to bioenergetics.

2. *Simple thermodynamics*

If thermodynamics would limit itself to the study of changes in the amount of energy (U) in a system, its application to biology would be rather dull. Energy (U) itself is a 'conserved' quantity, i.e., it can neither be destroyed nor created. The interesting part comes when it is realized that energy can appear in different forms which generally have different capacities to do useful work. In turn, these capacities depend on the type of system we are concerned with, e.g., heat has a large capacity to do useful work (only) if there is a large temperature difference between different parts of the system. Flux of a substance has a high capacity to do work, only if there is a high difference in its chemical potential (i.e., concentration) between different parts of the system. In the usual biological systems, there are no significant temperature gradients, so that pure heat is not a very useful form of energy, at least not in terms of doing work.

In an isothermal, isobaric system, the amount of energy that can be used to do useful work is equal to the Gibbs Free energy, G (defined as $U + PV - TS$). Pure heat in isothermal systems is a form of energy, the free-energy content of which is zero. All spontaneous chemical and physical processes proceed in such a way that some free energy is destroyed ('dissipated'). It should be noted that this does not imply that the free energy of a system has to decrease. When the system is in steady state, its free energy is constant. The correct implication then is that more free energy is imported into the system than is exported, such that the free energy of the system plus its surroundings does decrease. The expenditure of free energy, as fatalistic as it has been discussed at times, can also be seen from the more positive side: free energy is the factor that makes processes run. For systems not too far from equilibrium, it can even be shown that the rate of processes increases with the amount of free energy spent in making them run. Moreover, such systems evolve in such a way that the rate at which they dissipate free energy decreases with time [9]. It is a simple consequence of this that evolution will stop, i.e., steady state will be attained, when the free-energy dissipation cannot decrease any further: in the steady state free-energy dissipation is minimal.

As stated above, free-energy (G) dissipation does not imply that energy (U) is dissipated: it only implies that energy is (partly) transformed from free energy to pure heat, which equals the product of temperature and entropy. Consequently, the rate of free-energy dissipation is equal to the rate of entropy production multiplied by the absolute temperature.

The so-called dissipation function (Φ) analyzes the rate of free energy dissipation in terms of the different processes in which free energy is dissipated. If, for instance, a chemical reaction occurs with a free-energy difference, ΔG_r, and at a rate, J_{chem}, whereas at the same time a substance S flows from a space in which it has a high concentration to one where it has a low concentration, the rate at which the free energy is dissipated is:

$$\Phi = J_{chem} \cdot \Delta G_r + J_S \cdot \Delta \mu_S \qquad (1)$$

It may be noted that we define ΔG such that it equals the chemical potential of the substrate minus the chemical potential of the product. We noted above that the possibility of free-energy dissipation drives a reaction. Free-energy differences like ΔG_r and $\Delta \mu_S$ in the above equation embody such a possibility: they act as forces that drive the reaction. Other examples are: the contractile force on a muscle; the voltage drop across an electrical resistance; the osmotic pressure on a semipermeable membrane. The dissipation function consists of the sum of the products of fluxes (currents) and the (thermodynamic) forces that drive them [4].

The dissipation function always has a positive value (according to the second law of thermodynamics), but the sign of each of the separate flux-force couples is not a priori defined. Thus, the negative contribution of a diffusion flux against a concentration gradient may be compensated by the positive contribution of a chemical reaction proceeding at a high free-energy difference. This can, however, only occur if the two processes are coupled in one way or another. Any independent (not coupled) set of fluxes and associated forces must conform to the criterion of positive entropy production [10].

Generally speaking, a flux (J) in a system can depend on each of the forces ('X') in that system. Close to equilibrium it can be made feasible [1] that this function is in general proportional, i.e.:

$$J_i = L_{i1} \cdot X_1 + L_{i2} \cdot X_2 + \ldots + L_{in} \cdot X_n \qquad (2)$$

The proportionality constant L_{ij} is constant in the sense that it is independent of the forces (X).

The complete set of equations, relating fluxes and forces can be written in a matrix form and is called the set of phenomenological equations. It was shown by Onsager [1] that the matrix of the phenomenological proportionality constants is symmetrical:

$$L_{ij} = L_{ji}$$

This reduces the number of independent constants in the system.

3. *(Thermo-)kinetics*

The phenomenological equations given above are limited in (demonstrated) validity to near-equilibrium systems. They belong to the near-equilibrium NET approach discussed in the introduction. They demonstrate some of the limitations of the near-equilibrium approach in the sense that the relationship between the phenomenological constants (L_{ij}) and the biochemical and biophysical mechanisms within the biological system are obscure, and that no enzyme-saturation effects are recognizable in the equations. Consequently it was here that the second extension of thermodynamics, to include the mechanism and kinetics of enzyme-catalyzed reactions, had to start. Rottenberg [11], Hill [12] and Van der Meer et al. [13] have generated such an extension by translating the concentration parameters present in enzyme kinetics into thermodynamic parameters. We shall now first demonstrate the principles of this for the simpler case of ordinary chemical kinetics.

3.1. The principle

To investigate enzyme kinetics in terms of thermodynamics, it is appropriate to start with a consideration of a simple chemical reaction [1,4,6,14]:

$$\mathrm{S} \underset{k_{-1}}{\overset{k_1}{\rightleftarrows}} \mathrm{P} \tag{3}$$

Here S stands for 'substrate' and P for 'product'. By writing the net velocity of the reaction as the difference between the forward and the backward reaction:

$$v = k_1 \cdot [\mathrm{S}] - k_{-1} \cdot [\mathrm{P}] \tag{4}$$

and substituting the chemical potentials of the reactants:

$$\mu_i = \mu_i^\theta + R \cdot T \cdot \ln[\mathrm{i}], \qquad \mathrm{i} = \mathrm{S \ or \ P} \tag{5}$$

the following rate equation is obtained:

$$v = k_1 \cdot [S]^\dagger \cdot e^{(\mu_S - \mu_S^\dagger)/RT} - k_{-1} \cdot [P]^\dagger \cdot e^{(\mu_P - \mu_P^\dagger)/RT} \tag{6}$$

Here † refers to any reference state. Now we are faced with two questions:

(*i*) Can v be written as a function of $\mu_S - \mu_P = \Delta G$ only?

(*ii*) If so, what does that function look like?

For $\mu_S - \mu_S^\dagger \ll RT$ and $\mu_P - \mu_P^\dagger \ll RT$ the above equation can be approximated by its first-order Taylor expansion:

$$v = k_1 \cdot [S]^\dagger \cdot (1 - \mu_S^\dagger/RT) - k_{-1} \cdot [P]^\dagger \cdot (1 - \mu_P^\dagger/RT) + (k_1 \cdot [S]^\dagger/RT) \cdot (\mu_S - \mu_P) + (k_1 \cdot [S]^\dagger - k_{-1} \cdot [P]^\dagger/RT) \cdot \mu_P \tag{7}$$

It appears that generally the answer to the former question is negative. v is a function of the two independent variables μ_S and μ_P. However, there are two situations where v can be written as a function of $\mu_S - \mu_P$ only.

(**1**) If † refers to equilibrium:

$$k_1[S]_{eq} - k_{-1} \cdot [P]_{eq} = 0 \tag{8}$$

so that:

$$v = k_1 \cdot [S]_{eq} \cdot (\mu_S - \mu_P)/RT \tag{9}$$

This equation shows that the flux (v or J) through a chemical reaction near equilibrium is proportional to its free-energy difference $\mu_S - \mu_P$:

$$v = L \cdot (\mu_S - \mu_P) \tag{10}$$

with:

$$L = k_1 \cdot [S]_{eq}/RT \tag{11}$$

Note especially that the proportionality constant contains the absolute concentration of the reactants, in the form of $[S]_{eq}$. Furthermore, the derivation is limited to a rather narrow region near-equilibrium [cf. Ref. 6].

(**2**) If μ_S and μ_P are interdependent through some physical constraint. Then the v can also be written as a function of $\mu_S - \mu_P$ only.

Essig and Caplan [15] have made the point that, since a priori the physical constraint is arbitrary, it may be chosen such that v varies linearly with $\mu_S - \mu_P$. After having shown that linear flow-force relationships may have great advantages for biological systems, Stucki [16] suggested that the latter may have evolved in such a manner that linearity resulted. Below we shall examine the case in which classical enzyme kinetics by itself gives rise to linear relationships between the rate of, and the free-energy difference across, a reaction. Linear flow-force relationships for chemical reactions will then follow as a special case.

The above derivation illustrates a number of important points:

– the proportionality constant L is dependent on the rate constant k_1 (and consequently on the amount of chemical catalyst that affects this constant).

– the proportionality constant L is dependent on the some standard (e.g., equilibrium) concentration of the reactants.

– the proportionality between flux and force depends on the approximation in Eqn. 7. The exponential terms will be linear only if they are much smaller than RT.

– though in principle the reaction rate is a function of the chemical potentials of the substrate and the product separately, this function can be written as a function of the difference of the two chemical potentials (and hence as a function of the free energy of reaction) if the reaction is close to equilibrium.

– because the latter limitation to near-equilibrium systems is not welcome, it is important that the reaction rate can also be written as a function of the free-energy difference of reaction, if the concentrations of the substrate and product are not independent variables.

3.2. The physical constraint [S] + [P] constant

The argument that flux-force relations can be taken as approximately linear, because they can be approximated as the first order term in the expansion series of the real flux-force relation is relatively trivial. It would be more satisfactory if the linear approximation could be shown to be better than that. Van der Meer et al. [13], stressing that a metabolically important physical constraint is that of conservation of substrate plus product concentration, demonstrated such a better linear approximation for the enzyme-catalyzed reaction (see also Ref. 11):

$$\mathrm{E} + \mathrm{S} \underset{k_{-1}^{\mathrm{S}}}{\overset{k_1^{\mathrm{S}}}{\rightleftarrows}} \mathrm{ES} \underset{k_2^{\mathrm{P}}}{\overset{k_2^{\mathrm{S}}}{\rightleftarrows}} \mathrm{EP} \underset{k_1^{\mathrm{P}}}{\overset{k_{-1}^{\mathrm{P}}}{\rightleftarrows}} \mathrm{E} + \mathrm{P} \tag{12}$$

Here E is the enzyme and ES and EP are the enzyme-substrate and the enzyme-product complex, respectively. The rate of such a reaction (e.g., if $[\mathrm{P}]=0$) approaches a maximum value, V_S, at increasing substrate concentrations. Similarly, the backward reaction (i.e., at $[\mathrm{S}]=0$) has a maximum rate V_P at high P concentrations. With the introduction of the Michaelis-Menten constants:

$$K_\mathrm{S} = \left(k_{-1}^{\mathrm{S}} + k_2^{\mathrm{S}}\right)/k_1^{\mathrm{S}}$$

$$K_\mathrm{P} = \left(k_{-1}^{\mathrm{P}} + k_2^{\mathrm{P}}\right)/k_1^{\mathrm{P}} \tag{14}$$

the net velocity of the reaction can be written as:

$$v = \frac{V_\mathrm{S}\cdot[\mathrm{S}]/K_\mathrm{S} - V_\mathrm{P}\cdot[\mathrm{P}]/K_\mathrm{P}}{1+[S]/K_\mathrm{S}+[\mathrm{P}]/K_\mathrm{P}} \tag{15}$$

Using the Haldane relation:

$$\left(\frac{[P]_{eq}}{[S]_{eq}}\right) = K_{eq} = \frac{K_P \cdot V_S}{K_S \cdot V_P} \tag{16}$$

and:

$$e^{(\mu_S - \mu_P)/RT} = \frac{[S]}{[S_{eq}]} \cdot \frac{[P_{eq}]}{[P]} \tag{17}$$

this equation can be rewritten as:

$$\frac{v}{V_S} = \frac{e^{(\mu_S - \mu_P)/RT} - 1}{\left(\frac{K_S}{[S]+[P]} + 1\right) \cdot e^{(\mu_S - \mu_P)/RT} + \frac{V_S}{V_P} \cdot \left(\frac{K_P}{[S]+[P]} + 1\right)} \tag{18}$$

If we choose as the natural [cf. Refs. 13, 17] physical constraint that:

$$[S] + [P] = \text{constant} = c \tag{19}$$

then the flux is a function of the free-energy of the reaction (ΔG) only. In Fig. 1.1 the dependence of the rate on the Gibbs free-energy difference, described by this equation, has been plotted for two choices of the kinetic constants.

In the literature a distinction is sometimes made between irreversible and reversible reactions. Such a distinction is in apparent contradiction with the principle of microscopic reversibility [1,18]; if substrate and product concentrations were chosen

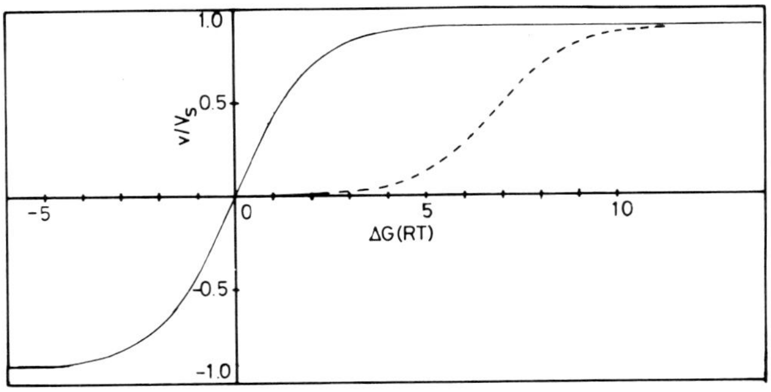

Fig. 1.1. Calculated plot of rate of an enzyme catalyzed reaction as a function of the free-energy difference across it, at constant sum concentration of substrate plus product. Calculations are according to Eqn. 18 with, for the 'kinetically reversible' reaction (———), $V_S = V_P = K_S = K_P = 10$ (i.e., $K_{eq} = 1$), $[S]+[P] = 1$, and, for the 'kinetically irreversible' reaction (------), $K_S = V_S = 10$, $V_P = 0.1$, $K_P = 100$ (i.e., $K_{eq} = 1000$), $[S]+[P] = 1$.

such that the free-energy of reaction (ΔG) was close to zero, then the reaction should proceed equally well either way, depending only on this ΔG, i.e., reactions are always reversible. Indeed, in accordance with microscopic reversibility, both lines in Fig. 1.1 pass through the origin of the plot, and in either case the reaction rate becomes negative at negative ΔG. Yet, the maximum reverse reaction rate of the dashed line is so much smaller than its maximum forward rate that it is effectively irreversible. Because the value of the kinetic parameters of the enzyme, and not the free-energy difference across the reaction, are responsible, enzymes have been distinguished in 'kinetically reversible' (full line in Fig. 1.1) and 'kinetically irreversible' (dashed curve in Fig. 1) reactions [13].

Until recently, non-equilibrium thermodynamic treatments always formulated the dependence of the rates of individual reactions on the free-energy difference (ΔG) across them in the same manner:

$$v = L \cdot (\Delta G) \tag{20}$$

Differences between different enzymes were supposed to be confined to differences in L; they were not supposed to affect the form, or position of the dependence of the reaction rate on ΔG. However, for the dashed curve in Fig. 1.1, the part where the approximation by this equation would be reasonable, is at very small reaction rates. In the presumably functionally more relevant part (v/V_S between 0.1 and 0.9), this equation (i.e., a straight line through the origin) is a completely unsatisfactory approximation of the actual relationship between v and ΔG. Yet, around $v/V_S = 0.5$ a different, though linear, approximation would seem possible. Below, we shall further substantiate this alternative approximation.

The finding that the dependence of the reaction rate on the free-energy difference across the reaction is a line traversing, in a quasi-linear fashion, the domain of the reaction rates between a maximum reverse rate and a maximum forward rate, is not astonishing. The conservation condition [S] + [P] = constant imposes a maximum to both [S] and [P] separately, and therefore also to both the maximum forward and the maximum backward rate. Since most kinetic equations [19] are monotonous functions of both [S] and [P] individually, a plot of v versus ΔG is bound to have three regions: at very high positive and negative values of ΔG the reaction rate is almost independent of ΔG, and between these regions there is a region where the rate changes smoothly from its lowest to its highest value. It may be noted that the presence of these general features does not depend on special values of the kinetic constants, they are mainly a consequence of the conservation condition. Consequently, they are also expected in the case of chemical kinetics.

When reformulating enzyme kinetics in terms of the dependence of reaction rates on ΔG, Rottenberg [11,20] did not use the conservation condition as the physical constraint, but rather that either the substrate, or the product concentration would be constant. He found similar appearances for the plots of reaction rates versus the free-energy difference across the reaction. Clearly, in this case the bounds to the maximum forward or reverse reaction rates are not due to the boundary condition chosen, but to saturation characteristics of enzyme kinetics alone.

The magnitudes of the maximum forward and backward rates, the position in the curve where ΔG and v are zero, the slope of the quasi-linear region and the curvature, are properties that depend on the magnitude of the kinetic constants and the sum concentration of substrate and product. We shall illustrate this for the case of Eqn. 18. As a consequence of its general appearance, a point of inflection occurs in the plot of reaction rate versus free-energy difference. In such an inflection point the second order derivative of v with respect to ΔG equals zero. Writing the reaction rate as a Taylor series around that inflection point one finds:

$$v = v^* + \left(\frac{\partial v}{\partial \Delta G}\right) \cdot (\Delta G - \Delta G^*) + \frac{1}{2}\left(\frac{\partial^2 v}{\partial \Delta G^2}\right) \cdot (\Delta G - \Delta G^*)^2 + \frac{1}{6}\left(\frac{\partial^3 v}{\partial \Delta G^3}\right) \cdot (\Delta G - \Delta G^*)^3 + \ldots\ldots \tag{21}$$

The usual approximation of the relationship between reaction rates is the first order approximation, i.e., all second and higher order terms in the Taylor series are neglected [6] (the actual curve is just replaced with the tangent to it). It has been shown that for (near-equilibrium) chemical reactions such approximations are generally valid only for ΔG ranges of less than 1.5 kJ/mol [6,14]. In the case where there is an inflection point, however, the linear approximation is much better than this: the neglection of the second order term in the Taylor series is no longer an approximation, it is equal to zero. In the present example (Eqn. 18) the linear approximation is satisfactory (i.e., error in predicted rate less than 15%) for a ΔG range of more than 7 kJ/mol [13]. Because the inflection point reflects a maximum of the slope, the quasi-linearity occurs where the dependence of the rate on ΔG is the strongest. Hence, the quality of the linear approximation is even more impressive in terms of the velocity range for which it is valid: with a deviation of less than 15% the following equation describes 75% of the range of the reaction rate:

$$v = \frac{1}{4RT}(\nu_S + \nu_P) \cdot (\mu_S - \mu_P) - \tfrac{1}{4}(\nu_S + \nu_P) \cdot \ln(\nu_S/\nu_P) + \tfrac{1}{2}(\nu_S - \nu_P) \tag{22}$$

Here:

$$\nu_S = V_S/(1 + K_S/c) \tag{23}$$

and

$$\nu_P = V_P/(1 + K_P/c) \tag{24}$$

According to enzyme kinetics ν_S and $-\nu_P$ are the reaction rates when [P] = 0, or [S] = 0, respectively.

Just as in the case of chemical kinetics, the slope of the linear approximation (i.e., $(\nu_S + \nu_P)/4RT$) depends on the kinetic constants and the sum concentration of substrate and product. It may also be noted that it is directly proportional to the amount of enzyme. Especially if the sum concentration [S] + [P] is much smaller than either Michaelis-Menten constant, its increase has the same effect as an increase in the concentration of the enzyme: the capacity of the enzyme to react is increased. If, on the contrary, the sum concentration vastly exceeds both K_m values, the dependence of the reaction rate on ΔG will be insensitive to changes in the sum concentration. Actually, this phenomenon may well explain why, at constant phosphate potential (ΔG_P), Van der Meer et al. [13] did observe a dependence of mitochondrial respiratory rate on the sum concentration of ATP and ADP, whereas Küster et al. [21], working at higher concentrations, did not.

Since a non-catalyzed reaction can be considered a limiting case of an enzyme-catalyzed reaction, this suggests that linearity might occur in both types of reaction. This can be shown by writing the rate equation for the non-catalyzed reaction as:

$$v = \nu_f \cdot \frac{e^{(\mu_S - \mu_P)/RT} - 1}{e^{(\mu_S - \mu_P)/RT} + \nu_f/\nu_b} \tag{25}$$

in which $\nu_f = k_1 \cdot ([S] + [P])$ and $\nu_b = k_{-1} \cdot ([S] + [P])$. This should be compared with the rearranged form of the rate equation for the enzyme-catalyzed case:

$$v = \nu_S \cdot \frac{e^{(\mu_S - \mu_P)/RT} - 1}{e^{(\mu_S - \mu_P)/RT} + \nu_S/\nu_P} \tag{26}$$

Apparently, with the boundary condition, [S] + [P] = constant, used here, the saturability of enzyme-catalyzed reactions has no marked effect on the linearity of the relation between rate and free-energy difference.

3.3. Short notation for the thermokinetic rate equations

The linear approximation such as given by Eqn. 22 could be summarized in either of two ways:

$$v = L^{\dagger} \cdot \Delta G + b \tag{27}$$

$$v = L^{\dagger} \cdot \Delta G^{\pounds} = L^{\dagger} \cdot (\Delta G - \Delta G^{\dagger}) \tag{28}$$

where b and $\Delta G^{\dagger}$ are constants, in the sense that they are independent of ΔG (at least, in the linear region). At a given ΔG, the reaction rate will always be directly proportional to the concentration of enzyme. As a consequence, b varies linearly with the enzyme concentration (cf., Eqn. 28). We prefer to have the effect of a variation in the enzyme concentration in one parameter (L) only, with $\Delta G^{\dagger}$ then being independent of the enzyme concentration.

In the kinetic example used here, the expression of $L^\dagger$ and $\Delta G^\dagger$ into the kinetic constants is:

$$L^\dagger = (\nu_S + \nu_P)/4RT \tag{29}$$

$$\Delta G^\dagger/RT = \ln(\nu_S) - \ln(\nu_P) - 2\cdot(\nu_S - \nu_P)/(\nu_S + \nu_P) \tag{30}$$

Only for certain kinetic constants, and sum concentration of substrate plus product, $\nu_S = \nu_P$ and $\Delta G^\dagger$ become zero, so that the flow-force relationship of Eqn. 20 which was used in near-equilibrium non-equilibrium thermodynamics [1,3–5,8], becomes valid. It should be noted that only at low magnitudes of [S] + [P] does the condition $\Delta G^\dagger = 0$ coincide with $K_{eq} = 1$.

In biological systems many reactions respond readily to changes in the free-energy difference across them. In terms of Fig. 1.1 they are in the steep range of the dependence of the reaction rate on ΔG. For them Eqn. 28 with $\Delta G^\dagger$ constant will be a good approximative description. Other reactions however, operate near their maximum rate in one of the two directions. One of the two horizontal parts of Fig. 1.1 would be representative for them. Eqn. 28 can also be made to describe these cases, provided that one takes $\Delta G^£$ rather than $\Delta G^\dagger$ constant [cf., 22]. It should be remembered that in between the region in which $\Delta G^\dagger$ is constant and the region in which $\Delta G^£$ is constant, there are (small) regions where neither approximation is valid.

4. *A mosaic in non-equilibrium thermodynamics (MNET)*

In phenomenological non-equilibrium thermodynamics a complicated system is described by taking each flux as dependent on each of the forces within the system. On the other hand, common sense tells us that some processes will in practice not depend noticeably on some of the forces. In fact, it could almost be considered a definition of an independent process that it is independent of driving forces other than its own driving force. Thus, we can simplify the description by splitting it up into independent processes. It should always be remembered, however, that such independence of processes remains a postulate of the description: failure to fit the properties of the system with the equations may be the result of an unnoticed coupling.

At this point it may be useful to emphasize that enzymes are designed to facilitate certain pathways of reaction. They may either catalyze a conversion of a single species or a coupled reaction, resulting in very different properties of the system. For instance, to a membrane across which a gradient of H^+ and K^+ ions exists we may add either nigericin or valinomycin plus a protonophore. In the first case, the system will come to a prolonged steady state in which the H^+ gradient is balanced by a K^+ gradient of opposite sign. In the second case, the system rapidly dissipates both gradients. Thus, the enzymatic complement of a system determines to a large extent its properties.

In mosaic non-equilibrium thermodynamics (MNET) [23,24] this phenomenon is accounted for: a complex system is considered as a mosaic of a number of independent building blocks. For each of the chemical species in the system the possible reactions in which it participates are sorted out and separately described. The total flux of the species is then defined as the sum of all the separate fluxes. Thus, the knowledge about the underlying biochemical and biophysical structure of the system is immediately included in the description. On the other hand, we may postulate a certain structure and derive a number of testable relations commensurate with this structure, which can then be used to assess the validity of the postulated structure.

We will now consider a few simple examples to illustrate the flux-force relations in a number of common elemental reactions.

4.1. Facilitated flux across a membrane

As a first approach the flux of solutes across membranes can be described by the simple linear equation:

$$J = L \cdot \Delta\tilde{\mu}_{\mathrm{S}} \tag{31}$$

The magnitude of the proportionality constant L is an indicator of the permeability of the membrane for the solute in question. Experimentally, we can manipulate this parameter, for instance by the addition of specific ionophores. Thus, uncouplers of oxidative phosphorylation increase the permeability of a membrane for H^+ ions, while valinomycin increases the permeability for K^+ ions. Addition of these ionophores will increase the flux of the respective ions (at equal electrochemical potential gradient) and, therefore, increase the dissipation of their gradient.

It may be instructive to note that the system will come to equilibrium with respect to the solute when the flux equals zero. From the above equation it is evident that this happens when the electrochemical gradient of the solute is zero or, in other words, when its electrochemical potential is equal on the two sides of the membrane.

4.2. Coupling between diffusion fluxes

The fluxes of solutes across membranes may be coupled. A clear example is the case of nigericin-induced K^+/H^+ exchange. This ionophore is a weak acid of which either the uncharged form or the potassium salt can cross the membrane. Thus, for each H^+ ion moving in one direction one K^+ ion has to move in the opposite direction.

In this case the flux-force relation for the elemental reaction may be written as (the reader should note that for simplicity we take here an example where the extra constant of Eqn. 28 is negligible):

$$J = L \cdot (\Delta\tilde{\mu}_{\mathrm{H}} - \Delta\tilde{\mu}_{\mathrm{K}}) \tag{32}$$

The flux of each of the two ions will depend on each of the two electrochemical gradients. Interestingly, equilibrium now is no longer reached at total dissipation of the gradients, but is attained at:

$$\Delta\tilde{\mu}_{H} = \Delta\tilde{\mu}_{K} \tag{33}$$

Since each of the two ions experiences the same membrane potential, this relation further simplifies to:

$$[H^{+}]_1/[H^{+}]_2 = [K^{+}]_1/[K^{+}]_2 \tag{34}$$

4.3. Coupling between chemical reaction and flux

The flux-force relations for (enzyme-catalyzed) chemical reactions were derived above. We consider now the case of a chemical reaction that is coupled to the vectorial movement of a solute across a membrane. An example is the ATP-driven H^+ pump present in the mitochondrial inner membrane. It catalyzes the hydrolysis of ATP to ADP and phosphate (P_i), with concomitant translocation of a number of H^+ ions from the mitochondrial matrix to the external medium. The reaction can be written down as:

$$ATP + n_H^P H^+_{left} \rightleftharpoons ADP + P_i + n_H^P \cdot H^+_{right} \tag{35}$$

We can predict the flux-force relation for this enzyme-catalyzed reaction in its general form:

$$J_P = L_P^{\dagger} \cdot \left(\Delta G_P - \Delta G_P^{\dagger} + n_H^P \cdot \gamma_H^P \cdot \Delta\tilde{\mu}_H\right) \tag{36}$$

The rate of the reaction will depend on both the free-energy difference of the chemical reaction and the electrochemical gradient of the H^+ ions, the latter weighted by the stoicheiometry of the reaction multiplied by a factor (γ) that may differ from 1 (see also section 5.2.1, and Refs. 24 and 25). The implications of this equation will be discussed in the section on application of MNET to oxidative phosphorylation (Section 5.2.1).

4.4 Leaks and slips

In coupled reactions there is always the possibility of a degree of imperfection in the coupling. This imperfection may be caused by either of two factors: parallel pathways of reaction or intrinsic uncoupling within the reaction pathway itself. It is useful to treat these two possibilities separately. They have been called leaks and slips, respectively.

An example of leaks can be found in the case of a solute symport system, for instance the proton-sugar symport in bacteria. In that case a protein specifically catalyzes the transport of protons and sugar across the membrane, by being able to cross the membrane either in the unloaded form or carrying both species simultaneously. Thus, the two solutes are transported in a strictly coupled way:

$$J_S = L_S \cdot (\Delta\mu_S + \Delta\tilde{\mu}_H) \tag{37}$$

$$J_H = J_S \tag{38}$$

If we now would add to the system a protonophore, we would introduce a parallel pathway for the movement of protons across the membrane:

$$J_H^l = L_H^l \cdot \Delta\tilde{\mu}_H \tag{39}$$

so that the strict stoicheiometric coupling between the solute fluxes is lost:

$$J_H = L_S \cdot \Delta\mu_S + \left(L_S + L_H^l\right) \cdot \Delta\tilde{\mu}_H \tag{40}$$

$$J_H/J_S = 1 + \frac{L_H}{L_S} \Bigg/ \left(1 + \frac{\Delta\mu_S}{\Delta\tilde{\mu}_H}\right) \tag{41}$$

This is a typical case of what we prefer to call a 'leak'. The two pathways (symport and leak) are completely independent and, therefore, can be described as two elemental reactions in the MNET treatment. In fact, the intrinsic passive permeability of the membrane for protons as well as for the sugar constitutes a leak that is always present; the importance of that leak can only be assessed by experiment.

A slip differs from a leak in that it is per se dependent on the machinery that couples two fluxes. Taking the same example as above, we can consider a slip in the proton-sugar symport to be present if the carrier sometimes crosses the membrane in association with the sugar only. This will practically result again in the loss of stoicheiometric coupling between the fluxes of protons and sugar. However, the rate of the 'side-reaction' is now no longer independent of the rate of the coupled reaction, since both are catalyzed by the same enzyme. Therefore, we can not treat the side reaction as an independent extra reaction in the MNET description. It should also be noted that a slip in, for instance, a symport system always involves more than one species. Thus, if our proton-sugar-symport carrier sometimes crosses the membrane in association with the the sugar only, this will result in a net flux of the sugar down its chemical gradient. At the same time, however, this flux together with the 'normal' movement of double-loaded carrier results in a flux of protons down their electrochemical gradient. Whereas the leak reactions are relatively simple to introduce in the MNET description, slip reactions are more complicated. Although detailed studies are in progress, the description of a slipping translocator is still relatively phenomenological. For the present example, the sugar and the proton flux would be given by (cf., Refs. 24, 26, 27]:

$$J_S = L_S \cdot \Delta\mu_S + L_S \cdot q \cdot Z \cdot \Delta\tilde{\mu}_H \tag{42}$$

$$J_H = L_S \cdot q \cdot Z \cdot \Delta\mu_S + L_S \cdot Z^2 \cdot \Delta\tilde{\mu}_H \tag{43}$$

with as criterion for slip that:

$$|q| < 1 \tag{44}$$

Z and q are called the phenomenological stoicheiometry and the degree of coupling, respectively. In case of slip they will differ from the mechanistic stoicheiometry (n) and ($-$)1, respectively.

5. *Application of MNET to biological free-energy converters*

We will now show how the application of MNET to a number of biological free-energy converters has not only led to an adequate description, but also to unexpected predictions about the properties of those systems. These examples will show how MNET can be applied, its use and also some of its limitations.

5.1. Bacteriorhodopsin liposomes

One of the simplest energy converters in biology is the light-driven proton pump bacteriorhodopsin. This protein can be isolated relatively easily from the halophilic bacterium *Halobacterium halobium* and reconstituted into phospholipid vesicles in a functional way. These vesicles take up protons from the medium upon illumination. The relevant elements of these bacteriorhodopsin liposomes are depicted in Fig. 1.2. Apart from the presence of the light-driven proton pump the vesicles are supposed to have a passive permeability for protons and other ions. This permeability can experimentally be manipulated by the addition of specific ionophores. The high heat conductance of lipid bilayers allows us to neglect any temperature gradient across the membrane (and any possible associated fluxes) [8,24]. Furthermore, the high

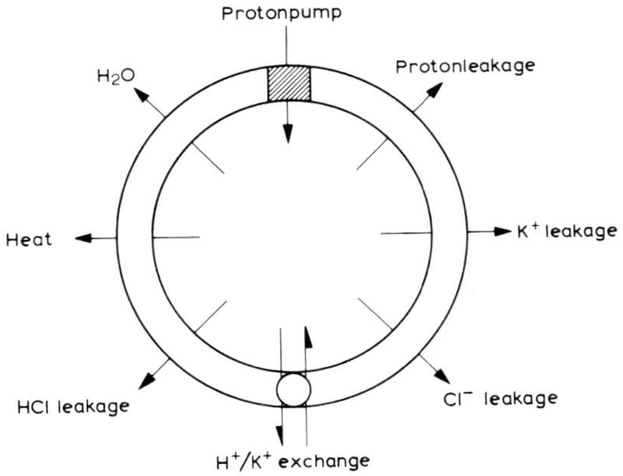

Fig. 1.2. The idealized bacteriorhodopsin liposome containing a light-driven proton pump in a membrane with some proton, K^+, Cl^-, HCl conductance and allowing some H^+/K^+ exchange.

permeability of lipid bilayers to water also allows us to neglect possible coupling of fluxes other than that of water to the difference between pressure and osmotic gradient.

According to the scheme of MNET we can now list the processes that can take place in the bacteriorhodopsin liposomes and write down the flux-force relation for each of the elemental processes. By adding the fluxes of each of the chemical species, we arrive at a set of equations, represented in matrix form:

$$\left.\begin{aligned}
J_\nu &= (1-\alpha)J_\nu^+ + \alpha J_\nu^- \\
J_\mathrm{H} &= (1-\alpha)J_\mathrm{H}^{\nu+} + \alpha J_\mathrm{H}^{\nu-} + J_\mathrm{H}^1 + J_\mathrm{H}^\mathrm{K} + J_\mathrm{H}^\mathrm{Cl} \\
J_\mathrm{K} &= J_\mathrm{K}^1 + J_\mathrm{K}^\mathrm{H} \\
J_\mathrm{Cl} &= J_\mathrm{Cl}^1 + J_\mathrm{Cl}^\mathrm{H}
\end{aligned}\right\} \tag{45}$$

$$\begin{pmatrix} J_\nu \\ J_\mathrm{H} \\ J_\mathrm{K} \\ J_\mathrm{Cl} \end{pmatrix} = \begin{pmatrix} L_\nu & n(1-2\alpha)L_\nu & 0 & 0 \\ n(1-2\alpha)L_\nu & n^2L_\nu + L_\mathrm{H}^1 + L_\mathrm{KOH} + L_\mathrm{HCl} & -L_\mathrm{KOH} & L_\mathrm{HCl} \\ 0 & -L_\mathrm{KOH} & L_\mathrm{K}^1 + L_\mathrm{KOH} & 0 \\ 0 & L_\mathrm{HCl} & 0 & L_\mathrm{Cl}^1 + L_\mathrm{HCl} \end{pmatrix} \cdot \begin{pmatrix} A_\nu \\ \Delta\tilde{\mu}_\mathrm{H} \\ \Delta\tilde{\mu}_\mathrm{K} \\ \Delta\tilde{\mu}_\mathrm{Cl} \end{pmatrix} \tag{46}$$

The parameter A_ν contains the free-energy of the absorbed photon, as well as the extra constant that may have to be added according to Eqn. 28 [24].

It is instructive to compare this result with that obtained by considering the bacteriorhodopsin liposomes as a black box, with gradients of H^+, K^+, Cl^- and their associated forces, plus the light-driven proton pump with its associated force. According to phenomenological thermodynamics, we would described such a system by the following set of equations in matrix form:

$$\begin{pmatrix} J_\nu \\ J_\mathrm{H} \\ J_\mathrm{K} \\ J_\mathrm{Cl} \end{pmatrix} = \begin{pmatrix} L_{\nu\nu} & L_{\nu\mathrm{H}} & L_{\nu\mathrm{K}} & L_{\nu\mathrm{Cl}} \\ L_{\mathrm{H}\nu} & L_\mathrm{HH} & L_\mathrm{HK} & L_\mathrm{HCl} \\ L_{\mathrm{K}\nu} & L_\mathrm{KH} & L_\mathrm{KK} & L_\mathrm{KCl} \\ L_{\mathrm{Cl}\nu} & L_\mathrm{ClH} & L_\mathrm{ClK} & L_\mathrm{ClCl} \end{pmatrix} \cdot \begin{pmatrix} A_\nu \\ \Delta\tilde{\mu}_\mathrm{H} \\ \Delta\tilde{\mu}_\mathrm{K} \\ \Delta\tilde{\mu}_\mathrm{Cl} \end{pmatrix} \tag{47}$$

The number of different proportionality constants in the latter set of equations is reduced by Onsager's reciprocity relation, which states that $L_\mathrm{ij} = L_\mathrm{ji}$. Such a symmetry is also present in these (though not in all) MNET equations.

The difference between the two sets of equations is [6] that only the MNET equations afford insight in the effect of changes in the underlying processes on the flux-force relations. For instance, a change in the permeability of the membrane towards protons will change the magnitude of L_H^1 and, since this constant only appears in one place, we can see directly that this change will only affect the rate of proton movement (if all the forces are kept equal). Such a statement could not be made in the phenomenological description. It should be stressed, however, that this extra information is a result of the proposed structure of the system described here. In other words: verification of the MNET relations can be used as evidence for the applicability of this proposed structure. The situation is strictly analogous to that encountered in enzyme kinetics: one devises a certain mechanism of action and derives the associated kinetic equations. An experimental test of these equations will allow one only to decide whether that particular mechanism is tenable.

In practice, it is not feasible to test the derived equations experimentally by varying all the forces and fluxes independently. Usually some simplification is gained by allowing the system to develop to a specific steady state. A useful steady state for illuminated bacteriorhodopsin liposomes is that of electroneutral total flow, i.e., the condition in which the net movement across the membrane of all chemical species adds up to no charge movement. It can be derived and shown that this condition is attained within seconds, considering the membrane resistance and electrical capacity in the usual salt media [28]. Electroneutral total flow is mathematically expressed as:

$$J_{\mathrm{H}} + J_{\mathrm{K}} - J_{\mathrm{Cl}} = 0 \tag{48}$$

Combining this equation with Eqn. 46 we obtain:

$$F\Delta\psi^{*e} = \frac{-\mathrm{n}L_\nu(1-2\alpha)A_\nu - \left(n^2L_\nu + L_{\mathrm{H}}^{\mathrm{l}}\right)\Delta\mu_{\mathrm{H}}^{*e} - L_{\mathrm{K}}^{\mathrm{l}}\,\Delta\mu_{\mathrm{K}}^{*e} + L_{\mathrm{Cl}}^{\mathrm{l}}\,\Delta\mu_{\mathrm{Cl}}^{*e}}{L_{\mathrm{eg}}} \tag{49}$$

with:

$$L_{\mathrm{eg}} = n^2L_\nu + L_{\mathrm{H}}^{\mathrm{l}} + L_{\mathrm{K}}^{\mathrm{l}} + L_{\mathrm{Cl}}^{\mathrm{l}} \tag{50}$$

Insertion of this expression ($*$e, electroneutral flow) into Eqn. 48 and some rearrangement leads to the following equations (from which one of the forces, $\Delta\psi$, has been eliminated):

$$\begin{Bmatrix} J_\nu^{*e} \\ J_{\mathrm{H}}^{*e} \\ J_{\mathrm{K}}^{*e} \end{Bmatrix} = 1/L_{\mathrm{eg}} \begin{Bmatrix} \begin{matrix}\left(L_{\mathrm{H}}^{\mathrm{l}} + L_e\right)L_\nu \\ +4\alpha(1-\alpha)n^2L_\nu^2\end{matrix} & nL_\nu L_e(1-2\alpha) & -nL_\nu L_{\mathrm{K}}^{\mathrm{l}}(1-2\alpha) \\ nL_\nu(1-2\alpha)L_e & \begin{matrix}L_e\left(n^2L_\nu + L_{\mathrm{H}}^{\mathrm{l}}\right) \\ +L_nL_{\mathrm{eg}}\end{matrix} & \begin{matrix}-L_{\mathrm{K}}^{\mathrm{l}}\left(n^2L_\nu + L_{\mathrm{H}}^{\mathrm{l}}\right) \\ -L_{\mathrm{KOH}}L_{\mathrm{eg}}\end{matrix} \\ -nL_\nu L_{\mathrm{K}}^{\mathrm{l}}(1-2\alpha) & \begin{matrix}-L_{\mathrm{K}}^{\mathrm{l}}\left(n^2L_\nu + L_{\mathrm{H}}^{\mathrm{l}}\right) \\ -L_{\mathrm{KOH}}L_{\mathrm{eg}}\end{matrix} & \begin{matrix}L_{\mathrm{K}}^{\mathrm{l}}\left(n^2L_\nu + L_{\mathrm{H}}^{\mathrm{l}} + L_{\mathrm{Cl}}^{\mathrm{l}}\right). \\ +L_{\mathrm{KOH}}L_{\mathrm{eg}}\end{matrix} \end{Bmatrix}$$

$$\times \begin{Bmatrix} A_\nu \\ \Delta\mu_{\mathrm{H}}^{*e} + \Delta\mu_{\mathrm{Cl}}^{*e} \\ \Delta\mu_{\mathrm{K}}^{*e} + \Delta\mu_{\mathrm{Cl}}^{*e} \end{Bmatrix} \tag{51}$$

with

$$L_e = L^1_K + L^1_{Cl} \qquad \text{and} \qquad L_n = L_{HCl} + L_{KOH}$$

These equations predict the effect of addition of ionophores on the rate of light-driven proton uptake by the bacteriorhodopsin liposomes. Such predictions have been experimentally tested and verified [23,28–30].

Also, interestingly, an implicit assumption of the description turned out to be applicable, namely that the light-driven pump is inhibited by the electrochemical gradient of protons which it develops. This results in an experimentally observed, hyperbolic rather than a linear dependence of the rate of proton pumping on the light intensity. The predicted inhibitory effect of $\Delta\tilde{\mu}_H$ on the proton pump was later shown more directly. Hellingwerf et al. [28,31], and later Quintanilha [32] and Dancshazy et al. [33], demonstrated that the presence of an electrochemical potential gradient for protons across the liposomal membrane inhibits the photocycle of bacteriorhodopsin. Arents et al. [34] demonstrated that light-driven proton pumping was indeed reduced in the presence of an opposing pH gradient. Bamberg et al. [35] showed that the photopotential, developed by bacteriorhodopsin in a black-lipid film, can be counteracted by an applied electrical counter-potential.

These results show that, for practical purposes, we can treat bacteriorhodopsin as a converter that utilizes light of a certain thermodynamic potential to create a gradient of protons of a certain electrochemical potential, and that the rate at which this converter operates is sensitive to a sort of 'respiratory control' phenomenon: it is inhibited by the proton gradient which it generates (for review see Ref. 36). Such apparently orthodox behaviour of a light-driven proton pump has been held improbable [12], because it would contradict the idea that photochemical reactions are 'irreversible'. At least (but see also Refs. 29, 30] in this sense the application of MNET to bacteriorhodopsin liposomes has had heuristic value.

5.2. Oxidative phosphorylation in mitochondria

As a second example of the application of MNET to biological energy converters we choose a more complicated system, namely the mitochondrion. This organelle is capable of converting the free energy of oxidation of substrates into the free energy of hydrolysis of ATP. Elsewhere in the cell this free energy of hydrolysis of ATP is utilized to drive free-energy-requiring processes.

5.2.1. Stoicheiometries

For the MNET description we take the presently most widely accepted model for this process as a basis: the chemiosmotic model of Mitchell [37]. It comprises three elemental reactions in the mitochondrial membrane (apart from the translocators bringing the substrates to the active site of the enzymes acting upon them) (cf., Fig. 1.3). The membrane itself is supposed to have a certain permeability to protons. The ATP synthase is a reversible H^+ pump, that pumps a certain number of protons across the membrane, coupled to the hydrolysis of ATP. The respiratory chain acts as a reversible pump, which couples the movement of electrons along the respiratory chain to the transmembrane movement of protons. The reversibility of the reactions allows the coupling of movement of electrons along the respiratory chain to synthesis of ATP, with a proton gradient across the mitochondrial membrane as the central coupling agent.

For each of the elemental processes we can write down the flux-force relation as follows:

$$J_{\mathrm{H}}^{1} = L_{\mathrm{H}}^{1} \cdot \Delta\tilde{\mu}_{\mathrm{H}} \tag{52}$$

$$J_0 = L_0^{\dagger} \cdot \left(\Delta G_0^{£} + \gamma_{\mathrm{H}}^{0} \cdot n_{\mathrm{H}}^{0} \cdot \Delta\tilde{\mu}_{\mathrm{H}} \right) \tag{53}$$

$$J_{\mathrm{P}} = L_{\mathrm{P}}^{\dagger} \left(\Delta G_{\mathrm{P}}^{£} + \gamma_{\mathrm{H}}^{\mathrm{P}} \cdot n_{\mathrm{H}}^{\mathrm{P}} \cdot \Delta\tilde{\mu}_{\mathrm{H}} \right) \tag{54}$$

The latter two equations, especially, deserve some comments. First, in $\Delta G^{£}((= \Delta G - \Delta G^{\dagger})$ we recognize the extra constant that may be present in the flow-force relations of enzyme catalyzed reactions (see Eqn. 28). Second, the γ factors in these equations extend the non-equilibrium thermodynamic description of enzyme-catalyzed reactions to the general case where the sensitivity of the reaction rate to changes in the concentrations of one substrate product couple is not equal to its sensitivity to changes in the other substrate-product couple. An example of this phenomenon is found in the reaction of the mitochondrial respiratory chain: whereas the reaction rate is extremely sensitive to changes in the free-energy difference across which it pumps the protons, it is almost insensitive to changes in the free-energy difference of the second substrate-product couple, i.e., O_2 and H_2O. This differential sensitivity is a characteristic of the enzyme and has to be reflected by the MNET description of the flow-force relationships of that enzyme. The γ factors see to this. The H^+ fluxes associated to the redox reaction and the ATP-synthase reaction are:

$$J_{\mathrm{H}}^{0} = n_{\mathrm{H}}^{0} J_0 = n_{\mathrm{H}}^{0} L_0^{\dagger} \left(\Delta G_0^{£} + \gamma_{\mathrm{H}}^{0} n_{\mathrm{H}}^{0} \, \Delta\tilde{\mu}_{\mathrm{H}} \right) \tag{55}$$

$$J_{\mathrm{H}}^{\mathrm{P}} = n_{\mathrm{H}}^{\mathrm{P}} J_{\mathrm{P}} = n_{\mathrm{H}}^{\mathrm{P}} L_{\mathrm{P}}^{\dagger} \left(\Delta G_{\mathrm{P}}^{£} + \gamma_{\mathrm{H}}^{\mathrm{P}} n_{\mathrm{H}}^{\mathrm{P}} \, \Delta\tilde{\mu}_{\mathrm{H}} \right) \tag{56}$$

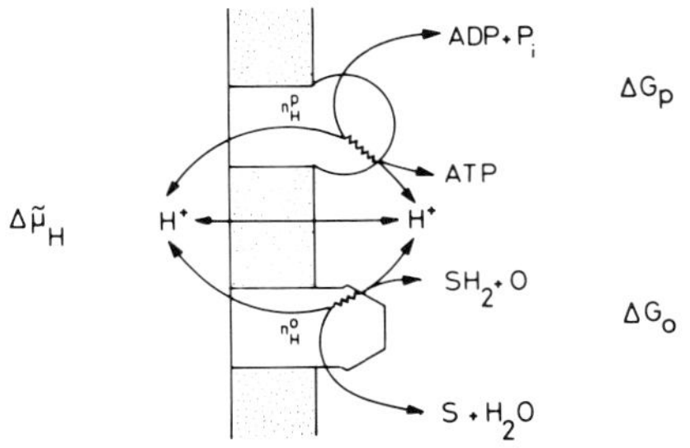

Fig. 1.3. A scheme of oxidative phosphorylation according to the chemiosmotic coupling hypothesis.

and the total H^+ flux is given by:

$$J_H = J_H^1 + J_H^0 + J_H^P \tag{57}$$

The complete set of equations, describing oxidative phosphorylation according to the chemiosmotic model, is as follows:

$$J_0 = L_0^* \,\Delta G_0^{ex} + \gamma_H^0 n_H^0 L_0^* \,\Delta\tilde{\mu}_H - L_0^* \,\Delta G_0^\dagger \tag{58}$$

$$J_H = n_H^0 L_0^* \,\Delta G_0^{ex} + \left(\gamma_H^0 \left(n_H^0\right)^2 L_0^* + \gamma_H^P \left(n_H^{P*}\right)^2 L_P^* + L_H^1\right) \Delta\tilde{\mu}_H$$

$$+ n_H^{P*} L_P^* \,\Delta G_P^{ex} - n_H^0 L_0^* \,\Delta G_0^\dagger - n_H^P L_P^* \,\Delta G_P^\dagger \tag{59}$$

$$J_P = \gamma_H^P n_H^{P*} L_P^* \,\Delta\tilde{\mu}_H + L_P^* \,\Delta G_P^{ex} - L_P^* \,\Delta G_P^\dagger \tag{60}$$

in which the asterisks have been included to account for the fact that the proportionality constants include the transport of the substrates. Furthermore, the number of protons translocated per ATP hydrolyzed (n_H^{P*}) includes any net proton movement coupled to the exchange of ATP for ADP and phosphate.

Again, it is convenient to select certain special cases to simplify the measurements. One obvious choice is the steady state of zero net proton movement across the membrane ($J_H = 0$). This state is usually reached rapidly, because of the small internal volume of the mitochondria: soon after an addition the maximal attainable proton gradient will have been established by the proton pumps and net movement of protons ceases. The steady-state condition ($J_H = 0$) allows us to mathematically eliminate one of the thermodynamic forces. Since $\Delta\tilde{\mu}_H$ is technically the most difficult to measure, we choose to eliminate this force. The elimination leads to the following relation between the rates of phosphorylation and oxidation:

$$J_P = -\left\{\frac{n_H^0}{n_H^{P*}} + \frac{L_H^1}{\gamma_H^0 n_H^0 n_H^{P*} L_0^*}\right\} \cdot J_0 + \frac{L_H^1}{\gamma_H^0 n_H^0 n_H^{P*}} \cdot \Delta G_0^{ex} - \frac{L_H^1 \Delta G_0^\dagger}{\gamma_H^0 n_H^0 n_H^{P*}} \tag{61}$$

The equation predicts that, at constant ΔG_0^{ex}, there should be a linear relation between rate of ATP synthesis ($-J_P$) and rate of oxygen uptake (J_0) by mitochondria. Furthermore, the slope of the line relating the two processes should increase when the membrane is made more permeable towards protons by addition of an uncoupler of oxidative phosphorylation (L_H^1 increases). Finally, the lines at different uncoupler concentrations should all intersect at a point where $-J_P/J_0 = n_H^0/n_H^{P*}$. Since the n values represent the molecular stoicheiometries, this intersection point gives us directly the 'theoretical P/O ratio'.

These predictions of Eqn. 61 were tested experimentally and were generally found to be consistent with the observations [38]. It is noteworthy that this method is one of the few available to allow an experimental determination of the P/O ratio, corrected for leakage of protons through the mitochondrial membrane. It should be noted also, however, that the procedure rests on two assumptions, i.e., that the delocalized chemiosmotic coupling scheme correctly describes mitochondrial free-energy transduction, and that the addition of FCCP does indeed solely increase the passive proton permeability of the inner mitochondrial membrane. Recently [27] it has been suggested that the main effect of FCCP is to increase slip in the redox driven proton pumps (cf., Section 5.2.3). Also the former assumption is sometimes challenged, e.g., as in the subsequent section.

5.2.2. Localization of the high-free-energy proton

A second special case is that where the ATP synthesis has come to a stop, because the maximal level of phosphorylation of the adenine nucleotides has been reached (State 4 condition). Assuming that no contaminating ATP hydrolyzing enzymes are present, this is the condition where the ATP-driven proton pump has come to a true equilibrium. From the relation $J_P = 0$, we derive easily:

$$\left(\frac{\Delta G_P^{ex}}{-\Delta\tilde{\mu}_H}\right)_{J_P=0} = n_H^{P*} \tag{62}$$

This relation was derived earlier on the basis of simple thermodynamic considerations. Experiments [39,40] show that the predicted constancy of the force ratio is not maintained if $\Delta\tilde{\mu}_H$ is decreased to low values. Interestingly, the experimentally determined ratio depends on $\Delta\tilde{\mu}_H$ in the same way, if $\Delta\tilde{\mu}_H$ is varied by an uncoupler or by an inhibitor of the respiratory chain [41].

The inconsistency between experiment and prediction must lead to the rejection of the model used to describe the system. In the case of oxidative phosphorylation this has led to a refined model, in which the chemiosmotic coupling is visualized as taking place within units of one (or a few) respiratory chain(s) plus ATP synthase, while the pumped protons have only limited access to the bulk phase inside and/or outside the mitochondrion [42]. This more refined model can again be tested by deriving from it flux-force relations according to the MNET approach. A discussion of the refined model can be found in Ref. 43.

5.2.3. Slipping proton pumps

The possibility that slip in the proton pumps, rather than passive leakage of protons across the inner mitochondrial membrane, is responsible for much of the uncoupling observed has been examined by Pietrobon et al. [44,45]. In analogy to the description of a slipping sugar transport system given above, a slipping redox driven proton pump may be described as (a very similar treatment has been presented earlier by Pietrobon and colleagues [26,27,45]):

$$J_0/L_0 = \Delta G_0 + q \cdot Z \cdot \Delta\tilde{\mu}_H \tag{63}$$

$$J_H^0/L_0 = q \cdot Z \cdot \Delta G_0 + Z^2 \cdot \Delta\tilde{\mu}_H \tag{64}$$

where the degree of coupling of that pump (q) and the phenomenological stoicheiometry (Z) will deviate from -1 and the mechanistic stochiometry, respectively, whenever there is slip in the pump. In addition to this pump there may be a finite proton back leakage across the membrane characterized by Eqn. 52. The steady-state condition of net zero proton movement, dictates that the two proton fluxes must be equal in magnitude, but opposite in sign. This extra equation allows us to eliminate L_0 (which is characteristic for the number of redox driven proton pumps) from the equations and thus derive the following relationship between the rate of electron transfer and $\Delta\tilde{\mu}_H$:

$$\frac{J_0 Z}{-L_H^1\,\Delta\tilde{\mu}_H} = \frac{1 + q \cdot Z \cdot \Delta\tilde{\mu}_H/\Delta G_0}{q + Z \cdot \Delta\tilde{\mu}_H/\Delta G_0} \tag{65}$$

In Fig. 1.4A we show a plot of what this relationship predicts for an experiment in which mitochondrial respiration is titrated away by the addition of respiratory inhibitor, for the case that $q = -1$ (no slip, dashed line) and cases in which $q > -1$ (slip, full lines). Pietrobon et al. [44, see also Refs. 46 and 47] have actually carried out this experiment and in Fig. 1.4B we reproduce their findings. It is clear that the experimental results are in line with the idea [44] that there is significant slip in the

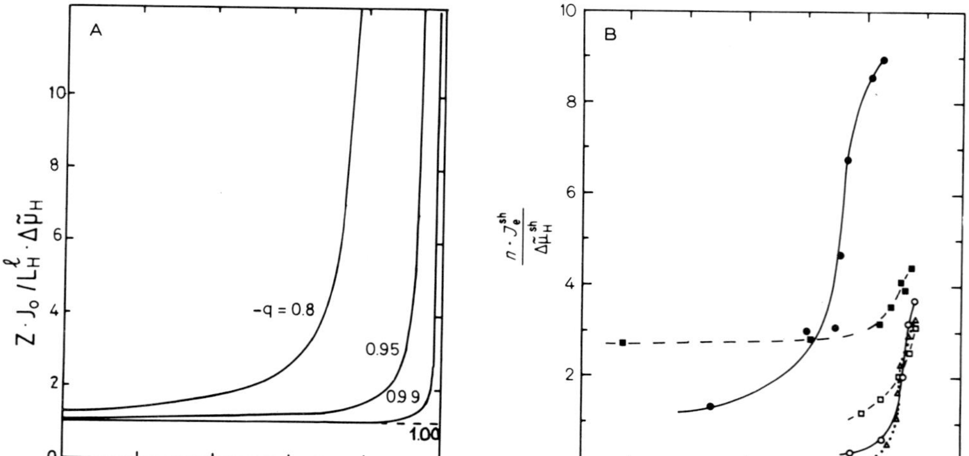

Fig. 1.4. Slipping proton pumps. (A) Simulation of the relationship between electron transfer rate (J_0) and $\Delta\tilde{\mu}_H$ for different degrees of coupling (q) of an electron transferring proton pump. $\Delta\tilde{\mu}_H$ expressed in units $Z \cdot \Delta G_0$. (B) Actual experimental results (from Ref. 44 with kind permission from the authors and the copywright owner) for different segments of the mitochondrial respiratory chain as proton pump. Both in (A) and in (B) the rate of electron transfer is varied through titration with an electron transfer inhibitor.

mitochondrial respiratory chain. The same authors were able to rule out the alternative possibility that 'nonohmicity' in the flow-force relationship of the proton leak itself was responsible for the experimental observations, by showing that the point at which $J_0/\Delta\tilde{\mu}_H$ begins its sharp increase, is different when different proton pumps are used. Only recently [43] it has been pointed out that the experimental observations can also be explained in a scheme where the proton pumps themselves do not slip, but where the high free-energy protons are not fully delocalized.

5.3. Bacterial growth

We will now demonstrate the utility of MNET by applying it to an even more complicated system, namely a population of duplicating cells. Thermodynamic consideration of microbial growth [e.g., Refs. 48 and 49] has led to the conclusion that the efficiency would be lower than 100%. Non-equilibrium thermodynamics with the inclusion of its symmetrical flow-force relations has only recently been applied to the description of the behaviour of a bacterial culture [22]. This allowed insight into the conversion of the free energy of catabolism into the free energy of biomass synthesis. However, phenomenological thermodynamics makes no use of the existing knowledge of the biochemical structure of the bacteria. MNET can, in principle, account for this knowledge in any desired detail. For the purpose of the present discussion, we [50] only consider catabolism as leading to the synthesis of intracellular ATP, and anabolism as leading to biomass synthesis at the expense of ATP, while there is also ATP hydrolysis that is not coupled to either process.

The simplified picture of a bacterium, defined in this way, is given in Fig. 1.5. By comparing it to Fig. 1.3 it is immediately clear that there is a strong analogy to the mitochondrial oxidative phosphorylation. We can set catabolism of the bacterium equivalent to oxidation in the mitochondrion, anabolism to ATP synthesis, and the free energy of ATP hydrolysis to the free energy of the proton gradient. Using this analogy, it is easy to derive the equations describing bacterial growth and metabolism:

$$J_a = L_a \cdot \left[\left(\Delta G_a - \Delta G_a^{\dagger} \right) + n_P^a \cdot \gamma_P^a \cdot \left(\Delta G_P - \Delta G_P^{\dagger} \right) \right] \tag{66}$$

$$J_P^a = n_P^a \cdot J_a \tag{67}$$

$$J_c = L_c \cdot \left[\left(\Delta G_c - \Delta G_c^{\dagger} \right) + n_P^c \cdot \gamma_P^c \cdot \left(\Delta G_P - \Delta G_P^{\dagger} \right) \right] \tag{68}$$

$$J_P^c = n_P^c \cdot J_c \tag{69}$$

To arrive at testable predictions, we go one step further by taking the steady state of constant intracellular ATP concentration, a condition that is reached very quickly [22,50]. This steady-state condition allows us to eliminate one of the forces, prefer-

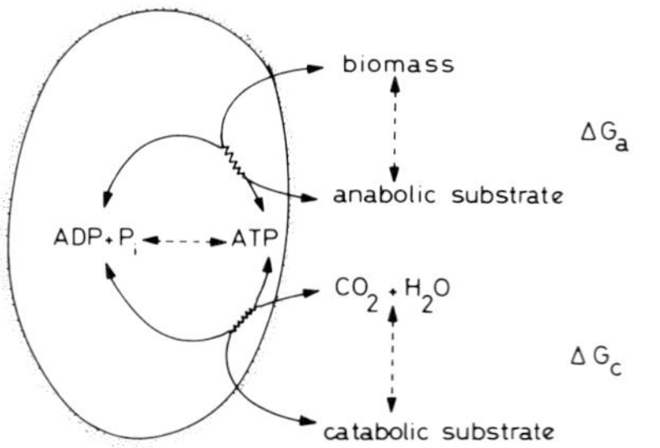

Fig. 1.5. A scheme of microbial growth stressing its analogy with other biological energy converters.

ably the intracellular ΔG_P. Depending on how we eliminate, we arrive at either of two equations, relating rate of substrate utilization and biomass synthesis:

$$J_c = \left\{ \frac{n_P^a}{n_P^c} \cdot \left(1 + \frac{L_P^l}{\gamma_P^a \cdot (n_P^a)^2 \cdot L_a} \right) \right\} \cdot (-J_a) + \left\{ \frac{L_P^l}{n_P^a \cdot n_P^c \cdot \gamma_P^a} \right\} \cdot \left(\Delta G_a - \Delta G_a^\dagger \right) \qquad (70)$$

$$J_c = \left\{ \frac{n_P^a}{n_P^c} \cdot \left(1 - \frac{L_P^l}{L_P^l + \gamma_P^c \cdot (n_P^c)^2 \cdot L_c} \right) \right\} (-J_a)$$

$$+ \left\{ \frac{L_P^l \cdot L_c}{L_P^l + \gamma_P^c \cdot (n_P^c)^2 \cdot L_c} \right\} \cdot \left(\Delta G_c - \Delta G_c^\dagger \right) \qquad (71)$$

Both equations are equally valid. It turns out, however, that the practical applicability differs according to the experimental conditions.

To study bacterial growth, microbiologists often use the so-called chemostat [51]. This is a closed vessel, containing a bacterial culture, into which fresh growth medium is pumped at a constant rate. At the same time, full-grown medium flows from the vessel at the same rate, such that the culture volume remains constant. After some time a steady state sets in, where one of the nutrients is limiting: its influx is almost exactly balanced by its use for growth and efflux of new bacteria. In principle, such a steady state will last indefinitely. Therefore, the chemostat offers an ideal instrument to determine the steady-state parameters during growth. The growth rate itself can be varied by varying the dilution rate in the chemostat. Of course the actual variation that induces an increase in growth rate is the concomitant variation of the growth limiting substrate.

In the simplest case, where catabolism and anabolism follow completely independent routes, we can define either catabolism or anabolism as 'limiting'. When one of the catabolic substrates is limiting, the free energy of anabolism is effectively constant (in biochemical terms: substrates for anabolism are always 'saturating')

[22,50]. This means that, under conditions of catabolite limitation, Eqn. 70 will give the most easily interpretable relation. Conversely, under anabolite limitation, Eqn. 71 will be the most practical. Both equations predict a linear relation between rate of substrate utilization and biomass formation. Furthermore, the relation between catabolism and anabolism has a positive intersection point with the ordinate. This positive catabolism at (extrapolated) zero growth rate has been interpreted as 'maintenance energy requirement' [52]. It follows naturally from the simple description of bacterial metabolism as we have used it here.

Some further information can be gained from a comparison of the predicted slopes of the lines relating catabolism and anabolism. Since n_P^a and n_P^c are the molecular stoicheiometries coupling anabolism to ATP synthesis or catabolism to ATP synthesis, respectively, the ratio n_P^a/n_P^c gives the 'theoretical stoicheiometry between growth and catabolism' (compare the 'theoretical P/O ratio'). Eqn. 70 shows that, under catabolite limitation, the slope of the rate of catabolism (J_c) versus growth rate ($-J_a$, or dilution rate) will be greater than that theoretical value. Conversely, under anabolite limitation conditions, the slope will be smaller than the theoretical value. Experiments reported in the literature confirm that the slope is higher under anabolite limitation than under catabolite limitation [53].

6. Prospects

The field of bioenergetics needs a rigorous quantitative description of processes to test proposed molecular mechanisms for these processes. Although one might think that kinetics alone would be sufficient to cover this need, in practice it is impossible to include all properties of the participating enzymes in a complex system in a complete kinetic description. Furthermore, kinetics alone will not reveal the thermodynamic limitations of the possible pathways. Therefore, a fruitful symbiosis of kinetics and thermodynamics has been developed over the past years.

In this chapter, we have only touched upon a number of examples, showing how this approach can be used in practice. For further details of these applications, the reader should consult the original papers or the more extensive treatment in Ref. 24. In doing so, it will be discovered that many further implications of the thermodynamic approach have been neglected here. For instance, the consideration of bioenergy converters as machines and their optimization with respect to certain output parameters has been extensively discussed by Stucki [54]. He also suggested that biological energy converters may have been designed in such a way that their flux-force relations are proportional over a very wide region, leading to a higher efficiency [16]. Finally, the introduction of thermodynamic notions in the description of control of bioenergy conversion has improved our insight in the complex regulation of biological systems [24,55].

We feel that the application of rigorous thermodynamic principles to biology has only just started bear practical fruit. We expect to see many more applications in the near future: if students do not fear to combine their physicochemical knowledge with their biochemical interest.

References

1 Onsager, L. (1931) Phys. Rev. 37, 405–426.
2 Morowitz, H.J. (1978) Foundations of Bioenergetics. Academic Press, New York.
3 Kedem, O. and Caplan, S.R. (1965) Trans. Faraday Soc. 21, 1897–1911.
4 Katchalsky, A. and Curran, P.F. (1967, 1974 2nd) Non-Equilibrium Thermodynamics in Biophysics. Harvard University Press, Cambridge, MA.
5 Rottenberg, H., Caplan, S.R. and Essig, A. (1970) in Membranes and Ion Transport (Bittar, E.E., ed.) pp. 165–191, Interscience, New York.
6 Wilson, D.F. and Westerhoff, H.V. (1982) Trends Biochem. Sci. 7, 275–279.
7 Blumenthal, R., Caplan, S.R. and Kedem, O. (1967) Biophys. J. 7, 735–757.
8 Westerhoff, H.V. and Van Dam, K. (1979) Curr. Top. Bioenerg. 9, 1–62.
9 Nicolis, G. and Prigogine, I. (1977) Self-Organization in Nonequilibrium Systems. Wiley and Sons, New York.
10 Keizer, J. (1975) J. Theor. Biol. 49, 323–335.
11 Rottenberg, H. (1973) Biophys. J. 13, 503–511.
12 Hill, T.L. (1977) Free Energy Transduction in Biology. Academic Press, New York.
13 Van der Meer, R., Westerhoff, H.V. and Van Dam, K. (1980) Biochim. Biophys. Acta 591, 488–493.
14 Prigogine, I., Outer, P. and Herbo, Cl. (1948) J. Phys. Coll. Chem. 52, 321–331.
15 Essig, A. and Caplan, S.R. (1981) Proc. Natl. Acad. Sci. U.S.A. 78, 1647–1651.
16 Stucki, J.W. (1983) In Biological Structure and Coupled Flows (Oplatka, A. and Balaban, M., eds.) pp. 33–41, Academic Press, New York.
17 Reich, J.G. and Sel'kov, E.E. (1981) Energy Metabolism of the Cell. A Theoretical Treatise. Academic Press, New York.
18 De Groot, S.R. and Mazur, P. (1962) Non-Equilibrium Thermodynamics. North-Holland, Amsterdam.
19 Cleland, W.W. (1963) Biochim. Biophys. Acta 67, 104–137.
20 Rottenberg, H. (1976) FEBS Lett. 66, 159–163.
21 Küster, U., Bohnensack, R. and Kunz, W. (1976) Biochim. Biophys. Acta 440, 391–402.
22 Westerhoff, H.V., Lolkema, J.S., Otto, R. and Hellingwerf, K.J. (1982) Biochim. Biophys. Acta 683, 181–220.
23 Westerhoff, H.V., Hellingwerf, K.J., Arents, J.C., Scholte, B.J. and Van Dam, K. (1981) Proc. Natl. Acad. Sci. U.S.A. 78, 3554–3558.
24 Westerhoff, H.V. and Van Dam, K. (1984) Mosaic Non-Equilibrium Thermodynamics of (the Control of) Energy Metabolism. Elsevier, Amsterdam, in the press.
25 Van Dam, K., Westerhoff, H.V., Rutgers, M., Bode, J.A., De Jonge, P.C., Bos, M.M. and Van den Berg, G. (1981) In Vectorial Reactions in Electron and Ion Transport in Mitochondria and Bacteria (Palmieri, F., Quagliariello, E. Siliprandi, N. and Slater, E.C., eds.) pp. 389–397, Elsevier, Amsterdam.
26 Pietrobon, D., Zoratti, M., Azzone, G.F., Stucki, J.W. & Walz, D. (1982) Eur. J. Biochem. 127, 483–494.
27 Walz, D. (1983) In Biological Structures and Coupled Flows (Oplatka, A. and Balaban, M., eds.) pp. 45–60, Academic Press, New York.
28 Hellingwerf, K.J., Arents, J.C., Scholte, B.J. and Westerhoff, H.V. (1979) Biochim. Biophys. Acta 547, 561–582.
29 Arents, J.C., Hellingwerf, K.J., Van Dam, K. and Westerhoff, H.V. (1981) J. Membrane Biol. 60, 95–104.
30 Westerhoff, H.V., Arents, J.C. and Hellingwerf, K.J. (1981) Biochim. Biophys. Acta 637, 69–79.
31 Hellingwerf, K.J., Schuurmans, J.J. and Westerhoff, H.V. (1978b) FEBS Lett. 92, 181–186.
32 Quintanilha, A.T. (1980) FEBS Lett. 117, 8–12.
33 Dancshazy, Zs., Helgerson, S.L. and Stoeckenius, W. (1983) Photobiochem. Photobiophys. 5, 347–358.
34 Arents, J.C., Van Dekken, H., Hellingwerf, K.J. and Westerhoff, H.V. (1981) Biochemistry 20, 5114–5123.

35 Bamberg, E., Dencher, N.A., Fahr, A. and Heyn, M.P. (1981) Proc. Natl. Acad. Sci. U.S.A. 78, 7502–7506.
36 Westerhoff, H.V. and Dancshazy, Zs. (1984) Trends Biochem. Sci. 9, 112–117.
37 Mitchell, P. (1961) Nature (London) 191, 144–148.
38 Van Dam, K., Westerhoff, H.V., Krab, K., Van der Meer, R. and Arents, J.C. (1980) Biochim. Biophys. Acta 591, 240–250.
39 Wiechmann, A.H.C.A., Beem, E.P. and Van Dam, K. (1975) In Electron Transfer Chains and Oxidative Phosphorylation (Quagliariello, E., Papa, S., Palmieri, F., Slater, E.C. and Siliprandi, N., eds.) pp. 335–342, North Holland, Amsterdam.
40 Azzone, G.F., Pozzan, T., Massari, S. and Pregnolato, L. (1978) Biochim. Biophys. Acta 501, 307–316.
41 Westerhoff, H.V., Simonetti, A.L.M. and Van Dam, K. (1981) Biochem. J. 200, 193–202.
42 Westerhoff, H.V., Colen, A.-M. and Van Dam, K. (1983) Biochem. Soc. Trans., 11, 81–85.
43 Westerhoff, H.V., Melandri, B.A., Venturoli, G., Azzone, G.F. and Kell, D.B. (1984) FEBS Lett. 165, 1–5.
44 Pietrobon, D., Azzone, G.F. and Walz, D. (1981) Eur. J. Biochem. 117, 389–394.
45 Pietrobon, D., Zoratti, M. and Azzone, G.F. (1983) Biochim. Biophys. Acta 723, 317–321.
46 Nicholls, D.G. (1974) Eur. J. Biochem. 50, 305–315.
47 Sorgato, M.C. and Ferguson, S.J. (1979) Biochemistry 18, 5737–5742.
48 Thauer, R.K., Jungermann, K. and Decker, K. (1977) Bacteriol. Rev. 41, 100–180.
49 Roels, J.A. (1983) Energetics and Kinetics in Biotechnology. Elsevier, Amsterdam.
50 Hellingwerf, K.J., Lolkema, J.S., Otto, R., Neijssel, O.M., Stouthamer, A.H., Harder, W., Van Dam, K. and Westerhoff, H.V. (1982) FEMS Microbiol. Lett. 15, 7–17.
51 Monod, J. (1950) Ann. Inst. Pasteur 79, 390–410.
52 Pirt, S.J.. (1965) Proc. Roy. Soc. Lond. (Biol.) 163, 224–231.
53 Neijssel, O.M. and Tempest, D.W. (1976) Arch. Microbiol. 107, 215–221.
54 Stucki, J.W. (1980) Eur. J. Biochem. 109, 269–283.
55 Westerhoff, H.V., Groen, A.K., Wanders, R.J.A. and Tager, J.M. (1983) Biosc. Rep. 4, 1–22.

Ernster (ed.) Bioenergetics
© 1984 Elsevier Science Publishers B.V.

CHAPTER 2

Mechanisms of energy transduction

DAVID NICHOLLS

Neurochemistry Laboratory, Department of Psychiatry, Ninewells Medical School, Dundee University, Dundee DD1 9SY, Scotland

1. Introduction

This chapter is concerned with the 'proton circuit' which in the chemiosmotic scheme links the generators and utilizers of proton electrochemical potential ($\Delta\mu_{H^+}$) [1,2]. Since the topic has been covered extensively in a recent monograph [3], this chapter will attempt to avoid repetition by concentrating on the ionic circuitry found in association with energy conserving organelles, and no attempt will be made to discuss the structures of the 'black boxes' of the membrane.

2. The basic features of the chemiosmotic theory

2.1. Principles

Energy-conserving (or energy-transducing) membranes possess the two classes of proton-pumping complex which are together required for the transduction of respiratory or photosynthetic energy into energy stored as ATP, or more precisely in the form of a displacement of the reactants and products of the ATPase reaction from equilibrium. The energy transducing membranes are the plasma membrane of prokaryotic bacteria and blue-green algae, the inner membrane of mitochondria, and the thylakoid membrane of chloroplasts.

One of these complexes, the ATP synthase, sometimes called the proton-translocating ATPase, is present in closely similar forms in all energy-transducing membranes. When the complex is assembled in the membrane it catalyzes a hydrolysis or synthesis of ATP which is coupled to the obligatory translocation of protons across the membrane.

The nature of the second proton-pumping complex in each membrane is dependent on the primary energy source utilized by the organelle. In the case of mitochondria or respiring bacteria a respiratory chain (also called an electron-transfer chain) transfers electrons from a donor substrate to an acceptor, often oxygen, at a more positive oxido-reduction potential, the process again being linked to obligatory proton translocation across the membrane. In the case of chloroplasts and photosynthetic bacteria, the obligatory linkage in the photosynthetic electron transfer chain is between photon capture and proton translocation.

The orientation of the two classes of complexes in each membrane is such that, in the absence of any thermodynamic back-pressure from a proton gradient, they would pump protons in the same direction. The means by which this assembly can be used for the continuous transduction of energy from substrate oxidation or photon capture to ATP synthesis can best be understood if the thermodynamic pressures operative on the ATP synthase are examined in detail.

If the electron transfer chain in the membrane is inhibited and an amount of ATP is made available, the ATP synthase would attempt to hydrolyze the ATP and pump protons across the membrane. This process would not continue indefinitely, since the ATP concentration would fall, and that of ADP and P_i would rise, lowering the Gibbs energy for the ATP hydrolysis. At the same time a $\Delta\mu_{H^+}$ would build up across the membrane, resulting in an increasing thermodynamic back-pressure against which further protons are pumped. Thermodynamic equilibrium will be attained when the Gibbs energy for ATP hydrolysis (ΔG_{ATP}) balances that required to pump further protons against the opposing $\Delta\mu_{H^+}$, i.e.,

$$-\Delta G_{ATP} = n \cdot \Delta\mu_{H^+} F$$

where n is the number of protons translocated for the hydrolysis of one ATP, and F is the Faraday constant.

Since the ATP synthase is reversible, any displacement from this equilibrium which increases $\Delta\mu_{H^+}$ or lowers ΔG_{ATP} would cause the complex to reverse, allowing protons to flow back down their electrochemical potential and re-synthesize ATP. In energy-transducing organelles the function of the second proton pump in the membrane (respiratory chain or photosynthetic electron transfer chain) is to create the conditions for such a reversal by continuously replenishing $\Delta\mu_{H^+}$. In this way ATP hydrolyzed by intra-cellular ATP-requiring reactions can be continuously replaced.

Protons thus continuously circulate across the membrane during ATP synthesis, driven against their electrochemical potential by the primary generator, and flowing down their electrochemical potential through the ATP synthase. This proton circuit is the most fundamental feature of the chemiosmotic theory, and the rest of this chapter is concerned with the evidence for its existence, and with its quantitation, control and coupling to other energy-requiring systems.

2.2. Energy flow pathways in mitochondria

The major pathways of energy flow in mitochondria (Fig. 2.1) were established prior to the general acceptance of the chemiosmotic theory [4] and thus provided a framework for the original chemiosmotic hypothesis [5–7]. In detail:

a. The mitochondrial respiratory chain possesses three regions where the ΔG

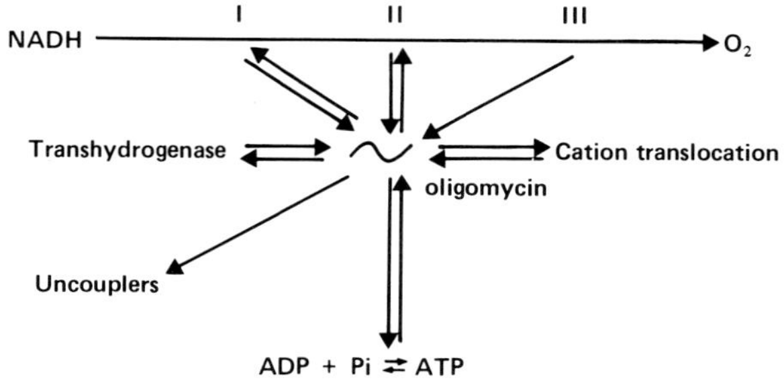

Fig. 2.1. Pathways of energy flow in mitochondria. After Ernster and Lee, 1964. The 'squiggle' represents the undefined energy-transducing intermediate. Note that not all energy-transfer pathways are reversible.

available from electron flow can be conserved for ATP synthesis, the accumulation of certain cations such as Ca into the matrix against their electrochemical potential gradient, or the thermodynamically unfavourable reduction of NAD and NADP.

b. In anaerobic mitochondria the hydrolysis of ATP can drive the other energy requiring processes.

c. Respiratory rate is controlled by ATP demand.

d. 'Uncouplers' abolish respiratory control and promote ATP hydrolysis.

e. Both the synthesis and uncoupler-stimulated hydrolysis of ATP are inhibited by oligomycin, an inhibitor of the ATP synthase.

2.3. The four postulates for the experimental verification of the chemiosmotic theory

2.3.1. The energy-transducing membrane is topologically closed and has a low proton permeability

In order to maintain a $\Delta\mu_{H^+}$ across a membrane, and to ensure that it is used for the synthesis of ATP and not dissipated by leakage, the membrane must be closed and not leaky to protons. From the rate at which a pH gradient across the membrane decayed, it was shown that the effective proton conductance of the mitochondrial inner membrane [8], bacterial plasma membrane [9], and chloroplast thylakoid membrane [10] have a value of only some 0.5 $\mu\Omega$/cm, or a million-fold less than the aqueous phases on either side.

The 'uncouplers' which abolish the coupling of respiratory rate to ATP synthesis act as proton translocators, inducing net proton translocation across the membranes. In this way the proton circuit can be 'short-circuited', allowing the protons translocated by the generator of $\Delta\mu_{H^+}$ to cross back across the membrane without passing through the ATP synthase and producing ATP. The majority of the uncouplers are protonatable, lipophilic compounds with an extensive pi-orbital system which allows the electron of the anionic, de-protonated form to be delocalized [11]. This enhances the permeability of the anionic form in the hydrophobic membrane, and allows the proton translocators to permeate in both their neutral (protonated) and anionic (deprotonated) forms. In this way they can catalyze the net transport of protons across the membrane down their electrochemical gradient. This can be seen in non-respiring mitochondria as an enhancement in the decay of a pH gradient [8], or as an increase in the rate of light scattering decrease due to swelling in media in which proton permeation is the limiting factor [12]. Note that in this latter example neither the respiratory chain nor the ATP synthase are functioning, eliminating the possibility that uncouplers interfere directly with the proton translocating complexes themselves.

2.3.2. There are proton- (or OH^-)-linked solute carrier systems for metabolite transport and osmotic stabilization

This postulate is not an absolute prerequisite for chemiosmotic coupling, but was suggested to reconcile the uptake of predominantly anionic metabolites into the mitochondrial matrix against an opposing membrane potential. Electroneutral proton-linked transport systems equilibrate with the pH gradient, the metabolites being accumulated in the matrix when, as is usual, the matrix is alkaline with respect to the incubation medium. The converse problem of excreting anionic metabolites from the matrix by proton-linked carriers is not encountered in practice, since the pH gradient is a relatively small component of the total $\Delta\mu_{H^+}$ [13].

A number of the metabolite transport systems are not electroneutral, but deliberately exploit the membrane potential to alter the distribution across the membrane. One example is the mitochondrial adenine nucleotide translocator, required because ATP is generated in the mitochondrial matrix and utilized in the cytosol. Since the translocator exchanges ADP^{3-} for ATP^{4-} the membrane potential will generate an asymmetric distribution of adenine nucleotides across the membrane [14]. The P_i required for phosphorylation must also be imported from the cytosol, and the combined effect of the electrogenic adenine nucleotide translocator and the electroneutral P_i/OH exchanger is to cause the influx of an additional proton per ATP synthesized (Fig. 2.2b). The potential of this proton is not wasted, however, since it leads to the accumulation of the substrates (ADP and P_i) of the ATP synthase, and to the export of the product (ATP). Thus a substantial proportion of the ΔG_{ATP} of the adenine nucleotide pool comes not from proton re-entry through the ATP synthase itself, but through these ancillary transport processes. As a result of this additional contribution, mitochondria can maintain a ΔG_{ATP} for the cytosolic adenine nucleotide pool which is considerably higher than found in the matrix, or in bacteria which have no need for such a transport system.

2.3.3. The ATP synthase is a reversible proton-translocating ATPase

The initial experiments which were important for the verification of the chemiosmotic hypothesis were those which showed that the complex was an autonomous proton pump when hydrolyzing ATP, and which showed that an artificial $\Delta\mu_{H^+}$ could cause the ATP synthase to generate ATP. Thus, if a limiting amount of ATP is injected into an anaerobic mitochondrial incubation a net expulsion of protons is observed, followed by a decay which is accelerated by proton translocators [15]. Less complications arise if the experiment is repeated with inverted sub-mitochondrial

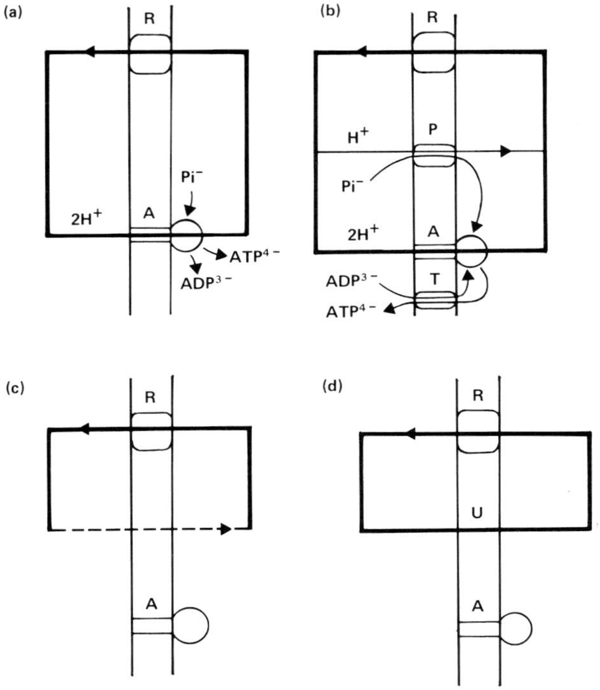

Fig. 2.2. The chemiosmotic proton circuit. a, during the synthesis of matrix ATP (State 3); b, during the synthesis of extra-mitochondrial ATP (State 3); c, during State 4 respiration; d, in the presence of proton translocator. R, respiratory chain; A, ATP synthase; P, phosphate carrier; U, uncoupler. The respiratory chain is simplified to a single proton pump.

particles [16], since the catalytic unit of the ATP synthase is now exposed on the outer face of the membrane. Purified ATP synthase reconstituted into phospholipid bilayers catalyzes a similar translocation [17].

In the synthetic direction, an artificially applied $\Delta\mu_{H^+}$ has been shown to result in net ATP synthesis when the purified complex from the thermophilic bacterium PS3 was incorporated into a synthetic lipid bilayer vesicle [18].

2.3.4. The respiratory and photosynthetic electron-transfer pathways are proton pumps operating with the same polarity as does the ATP synthase when hydrolyzing ATP

Since it is difficult to detect protons circulating in the steady-state, Mitchell and Moyle [19] studied the transient extrusion of protons when a small amount of oxygen is injected into an anaerobic incubation of mitochondria in the presence of substrate. Prior to this, Neumann and Jagendorf [20] had observed a light-dependent proton uptake into chloroplast thylakoid membranes.

The polarity of proton translocation during these transients is as predicted, namely extrusion from mitochondria and uptake into thylakoid membranes and inverted sub-mitochondrial particles. The polarity of the membrane potential maintained by continuous electron flow is consistent with the proton movements during transients, the mitochondrial matrix being negative with respect to the incubation, while the thylakoid space and the inner space of inverted sub-mitochondrial particles are positive.

As in the case of the ATP synthase, the most convincing evidence for the function of the respiratory chain as an autonomous proton pump comes from the ability to purify 'energy-conserving' segments of the chain and reconstitute them into artificial bilayers with the recovery of their proton translocating capacity [21,22].

3. The proton circuit

As discussed in Chapter 1, an accurate treatment of energy flows in a chemiosmotic system demands the use of non-equilibrium thermodynamics to relate flux to net driving force. However, since the same laws control the flow of energy in complex systems such as mitochondria and simple systems such as electrical or hydraulic circuits, the use of the latter as analogues for the proton circuit has considerable attractions to the non-mathematical bioenergeticist.

The mitochondrial respiratory chain consists of three proton pumps which act in series with respect to the electron flow and in parallel with respect to the proton circuit (Fig. 2.2a). Two limiting 'states' are frequently referred to for isolated mitochondria – State 4 in which the proton current is limited by the inhibition of proton re-entry through the ATP synthase (due to either actual inhibition of the synthase or to the attainment of equilibrium), and State 3 in which there is ready proton re-entry into the matrix and hence brisk respiration. The State 3 condition can be due to an induced proton leak in the membrane or to the maintenance of ΔG_{ATP} below that required to equilibrate with $\Delta\mu_{\mathrm{H}^+}$ (by either removing ATP, or following the addition of ADP.

Complex I, which transfers electrons from the matrix $\mathrm{NAD}^+/\mathrm{NADH}$ pool to the membrane-bound pool of ubiquinone/ubiquinol operates close to equilibrium such that the energy lost by the electrons (some 300 meV) is conserved in the proton electrochemical gradient. As a result, three protons can be extruded by this complex for the passage of two electrons [23].

Complex III transfers electrons from the quinone pool dissolved in the membrane to the pool of cytochrome *c* loosely associated with the cytosolic face of the membrane. This complex also operates with a near equilibrium between $\Delta\mu_{\mathrm{H}^+}$ and the oxido-reduction span of the electrons. In contrast the final complex of the mitochondrial respiratory chain, cytochrome *c* oxidase, transferring electrons from cytochrome *c* to oxygen, operates under non-equilibrium conditions and is strictly irreversible.

It is possible to introduce or remove electrons at the interfaces between the complexes. Thus, electrons may be added to the quinone pool from Complex II, which oxidizes succinate, while cytochrome *c* may donate electrons to, or accept electrons from, a variety of artificial redox mediators.

As with any other energy flow circuit, the proton circuit has inter-related potential, flow and conductance terms. These will now be considered in turn.

3.1. The potential term – proton electrochemical potential

The driving force for a proton crossing a membrane is a combination of the electrical and concentration gradients experienced by the ion. At 30°C

$$\Delta\mu_{H^+} = \Delta\psi - 59\Delta pH$$

Accurate quantification of $\Delta\mu_{H^+}$ has been important to establish the general validity of the chemiosmotic theory. Currently there is a debate concerning the extent to which the 'bulk-phase' proton electrochemical potential equilibrates with the proton circuit connecting the respiratory chain and the ATP synthase, and in a subsequent section we shall discuss the extent to which anomalies are real or artifactual. In the following, however, the potential refer to the 'bulk phase', regardless of whether this represents the true 'energy-coupling intermediate'.

3.1.1. Membrane potential

Membrane potential and pH gradient are always determined separately. Since a membrane potential is a delocalized parameter for a given membrane, it follows that the membrane potential generated by the translocation of protons across the membrane will be felt by all ions distributed across the membrane. If an electrical uniport pathway exists for one of these ions, then it will tend to come to an equilibrium when its electrochemical potential gradient is zero:

$$0 = \Delta\mu_{X^{n+}} = n\Delta\psi - \frac{2.3\,RT}{F}\log_{10}\frac{[X^{n+}]_{in}}{[X^{n+}]_{out}}$$

Note that this equilibrium involves only the membrane potential, and is independent of the pH gradient. Estimation of membrane potential from this Nernst equation involves the determination of the equilibrium concentration gradient of the indicator ion across the membrane, either by the use of isotopes, or by using an electrode in the medium responsive to the decrease in external concentration as the ion is accumulated by the organelle [13,24].

In both cases it is necessary to convert the concentrations of indicator in the two compartments into chemical activities by allowing for binding, or non-ideal behaviour of the indicator ion in the two compartments. It is here that the major problems arise for precise quantitation. The mitochondrial matrix compartment (or the bacterial cytosol) is about as far removed from an ideal solution as it is possible to be. If some 50% of the mitochondrial protein is soluble in the matrix, which typically has a volume of 1 μl/mg total protein, then this implies that protein in the matrix is present as a 50% solution, ignoring the additional metabolites and nucleotides. Little is known about activity coefficients under these conditions. The normal recourse is to make the assumption that K or Rb display ideal behaviour in the matrix, and to correct other ion distributions accordingly.

Apart from the necessity to correct for the activity coefficient, the ion should be of the correct charge to be accumulated rather than excluded from the internal compartment. Thus cations are required for mitochondria and bacteria, and anions for inverted sub-mitochondrial particles, chromatophores and thylakoid membranes. Additionally the ion must rapidly come to electrochemical equilibrium, which implies that it must be highly permeable by a uniport mechanism, but not able to permeate by any other mechanism. particularly if isotopes are used, it is essential that the ion is not metabolised.

The ions which are most frequently employed to estimate mitochondrial membrane potential are Rb in the presence of valinomycin, and lipophilic phosphonium cations. The ionophore valinomycin creates a uniport pathway for K or Rb in a variety of membranes [25]. The use of ^{86}Rb enables the concentration gradient to be determined isotopically after rapid separation of the mitochondria from the medium ^{86}Rb is usually considered to behave ideally in the matrix (i.e., not to be bound). Indeed, other methods of membrane potential measurement are usually referred back to the K or Rb gradient in the presence of valinomycin for their calibration. However, a disadvantage is that the ionophore will also transport K, which means that when the ionophore is added the membrane potential will tend to shift towards the value given by any K-gradient which happens to be pre-existing across the membrane.

The lipophilic phosphonium cations, TPP^+ (tetraphenyl phosphonium) and $TPMP^+$ (triphenylmethyl phosphonium) can cross bilayer membranes as charged species and distribute according to the membrane potential [26]. Since they can be used at very low concentration they tend to disturb the potential less than the use of valinomycin. The gradients established can be estimated either isotopically, or by the use of an electrode specific to the cation [27]. The latter allows the membrane potential in mitochondrial incubations to be monitored continuously. However, a disadvantage of these cations, particularly in the case of TPP^+, is that they do not behave ideally but bind to components in the matrix. A number of techniques have been described for correcting the accumulation ratios of the cations [28,29].

A less direct determination of membrane potential involves the use of endogenous or added lipophilic charged spectroscopic indicators which equilibrate with the membrane potential, and change their spectral properties when accumulated within the matrix, either by exhibiting a decreased extinction due to 'stacking' of the planar molecules in the matrix, or by showing a changed fluorescence. In either case the response is usually calibrated by applying a known K gradient across the membrane in the presence of valinomycin in order to generate a diffusion potential. These indicators have the advantage that membrane potential can be monitored directly, without the need to separate mitochondria from the incubation, but have the added complexity that the precise relationship between spectral change and concentration must be established, as must the existence of any interfering factors which influence for example the fluorescent yield [30].

3.1.2. Intrinsic indicators of membrane potential

Many photosynthetic membranes contain carotenoids which function as light harvesters [31]. These molecules contain long sequences of conjugated double bonds, and the energy levels of the electrons respond to an imposed electrical field with a shift of a few nm in their spectra (an electrochromic shift). The carotenoids have therefore been used as intrinsic indicators of membrane potential in photosynthetic membranes. They have the advantages that, being endogenous, they cause no artifacts, while as they are fixed in the membrane, rather than having to be accumulated into compartments, they respond rapidly. Indeed, during the primary events occurring in the reaction centre, a potential can be registered in a few ns. The problem with these indicators lies in their calibration, and in agreeing as to the relation between the field that they report and the bulk phase membrane potential measured by permeant ions.

The electrochromic shift of the carotenoids is usually calibrated with K-diffusion potential in the presence of valinomycin. One problem is that the shifts observed in respiring chromatophores (where the proton electrochemical potential is predominantly in the form of a membrane potential) are much larger than those induced by the calibrating diffusion potential, so that an extensive extrapolation is required. Thus, the carotenoids in illuminated chromatophores may indicate a membrane potential in excess of 300 mV, whereas the distribution of CNS^-, an electrically permeant anion, in the same system only indicates 140 mV [32]. The extent of this discrepancy, and the uncertainty as to whether the carotenoids see the bulk-phase potential, or only the local electrical field within the membrane, limits the confidence with which carotenoids may be used for quantitative as opposed to qualitative potential measurements.

3.1.3. pH gradient

The pH gradient across the membrane is usually estimated from the distribution of protonatable species which are permeant only in their uncharged form, i.e., as protonated weak acids or as deprotonated weak bases, even though their charged forms may be present in considerable molar excess. Since the indicators permeate electroneutrally, their distribution will be independent of the membrane potential, in fact the uncharged form will seek to achieve an equal concentration in the two compartments. In the two compartments the indicators will protonate or depronate according to the requirements of the Henderson-Hasselbach equation, and the final distribution of the ionized species will be given by the relationships:

$$[H^+]_{in}/[H^+]_{out} = [A^-]_{out}/[A^-]_{in}$$

for weak acids, which are therefore accumulation in alkaline compartments, and:

$$[H^+]_{in}/[H^+]_{out} = [BH^+]_{in}/[BH^+]_{out}$$

for weak bases, accumulated into acidic compartments.

All the limitations inherent in the membrane potential determinations are applicable also to the measurement of ΔpH.

3.2. Proton conductance

After discussing the generation and quantitation of the potential term of the proton circuit, we shall now turn to the proton current, and examine the factors which control the flux of protons around the circuit. Although it is not possible to determine the proton current under steady-state conditions directly, the parameter may be calculated indirectly from the respiratory rate and the stoicheiometry of proton extrusion by the respiratory chain. It is outside the scope of this chapter to discuss the contentious issue of the proton stoicheiometries of the complexes, but the important feature is that, unless the complexities of variable stoicheiometry are invoked, respiration and proton current vary in parallel.

If the only pathway for proton re-entry into the mitochondrial matrix were through the ATP synthase, then in conditions under which there is no net ATP synthesis, either due to inhibition of the complex itself, for example with oligomycin, or due to the attainment of thermodynamic equilibrium between the ATP pool and the proton electrochemical potential, there should be a complete inhibition of respiration. In practice mitochondria continue to respire at a finite rate even in this 'State 4' condition, since protons can slowly leak back across the membrane (Fig. 2.2).

As in an electrical circuit, where the current of electrons flowing through a resistive element is related to the electrical potential difference and the resistance by Ohm's law, the proton current flowing back into the mitochondrial matrix through a leak pathway will be given by the product of the membrane proton conductance and the proton electrochemical potential:

$$J_{H^+} = Cm_{H^+} \cdot \Delta\mu_{H^+}$$

In an experiment where the state 4 respiration and $\Delta\mu_{H^+}$ are determined, it is possible therefore to obtain a value for the leak proton conductance of the membrane. Typical values for mitochondria are in the range 0.3–1 nmol $H^+ \cdot min^{-1} \cdot mg^{-1}$ protein $\cdot$ mV^{-1} of $\Delta\mu_{H^+}$.

3.2.1. The special case of brown fat mitochondria

Brown fat is an exception to the optimization of thermodynamic efficiency characteristic of the chemiosmotic proton circuit. Since the tissue will be considered in Chapter 10, only those general aspects concerning the proton circuit will be considered here. Brown fat is a heat producing tissue in which the respiratory rate of the mitochondria far outstrips the ATP turnover in the cell [33]. This is possible because the mitochondria possess a 32 000 M_r protein in their inner membrane which functions as a regulatable uniport allowing protons to re-enter the matrix directly without passing through the ATP synthase [34]. The result is a physiological uncoupling of the mitochondria. The conductance of the protein appears to be regulated in situ by the fatty acids liberated by lipolysis of the triglyceride stores when the tissue is thermogenic, and can be increased by sub-micromolar concentrations of the fatty acids sufficiently to allow uncontrolled respiration to occur.

The fatty acids modulate the conductance of the protein in a rather unusual way [35]. The cytosolic purine nucleotides maintain the 32 000 M_r protein in a low conductance state in which the protein displays non-ohmic conductance characteristics. In other words the proton current through the membrane increases disproportionately when $\Delta\mu_{H^+}$ rises above a critical potential. This critical potential is so high that it is normally not exceeded during the oxidation of physiologically relevant substrates, and so respiratory control is not lost. However micromolar concentrations of fatty acids lower the critical potential to values which are insufficient to inhibit the proton circuit, the result being that mitochondrial respiration can now be freed from the constraints of respiratory control [35]. We have suggested that this is the mechanism by which the conductance of the 32 000 M_r protein is regulated in vivo.

3.3. Respiratory control

When the mitochondria are allowed to produce ATP, for example by adding ADP and P_i to an incubation, the potential and current parameters of the proton circuit both change in the direction which indicates an increase in the proton conductance of the membrane due to proton re-entry through the ATP synthase. Thus, $\Delta\mu_{H^+}$ falls and this lowers the thermodynamic back pressure upon the respiratory chain, which therefore respires more rapidly.

There is a controversy as to whether a drop in $\Delta\mu_{H^+}$ created by net ATP synthesis and by the addition of proton translocator create equal respiratory stimulations; this will be discussed in a later section. However, brown fat mitochondria investigated in our laboratory [36,37] show the same relationships between respiratory stimulation and $\Delta\mu_{H^+}$ decrease with a proton translocator and the 32 000 M_r uncoupling protein [36] (Fig. 2.3), or with a proton translocator and during state 3 [37].

Fig. 2.3 also shows that the transfer of energy from the respiratory chain to the proton circuit can be extremely efficient, in that a slight thermodynamic disequilibrium results in a considerable energy flux. The actual disequilibrium between the respiratory chain and the proton electrochemical potential is even less than appears from the drop in the latter, since the redox span across the respiratory chain proton pumps also contracts [24].

3.4. Reversed electron transfer and the proton circuit driven by ATP hydrolysis

The three proton pumps of the mitochondrial respiratory chain normally function in parallel with respect to the proton circuit. It is however possible to manipulate the conditions such that the proton electrochemical potential generated by two of the pumps can be used to reverse the third proton pump (but not cytochrome oxidase).

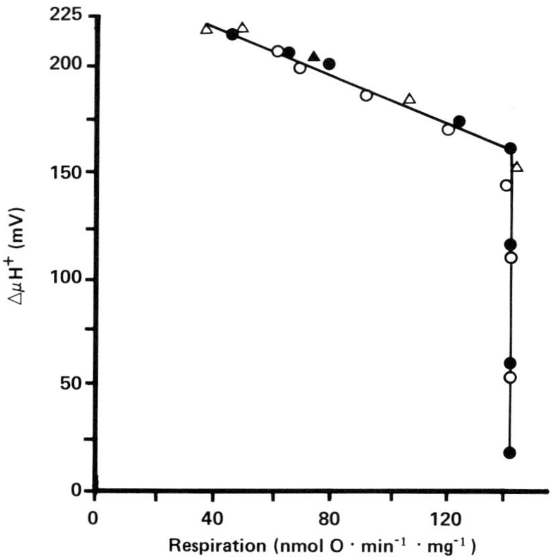

Fig. 2.3. Relationship between respiratory rate and $\Delta\mu_{H^+}$ for hamster brown fat mitochondria when the proton conductance is increased by titrating with proton translocator (●), by activating the 32000 M_r uncoupling protein (○), by varying the pH (△) or by adding fatty acids (▲). Data from Ref. 36.

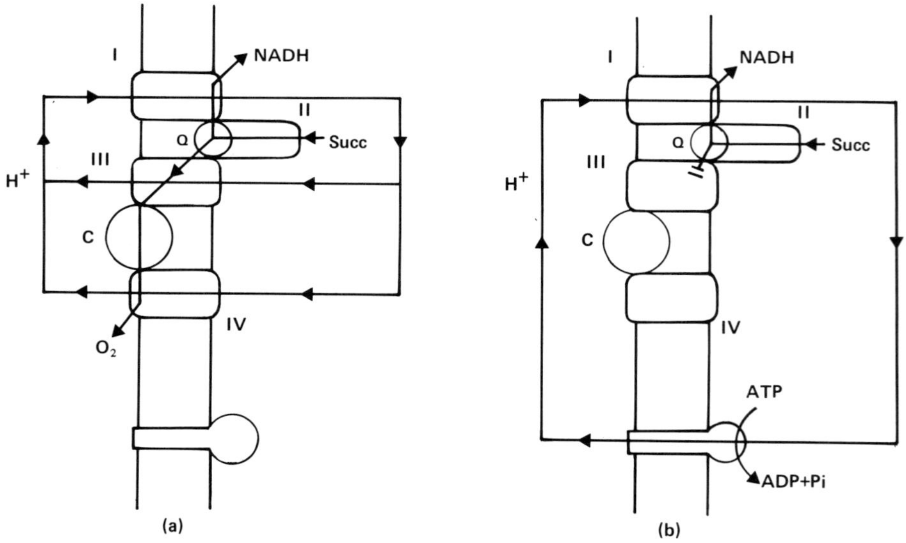

Fig. 2.4. Further variants of the proton circuit. a, reversed electron transfer in Complex I driven by $\Delta\mu_{H^+}$ from succinate oxidation. b, reversed electron transfer in Complex I driven by $\Delta\mu_{H^+}$ from ATP hydrolysis.

One way in which this may be achieved is shown in Fig. 2.4. The proton electrochemical potential generated by the 'downhill' flow of electron from Complex III to oxygen can be used to drive Complex I in reverse, protons entering the matrix through this complex and driving electrons 'uphill' from the ubiquinone pool to the $NADH/NAD^+$ pool, with a redox potential some 300 mV more negative.

There is one bacterial system where such reversed electron transfer is of great importance. In *Rps. sphaeroides* the $\Delta\mu_{H^+}$ generated by cyclic electron flow through the reaction centre and cytochrome system is used to induce reversed electron flow from the level of the ubiquinone pool to the $NADH/NAD^+$ pool, in a manner analogous to that described for mitochondria. The role of this is to supply low potential electrons for the biosynthetic functions of the cell [38].

Since the ATP synthase is reversible, it is also possible in the absence of respiration to drive a proton circuit by ATP hydrolysis (Fig. 2.4). The $\Delta\mu_{H^+}$ achievable by this means is identical to that supported by respiration [39] and can be utilized to drive ion transport, as well as to reverse electron flow through Complex I (Fig. 2.4).

While an ATP-driven proton circuit is an experimental device with mitochondria, anaerobic mitochondria such as *Strep. faecalis* use the hydrolytic mode of the ATP synthase to maintain a $\Delta\mu_{H^+}$ across their membrane for metabolite transport, the ATP being supplied by anaerobic glycolysis.

4. Coupling of the proton circuit to the transport of divalent cations

Since mitochondrial Ca transport is discussed in depth in Chapter 9, this section will be restrict to a brief summary of the way in which the proton circuit can be diverted into accumulating and regulating the transport of the cation, and how the permeation of weak acids is linked indirectly to net Ca transport.

The driving force (the electrochemical potential difference) for the uptake of a divalent cation by a uniport mechanism across a membrane is given by:

$$\Delta\mu_{Ca^{2+}} = 2\Delta\psi - 59 \log_{10} \frac{[Ca^{2+}]_{in}}{[Ca^{2+}]_{out}}$$

If a pathway exists, a divalent cation can be accumulated until electrochemical equilibrium is attained, i.e.

$$\frac{[Ca^{2+}]_{in}}{[Ca^{2+}]_{out}} = 10^{\Delta\psi/29.5}$$

It is apparent from this equation that the mitochondrion, with a membrane potential of up to 180 mV could, if a suitable pathway existed, accumulate divalent cations until equilibrium is attained with a 10^6 gradient of the free cation across the membrane. Mitochondria possess a pathway for divalent cation uniport which allows the passage of Ca, Sr, Ba and, at a lesser rate, Mn [40–42].

Since the electrical capacitance of the mitochondrial inner membrane is low, the membrane potential would be discharged by the uptake into the matrix of only about 1 nmol of Ca/mg protein if no other ions were to move across the membrane (Fig. 5). In order for further net accumulation of the cation to occur, a net extrusion of protons by either the respiratory chain or the ATP synthase is required.

This net proton extrusion results in a net acidification of the external medium and an alkalinization of the matrix (Fig. 2.5), and should be distinguished from the steady-state cycling which occurs during the operation of the proton circuit for ATP synthesis. Clear, alkalinization of the matrix cannot continue indefinitely, and the limitation which is set is essentially thermodynamic – the respiratory chain is incapable of maintaining a proton electrochemical potential in excess of 200–230

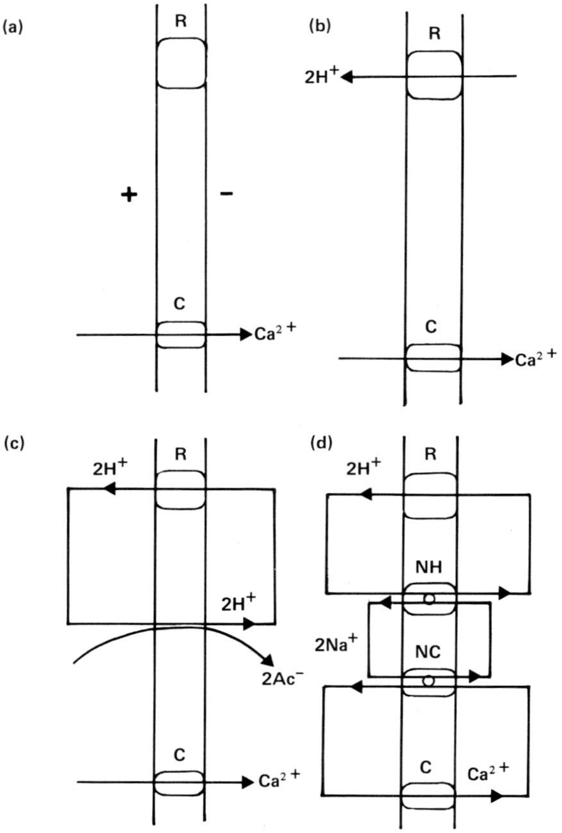

Fig. 2.5. Interaction of Ca transport with the proton circuit. a, Ca uptake alone discharges the membrane potential; b, Ca uptake in exchange for protons extruded by the respiratory chain generates a pH gradient; c, Ca uptake together with a weak acid such as acetate does not build up a pH gradient; d, Ca cycling in heart mitochondria driven by the proton circuit. R, respiratory chain; C, calcium uniport; NH, sodium/proton antiport; NC, sodium/calcium antiport.

mV; therefore as the pH component increases, the membrane potential must decrease. Since the pH-buffering capacity of the matrix is of the order of 20 nmol H^+/pH unit per mg protein, the uptake of about 20 nmol of Ca/mg protein, and the compensatory efflux of a bit less than 40 nmol H^+/mg protein would increase ΔpH by 2 units and lower the membrane potential by 120 mV. The lowered potential persists after net Ca accumulation has ceased. When these are the only ion movements two protons appear in the medium for each Ca accumulated.

In order to obtain more extensive Ca uptake, the independent but parallel movement of an electroneutrally permeant weak acid into the matrix is required. Fig. 2.5 shows the simplest case where a weak acid such as acetate is accumulated. The driving force for the acetate to enter the matrix is the pH gradient generated as above. Since the weak acids are accumulated electroneutrally, protons enter the matrix together with the anion and there is no net change in internal or external pH. Although there is no theoretical limit to the amount of Ca which can be accumulated by the matrix under these conditions, the osmotic activity of the soluble Ca acetate in the matrix can lead to swelling.

In the intact cell it is likely that P_i accompanies Ca during net accumulation of the cation. The interpretation of ionic movements in this condition is complicated by the formation of a $Ca_3(PO_4)_2$ complex in the matrix which is found in the form of Ca hydroxyapatite after ashing. For each Ca taken up, there is the appearance of about one proton in the medium, even though there is no change in matrix pH. This paradoxical imbalance is because the extruded protons originate from the phosphate as it is deprotonated on formation of the matrix complex.

In contrast to the permanent depression of membrane potential when Ca is accumulated in the absence of permeant cation, uptake in the presence of acetate or phosphate leads to only a transient drop in potential, similar to that seen during State 3.

So far we have limited discussion to the effects on the mitochondrial bioenergetics of the net uptake of a divalent cation. However, as will be discussed in a subsequent chapter, mitochondria possess a second class of pathway which catalyzes the exchange of matrix Ca for external Na or protons, the function of this being to allow a slow regulatory cycling of Ca across the membrane. During this steady-state cycling, the net transport of anions is not required (Fig. 2.5). In mitochondria in which Ca exchanges for protons, the Ca and proton cycles can be linked directly, whereas in mitochondria in which the Ca efflux pathway exchanges Ca for Na, there is an intermediate cycle of Na, linked to the proton circuit by the native Na/H^+ exchange activity in the membrane.

5. Is the proton circuit in equilibrium with the bulk aqueous phases on either side of the membrane?

In the proton circuit discussed so far in this chapter, it has been assumed that the protons, after being translocated across the membrane by the respiratory chain or

ATP synthase equilibrate with the aqueous phases of the matrix and cytosol (or incubation medium). While this may seem intuitive, in view of the extremely high conductance of water to protons, there have in recent years been a number of reports of significant anomalies in the thermodynamic and kinetic relationships between respiration, ATP synthesis and the measured $\Delta\mu_{H^+}$ sufficient to lead some investigators to suggest that this last parameter does not represent the true intermediate in the coupling of respiration to phosphorylation [24].

The majority of these reports still propose that the basic chemiosmotic feature of proton electrochemical coupling is still present, but hypothesize that microcircuits

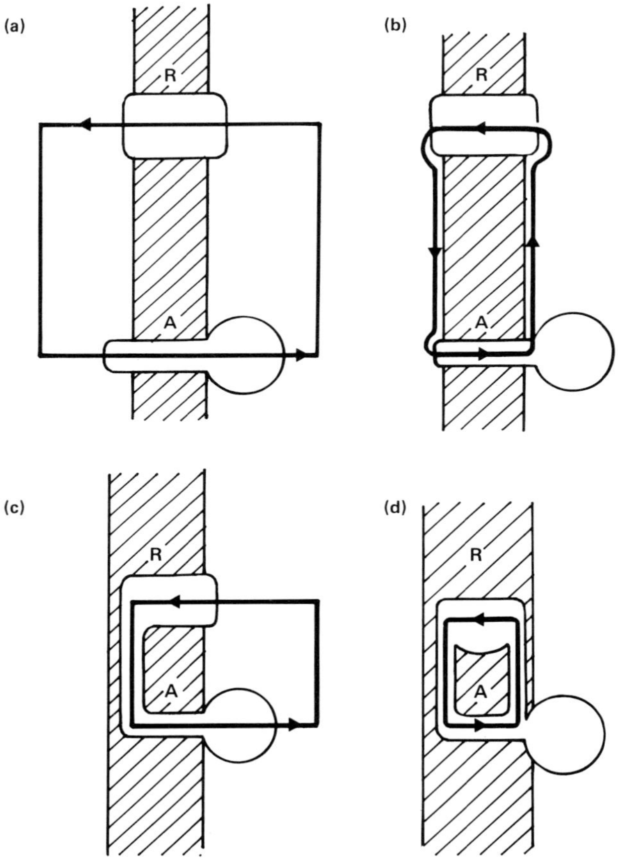

Fig. 2.6. Hypothetical localized variants on the proton circuit. a, fully delocalized circuit; b, proton current flows along surface of membranes (note that circuit is still in equilibrium with the bulk phases); c, one leg of the proton circuit is conducted through a lateral channel insulated from the bulk phase; d, both legs of the proton circuit are conducted through lateral channel insulated from the bulk phases; R, respiratory chain; A, ATP synthase. Note the necessity for both outward and return legs in all models.

exist between individual respiratory chains and ATP synthase units, which are significantly out of equilibrium with the bulk phase $\Delta\mu_{H^+}$ as measured by the techniques discussed earlier in this chapter.

Before considering this information in detail, it is worthwhile to summarize briefly the implications of a localized proton circuit. One possibility is that the major part of the proton current flows not through the bulk aqueous phase (Fig. 2.6a) but along the two surfaces of the membrane (Fig. 2.6b). Note that in this model there is no insulating barrier between the surfaces of the membrane and the bulk phases. Therefore, under steady-state conditions, the electrochemical potential of the protons on the surfaces of the membrane must be the same as in the bulk phases, since otherwise there would be a net flow of protons down the supposed gradient from surface to bulk. This model does not therefore represent true localized chemiosmosis, since the bulk-phase potential measured experimentally will accurately reflect the true potential driving ATP synthesis.

The next stage in developing a model of localized chemiosmosis is to assume that there is a substantial resistance between the local circuit and the bulk phase [43,44]. One could devise hypothetical models in which one (Fig. 2.6c) or both (Fig. 2.6d) of the portions of the proton circuit parallel to the membrane were insulated from the bulk phases. It is important, however, to appreciate that such models still require the presence of a highly insulating phase separating the 'outward' and 'return' limbs of the proton circuit, with the addition of one or two further substantial resistances to protons.

The evidence which has been put forward to favour such 'localized' schemes is largely thermodynamic and will be summarized below. It must be borne in mind, however, that any localized circuit must be consistent, not only with the deviations in the thermodynamic relationships which led to the particular model, but also with the wealth of information available about chemiosmotic coupling in general. Some of these have been discussed above and will now be restated:

a. It is possible to reconstitute individual respiratory chain complexes and the ATP synthase into artificial bilayer membranes such that they individually pump protons into the bulk phase, or when linked together synthesize ATP. In a well known experiment, Racker and Stoekenius [45] paired the bacteriorhodopsin of the archaeobacterium *Halobacterium halobium* with the beef heart ATP synthase and observed ATP synthesis. Such an artificial system could hardly have the components to assemble an insulated proton conduction pathway between the complexes.

b. An artificial, bulk-phase proton electrochemical potential can drive ATP synthesis by a reconstituted ATP synthase from the thermophilic bacterium PS3 [18].

c. When mitochondria are presented with cations which are either non-specifically permeable through lipid bilayers, such as tetraphenyl phosphonium or triphenyl methyl phosphonium, or which possess specific transport pathways for electrical uniport, as is the case for Ca, there is a rapid uptake of the cation and a compensatory net efflux of protons from the matrix to the medium, as the respiratory chain responds to the depolarizing influence of the cation uptake by extruding more protons. These protons are evidently translocated from the matrix, which becomes alkaline, to the bulk external phase, where they can be detected by a pH electrode, with no significant kinetic barrier.

d. Returning to the case of *Halobacterium halobium*, the bacteriorhodopsin aggregates in the membrane into two-dimensional arrays of 'purple patches', which exclude other proteins. Thus, many of these light-driven proton pumps must be relatively distant from the nearest ATP synthase, and presumably incapable of forming a local circuit.

With these observations in mind we shall now consider the nature of the thermodynamic and kinetic discrepancies which have led a number of groups to consider the possibility of localized proton circuits.

5.1. Kinetic evidence

Respiratory control is a consequence of the thermodynamic back-pressure exerted by the proton electrochemical potential upon the respiratory chain, limiting the rate at which further protons can be extruded. Since the chemiosmotic theory states that the only connection between the respiratory chain and the other proton-utilizing components in the membrane is through the proton electrochemical gradient, the respiratory chain should be incapable of distinguishing between the different ways of producing a given depression in $\Delta\mu_{H^+}$.

Testing for such an identity provides a severe test for the chemiosmotic theory. While some groups [e.g., 36,37,46] show that the transition to State 3 or addition of small concentrations of proton translocator cause similar respiratory stimulations, other groups have reported major discrepancies. Thus, Padan and Rottenberg [47] found that a much greater depression of $\Delta\mu_{H^+}$ was required to produce a given respiratory stimulation if produced by addition of proton translocator than during the transition to State 3. Similar results were obtained by Azzone et al. [48]. Wilson and Forman [29] and Zoratti et al. [49] could find no correlation between $\Delta\mu_{H^+}$ and respiration when the former is varied by proton translocator, by changing osmolarity, or by selectively lowering either pH gradient or membrane potential. Finally, Rottenberg [50] reported that local anaesthetics could increase State 4 respiration without detectably lowering $\Delta\mu_{H^+}$. In an extreme case, Michel and Oesterhelt [51] were able to observe ATP synthesis in *Halobacterium halobium* in the absence of any detectable $\Delta\mu_{H^+}$.

A similar argument applies to the rate of ATP synthesis. From the chemiosmotic standpoint, this should be regulated by the relation between $\Delta\mu_{H^+}$ and ΔG_{ATP}, and should not depend on the way in which $\Delta\mu_{H^+}$ was varied. However, Baccarini-Melandri et al. [52] found that the rate of ATP synthesis by bacterial chromatophores varies with $\Delta\mu_{H^+}$ depending on whether electron flow is limited or proton translocator added.

These discrepancies could be due to experimental problems in the measurement of the potentials, or to the existence of micro-circuits out of equilibrium with the bulk phases. These would possess a significant resistance between the localized proton circuit and the bulk phase. The respiratory chain would see the localized circuit linked to the ATP synthase directly, but the bulk-phase potential only indirectly and through the resistance. Proton translocators would only act on the bulk phase potential, and only the bulk-phase potential would be measurable experimentally.

5.2. Thermodynamic anomalies

When there is no net synthesis of ATP in the matrix by the ATP synthase, then according to the chemiosmotic theory the mitochondrion should maintain an equilibrium between $\Delta\mu_{H^+}$ and the ΔG_{ATP} such that the ratio between the two thermodynamic terms should be invariant when $\Delta\mu_{H^+}$ is altered, regardless of how this alteration is brought about. While this has been observed by some groups [27,37], other have found large discrepancies which, unless the somewhat unattractive concept of a variable stoicheiometry for the ATP synthase is resorted to, imply that the $\Delta\mu_{H^+}$ being measured is not the true energy-transducing intermediate.

In a localized scheme, the measured bulk-phase $\Delta\mu_{H^+}$ is not the true 'energy-transducing intermediate' and so deviations are possible. Generally, the deviation which is found is that ΔG_{ATP} remains high when $\Delta\mu_{H^+}$ is lowered by titration with proton translocator [e.g., 48,53,54]. Putting aside questions as to whether all investigators have ensured that their ATP synthase is at a true equilibrium, these results would predict that slowing the proton circuit by titration with respiratory inhibitor would give more time for the local and bulk phases to equilibrate, such that the discrepancies would decrease. However, Westerhoff et al. [54] reported that the ΔG_{ATP} remained above that predicted from the $\Delta\mu_{H^+}$ even in this latter case, and so were forced to invoke a parallel coupling (both direct and via the proton circuit) between the respiratory chain and the ATP synthase of the type initially proposed by Padan and Rottenberg [47].

6. Conclusion

The next few years will decide whether the present anomalies in the quantitation of the proton circuit are merely due to experimental problems in the precise measurement of the parameters, or whether they will demand a wholesale revision in our understanding of the proton electrochemical coupling of respiration to ATP synthesis. In the author's opinion the evidence is not yet adequate for the second alternative, but there is a need for proponents of 'delocalized' chemiosmosis to propose straightforward and unambiguous explanations for these phenomena.

References

1 Mitchell, P. (1976) Biochem. Soc. Trans. 4, 399–430.
2 Mitchell, P. (1979) Eur. J. Biochem. 95, 1–20.
3 Nicholls, D.G. (1982) Bioenergetics. Academic Press, London.
4 Ernster, L. and Lee, C.P. (1964) Annu. Rev. Biochem. 33, 729–788.
5 Mitchell, P. (1961) Nature (London) 191, 423–427.

6 Mitchell, P. (1966) Chemiosmotic Coupling in Oxidative and Photosynthetic Coupling. Glynn Research, Bodmin, UK.
7 Mitchell, P. (1968) Chemiosmotic Coupling and Energy Transduction. Glynn Research, Bodmin, UK.
8 Mitchell, P. and Moyle, J. (1967) Biochem. J. 104, 588–600.
9 Scholes, P. and Mitchell, P. (1970) J. Bioenerg. 1, 67–72.
10 Jagendorf, A.T. (1975) In Bioenergetics of Photosynthesis (Govindjee, ed.) pp. 413–492, Academic Press, New York.
11 Terada, H. (1981) Biochim. Biophys. Acta 639, 225–242.
12 Mitchell, P. and Moyle, J. (1969) Eur. J. Biochem. 9, 149–155.
13 Rottenberg, H. (1979) Methods Enzymol. 55, 547–569.
14 Klingenberg, M. and Rottenberg, H. (1977) Eur. J. Biochem. 73, 125–130.
15 Mitchell, P. and Moyle, J. (1968) Eur. J. Biochem. 4, 530–539.
16 Thayer, W.S. and Hinkle, P. (1973) J. Biol. Chem. 250, 5330–5335.
17 Kagawa, Y., Kandrach, A. and Racker, E. (1973) J. Biol. Chem. 248, 676–684.
18 Sone, N., Yoshida, M., Hirata, H. and Kagawa, Y. (1977) J. Biol. Chem. 252, 2956–2960.
19 Mitchell, P. and Moyle, J. (1967) Biochem. J. 105, 1147–1162.
20 Neumann, J. and Jagendorf, A.T. (1964) Arch. Biochem. Biophys. 107, 109–119.
21 Kagawa, Y. (1972) Biochim. Biophys. Acta 265, 297–338.
22 Racker, E. (1979) Methods Enzymol. 55, 699–711.
23 De Jonge, P.C. and Westerhoff, H.V. (1982) Biochem. J. 204, 515–523.
24 Ferguson, S.J. and Sorgato, M.C. (1982) Annu. Rev. Biochem. 51, 185–217.
25 Pressman, B.C. (1976) Annu. Rev. Biochem. 45, 501–530.
26 Skulachev, V.P. (1971) Curr. Topics Bioenerg. 4, 127–190.
27 Kamo, N., Muratsugo, M., Hongoh, R. and Kobatake, Y. (1979) J. Membr. Biol. 49, 105–121.
28 Scott, I.D. and Nicholls, D.G. (1980) Biochem. J. 21–33.
29 Wilson, D.F. and Forman, N.G. (1982) Biochemistry 21, 1438–1444.
30 Bashford, C.L. and Smith, J.C. (1979) Methods Enzymol. 55, 569–586.
31 Baccarini-Melandri, A., Casadio, R. and Melandri, B.A. (1981) Curr. Topics Bioenerg. 12, 197–258.
32 Ferguson, S.J., Jones, O.T.G., Kell, D.B. and Sorgato, M.C. (1979) Biochem. J. 180, 75–85.
33 Nicholls, D.G. and Locke, R.M. (1984) Physiol. Rev. 64, 1–64.
34 Heaton, G.M., Wagenvoord, R., Kemp, A. and Nicholls, D.G. (1978) Eur. J. Biochem. 82, 515–521.
35 Rial, E., Poustie, A. and Nicholls, D.G. (1983) Eur. J. Biochem. 137, 197–203.
36 Nicholls, D.G. (1974) Eur. J. Biochem. 49, 573–583.
37 Nicholls, D.G. and Bernson, V.S.M. (1977) Eur. J. Biochem. 75, 601–612.
38 Jones, O.T.G. (1977) In Microbial Energetics (Haddock, B.A. and Hamilton, W.A., eds.) pp. 151–183. Cambridge Univ. Press, Cambridge.
39 Nicholls, D.G. (1974b) Eur. J. Biochem. 50, 305–315.
40 Carafoli, E. and Crompton, M. (1978) Curr. Top. Membr. Trans. 10, 151–216.
41 Nicholls, D.G. and Crompton, M. (1980) FEBS Lett. 111, 261–268.
42 Nicholls, D.G. and Akerman, K.E.O. (1982) Biochim. Biophys. Acta, 683, 57–88.
43 Williams, R.J.P. (1978) Biochim. Biophys. Acta 505, 1–44.
44 Kell, D.B. (1979) Biochim. Biophys. Acta 549, 55–99.
45 Racker, E. and Stoekenius, W. (1974) J. Biol. Chem. 249, 662–663.
46 Kuster, U., Letko, G., Kunz, W., Duszynski, J., Bogucka, K. and Wojtczak, L. (1981) Biochim. Biophys. Acta 636, 32–38.
47 Padan, E. and Rottenberg, H. (1973) Eur. J. Biochem. 40, 431–437.
48 Azzone, G.F., Pozzan, T., Massari, S. and Bragadin, M. (1978) Biochim. Biophys. Acta 501, 296–306.
49 Zoratti, M., Pietrobon, D. and Azzone, G.F. (1982) Eur. J. Biochem. 126, 443–451.
50 Rottenberg, H. (1983) Proc. Natl. Acad. Sci. U.S.A. 80, 3313–3317.
51 Michel, H. and Oesterheldt, D. (1980) Biochemistry 19, 4615–4619.
52 Baccarini-Melandri, A., Cassadio, R. and Melandri, B.A. (1977) Eur. J. Biochem. 78, 389–402.
53 Sorgato, M.C. and Ferguson, S.J. (1980) Biochem. J. 188, 945–948.
54 Westerhoff, H.V., Simonetti, A.L.M. and van Dam, K. (1981) Biochem. J. 200, 193–202.

Ernster (ed.) Bioenergetics

© 1984 Elsevier Science Publishers B.V.

CHAPTER 3

The mitochondrial respiratory chain

MÅRTEN WIKSTRÖM and MATTI SARASTE

Department of Medical Chemistry, University of Helsinki, Siltavuorenpenger 10A, SF-00170 Helsinki, Finland

1. Introduction

The term respiratory chain may actually be a misnomer in the sense that the respiratory components of the inner mitochondrial membrane probably do not form chain-like physical complexes with one another, at least not with a long life-time. Yet, a *chain* is still a good collective description of the catalysts of mitochondrial respiration, which transfer reducing equivalents in a very specific and functionally chain-like sequence from hydrogen-donating substrates to dioxygen.

The mitochondrial respiratory chain (Fig. 3.1A) consists of three catalytic complexes, viz. NADH:ubiquinone (Q) oxidoreductase (Complex I), ubiquinol:ferricytochrome *c* oxidoreductase (Complex III) and ferrocytochrome c:O_2 oxidoreductase (Complex IV), *plus* two lower molecular weight redox carriers, cytochrome *c* and ubiquinone-10 (Q). Succinate:ubiquinone reductase (Complex II) is not included here (although this is often done) because it may be considered only auxiliary to the energy-conserving respiratory chain. Its function is analogous to some other enzymic activities in the inner mitochondrial membrane: They feed reducing equivalents into ubiquinone from various substrates, without being directly engaged in the conservation of energy.

Cytochrome *c* carries electrons from the cytochrome bc_1 complex, but also from auxiliary redox enzymes (see e.g., Refs. 1,2) to cytochrome oxidase. Ubiquinone is the only non-protein member of the chain. This highly hydrophobic substituted

---

Abbreviations: AA, antimycin; BAL, British Anti-Lewisite (2,3-dimercaptopropanol); DCCD, dicyclohexylcarbodiimide; DTNB, 5,5′-dithiobis(2-nitrobenzoate); E_h, oxidoreduction potential relative to the Normal Hydrogen Electrode; E_m, midpoint oxidoreduction potential; $E_{m,x}$, midpoint oxidoreduction potential at pH = x; FeS, iron-sulphur (centre or protein); FMN, flavin mononucleotide; HMHQQ, 7-(*n*-heptadecyl)mercapto-6-hydroxy-5,8-quinolinequinone; HOQNO, 2-*n*-heptyl-4-hydroxyquinoline *N*-oxide; Lb, leghaemoglobin; MX, myxothiazol; NEM, *N*-ethylmaleimide; pmf, protonmotive force, electrochemical proton gradient; Q, ubiquinone; Q_N, ubiquinone bound to Complex I; SQ, ubisemiquinone; SQ^-, ubisemiquinone anion; UHDBT, 5-*n*-undecyl-6-hydroxy-4,7-dioxobenzothiazol.

A

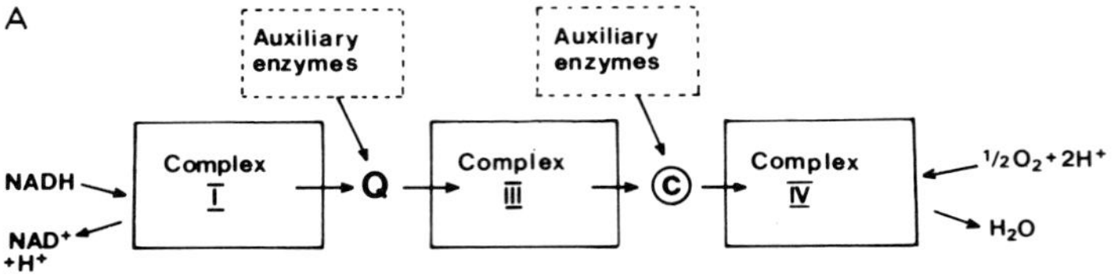

B

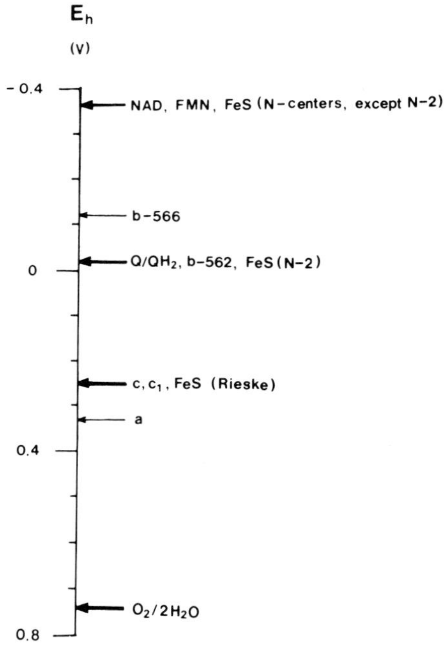

Fig. 3.1. A, The respiratory chain. Q and c stand for ubiquinone and cytochrome *c*, respectively. Auxiliary enzymes that reduce ubiquinone include succinate dehydrogenase (Complex II), α-glycerophosphate dehydrogenase and the electron-transferring flavoprotein (ETF) of fatty acid oxidation. Auxiliary enzymes that reduce cytochrome *c* include sulphite oxidase. B, Thermodynamic view of the respiratory chain in the resting state (State 4). Approximate E_h values are calculated according to the Nernst equation using oxidoreduction states from work by Muraoka and Slater, (NAD, Q, cytochromes $c + c_1$ and *a*; oxidation of succinate [6]), and Wilson and Erecinska (*b*-562 and *b*-566 [7]). The NAD, Q, cytochrome *b*-562 and oxygen/water couples are assumed to equilibrate protonically with the M phase at pH 8 [7,8]. $E_{m,7}$ ($\Delta E_m/\Delta$pH) for NAD, Q, *b*-562, and oxygen/water are taken as −320 mV (−30 mV/pH), 66 mV (−60 mV/pH), 40 mV (−60 mV/pH), and 800 mV (−60 mV/pH) [7–10]. FMN and the FeS centres of Complex I (except N-2) are assumed to be in redox equilibrium with the NAD/NADH couple, FeS(N-2) with ubiquinone [11], and cytochrome c_1 and the Rieske FeS centre with cytochrome *c* [10]. The position of cytochrome *a* in the figure stems from its redox state [6] and its apparent effective $E_{m,7.2}$ of 285 mV in aerobic steady states [12].

benzoquinone collects reducing equivalents from various dehydrogenases (including Complex I), carrying them to the cytochrome bc_1 complex. In addition, ubiquinone apparently has more specific functions, at least within the latter complex.

All three respiratory complexes are typical integral membrane proteins that span the inner mitochondrial membrane. Each consists of several different subunits, the exact number of which is still under debate. The genes of some subunits of cytochrome oxidase and the bc_1 complex are in mitochondrial DNA (mtDNA). These proteins are synthesised inside the mitochondrion. However, most proteins of these complexes, as well as cytochrome *c*, are synthesised on cytoplasmic ribosomes and coded by the nuclear genome. This raises intriguing questions of how the latter are imported into the mitochondrion and inserted into the mitochondrial membrane, as well as of how mitochondrial and cytoplasmic transcription and translation are synchronised [3–5].

In this account we will review the structural and functional properties of the respiratory chain components, as they are known from studies with intact mitochondria, vesicles of the inner mitochondrial membrane (submitochondrial particles), or isolated complexes. The latter may additionally be 'reconstituted' into liposomal membranes. To some extent we will also review the knowledge on the integrated functions of the respiratory chain with main emphasis on proton translocation and essential thermodynamic and kinetic properties.

2. *General survey*

2.1. *The central dogma*

The respiratory chain catalyses transfer of reducing equivalents from NADH generated in the mitochondrial matrix or M space, to dioxygen (Fig. 2.1A). Fig. 2.1B shows a thermodynamic view, giving the operational redox potentials (E_h) for the main individual components (for details, see below). The total redox span is about 1.11 V for oxidation of NADH, and about 760 mV for oxidation of ubiquinol (or succinate).

The exergonic respiratory chain activity is utilised to drive proton translocation from the matrix (M) to the cytoplasmic (C) side of the membrane with generation of an electrochemical proton gradient (protonmotive force (pmf) [13],

$$\text{pmf} = \Delta\psi - 2.303\ RT/F \quad \Delta\text{pH} \tag{1}$$

where $\Delta\psi$ is the electrical membrane potential (C side minus M side) and ΔpH is the pH differential (C side minus M side), R, T and F having their usual meanings.

Thus, redox energy is conserved as pmf, which may subsequently be utilised to drive ATP synthesis or other energy-requiring reactions of the mitochondrion, such as ion transport. This principle of respiratory chain function is generally accepted. It may be called the *central dogma of membrane bioenergetics,* or the principle of protonic coupling, and may be visualised by the following scheme

$$\text{RESPIRATION} \rightleftharpoons \text{pmf} \rightleftharpoons \text{ATP}$$

The molecular details of this principle, e.g. the mechanism of proton translocation by the respiratory complexes, and the pathways of the proton circuitry connecting respiration to phosphorylation, are still open questions [8,14,15]. Mitchell's chemiosmotic theory [13,16], though instrumental in the development of the dogma, stresses such mechanistic details [17,18]. Hence, it is a special case of the more general dogma, which also encompasses the early proposals by Williams [19,20] of protonic coupling within the membrane. The latter idea originally caused less impact than the chemiosmotic theory, probably due to its lack of details amenable to experimental scrutiny (but see, e.g., Refs. 21,22).

2.2. Thermodynamic limits for mechanisms

Any considered mechanism must, first of all, be consistent with the thermodynamic constraints of the system. Such limits are set by the span of oxidoreduction potentials in the respiratory chain and by the protonmotive force that opposes the proton movement. The relative magnitudes of these two forces set an absolute upper limit for H^+/e^- stoicheiometry of proton translocation. The stoicheiometry in turn, puts limits on the underlying mechanisms. Analogous limits for the H^+/ATP stoicheiometry of ATP synthesis are obtained from the relative magnitudes of phosphorylation potential and pmf. An elementary thermodynamic analysis of the system can therefore be helpful in defining the degree of freedom in discussions of chemical mechanisms (see Ref. 8).

The redox centres of the chain are arranged roughly in three isopotential groups with the dioxygen/water couple as the fourth relevant redox pair (Fig. 3.1B). Hence, under normal operating conditions there are three large drops in oxidoreduction potential, corresponding roughly to the 'sites' of energy conservation, or the segments where flux of reducing equivalents is linked to generation of pmf. The magnitudes of these three 'redox spans' are unequal. Therefore, neither the ATP/$2e^-$ ratio nor the H^+/e^- ratio at these 'sites' are necessarily equal. In fact, they are likely to be unequal, and need not even be integer numbers.

Mitochondrial respiration has certain characteristic states, defined originally by the classical work of Chance and Williams [23]. States 3 and 4 are of particular relevance here. In State 4 the rate of respiration is minimal ('resting state') due to a maximal back-pressure from the pmf (or from the phosphorylation potential via the pmf). This is a state termed 'static head' in the theory of thermodynamics of irreversible processes (see, e.g., Refs. 24, 25).

State 3 ('active', phosphorylating state) is characterised by high respiratory activity due to a lowered phosphorylation potential (addition of ADP). However, respiration is usually not at its maximum, and the measured pmf is only little decreased from its State 4 value [26]. Maximal respiratory activity is obtained in the presence of uncoupling agents that rapidly dissipate the pmf by making the coupling membrane permeable to protons. The latter state (State 3u) is equivalent (or nearly so) to the thermodynamicist's 'level flow' [24,25], where the pmf is zero.

The maximum pmf generated in State 4 has been measured by various groups using several techniques (see Refs. 8,26,27). The results vary between about 160 and 240 mV. However, from the observed shifts of the redox equilibria in the respiratory chain imposed by the pmf (or by ATP via the pmf), it can be concluded that the functionally relevant pmf must be at least 200 mV in State 4 of well-coupled mitochondria [8]. (This need not be equivalent with the pmf between the bulk aqueous C and M phases). This limits the H^+/e^- ratio in State 4 to *maximal* values of 5.6 and 3.8 for the spans from NADH and ubiquinol to oxygen, respectively [8]. These are indeed upper limit stoicheiometries, attainable, of course, only at thermodynamic equilibrium between respiration and proton translocation. The true values must therefore be lower, particularly as the terminal step in the chain, i.e., oxidation of cytochrome *c* by O_2 is irreversible (see Refs. 28, 29 and below).

Recently Forman and Wilson [30] presented strong evidence for the notion [28] that the respiratory chain between NADH and cytochrome *c* is very close to thermodynamic equilibrium with the phosphorylation potential in State 4 mitochondria. If this and the *central dogma* are accepted, the observed ATP/2e ratio very near 2 [30] means that H^+/ATP equals the effective H^+/e^- for the 'site 1 + 2' segment. Since the highest effective H^+/e^- ratio observed for this segment is 3 (Table 3.1), the maximum H^+/ATP ratio would be limited to this value and not be as high as 4 as proposed by some authors (cf., Refs. 26, 31–34, 43). It would also follow that the effective pmf must be in near equilibrium with the 'site 1 + 2' segment of the respiratory chain in such conditions. From the observed redox span [30] and the maximal effective H^+/e^- ratio (Table 3.1) it can be concluded also from these results that the effective pmf in State 4 must be at least 200 mV (cf. above). Interestingly, the *measured* pmf was only of the order of 170 mV [30].

The above estimates of the minimum pmf in State 4 are important also insofar as they are independent of the present uncertainty of whether protonic coupling is localised [15,19–22,35–38] with direct protonic circuits between respiratory complexes and ATP-synthase, or entirely delocalised between the bulk aqueous compartments on each side of the membrane. Clearly, any reliable measurement of a bulk phase pmf in State 4 that yields values lower than 200 mV is indicative of localised coupling (as may be the case above).

Table 3.1 summarises the proposed stoicheiometries of proton and charge translocation in the different segments of the respiratory chain. It shows that there is an embarassing disagreement about these fundamental parameters between different laboratories, and even within single research groups. Note, however, that there is full agreement on 'segment 2' between ubiquinol and ferricytochrome *c*.

According to the *dogma*, the P/O (or $ATP/2e^-$) ratio equals the $H^+/2e^-$ ratio divided by the H^+/ATP ratio. The latter is minimally 2 for the F_1F_0 system [13]. Since the ADP/ATP translocase is electrogenic (but see Ref. 70) and inorganic

TABLE 3.1

Stoicheiometry of proton translocation in the respiratory chain

Segment		H^+/e^- [a]	q^+/e^- [a]
1	NADH-ubiquinone		
a.	(Refs. 39–42)	1	1
b.	(Refs. 32, 43, 44)	2	2
2	Ubiquinol-cyt. *c*		
	(Refs. 8, 14, 40, 43–47)	2 [b]	1
3	Cytochrome *c*-O_2 [c]		
a.	(Refs. 13, 45, 48–51)	0	1
b.	(Refs. 14, 47, 52–55)	1	2
c.	(Refs. 44, 56, 57)	2	3
2+3	Ubiquinol-O_2 [d]		
a.	(Refs. 17, 40, 45, 58, 59, 60)	2	2
b.	(Ref. 45)	2.5	2.5
c.	(Refs. 8, 14, 46, 47, 52, 61–64)	3	3
d.	(Refs. 34, 44, 65, 66)	4	4

See page 55 for notes to Table 3.1.

phosphate (P_i^-) is translocated in symport with H^+ [71], the effective H^+/ATP ratio for intact mitochondria is minimally 3.0. Therefore, the lowest proposed proton stoicheiometries of the respiratory chain (Table 3.1) are probably incorrect, as they would yield $ATP/2e^-$ ratios considerably lower than those routinely measured. Table 3.1 also shows that the highest proposed H^+/e^- stoicheiometries exceed the limit values for equilibrium calculated above for State 4. Such high ratios would therefore be thermodynamically possible only in states other than State 4. If these high ratios are correctly measured (see Refs. 14, 72 for criticism), the proton stoicheiometry must be *variable*, i.e., higher in other states than State 4 where decoupling or 'slipping' of the proton pumps [26,73] would decrease it.

2.3. *Occupancy and mobility of the respiratory chain in the membrane*

Table 3.2 gives the amounts of the different respiratory chain constituents in rat liver and beef heart mitochondria. It may be seen that for each cytochrome bc_1 unit (or monomer), there are two cytochrome aa_3 units (monomers), two to three cytochrome *c* molecules, and about 16 ubiquinone molecules, but only about $\frac{1}{4}$ of a Complex I molecule.

The area of the inner membrane of rat liver mitochondria has been estimated to be about 40 $m^2 \cdot g^{-1}$ of mitochondrial protein [13,80]. Together with the data of Table 3.2, this means that, on average, the above number of respiratory chain components occupies a membrane area of approx. 900 nm^2 (e.g., a square with a side

Notes to Table 3.1

[a] H^+/e^- and q^+/e^- refer to the stoicheiometries of protons released on the outside of the mitochondrion and the number of electrical charge equivalents translocated, respectively (see Ref. 8). Note that the values for segment 1 are usually measured as the difference between segments 1 + 2 and 2.

[b] Though two protons are released from mitochondria, only one is translocated electrogenically. The thermodynamically effective proton/electron ratio is therefore 1 (see Refs. 8, 26).

[c] One less proton is released on the C side than the number of translocated electrical charges. In mitochondria the thermodynamically effective proton translocation is that indicated by the q^+/e^- ratio (see Ref. 8, 26).

[d] The wide diversity of observed proton/electron ratios is surprising, especially for the uninitiated reader. We therefore include here a brief historical survey of experimental reports on the much studied segment 2 + 3 as a guide into the complexity of the literature.

Mitchell and Moyle [58] reported in 1967 that H^+/e^- was very near 2.0 for rat liver mitochondria, and practically constant between pH 5.5 and 8.5, or in various media containing, e.g., KCl or choline chloride. This view still prevailed in 1979 [17]. Then Mitchell also reported, contrary to findings of Brand et al. [64] in 1976, that *N*-ethylmaleimide (NEM) had no effect on the ratio in a sucrose/$MgCl_2$ medium [17]. But later in the same year Mitchell and Moyle [45] communicated that the ratio was enhanced to 2.5 at low extramitochondrial pH, or in media containing choline chloride, or in all tested media when NEM was present. This rise of the ratio was interpreted as due to a very special anomalous reaction sequence of the cytochrome bc_1 complex [67]. The phenomenon was nevertheless not observed when segment 2 was studied alone.

On the basis of extensive fast kinetics studies Papa et al. [62,63] showed in 1973–1974 that H^+/e^- of proton uptake is 3 or slightly higher in submitochondrial particles oxidising succinate or ubiquinol. In a paper that appeared in 1975 [60], the same authors found a ratio of 2. The latter conclusion has been maintained to date.

In 1976 Brand et al. [46,64] reported a ratio very near 3 in oxygen pulse experiments with rat liver mitochondria. This result has been supported by data from this laboratory [8,14,47,52,61]. Subsequently, Lehninger et al. [44,65,66] and Azzone et al. [43,56] obtained results that were interpreted to suggest a ratio of 4.

Since all groups agree on the stoicheiometry for segment 2, the above diversity obviously mirrors a profound experimental disagreement on segment 3. Mitchell [18] has recently stated that the proposal that cytochrome oxidase is a proton pump suffers from "much the same excessive hypothesis-building and lack of scientific realism as was responsible for the downfall of the chemical coupling theory of oxidative and photosynthetic phosphorylation." The first part of this statement is in poor accord with the factual experimental development just related. The second part is an unfortunate underrating of the so-called chemical coupling theory. It was, in fact, more an abstract framework for linking experimental results logically together (see, e.g., Ref. 68). This frame is still valid though perhaps no longer very useful. Its importance in the earlier development of the field cannot be diminished by the present deeper level of understanding. Even though we now write the year 1984 one would prefer the past to be related more fairly than in the case of Comrade Ogilvy [69]. It may indeed be "especially interesting for scientific historians to observe the outcome in the case of the 'proton pump' of cytochrome oxidase" [18].

of 300 Å). On average, this segment also contains a pair of ATP synthase molecules (see Refs. 75, 79), and approx. 1500 phospholipid molecules.

The respiratory complexes diffuse laterally in the membrane with diffusion coefficients of 8×10^{-10}–2×10^{-9} $cm^2 \cdot s^{-1}$ [81–83]. Cytochrome *c* and ubiquinone have been quoted to diffuse at a velocity (10^{-8} $cm^2 \cdot s^{-1}$) comparable to that of phospholipid molecules [84,85]. However, the bulky isoprenoid side chain of Q may slow down its mobility [86]; in chromatophores, which may be compared with mitochondria, a mobility of 10^{-9} $cm^2 \cdot s^{-1}$ has been estimated [87]. Overfield and

TABLE 3.2

Content of respiratory carriers in mitochondria

The data given is collected from Refs. 74–79. Cytochrome aa_3 contains two haems and two coppers per monomer. Considerable variations in the literature are due in part to differences in protein determination, and in part to use of erroneous extinction coefficients. The ratio of Complex I : Complex III : Cytochrome c : Complex IV : Ubiquinone is about 1 : 4 : 8 : 8 : 64 in most mitochondria. The content of cytochrome c is somewhat variable, however, and is lowered by extensive washing of mitochondria is salt solutions.

Component	Liver (nmol/mg protein)	Heart (nmol/mg protein)
Cytochrome		
aa_3	0.14	0.50
b	0.14	0.50
c_1	0.07	0.25
c	0.20	> 0.35
Rieske's FeS centre	0.07	0.25
Ubiquinone	2.5	4
Complex I	0.02	0.05

Wraight [88] have shown that both the ionic strength and the temperature are important mobility-determining parameters for cytochrome c. Their estimated lateral diffusion coefficients above 25°C were $0.6–2 \times 10^{-10}$ $cm^2 \cdot s^{-1}$ and $0.3–1 \times 10^{-9}$ $cm^2 \cdot s^{-1}$ for no salt and 0.1 M NaCl, respectively, in membranes composed of charged phospholipids. Thus, the mobilities of c and Q may not, in fact, be much higher than those of the integral membrane complexes. The time necessary for a molecule of cytochrome c or of ubiquinone to sweep the area of 900 nm^2 would then be about 10 ms at physiological temperature and ionic strength.

It is remarkable that reported rates of respiration per mg of protein can vary between different laboratories for mitochondria from the same source by a factor as large as 3–4. This may be partially due to differences in protein determination. A more appropriate point of reference is, therefore, the content of respiratory chains. The respiratory rate with natural substrates is maximally approx. 30 $e^-/s \cdot aa_3$ in rat liver mitochondria (State 3u with succinate; M. Wikström, unpublished results). The rate in State 4 is lower by about one order of magnitude. The maximum respiratory velocity of a 900 nm^2 membrane segment is then 60 e^-/s and the average turnover time of cytochrome c and ubiquinone, 50 and 500 ms, respectively (cf., Ref. 84). In spite of the lower mobilities applied here, it seems that lateral diffusion of cytochrome c and ubiquinone is still sufficiently fast to be compatible even with maximal rates of respiration with natural substrates.

Rotational correlation measurements have also indicated that about 40% of cytochrome oxidase is immobile in the mitochondrial membrane in situ [82,83]. Capaldi [85] suggested that this could be due to an immobilising effect of the high protein content of the M phase on integral membrane proteins, a significant fraction of which protrudes into the matrix. This is not expected to significantly affect the rate of respiration since this effect should not decrease the mobilities of *c* and Q [85].

2.4. *Functional domains in the membrane*

At present much evidence is quoted in favour of the idea [19–22,25,26,38] that the proton circuitry between respiration and ATP synthesis is confined to the membrane proper or its interphases with the aqueous media (*transversal localisation*). Special high conductance pathways of the protons have been postulated along the membrane, with resistive and capacitative barriers against delocalisation of translocated protons into the bulk media. Typical of this idea is the prediction that the functionally relevant pmf (across a restricted membrane domain) is higher than that between the bulk aqueous phases (cf., above). However, other sets of experiments based on inhibitor titrations [35–37,89,90] suggest *lateral localisation* of protonic circuits. This implies that a particular respiratory chain complex would be able to drive ATP synthesis only in a limited membrane domain containing one or very few ATP synthase complexes. These two modes of localisation are, of course, not mutually exclusive. Transversal localisation does not necessarily require lateral localisation, but the latter is difficult to envisage unless the former is also true.

3. *Cytochrome oxidase or complex IV*

3.1. *Composition*

Mitochondrial cytochrome oxidase (EC 1.9.3.1) consists of three polypeptides synthesised in the mitochondrion, and several others synthesised in the cytoplasm (Table 3.3). Preparations of the mammalian enzyme have been shown to contain up to 13 different polypeptides, and there is uncertainty at present as to which are true constituents of the enzyme. The present opposing views are those of Capaldi et al. [85,96], who regard eight polypeptides as unique constituents, and Kadenbach et al. [93,94], who suggest that all 13 are true parts of the enzyme.

All polypeptides involved in one way or another have been sequenced (Table 3.3); in some cases the sequence is known only from the corresponding nucleotide sequence of mtDNA.

Cytochrome oxidase contains two haem groups and two protein-bound coppers per minimal functional unit, i.e., the monomer (Table 3.4). Apparently a single copy of most polypeptides is present in this monomer. The haems are chemically identical (haem A), but are bound quite differently to the protein, which gives them widely different functions and spectroscopic properties (Table 3.4). The same is true of the two coppers, Cu_A and Cu_B. The haems will be called *a* and a_3, respectively. As discussed below, it is not certain whether they have different apoprotein parts, i.e., whether they are formally different 'cytochromes'.

TABLE 3.3

Polypeptides of cytochrome oxidase from beef heart

After Refs. 91, 92. The main nomenclature is that of Buse et al. [91].

Polypeptide	M_r	Synthesis	Stoich-eiometry	*N*-terminal
I	56993	mito	1	f-Met-Phe-Ile-Asn
II	26049	mito	1	f-Met-Ala-Tyr-Pro
III	29918	mito	1(?)	(Met)-Thr-His-Gln
IV	17153	cyto	1	Ala-His-Gly-Ser
V (Va [a])	12436	cyto	1	Ser-His-Gly-Ser
VIa (Vb [a], a [b])	10670	cyto	1	Ala-Ser-Gly-Gly
VIb (VIa [a], b [b])	9419	cyto	1	Ala-Ser-Ala-Ala
VIc (VIb [a], c [b])	8480	cyto	1	Ser-Thr-Ala-Leu
VII (VIc [a], VI [b]) (VIIa [a])	10068	cyto	1	acetyl-Ala-Glu-Asp-Ile
VIIIa (VIIb [a], VIIser [b])	5541	cyto	1	Ser-His-Tyr-Glu
VIIIb (VIIc [a], VIIile [b])	4962	cyto	2	Ile-Thr-Ala-Lys
VIIIc (VIII [a], VIIphe [b])	6244	cyto	1	Phe-Glu-Asn-Arg

[a] Nomenclature of Kadenbach et al. [93,94].
[b] Nomenclature of Azzi [95] and Capaldi [85].

TABLE 3.4

Properties of redox centres in cytochrome oxidase

The haems exhibit considerable redox interactions making differentiation between them difficult. 'Optical' refers to commonly used wavelength couples for measuring oxidoreduction of the haems. Extinction coefficients (for reduced *minus* oxidised per aa_3 unit) are given in parentheses. The approximate contributions of haem a to the reduced *minus* oxidised difference at 605–630 and 445–460 nm are 75–80% and 33%, respectively, the rest being attributed to reduced *minus* oxidised haem a_3. The EPR data refer to low-spin ferric haem a and high-spin ferric a_3, respectively, of which only the former is observed in the oxidised resting enzyme. The two $E_{m,7}$ values for aa_3 represent midpoints of two redox transitions of the haem system, to which both haem residues contribute. The last column indicates that both these midpoints are pH-dependent. All parameters refer to the native enzyme as isolated or in mitochondria. In the presence of ligands to haem a_3 there may be drastical changes (see reviews in Refs. 12, 97–99). Cu_B is generally undetectable by EPR due to its proximity to haem iron of a_3 (but see Refs. 100, 101 for exceptions). It has no known optical transitions.

Component	Optical	EPR	$E_{m,7}$ (mV)	pH-dep
Haems				
a and a_3				
α-band	605–630 nm (27 $mM^{-1} \cdot cm^{-1}$)			
γ-band	445–460 nm (148 $mM^{-1} \cdot cm^{-1}$)			
Ferric a		$g = 3, 2, 1.5$		
Ferric a_3		$g = 6$		
High potential			380 mV	+
Low potential			220 mV	+
Cu_A	830 nm (2 $mM^{-1} \cdot cm^{-1}$)	$g = 2$	240 mV	–
Cu_B	–	–	340 mV	–

3.2. Topography and image reconstruction

The topography of Complex IV with respect to the membrane has been studied both by the aid of several kinds of labelling and cross-linking techniques (see Refs. 85, 92, 95, 96, 99), and by image reconstruction of two-dimensional crystal structures [102–106]. The wealth of this information may be summarised as the model in Fig. 3.2, which also includes information on the positions of the redox centres (see below).

3.3. Catalytic activity

Cytochrome oxidase catalyses electron transfer from cytochrome *c* to dioxygen, reducing the latter to water without release of intermediates. The maximal electron transfer activity may reach 400 (moles of cytochrome *c* oxidised per second per mole of cytochrome aa_3) in optimal conditions at pH 7 and 25°C, both in situ and in detergent solution, although much lower activities are often encountered in the latter case (see Ref. 99). In mitochondria, respiration with natural substrates proceeds at much lower rates (cf., Section 2.3). The kinetic capacity of cytochrome oxidase greatly exceeds demands for reasons not understood at present.

3.4. Interaction with cytochrome c

Cytochrome *c* delivers electrons to the oxidase after binding to a specific site in subunit II (K_d approx. 10^{-8} M) that is located on the C side of the membrane

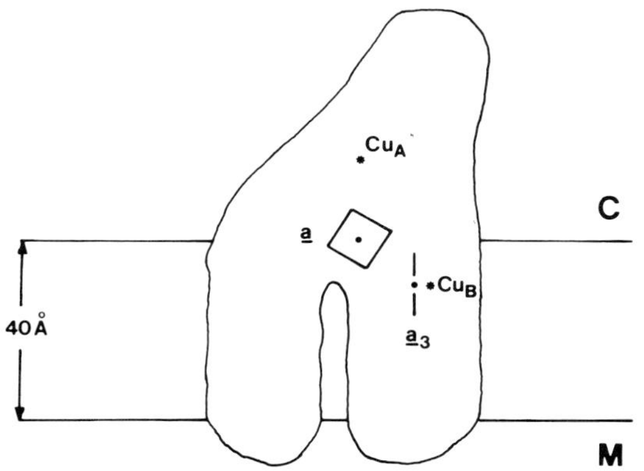

Fig. 3.2. Model of the cytochrome oxidase monomer. The structural model of the membrane-bound cytochrome oxidase monomer that has emerged from image reconstruction studies of two-dimensional 'crystals' [102–106] is an asymmetric 'Y'-shaped molecule (see also Ref. 99). A considerable part protrudes on the cytoplasmic side of the membrane, but only little on the M side. In the membrane there are two separated domains. Chemical labelling and cross-linking studies (see Refs. 85, 92, 95, 96, 99 for reviews) have given at least a rough topography of the individual subunits. The protruding C-domain is made up mainly of subunits I, II, III and V, while the aqueous M-domain is largely due to subunit IV. The two membranous domains are formed by subunits I and III, but also subunits II, IV, VIIIa and VIIIb (Table 3.3) probably contribute with 2, 1, 1 and 1 transmembranous polypeptide segments, respectively. The positions of the redox centres with respect to the membrane are also shown (see the text).

[107–111]. The structure of cytochrome *c* is known from X-ray studies to < 2 Å resolution [112]. It binds to subunit II, and also to cytochrome c_1 of Complex III [113,114], with a lysine-rich domain near the solvent-accessible haem edge. This domain mainly involves lysines 13, 72, 86, possibly 79, and 27 to a lesser extent [115–118]. When bound to the dimeric oxidase cytochrome *c* may lie in a cleft between the monomers so that the side opposite to the lysine-rich region interacts weakly with subunit III of the second monomer [119]. The binding site on subunit II has recently been mapped more accurately. Four acidic residues have been suggested to be involved, i.e., Asp-158 [120], Asp-112, Glu-198 [121,122] and Glu-114 [123] (numbering refers to the bovine subunit; see also Fig. 3.4 below). Of these, Glu-114 is not conserved in subunits II from human, yeast and maize cytochrome oxidase (see Ref. 92).

Cytochrome *c* also binds to low-affinity sites (K_d approx. 10^{-6} M), possibly composed of bound cardiolipin molecules [124,125]. The function of these is not understood, but the negatively charged cardiolipin could help to orientate the cytochrome *c* dipole [115] for subsequent proper interaction with the high-affinity site [126].

3.5. Mechanism of electron transfer and reduction of O_2

Electrons from cytochrome *c* are transferred rapidly ($k = 8 \times 10^6$ $M^{-1} \cdot s^{-1}$) to haem *a* and Cu_A. Although the former has been suggested to be the primary electron acceptor, this view is presently uncertain due to the very fast electron equilibration between the two receiving centres [127,128].

Subsequently, the electrons are transferred to haem a_3 and Cu_B, which are closely apposed and function as the enzyme's binuclear O_2-reducing centre (see Refs. 8, 92, 97–99). It is not known whether haem *a* or Cu_A (or both) is the principal electron donor. Nor is it known whether haem a_3 or Cu_B is the acceptor. However, in practice the acceptor is in any case a haem a_3/Cu_B-oxygen complex during turnover.

One factor that has greatly complicated elucidation of electron transfer in cytochrome oxidase is the strong redox interaction between the haems (see Refs. 12, 99). Although this phenomenon is relatively well elucidated, its functional relevance is not yet understood.

The mechanism of reduction of dioxygen has been partially clarified, mainly thanks to the low temperature trapping technique designed by Chance et al. [129]. Although the precise mechanism is still not understood, it seems probable that dioxygen is reduced to water in two concerted two-electron steps (Fig. 3.3) in which certain intermediates have been identified (see Refs. 8, 92, 97–100, 129–133). One important feature of this particular mechanism is that it affords a switch from one-electron transfer reactions (of cytochromes *c*, *a* and Cu_A) to effective two-electron steps in the reduction of O_2. The latter is necessary for thermodynamic and kinetic reasons and to effectively prevent release of toxic oxygen radicals from the active site (see Ref. 99).

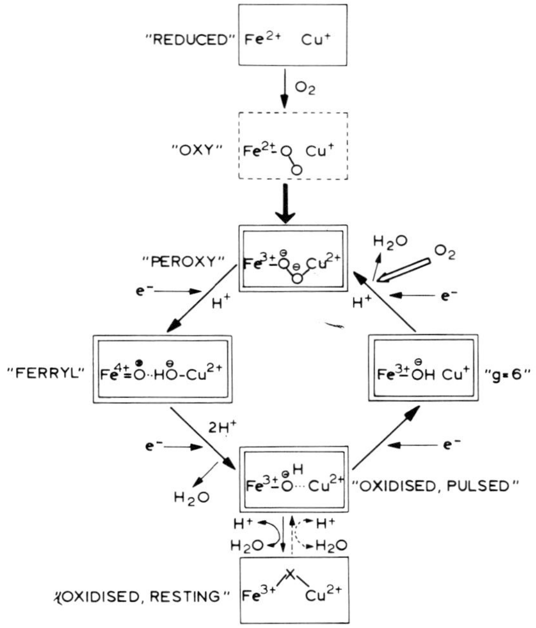

Fig. 3.3. Tentative mechanism of reduction of dioxygen. The scheme shows some of the more significant reaction steps at the haem a_3 iron-Cu_B centre of cytochrome oxidase. The reaction may be initiated by delivery of dioxygen to the reduced enzyme (in anaerobiosis; top of figure). An initially formed 'oxy' intermediate is normally extremely short-lived, but can be stabilised and identified in artificial conditions (see Refs. 92, 99, 129, 134). Concerted transfer of two electrons from Fe and Cu to bound dioxygen yields a 'peroxy' intermediate. This, or its electronic analogue, is stabilised in the absence of electron donors (ferrocytochrome *a* and/or reduced Cu_A), and has been termed 'Compound C' [129,130,132]. It may also be observed at room temperature, and is then probably generated from the 'oxidised' state by partial oxidation of water in the active site, in an energy-linked reversed electron transfer reaction [29] (see also Refs. 92, 99). Also the 'ferryl' intermediate [92,99,100] has been tentatively observed in such conditions [29]. In aerobic steady states the reaction is thought to involve the cycle of intermediates in the centre of the figure (dark frames). The irreversible step is probably the conversion of '$g = 6$' (see Refs. 98, 133) to 'peroxy'.

The conversion of one- to two-electron transfer is an essential feature of this model. It is made possible by the electronic flexibility of haem iron and by the presence of two metals in the centre. The input of electrons shown is thought to arise from haem *a*, or possibly Cu_A [99]. For further details and discussion, see Refs. 29, 97–100, 129–133.

The 'oxygen intermediates' of the haem a_3 centre have generally been observed at low temperatures during the forward reaction. However, partial reversal of electron transfer (i.e., from water to cytochrome *c*) may take place at a high pmf (or a high phosphorylation potential) [29] with accumulation of some of the 'oxygen intermediates' at room temperature. This reaction apparently proceeds only to the stage of bound peroxide (Fig. 3.3). Complete oxidation of water to O_2 has not been achieved, underlining the irreversibility of the cytochrome oxidase reaction.

The actual mechanisms of electron transfer from cytochrome *c* to the oxidase, or within the oxidase, are not known. Outer sphere electron transfer or thermally assisted electron tunneling are mechanisms to be considered [135]. Conserved clusters of aromatic amino acids have been observed in the primary structures of the subunits, and might be involved in electron transfer [92,122,136] (cf., Fig. 3.4).

3.6. The redox centres and their location

It is generally accepted that the four redox centres are located in subunits I and II (see Refs. 85, 92, 96, 99). Cu_A is almost certainly in subunit II which, near the carboxy terminus, has a segment that is homologous to the copper-binding site of copper proteins [92,122,136] (Fig. 3.4). Moreover, Chan et al. [137] have recently

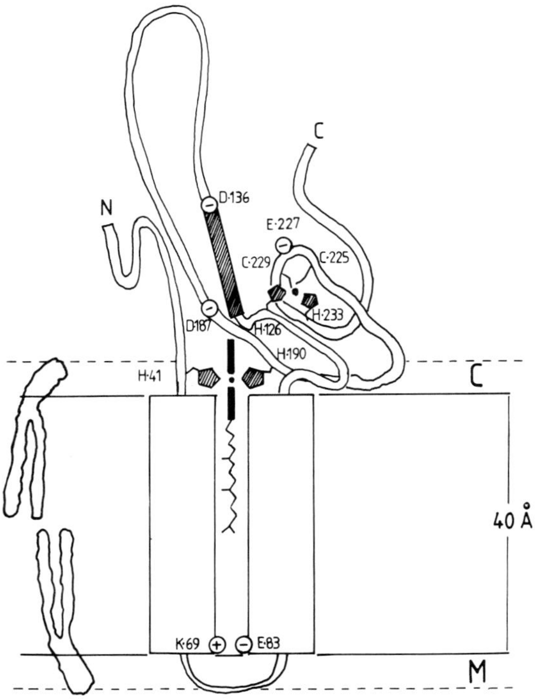

Fig. 3.4. Model of subunit II in the membrane. (From Ref. 92; modified from Refs. 122, 138). The amino acid numbering derives from an alignment of subunits II from seven organisms. His-41 (24 in the bovine subunit) and possibly His-190 (161) may be haem ligands if this subunit binds the haem of cytochrome *a*. Cys-225 (196), Cys-229 (200) and His-223 (204) are most probable Cu_A ligands. His-126 (102) is not fully conserved, and is uncertain. His-190 (161) is a likely copper ligand if subunit II does not bind haem. Asp-136 (112), Asp-187 (158), Glu-227 (198) and Glu-138 (114) (not indicated in the figure) may be involved in the binding of cytochrome *c* (Section 3.4.). The cross-hatched segment near Asp-136 (and Glu-138) is a cluster of aromatic amino acids, possibly involved in electron transfer (Section 3.5.).

shown that the ligands of Cu_A are at least one histidine and one cysteine. The two cysteines of subunit II (Fig. 3.4) are the only ones fully conserved in subunits I–III of cytochrome oxidase from seven different species (see Ref. 92). The folding of subunit II in the membrane predicted from the primary structure is in excellent agreement with labelling data [138] and places this copper site on the C side of the membrane (Fig. 3.4). This also agrees with observed magnetic interactions between aqueous dysprosium probes in the C phase and Cu_A [139].

Winter et al. [140] suggested that haem *a* is bound to subunit II. If so, it may be uniquely located because it is a bisimidazole complex and the number of conserved histidines is limited. It would then be expected to be near the C domain, 'sandwiched' in part between transmembranous helices [92] (Fig. 3.4). This would agree with the perpendicularity between the haem and membrane planes observed by polarised spectroscopy of orientated enzyme specimens [141–143].

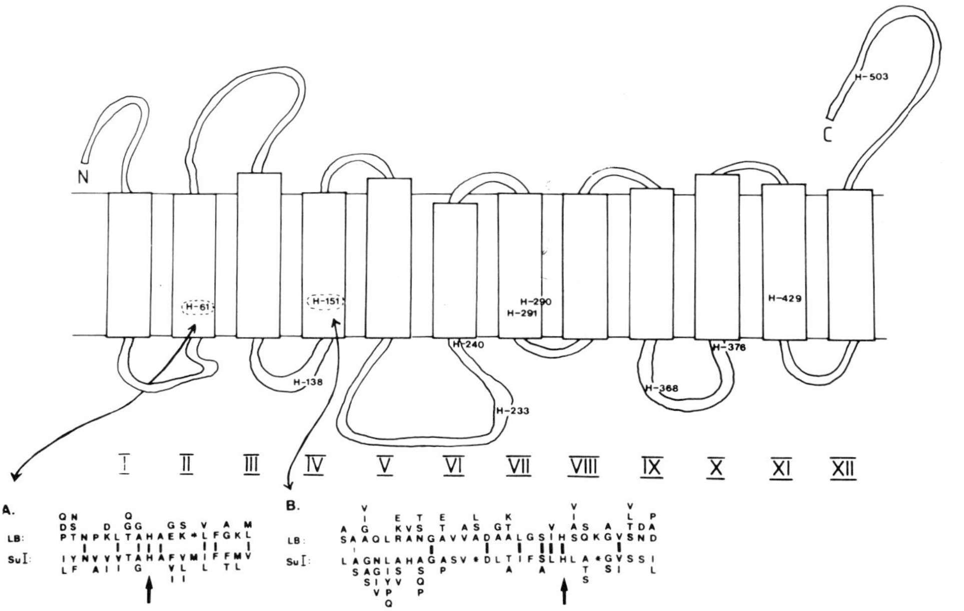

Fig. 3.5. Subunit I of cytochrome oxidase. The alternating pattern of hydrophobic and hydrophilic segments in the primary structure suggests a model of how this subunit might fold in the membrane. 12 hydrophobic α-helices are shown as boxes traversing the membrane. Ribbons on the membrane surfaces are hydrophilic sequences (See Ref. 92). The location of the 11 invariant histidines [92,152] is shown (numbering refers to the bovine subunit [153]). Their location towards the lower aspect of the figure suggests that this is the cytoplasmic side of the membrane (see the text). In the lower part of the figure the sequences around His-61 and His-151 are compared with the amino acid sequences around distal (A) and proximal (B) histidines in leghaemoglobin (Lb) [92]. Amino acid substitutions in four Lb [154] and six subunit I sequences are included in the alignment. A single residue and a black bar denotes invariance; * denotes a deletion. The arrows in A and B show the distal and proximal histidines of Lb, respectively.

However, at present it cannot be excluded that both haem groups may be linked to the same protein, viz. subunit I. This subunit contains 11 invariant histidines, of which four are predicted to be located in the aqueous domains, and seven in the membrane (Fig. 3.5).

Haem a_3 is structurally very similar to the haems of myoglobin and haemoglobin [144–146]. The proximal (5th) ligand is histidine [147] (see also review in Ref. 99). It is therefore of interest that the sequences around two of the 'membranous' histidines of subunit I show homology to the regions the proximal and distal histidines of myoglobin and leghaemoglobin [92] (Fig. 3.5). These histidines are predicted to be located at similar depth from the membrane surface and in different transmembranous helices. It is possible, therefore, that haem a_3 is sandwiched between these helices. However, Welinder and Mikkelsen [148] favour a different site involving His-233 and His-376 (see Fig. 3.5).

Of the nine other conserved histidines in subunit I, two may be potential ligands of haem *a* (in case this haem is not associated with subunit II). In such a case subunit I would be similar to cytochrome *b* of the bc_1 complex, which binds two haems (see below). Note that the conserved membranous histidines of subunit I (Fig. 3.5) are all predicted to be located in the lower aspect of the membrane.

Recently an enhancement of spin relaxation of ferrous haem a_3-NO by ferric haem *a* has been demonstrated [149,150]. From this, and the known location of haem *a* near the C side of the membrane [99], it can be concluded that both haems are near this side with an Fe-Fe distance of approx. 15 Å (about 10 Å when projected on the same normal to the membrane). This would identify the lower aspect of Fig. 3.5 as the cytoplasmic side.

Cu_B is known to reside within 3–5 Å from the iron of haem a_3 [146,151], almost certainly on the distal side of the haem. It seems probable, then, that this copper lies in subunit I in the region near the putative distal histidine (Figs. 3.2 and 5; see Ref. 92).

3.7. Energy conservation

The cytochrome *c* oxidase reaction encompasses the so-called third site of oxidative phosphorylation. There is no doubt that oxidation of cytochrome *c* by dioxygen results in generation of pmf. Cytochrome oxidase was long believed to do so simply by catalysing transmembranous electron transfer, with uptake of the protons required in reduction of O_2 to water from the M phase. Such a function is thermodynamically equivalent to translocation of one proton per transferred electron, although no protons appear on the C side [8].

However, today a wealth of evidence suggests that cytochrome oxidase functions as a redox-linked proton pump [155] (reviewed in Refs. 8, 14, 47, 52, 92, 99). Two electrical charge equivalents are proposed to cross the membrane per transferred electron; two protons are taken up from the M side, one of which is transported to the C side (Fig. 3.6). This function has been confirmed also for cytochrome oxidase liposomes [52,54,55,156–161], and includes work with bacterial cytochrome oxidase

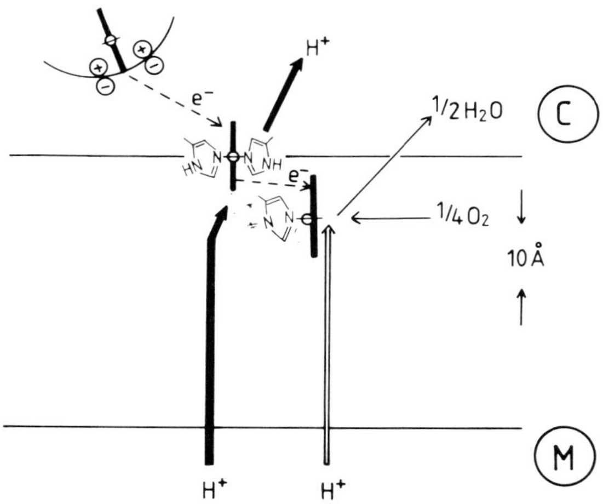

Fig. 3.6. Principle of energy conservation by cytochrome oxidase. The approximate location of haems *a* and a_3 are indicated with respect to the membrane (see the text). Cytochrome *c* is also shown (upper left) at the approximately correct distance from haem *a*. The 40 Å lipid domain of the membrane is indicated by horizontal lines. The pathway of electron transfer is shown by dotted arrows; reduction of dioxygen by thin arrows. Thick black arrows symbolise the redox-linked proton-pumping function. The thick white arrow shows the uptake of protons into the haem a_3/Cu_B site. See the text for details. (From Ref. 92).

[162–164]. It is thermodynamically equivalent to translocation of $2H^+/e^-$ [8]. Despite the comparatively strong evidence, two groups disagree with the proposal of cytochrome oxidase as a proton pump [17,18,48–51] (Table 3.1). Two other groups suggest a proton translocation stoicheiometry higher than that in Fig. 3.6 (Table 3.1) [44,56,57].

Fig. 3.6 implies that cytochrome oxidase generates pmf by two different mechanisms coupled in series. The first is the pure proton-translocating function, which may specifically involve haem *a* (see below). The second and thermodynamically equivalent function is the 'annihilation' of electrical charges when electrons deriving from cytochrome *c* on the C side of the membrane 'meet' with protons deriving from the M side, in the reduction of O_2 to water at the haem a_3 centre.

3.7.1. On the mechanism of proton/electron 'annihilation'

This energy-conserving function is simply a consequence of two topological features; electron donation takes place from the C side, and proton donation for conversion of O_2 to water from the M side (for the evidence, see below). The position of the haem a_3 centre with respect to the membrane surfaces then determines to what extent the reaction requires electron and proton translocation. Since the evidence points strongly towards the positioning of haem a_3 near the C side (Section 3.6), the mechanism involves mainly proton translocation, and electron translocation only to a lesser degree. Hence, not only the thermodynamical, but also the structural features of this mechanism resemble those of a 'proton pump'. It differs from the latter only by lacking the (thermodynamically unimportant) proton ejection step to the C phase. Instead the proton is consumed at the haem a_3/Cu_B centre within the membrane.

A proton-conducting pathway from the haem a_3 domain to the M phase must be postulated. The evidence for uptake of the 'water protons' from the M side stems from the finding that, when the proton pump is inoperative after removal of subunit III (see below), the oxidase still conserves energy, but with only half efficiency [55,92,99]. If the protons were taken from the C side, there could be no generation of pmf in conditions where the proton pump is inoperative.

3.7.2. On the mechanism of proton translocation

Of the four redox centres, haem *a* is the most likely one to be coupled to proton translocation (Fig. 3.6). It exhibits heterogeneous oxidoreduction kinetics, membrane sidedness with respect to proton dependence, and looses the proton-dependence of oxidoreduction simultaneously with loss of proton pumping when subunit III is removed from the enzyme (see below, and discussion in Refs. 92, 99).

However, the details of the molecular mechanism of redox-linked proton translocation are still largely unknown, not only for cytochrome oxidase but for the entire respiratory chain. In considering various possibilities it is important to distinguish between basically different elementary steps of the process. *Proton conduction* through the protein is but one such element. Several possible mechanisms have been proposed for this function, based on conduction along hydrogen-bonded networks of amino acid residues within the membrane (reviewed in Ref. 8). In redox-linked

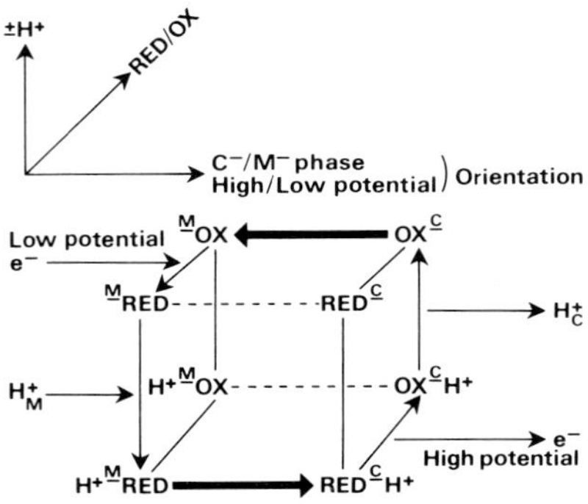

Fig. 3.7. Cubic model of a redox-linked proton pump. OX and RED denote a redox centre in the oxidised and reduced state. The bar marked M or C next to OX and RED indicates an acidic group, the function of which is linked to the redox centre. M and C mean that the group is connected protonically either with the aqueous matrix or cytoplasmic phases, respectively. When the group is protonated the bar is supplemented with H^+. Left and right faces of the cube separate states in electronic and protonic contact with the input and output sides of the transducer, respectively. 'Allowed' transitions between these are indicated by thick arrows. Dotted lines denote 'forbidden' transitions. If the latter gain significant probability relative to 'allowed' transitions proton transport becomes decoupled from electron transfer (so-called 'slipping'). (From Ref. 8.)

proton pumping such 'proton conductors' are necessary but not sufficient elements. The minimum model of a redox-linked proton pump requires eight different states. This may be presented as a cubic scheme (Fig. 3.7).

Fig. 3.7 is strongly related to the 'cycles' of energy-transducing enzyme systems analysed in detail by Hill [165], and to the energy transducer described by DeVault [166]. The essential elements of a redox-linked proton pump minimally include one *acidic group,* the function of which is specifically coupled to the redox state changes in a *redox centre* (probably haem *a* in cytochrome oxidase). Both can exist in two states (protonated and deprotonated; reduced and oxidised, respectively). In addition, both the *redox centre* and the *acidic group* must be able to exist in two 'orientations' in which electronic and protonic contact, respectively, are established with the input and output sides of the transducer (cf. Refs. 8, 52). States ascribed to these two latter configurations have been observed for the haem of cytochrome *a* [99].

The most intriguing question regarding molecular mechanism concerns the nature of the *molecular linkage* between the *acidic group* and the *redox centre.* In cytochrome oxidase the formyl carbonyl and the propionate carboxyls of haem *a* have specific properties that make them potential candidates for providing such a linkage [92,99,167–169].

3.7.3. Role of subunit III in proton translocation

There are three main reasons to suggest a specific function of subunit III in proton translocation. First, Casey et al. [171] showed that modification of this subunit with dicyclohexylcarbodiimide (DCCD) blocks proton translocation, but has little effect on electron transfer. Similar results have been obtained with the reconstituted oxidase from the thermophilic bacterium PS3 [164]. Prochaska et al. [160] showed that DCCD binds mainly to Glu-90 of the bovine subunit III, which is predicted to lie within the membrane domain and hence to be a site analogous to the DCCD binding site in the membranous F_0 sector of the ATP-synthase (Fig. 3.8; see also Ref. 85). Since the latter is a part of a proton-conducting channel in ATP synthase, subunit III was thought to have the same function. However, there is one essential difference between the two phenomena. Modification of the membranous glutamic residue in F_0 by DCCD leads also to inhibition of ATP hydrolysis in the F_0F_1 complex, as expected for two linked reactions. In contrast, DCCD has little or no effect on electron transfer in cytochrome oxidase under conditions where H^+ translocation is abolished. Hence, DCCD cannot simply be judged to 'block a proton channel' in the oxidase. More appropriately, it 'decouples' proton translocation from electron transfer.

The second reason is the finding [55,170,172] that removal of subunit III abolishes all attributes of the proton pump after reconstitution of the enzyme into liposomes. Again, electron transfer is not significantly affected, implicating 'decoupling', rather than primary inhibition of the proton-translocating function. Notably, the reconstituted enzyme retains about 50% of the energy-conserving capacity, suggesting that the proton pump is not the only energy-transducing mechanism (cf.,

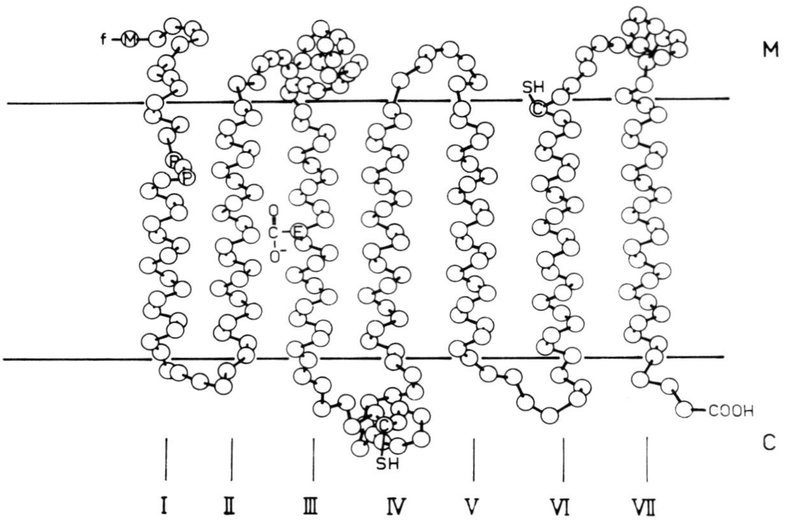

Fig. 3.8. Model of subunit III in the membrane. The folding is a model suggested by the alignment of primary structures of the subunit from six different species and the resulting conserved pattern of hydrophobic and hydrophilic stretches of amino acids [92]. The DCCD-sensitive Glu-90 (bovine numbering) is shown in the middle of the membrane (see text). Two adjacent prolines are shown in the transmembranous segment I. Two cysteines of the bovine subunit are also shown. Of these Cys-115 (the SH-group between segments III and IV) is almost certainly located in the aqueous C phase [92,119], which defines the sidedness of the figure. (Modified from 92; courtesy of Dr. Timo Penttilä.)

above). Unfortunately, it has not yet been possible to reconstitute subunit III with the depleted enzyme.

Recently, a third piece of evidence was added by Chan and Freedman [161], who showed that an antibody towards subunit III specifically blocked proton translocation in cytochrome oxidase vesicles. 'Resting' or coupled respiration was stimulated so that the respiratory control index fell by a factor of two. These findings exactly parallel those obtained by removal of subunit III [55,172].

The role of subunit III in proton translocation became more enigmatic after the demonstration [163] that the reconstituted cytochrome oxidase from *Paracoccus denitrificans* (which lacks the equivalent of subunit III [173]) translocates protons, albeit with an apparently lower efficiency. The proton translocation is insensitive to DCCD, which does not bind covalently to this enzyme [174].

Two main possibilities are apparent. Subunit III might be an essential functional part of the proton pump, in which case its most essential structures may be expected to be built into the two subunits of *Paracoccus*. Unfortunately, the primary structures of the latter are not yet known. The absence of DCCD-sensitivity does not discount this possibility. Bacterial mutants are known, for example, in which the H^+-ATPase functions properly but has lost the sensitivity towards DCCD, although the potential DCCD-binding residue is retained in the primary structure [175]. The second possibility is that the role of subunit III in proton translocation is more indirect. It is clear, in any case, that subunit III can have no important role in electron transfer or binding of the redox centres, since optical spectra and redox activity are very little affected by its removal [55]. Phenomenologically, the role of subunit III is to somehow facilitate the coupling between electron transfer and proton translocation.

4. *The cytochrome* bc$_1$ *complex*

It has become clear in the recent years that electron transfer chains of mitochondria, chloroplasts and some bacteria all contain a cytochrome *bc* complex with very similar structural and functional properties (see Refs. 87, 176–180). Although we focus here on the mitochondrial Complex III, much information has, in particular, come from studies on the bacterial chromatophore system [8,87,176,178].

4.1. Composition and structure

Complex III (ubiquinol:ferricytochrome *c* oxidoreductase; also cytochrome *c* reductase; EC 1.10.2.2) contains one haem of cytochrome c_1, two of cytochrome *b* and one iron-sulphur centre (Rieske's centre) per 200–250 kDa of protein. It is a multisubunit complex (reviewed in Refs. 8, 14, 85, 177–180) consisting of eight or more different polypeptides (Table 3.5). At present a functional property can be assigned only to a few of these. The apoproteins of cytochromes c_1 and *b*, and the Rieske FeS protein have been isolated, and thus identified, among the subunits. The proteins with the highest molecular weight are called 'core proteins' I and II, although recent evidence suggests that they may rather be peripheral in the mem-

TABLE 3.5

Subunit composition of Complex III

Number	Name/function	M_r	Stoicheiometry
I	'core protein' I	46–52000	1
II	'core protein' II	43–45000	1 or 2
III	cytochrome b	43–44000 [a]	1
IV	cytochrome c_1	28 [a] –31000	1
V	Rieske's FeS protein	24–25000	1
VI	–	12–14000	1 or 2
VII	–	8–12000	1 or 2
VIII	–	6–9000	heterogeneous

[a] From the primary structures; bovine cytochrome c_1 [181], human [182], bovine [153], mouse [183], *Saccharomyces* [184], *Aspergillus* [185], and *Neurospora* [186] cytochrome *b*. Studies on the bovine [187] and *Neurospora* [188,189] Complex III are summarised. Complex III from rat [190] and *Saccharomyces* [191] are very similar but the latter may lack the smallest 'subunit', band VIII [192]. The complex forms dimers in Triton X-100 [189,192,193].

branous enzyme (see below). Their functional role is unknown. The function of the smallest proteins (VI–VIII; Table 3.5) is also unknown, though some of them might bind ubiquinone specifically [194–196]. At least band VIII is heterogeneous and contains three different polypeptides in preparations of the bovine enzyme [187].

The stoicheiometry of the polypeptides in the complex is somewhat controversial (Table 3.5). Complex III can apparently take two oligomeric forms. Hydrodynamic experiments of Tzagoloff et al. [197], carried out in the presence of taurocholate or taurodeoxycholate, suggested a monomeric complex. In contrast, when solubilised with Triton X-100 it appears to be a dimer. Both these preparations are enzymically active [see 178].

4.2. Cytochrome b

It was thought originally that there may be two copies of apocytochrome *b* in monomeric Complex III. However, the earlier estimates of molecular weight of around 30 000 are probably in error. As judged from the nucleotide sequence of the cytochrome *b* gene, apocytochrome *b* has a molecular weight of approx. 45 000 (Table 3.5). When the older data are corrected for this, it appears that there is, in fact, only one apocytochrome *b* molecule per monomeric complex. This means that the single polypeptide must accommodate two protohaem prosthetic groups [198–200].

Most spectroscopic properties of cytochrome *b* suggest by comparison with model compounds and haemoproteins that the axial haem ligands are histidines [200]. If so, the ferric haems have very unusual EPR properties with exceptionally low-field g_z values and unusual line shape [191,200–202]. Comparison with model compounds led Carter et al. [203] to suggest that this may be due tosteric strain on the histidines. Comparison of the amino acid sequences of the mitochondrial cytochromes *b* shows that there are only six invariant histidines [204,204a]. Only four of these are conserved in the otherwise highly homologous cytochrome b_6 from chloroplasts [205]. These four histidines are the best candidates for axial haem ligands (Fig. 3.9).

Cytochrome *b* is a very hydrophobic protein. A model of how it may be folded in the inner mitochondrial membrane may be obtained from its primary structure [204,204a,205] (Fig. 3.9). The haem-binding histidines are positioned pairwise in two segments that traverse the membrane. The haems may therefore be ‘sandwiched’ between transmembranous helices (cf., cytochrome oxidase haem groups, Section 3.6). This agrees with the perpendicularity between the haem and the membrane planes [206], and with proposed transmembranous electron transfer catalysed by the cytochrome *b* haems [207,208].

In most potentiometric titrations only two cytochrome *b* species are distinguished [208–210], i.e., *b*-562 and *b*-566, with $E_{m,7}$ values of about +40 and −40 mV, respectively (both pH dependent). However, proposals have been made of up to four different species (see Refs. 202, 208, 211), i.e., two components *b*-562 and separate identity of *b*-558 and *b*-566. Although it now seems clear that there are two haems *b* per monomer, a functional dimer [211] or different functional states of the monomer

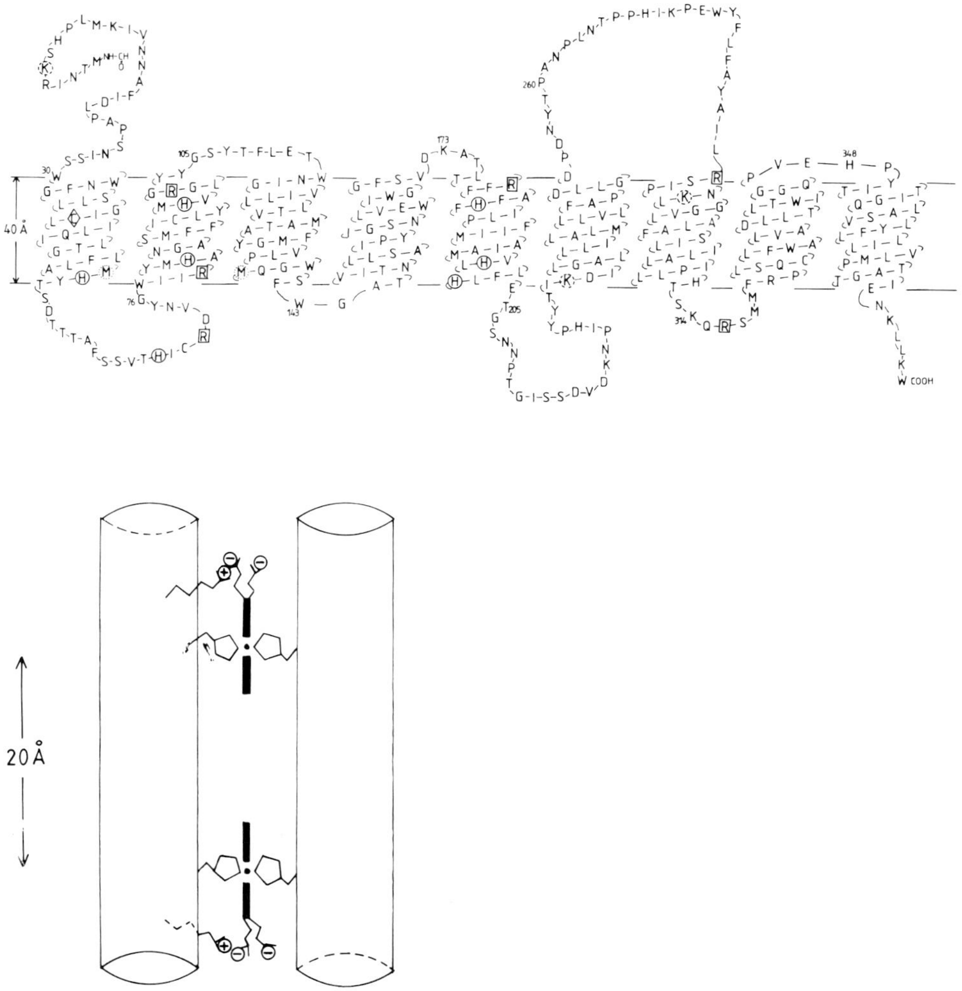

Fig. 3.9. Model of cytochrome *b* in the membrane. The amino acid sequences of six cytochromes *b* (Table 3.5) are highly homologous. Nine hydrophobic segments are predicted to traverse the membrane as α-helices (the sequence shown and the numbering refer to the bovine protein). Invariant histidines (H), arginines (R) and lysines (K) are indicated [204]. His-68 is not conserved in cytochrome *b* from maize mitochondria (A. Dawson, V. Jones and C.J. Leaver, personal communication). In two transmembranous segments (2nd and 5th from the left) the invariant histidines are found in pairs, separated in each case by 13 residues. This places them on the same helix surface. Thus, two protohaems could be bound between the helices, as indicated in the enlargement below. Two invariant arginines (indicated by '+' signs) could form salt bridges to propionic acid side chains of the haems. (From Ref. 204a.)

could explain further diversity. Such diversity has also been described for the Rieske FeS centre [211]. Yet, it is felt that, at the moment, the concept of more than two functionally different haem *b* groups or more than one FeS centre is not yet sufficiently documented, and that other more trivial explanations of apparent diversity are not fully excluded. We therefore restrict our analysis of electron transfer models to consideration of the simplest case only.

4.3. *Cytochrome* c_1

Cytochrome c_1 is an amphiphilic protein with a molecular weight of 28–31 000. Weiss et al. [212] found that it may be isolated only with the help of a detergent. However, by mild proteolysis they could release the haem-binding domain from the rest of the protein. This segment is soluble in aqueous solutions. The amino acid sequence of the bovine protein [181] shows that it has only one continuous hydrophobic segment that is close to the C-terminus. This segment probably acts as an anchor to the membrane. The covalently bound haem is located in the water-soluble part, with its plane perpendicular to the membrane plane [206]. The two-domain structure makes the architecture of cytochrome c_1 very similar to that of microsomal cytochrome b_5 [213].

Cytochrome c_1 is the site in Complex III that interacts with high affinity with cytochrome *c* [113,114] (see also Section 3.4). The E_m is 225–245 mV and is independent of pH in the physiological range [210,214].

4.4. *The Rieske FeS protein*

After suggestions on an iron-sulphur protein in the bc_1 complex [215], this subunit was isolated and characterised by Rieske et al. [216,218], but in a form that was not active in reconstitution. Isolation in a reconstitutively active form was pioneered by Racker et al. [219], who showed that a soluble 'oxidation factor' was required for activity. Subsequently, Trumpower and Edwards [220] purified 'oxidation factor' and identified it as a form of the FeS protein that is active in reconstitution. An excellent review on the structure and function of the FeS protein is available [221].

The iron-sulphur protein has an apparent molecular weight of 24–25 000. Li et al. [222] were able to split it into two parts by limited proteolysis. While the native protein is soluble only in the presence of detergent, a 16 kDa water-soluble fragment was released, which apparently contains the FeS centre. From this it appears that the Rieske protein is amphiphilic (cf., cytochrome c_1), containing a membrane anchor and a catalytic centre in different domains.

The iron-sulphur centre is probably of the 2Fe-2S type [191,223]. It is a one-electron donor/acceptor with $E_{m,7}$ of approx. 280 mV in mitochondria (pH independent below pH 8; [224,225]). It exhibits an EPR spectrum in the reduced state that is somewhat anomalous for 2Fe-2S clusters (see Ref. 221). This, as well as the high midpoint redox potential, suggest that the iron ligands may be less electronegative than the four cysteine sulphurs of the plant ferredoxin model (see Ref. 226). The EPR spectrum of the FeS cluster is affected by the redox state of ubiquinone

[191,201,202,211], and by binding of ubiquinone analogue inhibitors to the complex [221,227,228]. Hence the FeS protein is believed to be close to, or form part of, one of the ubiquinone-binding sites of the enzyme (cf., below).

The FeS protein functions as a ubiquinol:ferricytochrome c_1 oxidoreductase. It also has a central role in the coupled oxidation of semiquinone by ferricytochrome *b* (see Refs. 179, 221, 229 and below).

4.5. Subcomplexes and image reconstruction of membrane crystals

The 'core proteins' and the Rieske FeS protein can be dissociated from Triton X-100-solubilised cytochrome *c* reductase in concentrated NaCl. The cytochrome bc_1 core was isolated from the dissociated Complex III of *Neurospora* by gel filtration [230]. It retains most of the hydrophobic character of the parent protein and seems to correspond to the membrane-embedded part of Complex III.

Both the intact Complex III and its hydrophobic 'core' can form two-dimensional crystals [230,231]. In both cases the crystals contain dimeric complexes. The monomeric entity is an elongated particle about 15 nm long that extends out of the predicted membrane layer on both of its sides. This is shown schematically in Fig. 3.10. Comparison of the two kinds of crystals has suggested that the 'core proteins' might be located in the larger extramembranous domain, while the hydrophilic parts of cytochrome c_1 and the FeS protein would contribute to the smaller one [222]. The latter domain should hence be the C side and the former the M side of the membrane. Cytochrome *b* seems to occupy a large part of the membrane-embedded domain (cf., Section 4.2), while the Rieske FeS protein is located "between cytochromes *b* and c_1" [222]. This topography of the complex is in general agreement with labelling data using hydrophilic [232] and hydrophobic [233] probes.

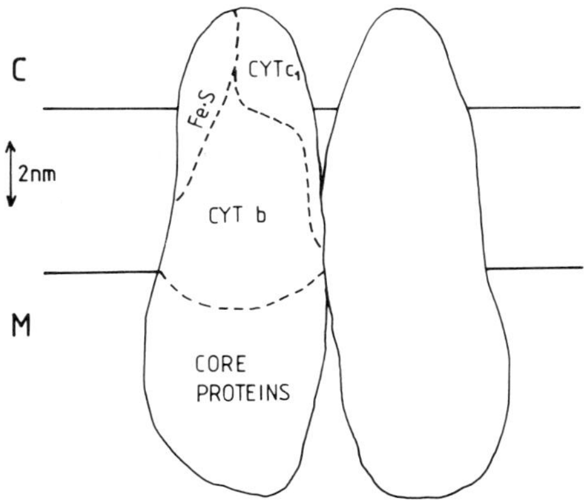

Fig. 3.10. Topography of Complex III. Complex III is a dimer in the two-dimensional crystal form studied by electron microscopy. The shape of the membrane-bound enzyme particle was resolved by image reconstruction of micrographs [230]. The location of various components was predicted by comparing crystals of Complex III with those of a subcomplex lacking the Rieske FeS protein and the 'core proteins' [222,231]. The schematic figure is adapted from Li et al. [222].

4.6. Topography of redox centres

Ohnishi et al. [234] studied paramagnetic interactions between water-soluble dysprosium probes and redox centres of Complex III in the isolated proteins, in the Complex, and in mitochondrial membranes. In Complex III the distance from the protein surface to haems *b*-562 and *b*-566 were 22 and 17 Å, respectively (cf., Figs. 3.9,10). Haem c_1 was about 10 Å from the surface of the isolated cytochrome. The FeS cluster is deeply buried within the isolated protein; about 19 Å from the protein surface. In Complex III, as well as in mitochondrial membranes the distance is 20 Å. However, the magnetic interactions between FeS and the dysprosium probe was similar, irrespective of whether the latter was added to mitochondria or submitochondrial particles [234,235]. This would place the FeS cluster near the centre of the membrane, somewhat in contrast to other topographical data (see Fig. 3.10 and above). However, the FeS protein may lack interactions with phospholipids by being shielded by other polypeptides in the Complex [233].

Studies with orientated membrane multilayers have suggested that the Fe-Fe axis of the FeS centre lies in the membrane plane [206,236]. Based on pH-dependent inhibition by a ubiquinone analogue, Harmon and Struble [237] placed the FeS centre on the C side of the membrane. However, their results would also be consistent with a location inside the membrane domain, provided that there is protonic communication with the aqueous C phase. The observed interactions between Q and the FeS cluster (above) suggest, together with the location of ubiquinone in the hydrophobic domain of the membrane (see below), that the FeS centre is buried within this domain.

4.7. Ubiquinone

Ubiquinone is a substituted (2,3-dimethoxy-5-methyl-(1,4)-)benzoquinone with a long isoprenoid side chain in position 6 (see Ref. 238). The fact that ubiquinol is a donor of two reducing equivalents, while cytochrome *c* is a one-electron acceptor, requires special arrangements of electron transfer (cf., the analogous but opposite problem in cytochrome oxidase). Although ubisemiquinone is very unstable in most circumstances, it can be stabilised by specific binding to a catalytic site. Two such sites have been identified in Complex III [236,239–244]. Quinone-binding proteins have also been described [194–196,245].

4.7.1. Redox properties

Semiquinone stability in solution is defined by the equilibrium constant (K_s) of the reverse of the semiquinone dismutation reaction

$$2\,SQ \rightleftharpoons QH_2 + Q \qquad (2)$$

where SQ, QH_2 and Q are, respectively, the sums of the activities of all protonation states of semiquinone, quinol, and quinone. Thus

$$K_s = (SQ)^2/(Q)(QH_2) \qquad (3)$$

In a hydrophobic milieu such as the inner membrane K_s has been estimated to be about 10^{-10} [246], which makes SQ of the membranous pool undetectable by EPR.

K_s also determines the difference between the redox midpoint potentials of the one-electron couples Q/SQ (E_{m1}) and SQ/QH_2 (E_{m2}), so that

$$E_{m1} - E_{m2} = RT/F \quad \ln K_s \tag{4}$$

Also usually,

$$E_{m1} + E_{m2} = 2E_m \tag{5}$$

where E_m is the midpoint potential of the two-electron Q/QH_2 couple (but see Ref. 247). Thus, the E_m values of the semiquinone couples are often displaced symmetrically above and below the E_m of the quinone/quinol couple.

The functionally relevant redox potentials of the semiquinone couples are obviously those of quinone-enzyme complexes, not of free quinone molecules. For instance, the effective E_m of the QH_2/Q couple bound to the enzyme is shifted from the corresponding value of free couple by an amount that is determined by the *relative* affinities of binding of quinone and quinol to the binding-site. The effect of binding on the E_m is given by the equation (see Ref. 247)

$$E_{mB} - E_{mF} = RT/nF \quad \ln(K_o/K_r) \tag{6}$$

where E_{mB} and E_{mF} are the midpoint potentials of the protein-bound and free couples, respectively, and K_o and K_r are the dissociation constants of the oxidised and reduced forms of the couple.

If K_r, K_o and K_{sq} are the dissociation constants of quinol, quinone and semiquinone, the $E_{m1(B)}$ and $E_{m2(B)}$ of the *bound* one-electron couples may be derived, as well as the 'stability constant' K_{sB} for the bound semiquinone.

$$K_{sB} = K_s(K_o)(K_r)/(K_{sq})^2 \tag{7}$$

$$E_{m1(B)} = E_{m(F)} + RT/F \quad \ln(K_s K_o/2K_{sq}) \tag{8}$$

$$E_{m2(B)} = E_{m(F)} - RT/F \quad \ln(K_s K_r/2K_{sq}) \tag{9}$$

where $E_{m(F)}$ is the midpoint potential of the free Q/QH_2 couple.

The stability of semiquinone is an equilibrium property of the reversed dismutation reaction (Eqn. 2). It is stressed that the above treatment refers to equilibrium conditions. Tight binding to a proteinaceous site could mean that a (semi)quinone molecule would not dissociate out of it before some catalytically important event has taken place. Kinetically determined midpoint potentials may in such cases be functionally more relevant, and could be considerably distorted from the equilibrium values. Lack of equilibration of protons between the proteinaceous site and the bulk phases in the kinetic time domain would be another related cause for such distortions (see, e.g., Ref. 87).

The $E_{m,7}$ of the ubiquinone pool [248] of the mitochondrial membrane has been reported to be 65 mV [9]. In mitochondria the E_m of the pool is a function of pH in the M phase [7]. The $E_{m,7}$ values of Q/QH_2, and the two semiquinone couples $Q^{\cdot -}/QH_2$ and $Q/Q^{\cdot -}$ *bound* to one of the two sites in Complex III (site 'i'; cf., below) have been estimated to about 105, 140 and 70 mV, respectively [229,241,249].

4.7.2. Ubiquinone in the membrane

Proton NMR studies of ubiquinone and ubiquinol in dimyristoylphosphatidylcholine vesicles (DMPC) [250] have confirmed the location of Q, including the benzoquinone ring, deep in the hydrophobic interior of the membrane. Chance [251] suggested a depth of about 15 Å for the quinone moiety based on quenching of probe fluorescence. The more hydrophilic quinol moiety may come closer to the membrane interphase. Based on perturbation of the proton NMR of the methoxy groups by membrane impermeable shift reagents, it was estimated that the minimum rate of transbilayer 'flip-flop' of the benzoquinone headgroup is 23 s^{-1} for ubiquinone-10 in DMPC vesicles [250]. This is in agreement with measured rates of transmembranous transfer of reducing equivalents catalysed by ubiquinone in model membranes [252]. These rates are fast enough to be consistent with such 'flip-flop' required in the Q cycle model (see below). Recent surface pressure and calorimetric studies of model membranes have suggested that the long ubiquinone side chain may lie parallel to the plane of the membrane [253,254]. Such an orientation makes a fast 'flip-flop' (of the head group) more understandable than would a phospholipid-like orientation. An axial rotation around the side chain axis in the plane of the membrane may be sufficient.

4.8. Pathway of electron transfer

It is now generally accepted that the oxidation of ubiquinol takes place in two distinct steps at different redox potentials, with semiquinone as the intermediate [255,256]. This yields a branching point in the electron transfer chain with interesting kinetic, mechanistic and thermodynamic implications (see Ref. 246). The important finding by Deul and Thorn [257] on the necessity of two different inhibitors (BAL(2,3-dimercaptopropanol) and antimycin to block the reduction of cytochrome *b* was long overlooked. This result, together with more recent related data using other combinations of inhibitors (Table 3.6), or on the effects of removal and reconstitution of the FeS protein (see Refs. 8, 14, 179, 199, 229, 238 for reviews), strongly indicates that there are two pathways of electron flow into cytochrome *b*. One is sensitive to one group of inhibitors of which antimycin is the prototype, and the other is blocked by BAL, myxothiazol, UHDBT, or by removal of the FeS protein (Table 3.6). It is, therefore, very likely that the electron transfer path is not only branched, but cyclic as well.

TABLE 3.6

Inhibitors of Complex III

The inhibitory site refers broadly to sites 'i' and 'o' of the Q cycle (see Ref. 246 and the text). Inhibitors of site 'i' prevent oxidation of cytochrome *b* via bound quinone and the FeS centre, whereby oxidant-induced reduction of the *b* cytochromes is 'stabilised' (see Refs. 208, 258). Reduction of the FeS cluster by ubiquinol is not affected. Site 'i' is intimately associated with cytochrome *b*-562. Inhibitors of site 'o' prevent oxidant-induced reduction of cytochrome *b* and reduction of FeS by quinol. Cytochrome *b* may still be reduced via site 'i'. Site 'o' is intimately associated with the FeS centre and with cytochrome *b*-566. Reduction of cytochrome *b* is completely blocked only by blocking both sites 'i' and 'o'. The Q analogue inhibitors can bind to both sites with different affinity. See also Fig. 11.

Compound	Inhibitory site	Comment	Refs.
Antimycin	i	red shift of *b*-562 spectrum	258
BAL	o	destroys FeS in the presence of oxygen	259
DTNB	o	irreversible; affects EPR of FeS	227
Funiculosin	i	like antimycin, but no spectral shift	229, 260
HMHQQ	o, i	Q analogue; changes EPR of FeS; binding depends on FeS redox state; different affinity for o and i	228
HOQNO	i	like antimycin, but no spectral shift	261, 262
Myxothiazol	o	blue shift of *b*-566 spectrum changes EPR of cyts. *b* and FeS	263–266
UHDBT	o (i)	Q analogue, changes EPR and $E_{m,7}$ of FeS; displaced by myxothiazol	265, 267, 268
Mucidin	o	like myxothiazol	229, 268

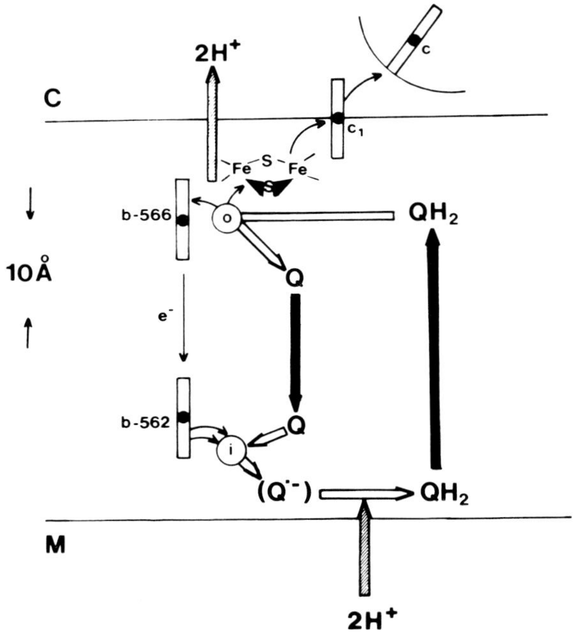

Fig. 3.11. The ubiquinone or Q cycle. The redox centres of Complex III are placed in their approximate positions with respect to the membrane (cf., the text and Fig. 12B). o and i represent the centres of interaction of ubiquinone (see text and Table 3.6). Thin arrows show electron transfer.

At present the most entertained model which incorporates the above principles is the 'Q cycle' proposed by Mitchell [246], shown in Fig. 3.11. Many discussions of the Q cycle or variants thereof are available [8,14,67,87,178,179,199,211,221,229,238,246]. In essence, ubiquinol is oxidised at 'centre o', which is closely associated with the FeS centre and the haem of *b*-566. This results usually in transient generation of SQ bound to this centre, as identified by EPR spectroscopy [239]. Then there is rapid electron transfer from SQ to *b*-566 with formation of Q. The electron in FeS is transferred rapidly to cytochrome *c* via cytochrome c_1. The electron in *b*-566 moves transmembranously to haem *b*-562 (Figs. 3.9,11). The latter may then reduce either SQ or Q at 'centre i', of which SQ has again been identified by EPR [239–244]. This reaction is a weak point of the Q cycle, and has been subject to modifications by several workers (see Refs. 270, 271 and Refs. above).

Outgoing from the model of Wikström and Berden [256], an alternative proposal termed the '*b* cycle' was made based on the more recent experimental developments (Fig. 3.12A) [8,14]. Also this model clearly requires separate SQ/quinol interactions with the FeS/haem *b*-566 and *b*-562 domains, respectively, corresponding to 'o' and 'i' in the Q cycle. Contrary to some statements [199,239], the *b* cycle is therefore consistent with the findings of two different SQ species. The most essential difference from the Q cycle is that the SQ/quinol couple is suggested to be able to shuttle reducing equivalents between sites 'i' and 'o'.

The Q cycle requires that following oxidation of cytochrome *c*, the oxidation of the *b* cytochromes must be preceded by their reduction (see Fig. 3.11). Yet, the velocity of oxidation of the *b* cytochromes is unimpeded even if they are completely reduced before the pulse of oxidant (Ref. 272 and M. Wikström, unpublished observations). To explain this, the Q cycle would require a special not normally operative mechanism of conducting SQ from site 'o' to site 'i' without equilibration with the ubiquinone pool. Interestingly, this is a normal pathway of the *b* cycle. Application of Occam's razor would hence favour the latter.

Fig. 3.12B shows a more detailed molecular interpretation of the *b* cycle. A single ubiquinone species in a 'ubiquinone pocket' of the Complex III monomer is shown to interact either with the quinol oxidase site (o), or with a semiquinone reductase site (i). 'Flip-flop' of the headgroup (usually of SQ^-) between the sites might take place, in principle as proposed for bulk phase ubiquinone (see Section 4.7.2). However, here it occurs within a specific proteinaceous pocket so that regulation by the enzyme is possible.

Following oxidation of ubiquinol at site 'o', with resultant transfer of electrons to cytochrome *c* (via FeS and c_1) and to *b*-562 (via *b*-566), ubiquinone may leave the pocket in exchange for another molecule of ubiquinol. The latter is similarly oxidised to SQ^- by FeS. But in this situation the SQ anion rotates so that contact with site 'i' is established. It is then reduced to ubiquinol by the ferrous *b*-562. The reaction at the 'i' site is thus unambiguous (contrast above).

The Q pocket may normally contain only a single Q molecule. However, it must probably additionally accommodate an inhibitory short-chain Q analogue (see Table 3.6). The *b* cycle is a compact 'enzymic' model which does not require translocation

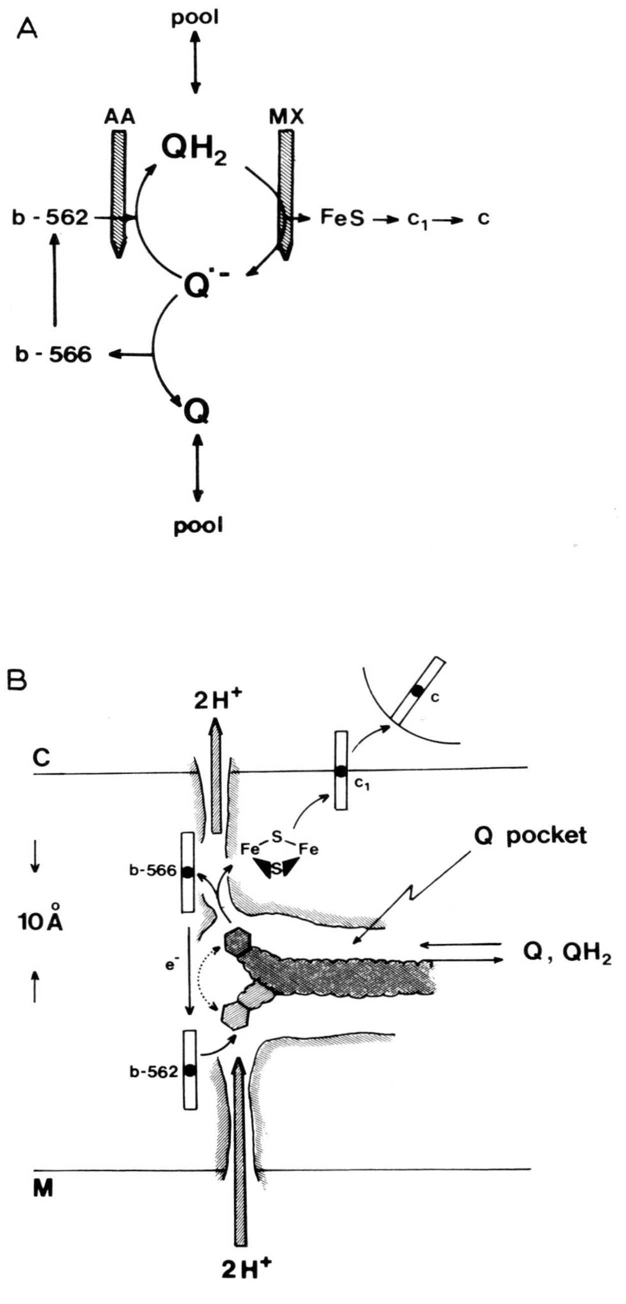
A
pool
AA
MX
QH2
b-562
FeS → c1 → c
Q·-
b-566
Q
pool
B
2H+
C
c
c1
Q pocket
Fe S Fe
S
b-566
10 Å
e-
Q, QH2
b-562
M
2H+

of bulk ubiquinone, but merely communication between ubiquinone in the 'pocket' and in the membrane pool. Obviously, also the Q cycle can be more 'compactly' interpreted so that the two models need not differ in this respect. However, in the Q cycle two ubiquinone molecules may be able to occupy the 'active site' simultaneously (at 'o' and 'i').

Rich and Bendall [273–275] have suggested that the transient complexes between ubiquinol and the protein involve the anionic QH^- and $Q^{\cdot -}$, which may be stabilised by positive charges on the protein and by hydrophobic bonding. Interestingly, mitochondrial apocytochrome *b* contains two invariant lysines and invariant arginines predicted to be located near opposite sides of the membrane [204]. The lysines and two of the arginines reside in the carboxyterminal half of the protein, which is split from apocytochrome b_6 in the chloroplast [205]. The function of this part of mitochondrial apocytochrome *b* could, therefore, be to form (part of) the Q-binding domains in the hydrophobic ubiquinone pocket.

4.9. Proton translocation

$2H^+$ are agreed to be released on the C side of the membrane, and one electrical charge equivalent translocated, per transferred electron when ubiquinol is oxidised by ferricytochrome *c*. In mitochondria this is thermodynamically equivalent to translocation of $1H^+/e^-$ because release of the second proton to the well-buffered C phase is thermodynamically futile (without corresponding charge translocation; Table 3.1).

Mechanisms of so-called direct or indirect coupling (see Refs. 14, 17, 18) are difficult to distinguish experimentally for this span of the chain, since oxidoreduction involves the hydrogen carrier ubiquinone (contrast cytochrome *c* oxidase; Section 3.7). The principle of proton translocation is straightforward in the Q cycle (Fig. 3.11). The main electrogenic event is the translocation of electrons between the haem *b* residues. Translocation of hydrogen is the result of transport of QH_2 one way and of Q the other way across the membrane (or the 'Q pocket'; cf., above).

An indirectly coupled proton pump principle was originally suggested in connection with the *b* cycle [14]. However, the directly coupled alternative of Fig. 3.12B is

---

Fig. 3.12. A, Schematic representation of the '*b*-cycle'. Modified from Refs. 5, 15. The reaction steps thought to be inhibited by antimycin (site 'i') and myxothiazol (site 'o') are indicated (cf., Table 3.6 and the text). B, Molecular representation of the *b* cycle. The redox centres are placed in the membrane-embedded Complex III (cf., text and Fig. 3.11). A ubiquinone 'pocket' is indicated with an attached ubiquinone molecule. The latter can take one of two configurations in which the headgroup interacts with the haem *b*-566/FeS ('o') or haem *b*-562 ('i') region of the pocket. A switch in configuration (dotted two-headed arrow) is thought to be effected by rotation of the Q molecule around the axis through the hydrophobic side chain. Quinone and quinol in the pocket can exchange with the ubiquinone pool in the membrane. QH^- and $Q^{\cdot -}$ may be stabilised in the pocket by interaction with positively charged arginines or lysines in the 'o' and 'i' regions (not shown; see text). Proton-conducting 'channels' are indicated, which connect the aqueous C and M phases with the sites 'o' and 'i', respectively. Modified from Refs. 8, 14 (Fig. 3.12A). For further details, see the text.

at least equally feasible. It differs from the Q cycle in that the electrogenic event is now not only electron transfer between the cytochrome *b* haems, but also translocation of the SQ anion in the active site.

In chromatophores light flash-induced reduction of the *b* cytochromes (e.g., in the presence of antimycin) is not associated with conservation of energy, as judged from the lack of an electrogenic spectral change in the carotenoids. It is the antimycin-sensitive reoxidation of cytochromes *b* (through site 'i') that appears electrogenic in this sense (see Refs. 87, 276, 277). This contrasts to models (cf., Fig. 3.9) of electron translocation by the *b* cytochromes, and the observations that the relative redox poise of the haems is distorted by $\Delta\psi$ [246,278]. However, the position of carotenoids may be such in the membrane that a local field between the *b* haems is not sensed until it is delocalised. Such a field could also be distorted by a bound SQ^- molecule.

4.10. Reconstitution of Complex III

Comparatively few papers describe the properties of Complex III after reconstitution into liposomal membranes [198,279,280]. The found stoicheiometries of proton and electrical charge translocation agree well with the data for mitochondria.

5. The NADH-ubiquinone reductase complex

NADH-ubiquinone reductase (EC 1.6.5.3) or Complex I is structurally by far the most complicated member of the respiratory chain. It is also the least known in terms of structure, electron transfer pathway or mechanism of proton translocation. Even the nomenclature of the isolated enzyme entities and of the FeS centres is problematic because it differs between research groups.

Complex I is the lipid-containing isolated preparation, with properties (Table 3.7) [39] that best correspond to the 'Site 1' region of the respiratory chain of intact mitochondria.

TABLE 3.7

Properties of Complex I (modified from Ref. 39)

1. Contains FMN, nonhaem Fe, acid-;abile sulphur, ubiquinone-10 and lipids (Table 3.8). The several constituent polypeptides probably include ubiquinone-binding proteins [281].
2. Catalyses rapid rotenone-sensitive reduction of Q-1 by NADH.
3. Catalyses rapid NADH-linked reduction of ferricyanide, but reacts slowly with other acceptors.
4. NADH induces multiple EPR signals of similar lineshape and E_m to those in mitochondria and submitochondrial particles (see Table 3.9).
5. Other types of NADH dehydrogenase may be prepared from Complex I.
6. Reconstitutes NADH oxidase activity with Complexes III, IV and cytochrome *c*.
7. Energisation and ATP synthesis can be coupled to NADH-Q oxidoreductase activity after 'reconstitution' of Complex I (and ATP synthase) into liposomes.

Two types of soluble NADH dehydrogenase preparations are also currently studied. The high molecular weight (Type I) NADH dehydrogenase [282] is in many respects similar to Complex I, but lacks phospholipid and ubiquinone and does not catalyse rotenone-sensitive reduction of added ubiquinone-1. In addition, it has a somewhat different content of non-haem Fe and acid-labile S per FMN.

The low molecular weight (Type II) NADH dehydrogenase is a subcomplex of the former [283,284], and may be prepared from Complex I by treatment with chaotropic agents [285]. Type II dehydrogenase catalyses oxidation of NADH by several acceptors such as cytochrome *c*, ubiquinone-1 and menadione. It contains approx. four non-haem Fe and four acid-labile S/FMN, and has a molecular weight between 70 and 80 000.

5.1. Structure

Complex I is traditionally isolated by precipitation with ammonium sulphate from partially solubilised mitochondria [286,287]. Such preparations contain 1.2–1.5 nmol FMN/mg protein. Thus the minimum molecular weight seems to be in the range 670–830 000 (see, e.g., Refs. 39, 288). Table 3.8 gives the composition of Complex I.

Electrophoretic analysis reveals that there are more than 20 different polypeptides in the preparation [39,289,290]. The status of these proteins as enzyme subunits is not clear. However, some are associated with FeS centres and may be isolated in defined subcomplexes (see below). Some may constitute specific binding sites for ubiquinone [281]. No success in isolation of rotenone-sensitive Complex I with a smaller number of constituent polypeptides has been reported.

The isolated Complex I contains a large amount of iron, which is arranged in bi- or tetranuclear FeS centres (Table 3.9; Refs. 295, 296; see below).

Treatment of Complex I with chaotropic agents dissociates two water-soluble subcomplexes from the enzyme. Both contain iron, and one also contains FMN. These are called 'iron-sulphur protein' (ISP) and 'flavoprotein' or NADH dehydrogenase (Type II), respectively [285]. The hydrophobic residue left behind still contains six or seven of the original 22–23 Fe atoms per FMN, while ISP contains 9.3, and the dehydrogenase 6.2 [296]. The ISP fraction may be further dissociated into three parts with trichloroacetate, each containing an FeS centre. Two of these

TABLE 3.8

Composition of Complex I (see Refs. 39, 287)

Component	Per mg of protein
Acid-extractable FMN	1.2–1.5 nmol
Nonhaem iron	23–26 nmol
Acid-labile sulphur	23–26 nmol
Ubiquinone-10	4.2–4.5 nmol
Cytochromes	$<$ 0.1 nmol
Lipids	0.22 mg

TABLE 3.9

FeS centres of Complex I (see Refs. 235, 291–294)

The centres are grouped in the order of their midpoint potentials. 'Centre' and 'Cluster' refer to the nomenclatures of Ohnishi et al. [235, 291–293] and Albracht and Beinert [294], respectively.

Nomenclature		$E_{m,7}$ (mV)	Amount per FMN	$\Delta E_m/\Delta pH$ (mV)	EPR $g_{x,y,z}$
Centre	Cluster				
N-2	2	−20	0.8–0.9	−60	1.92 1.92 2.05
N-1b	1	−245	0.4–0.8	0	1.92 1.94 2.02
N-3	4	−245	0.9–1.1	0	1.86 1.93 2.04
N-4	3	−245	0.6–0.7	0	1.88 1.93 2.10
N-5	?	−270	0.06–0.25	0	1.89 1.92 2.06
(N-1a)	??	−370	about 1	−60	1.94 1.94 2.03

contain only a single apoprotein while the third has two stoicheiometric polypeptides [296]. This approach to dissect the structure of Complex I, as recently updated by Ragan et al. [295,296], yields the rough map shown in Fig. 3.13A.

An interesting topographic feature of Complex I is that although the flavoprotein and FeS protein subcomplexes are soluble in water, each appears to be buried in the intact membrane-bound enzyme. Thus, the three flavoprotein subunits are inaccessible to surface labelling, and some of the proteins in the ISP fraction are probably transmembranous [289,299]. The corollary of this would be that the enzyme may have a hydrophilic core within the membrane, which is surrounded by hydrophobic protein components (Fig. 3.13B).

The flavoprotein fraction or Type II NADH dehydrogenase contains the NADH binding site, which is located in the 51 000 protein [298]. The so-called subunit II of the NADH dehydrogenase has recently been sequenced [299]. It has no long hydrophobic segments. Two pairs of adjacent cysteines are found in the sequence.

Recently, an image reconstruction of Complex I has been reported [300]. The unit in the two-dimensional crystals studied is apparently a tetramer with a mass of about 750 kDa. Hence, it should in principle contain only a single unit of FMN-binding Complex I. But the tetrameric structure of the unit does not easily fit to this idea, and further biochemical characterisation is needed before the identity of the crystallised protein with Complex I (or with its subfragment) can be accepted.

5.2. *Iron-sulphur centres*

Four FeS centres were originally described by Orme-Johnson et al. [301] and termed centres 1–4. These are rapidly reduced by NADH and were reported to be present in amounts stoicheiometric with FMN. Potentiometric titrations of mitochondria combined with EPR spectroscopy at extremely low temperatures later revealed possible further centres [291]. Alternatively, the multiplicity of EPR signals observed could be due to centre-centre interactions. This subject has been highly controversial.

The FeS centres considered today and their properties are listed in Table 3.9. Of

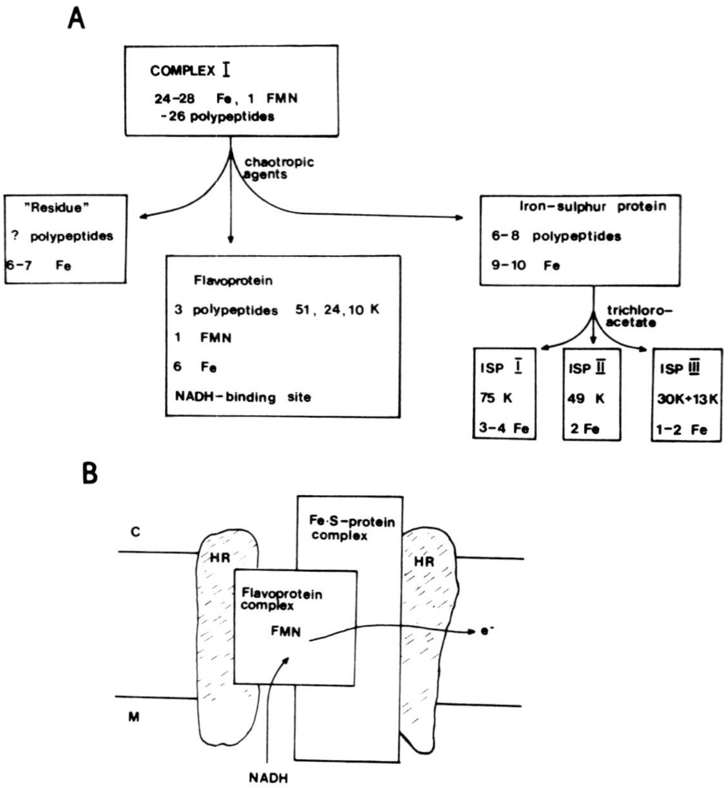

Fig. 3.13. Complex I. A, the map shows how Complex I can be dissected into subcomplexes. Treatment with chaotropic agents splits the Complex into three fractions: an iron-sulphur protein (ISP), a flavoprotein (NADH dehydrogenase) complex, and a hydrophobic residue. The figure shows the polypeptide composition and the flavin and iron content of these subcomplexes. ISP can be further split into three FeS-containing fractions by trichloroacetate. See text for details and references. B, the topography of the components mapped in A has been studied by Ragan et al. [290,296]. The NADH-binding site is in the flavoprotein fraction [297]. This, as well as the FeS protein complex (ISP), are possibly buried in the membrane and 'shielded' from phospholipids by the 'hydrophobic residue' (HR). Adapted from Refs. 290, 296.

these, the N-1a cluster is still controversial. N-1b (or cluster 1) is probably in the NADH dehydrogenase part of Complex I [302]. The N-1 type centre is most probably of the binuclear type [223,303]. Centre N-3 may interact with FMN [292]. On the basis of its $E_{m,7}$ value, centre N-2 probably lies closest to the electron output side of Complex I. The involvement of N-5 in Complex I is uncertain.

Paech et al. [304] used the cluster extrusion technique to detect and quantify FeS clusters in high molecular weight NADH dehydrogenase (type I). Bi- and tetranuclear clusters were found in the ratio 2 : 1, and the enzyme was suggested to contain four (2Fe-2S) and two (4Fe-4S) clusters per FMN. However, the type I preparation differs from Complex I in its slightly lower Fe/FMN content and higher acid-labile S/FMN content. This difference has not been resolved.

A recent penetrating review on the mitochondrial FeS centres is available [294].

5.3. Inhibitors and electron transfer pathway

NADH-Q-10 reductase activity can be inhibited by several compounds, of which thiol reagents, barbiturates, rotenone and piericidin are the most common.

Singer et al. [305] characterised five different thiol groups in the 'Site 1' segment of the respiratory chain on the basis of different reactivity towards thiol reagents (see also Ref. 306). An interesting finding is that the reaction of mersalyl halves the apparent number of binding sites of rotenone or piericidin [306].

Barbiturates, rotenone and piericidin block electron transfer from the FeS centres to ubiquinone-10, but do not block the reduction of artificial acceptors by NADH in Type II NADH dehydrogenase. Both rotenone [282,286] and piericidin [307] bind to the enzyme with an apparent 1 : 1 stoicheiometry to the content of FMN [39]. The binding of these inhibitors is non-covalent which has so far prevented identification of the binding site(s). All EPR- or optically detectable redox centres (possibly with the exception of centre N-1a) are reducible by NADH in the presence of these inhibitors [308,309]. The structural analogy between piericidin and Q-10 provides further evidence that these inhibitors interact at the point where the enzyme delivers reducing equivalents to ubiquinone (Fig. 3.14).

Rhein (4,5-dihydroxyanthraquinone-2-carboxylate) inhibits NADH oxidation competitively with NADH, with an apparent K_i of 2 μM [310].

From the above data a rough picture of the electron transfer sequence of Complex I may be constructed (Fig. 3.14).

5.4. Energy conservation

Oxidation of NADH by ubiquinone is linked to proton translocation. But also here the stoicheiometry is under debate; the proposals range between 1 and $2H^+/e^-$ (see Table 3.1 and Section 2).

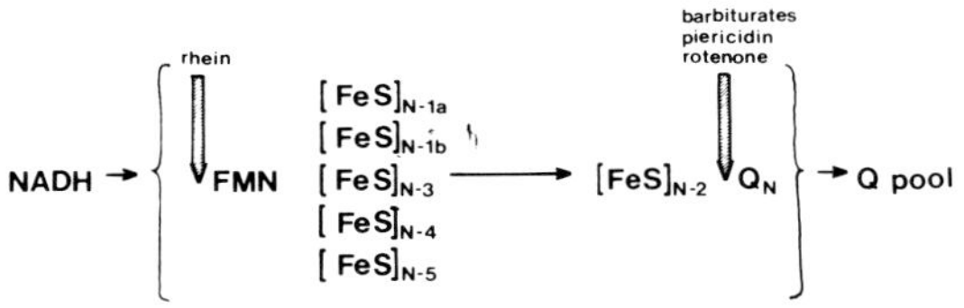

Fig. 3.14. Electron transfer in Complex I. Rough arrangement of the redox centres in Complex I with respect to electron transfer. This arrangement is based mainly on the measured E_m values (Table 3.9). Apparent sites of inhibitor action are also shown. Q_N is a postulated form of ubiquinone bound to the complex [281]. See the text for details.

Ragan and Hinkle [42] reported on proton translocation by Complex I reconstituted into liposomes. During oxidation of added NADH by ubiquinone-1 they observed an uptake of about 0.7 H^+/e^- in addition to the trivial proton uptake linked to reduction of quinone by NADH. Liposomes reconstituted with Complex I *plus* ATP synthase exhibited ATP synthesis linked to oxidation of NADH by Q-1 [311].

Energisation by ATP has been shown to shift the apparent midpoint redox potentials of some of the FeS centres [293,312,313]. It is, however, difficult at the present time to draw any firm conclusions on the basis of these phenomena (but see discussions in Refs. 235,292,314).

The mechanism of proton translocation is not understood. Since the stoicheiometry is almost certainly higher that one $H^+/e-$ (Table 3.1; Section 2.2), the prototype of a Mitchellian redox loop (see Refs. 39. 41) may be rather safely excluded.

6. *Epilogue*

From the above account of the mitochondrial respiratory chain it should be clear that the research has, during recent years, been revolutionised by an extensive structural attack, which has included the elucidation of the primary structures of the constituent polypeptides. The structural approach has given an entirely new dimension and depth to mitochondrial bioenergetics. It is hoped, however, that this progress will not tend to blur the importance of results obtained by using more conventional bioenergetic approaches. It is the synthesis of structural, dynamic and thermodynamic information that will most effectively contribute to our understanding of cellular respiration and energy conservation in the future.

Acknowledgements

Work in this laboratory has been supported by grants from the Sigrid Juselius Foundation and the Academy of Finland (Medical and Science Research Councils). We are grateful to a large number of colleagues for freely communicating their opinions, and for criticism. However, the responsibility of the final outcome is of course the authors'. We have felt serious insufficiency in giving proper credits for the vast amount of work in this field, and offer our apologies for the many omissions that are bound to have occurred. It is hoped that our practise to refer to several excellent specialised review articles may help to make these shortcomings less serious.

We thank Ms. Hilkka Vuorenmaa for drawing most of the figures, and for her help with preparation of the manuscript.

References

1 Oshino, N. and Chance, B. (1975) Arch. Biochem. Biophys. 170, 514–528.
2 Matlib, M.A. and O'Brien, P.J. (1976) Arch. Biochem. Biophys. 173, 27–33.
3 Schatz, G. (1979) FEBS Lett. 103, 203–211.
4 Wallace, D.C. (1982) Microbiol. Rev. 46, 208–240.
5 Tzagoloff, A. (1982) Mitochondria. Plenum Press, New York.
7 Muraoka, S. and Slater, E.C. (1969) Biochim. Biophys. Acta 180, 227–236.
7 Wilson, D.F. and Erecinska, M. (1975) Arch. Biochem. Biophys. 167, 116–128.
8 Wikström, M., Krab, K. and Saraste, M. (1981) Annu. Rev. Biochem. 50, 623–655.
9 Urban, P.F. and Klingenberg, M. (1969) Eur. J. Biochem. 9, 519–525.
10 Slater, E.C., Rosing, J. and Mol, A. (1973) Biochim. Biophys. Acta 292, 534–553.
11 Ohnishi, T., Wilson, D.F., Asakura, T. and Chance, B. (1972) Biochem. Biophys. Res. Commun. 46, 1631–1638.
12 Wikström, M., Harmon, H.J., Ingledew, W.J. and Chance, B. (1976) FEBS Lett. 65, 259–277.
13 Mitchell, P. (1966) Chemiosmotic Coupling in Oxidative and Photosynthetic Phosphorylation. Glynn Research Ltd., Bodmin, U.K.
14 Wikström, M. and Krab, K. (1980) Current Top. Bioenerg. 10, 51–101.
15 Skulachev, V.P. (1982) FEBS Lett. 146, 1–4.
16 Mitchell, P. (1961) Nature 191, 144–148.
17 Mitchell, P. (1979) Eur. J. Biochem. 95, 1–20.
18 Mitchell, P. (1981) In Of Oxygen, Fuels and Living Matter, Part 1 (Semenza, G., ed.) pp. 1–160, John Wiley & Sons, New York.
19 Williams, R.J.P. (1961) J. Theor. Biol. 1, 1–17.
20 Williams, R.J.P. (1962) J. Theor. Biol. 3, 209–229.
21 Williams, R.J.P. (1978) Biochim. Biophys. Acta 505, 1–44.
22 Kell, D.B. (1979) Biochim. Biophys. Acta 549, 55–99.
23 Chance, B. and Williams, G.R. (1955) J. Biol. Chem. 217, 409–427.
24 Rottenberg, H. (1979) Biochim. Biophys. Acta 549, 225–253.
25 Van Dam, K. and Westerhoff, H.V. (1984) this volume, pp. 1–27.
26 Ferguson, S.J. and Sorgato, M.C. (1982) Annu. Rev. Biochem. 51, 185–217.
27 Rottenberg, H. (1979) Methods Enzymol. 55, 547–569.
28 Wilson, D.F., Owen, C.S. and Holian, A. (1977) Arch. Biochem. Biophys. 182, 749–762.
29 Wikström, M. (1981) Proc. Natl. Acad. Sci. U.S.A. 78, 4051–4054.
30 Forman, N.G. and Wilson, D.F. (1982) J. Biol. Chem. 257, 12908–12915.
31 Berry, E.A. and Hinkle, P.C. (1983) J. Biol. Chem. 258, 1474–1486.
32 Rottenberg, H. and Gutman, M. (1977) Biochemistry 16, 3220–3226.
33 Alexandre, A., Reynafarje, B. and Lehninger, A.L. (1978) Proc. Natl. Acad. Sci. U.S.A. 75, 5296–5300.
34 Lemasters, J.J. and Billica, W.H. (1981) J. Biol. Chem. 256, 12949–12957.
35 Baum, H., Hall, G.S., Nalder, J. and Beechey, R.B. (1971) In Energy Transduction in Respiration and Photosynthesis (Quagliariello, E. et al., eds.) pp. 747–755, Adriatica Editrice, Bari, Italy.
36 Ernster, L., Juntti, K. and Asami, K. (1973) J. Bioenerg. 4, 148–159.
37 Hitchens, G.D. and Kell, D.B. (1982) Biochem. J. 206, 351–357.
38 Van Dam, K., Wiechmann, A.H.C.A., Hellingwerf, K.J., Arents, J.C. and Westerhoff, H.V. (1978) Fed. Eur. Biochem. Soc. Symp. 45, 121–132.
39 Ragan, C.I. (1976) Biochim. Biophys. Acta 456, 249–290.
40 Mitchell, P. and Moyle, J. (1967) In Biochemistry of Mitochondria (Slater, E.C. et al., eds.) pp. 53–74, Academic Press, New York.
41 Lawford, H.G. and Garland, P.B. (1972) Biochem. J. 130, 1029–1044.
42 Ragan, C.I. and Hinkle, P.C. (1975) J. Biol. Chem. 250, 8472–8476.
43 Pozzan, T., Miconi, V., Di Virgilio, F. and Azzone, G.F. (1979) J. Biol. Chem. 254, 10200–10205.

44 Lehninger, A.L., Reynafarje, B., Alexandre, A. and Villalobo, A. (1979) In Membrane Bioenergetics (Lee, C.P. et al., eds.) pp. 393–404, Addison-Wesley, London and Massachusetts.
45 Mitchell, P. and Moyle, J. (1979) Biochem. Soc. Trans. 7, 887–894.
46 Brand, M.D. (1977) Biochem. Soc. Trans. 5, 1615–1620.
47 Wikström, M. and Krab, K. (1979) Biochem. Soc. Trans. 7, 880–887.
48 Moyle, J. and Mitchell, P. (1978) FEBS Lett. 88, 268–272.
49 Moyle, J. and Mitchell, P. (1978) FEBS Lett. 90, 361–365.
50 Lorusso, M., Capuano, F., Boffoli, D., Stefanelli, R. and Papa, S. (1979) Biochem. J. 182, 133–147.
51 Papa, S., Guerrieri, F., Lorusso, M., Izzo, G., Boffoli, D., Capuano, F., Capitanio, N. and Altamura, N. (1980) Biochem. J. 192, 203–218.
52 Wikström, M. and Krab (1979) Biochem. Biophys. Acta 549, 177–222.
53 Wikström, M. and Penttilä, T. (1982) FEBS Lett. 144, 183–189.
54 Casey, R.P., O'Shea, P.S., Chappell, J.B. and Azzi, A. (1984) Biochim. Biophys. Acta 765(1), 30–37.
55 Penttilä, T. (1983) Eur. J. Biochem. 133, 355–361.
56 Azzone, G.F., Pozzan, T. and Di Virgilio, F. (1979) J. Biol. Chem. 254, 10206–10212.
57 Reynafarje, B., Alexandre, A., Davies, P. and Lehninger, A.L. (1982) Proc. Natl. Acad. Sci. U.S.A. 79, 7218–7222.
58 Mitchell, P. and Moyle, J. (1967) Biochem. J. 105, 1147–1162.
59 Papa, S., Capuano, F., Markert, M. and Altamura, N. (1980) FEBS Lett. 111, 243–248.
60 Papa, S., Lorusso, M. and Guerrieri, F. (1975) Biocheim. Biophys. Acta 387, 425–440.
61 Wikström, M. (1982) EBEC Reports, Vol. 2, pp. 413–414.
62 Papa, S., Guerrieri, F. and Lorusso, M. (1974) In Dynamics of Energy-Transducing Membranes (Ernster, L. et al., eds.) pp. 417–432, Amsterdam. Elsevier, (B.B.A. Library, Vol. 13).
63 Papa, S., Guerrieri, F., Simone, S. and Lorusso, M. (1973) In Mechanisms in Bioenergetics (Azzone, G.F. et al., eds.) pp. 451–472, Academic Press, London and New York.
64 Brand, M.D., Reynafarje, B. and Lehninger, A.L. (1976) J. Biol. Chem. 251, 5670–5679.
65 Lehninger, A.L., Reynafarje, B., Davies, P., Alexandre, A., Villalobo, A. and Beavis, A. (1981) In Mitochondria and Microsomes (Lee, C.P. et al., eds.) pp. 459–479, Addison-Wesley, London and Massachusetts.
66 Lehninger, A.L., Reynafarje, B. and Alexandre, A. (1977) In Structure and Function of Energy-Transducing Membranes (van Dam, K. and van Gelder, B.F., eds.) pp. 95–106, Elsevier/North-Holland, Amsterdam.
67 Mitchell, P. and Moyle, J. (1982) In Function of Quinones in Energy Conserving Systems (Trumpower, B.L. ed.) pp. 553–575, Academic Press, New York and London.
68 Slater, E.C. (1966) In Comprehensive Biochemistry, Vol. 14 (Florkin, M. and Stotz, E.H., eds.) pp. 327–396, Elsevier, Amsterdam.
69 Orwell, G. (1949) Nineteen Eighty-Four. Penguin Books Ltd.
70 Wilson, D.F., Erecinska, M. and Schramm, V.L. (1983) J. Biol. Chem. 258, 10464–10473.
71 LaNoue, K.F. and Schoolwerth, A.C. (1984) this volume, pp. 221–268.
72 Mitchell, P. (1982) In Oxidases and Related Redox Systems (King, T.E. et al., eds.) pp. 1247–1263, Pergamon Press, Oxford.
73 Pietrobon, D., Zoratti, M. and Azzone, G.F. (1983) Biochim. Biophys. Acta 723, 317–321.
74 Slater, E.C. (1966) In Biochemistry of Mitochondria (Slater, E.C. et al., eds.) pp. 1–10, Academic Press, New York and London.
75 Bertina, R.M., Schrier, P.I. and Slater, E.C. (1973) Biochim. Biophys. Acta 305, 503–518.
76 Gutman, M., Singer, T.P. and Casida, J.E. (1976) J. Biol. Chem. 245, 1992–1997.
77 Smith, S., Collingham, I.R. and Ragan, C.I. (1980) FEBS Lett. 110, 279–282.
78 Papa, S., Guerrieri, F. and Lorusso, M. (1974) Biocheim. Biophys. Acta 357, 181–192.
79 Slater, E.C. (1974) In Dynamics of Energy-Transducing Membranes (Ernster, L. et al., eds.) pp. 1–20, Elsevier (B.B.A. Library, Vol. 13), Amsterdam.
80 Klingenberg, M. (1967) In Mitochondrial Structure and Compartmentation (Quagliariello, E. et al., eds.) pp. 124–125, Adsriatica Editrice, Bari, Italy.

81 Sowers, A. and Hackenbrock, C.R. (1980) Fed. Proc. 39, 1955.
82 Kawato, S., Sigel, E., Carafoli, E. and Cherry, R.J. (1980) J. Biol. Chem. 255, 5508–5510.
83 Muller, M., Krebs, J.J.R., Cherry, R.J. and Kawato, S. (1982) J. Biol. Chem. 257, 1117–1120.
84 Hackenbrock, C.R. (1981) Trends Biochem. Sci. 6, 151–154.
85 Capaldi, R.A. (1982) Biochim. Biophys. Acta 694, 291–306.
86 Trumpower, B.L. and Katki, A.G. (1979) In Membrane Proteins in Energy Transduction (Capaldi, R.A., ed.) pp. 89–200, Academic Press, New York and London.
87 Crofts, A.R. and Wraight, C.A. (1983) Biochim. Biophys. Acta 726, 149–185.
88 Overfield, R.E. and Wraight, C.A. (1980) Biochemistry 19, 3328–3334.
89 Ferguson, S.J., Lloyd, W.J. and Radda, G.K. (1976) Biochim. Biophys. Acta 423, 174–188.
90 Wikström, M. (1981) Trends Biochem. Sci. 6, 166–170.
91 Buse, G., Steffens, G.C.M., Steffens, G.J., Meinecke, L., Biewald, R. and Erdweg, M. (1982) EBEC Reports, Vol. 2, pp. 163–164.
92 Wikström, M., Saraste, M. and Penttilä, T. (1984) In The Enzymes of Biological Membranes (Martonosi, A.N., ed.) 2nd edition, in the press.
93 Kadenbach, B. and Merle, P. (1981) FEBS Lett. 135, 1–11.
94 Kadenbach, B., Jarausch, J., Hartman, R. and Merle, P. (1983) Anal. Biochem. 129, 517–521.
95 Azzi, A. (1980) Biochim. Biophys. Acta 594, 231–252.
96 Capaldi, R.A., Malatesta, F. and Darley-Usmar, V.M. (1983) Biochim. Biophys. Acta 726, 135–148.
97 Malmström, B.G. (1979) Biochim. Biophys. Acta 549, 281–303.
98 Malmström, B.G. (1982) Annu. Rev. Biochem. 51, 21–59.
99 Wikström, M., Krab, K. and Saraste, M. (1981) Cytochrome Oxidase-A Synthesis. Academic Press, London and New York.
100 Reinhammar, B., Malkin, R., Jensen, P., Karlsson, B., Andreasson, L.-E., Aasa, R., Vänngård, T. and Malmström, B.G. (1980) J. Biol. Chem. 255, 5000–5003.
101 Stevens, T.H., Brudwig, G.W., Bocian, D.F. and Chan, S.I. (1979) Proc. Natl. Acad. Sci. U.S.A. 76, 3320–3324.
102 Henderson, R., Capaldi, R.A. and Leigh, J.S., Jr. (1977) J. Mol. Biol. 112, 631–648.
103 Fuller, S., Capaldi, R.A. and Henderson, R. (1979) J. Mol. Biol. 134, 305–327.
104 Deatherage, J.F., Henderson, R. and Capaldi, R.A. (1982) J. Mol. Biol. 158, 487–499.
105 Deatherage, J.F., Henderson, R. and Capaldi, R.A. (1982) J. Mol. Biol. 158, 501–514.
106 Frey, T.G., Costello, M.J., Karlsson, B., Haselgrove, J.C. and Leigh, J.S., Jr. (1982) J. Mol. Biol. 162, 113–130.
107 Bisson, R., Gutweniger, H., Montecucco, C., Colonna, R., Zanotti, A. and Azzi, A. (1977) FEBS Lett. 81, 147–150.
108 Bisson, R., Azzi, A., Gutweniger, H., Colonna, R., Montecucco, C. and Zanotti, A. (1978) J. Biol. Chem. 253, 1874–1880.
109 Briggs, M.M. and Capaldi, R.A. (1978) Biochem. Biophys. Res. Commun. 80, 553–559.
110 Bisson, R., Jacobs, B. and Capaldi, R.A. (1980) Biochemistry 19, 4173–4178.
111 Millett, F., Darley-Usmar, V.M. and Capaldi, R.A. (1982) Biochemistry 21, 3857–3862.
112 Takano, T. and Dickerson, R.E. (1980) Proc. Natl. Acad. Sci. U.S.A. 77, 6371–6375.
113 Broger, C., Nalecz, M.J. and Azzi, A. (1980) Biocheim. Biophys. Acta 592, 519–527.
114 Broger, C., Salardi, S. and Azzi. A. (1983) Eur. J. Biochem. 131, 349–352.
115 Koppenol, W.H. and Margoliash, E. (1982) J. Biol. Chem. 257, 4426–4437.
116 Ferguson-Miller, S., Brautigan, D.L. and Margoliash, E. (1978) J. Biol. Chem. 253, 149–159.
117 Rieder, R. and Bosshard, H.R. (1980) J. Biol. Chem. 255, 4732–4739.
118 Smith, H.T., Ahmed, A.J. and Millett, F. (1981) J. Biol. Chem. 256, 4984–4990.
119 Capaldi, R.A., Darley-Usmar, V., Fuller, S. and Millett, F. (1982) FEBS Lett. 138, 1–7.
120 Bisson, R., Steffens, G.C.M., Capaldi, R.A. and Buse, G. (1982) FEBS Lett. 144, 359–363.
121 Bisson, R. and Montecucco, C. (1982) FEBS Lett. 150, 49–53.
122 Millett, F., De Jong, C., Paulson, L. and Capaldi, R.A. (1983) Biochemistry 22, 546–552.
123 Millett, F., De Jong, C., Stonehuerner, J., O'Brian, P. and Capaldi, R.A. (1983) Fed. Proc. 42, 1781.

124 Vik, S.B., Georgevich, G. and Capaldi, R.A. (1981) Proc. Natl. Acad. Sci. U.S.A. 78, 1456–1460.
125 Robinson, N.C., Strey, F. and Talbert, L. (1980) Biochemistry 19, 3656–3660.
126 Semin, B.K., Saraste, M. and Wikström, M. (1984) Biochim. Biophys. Acta 769, 15–22.
127 Wilson, M.T., Greenwood, C., Brunori, M. and Antonini, E. (1975) Biochem. J. 147, 145–153.
128 Antalis, T.M. and Palmer, G. (1982) J. Biol. Chem. 257, 6194–6206.
129 Chance, B., Saronio, C. and Leigh, J.S., Jr. (1975) J. Biol. Chem. 250, 9226–9237.
130 Chance, B., Saronio, C. and Leigh, J.S., Jr. (1979) Biochem. J. 177, 931–941.
131 Clore, G.M., Andreasson, L.-E., Karlsson, B., Aasa, R. and Malmström, B.G. (1980) Biochem. J. 185, 139–154.
132 Clore, G.M., Andreasson, L.-E., Karlsson, B., Aasa, R. and Malmström, B.G. (1980) Biochem. J. 185, 155–167.
133 Wilson, M.T., Jensen, P., Aasa, R., Malmström, B.G. and Vänngård, T. (1982) Biochem. J. 203, 483–492.
134 Fiamingo, F.G., Altschuld, R.A., Moh, P.P. and Alben, J.O. (1982) J. Biol. Chem. 257, 1639–1650.
135 DeVault, D. (1980) Q. Rev. Biophys. 13, 387–564.
136 Steffens, G.J. and Buse, G. (1979) Hoppe-Seyler's Z. Physiol. Chem. 360, 613–619.
137 Stevens, T.H., Martin, G.T., Wang, H., Brudvig, G.W., Scholes, C.P. and Chan, S.I. (1982) J. Biol. Chem. 257, 12106–121113.
138 Bisson, R., Steffens, G.C.M. and Buse, G. (1982) J. Biol. Chem. 257, 6716–6720.
139 Ohnishi, T., Blum, H., Leigh, J.S., Jr. and Salerno, J.C. (1979) In Membrane Bioenergetics (Lee, C.P. et al., eds.) pp. 21–30, Addison-Wesley, London and Massachusetts.
140 Winter, D.B., Bruynincks, W.J., Foulke, F.G., Grinich, N.P. and Mason, H.S. (1980) J. Biol. Chem. 255, 11408–11414.
141 Erecinska, M., Wilson, D.F. and Blasie, J.K. (1978) Biochim. Biophys. Acta 501, 53–62.
142 Erecinska, M., Wilson, D.F. and Blasie, J.K. (1978) Biochim. Biophys. Acta 501, 63–71.
143 Blum, H., Harmon, H.J., Leigh, J.S., Jr., Salerno, J.C. and Chance, B. (1978) Biochim. Biophys. Acta 502, 1–10.
144 Babcock, G.T., Vickery, L.E. and Palmer, G. (1976) J. Biol. Chem. 251, 7907–7919.
145 Blumberg, W.E. and Peisach, J. (1979) In Cytochrome Oxidase (King, T.E. et al., eds.) pp. 153–159, Elsevier/North-Holland, Amsterdam.
146 Powers, L., Chance, B., Ching, Y. and Angiolillo, P. (1981) Biophys. J. 34, 465–498.
147 Stevens, T.H. and Chan, S.I. (1981) J. Biol. Che,m. 256, 1069–1071.
148 Welinder, K.G. and Mikkelsen, L. (1983) FEBS Lett. 157, 233–239.
149 Ohnishi, T., LoBrutto, R., Salerno, J.C., Bruckner, R.C. and Frey, T.G. (1982) J. Biol. Chem. 257, 14821–14825.
150 Mascarenhas, R., Wei, Y.-H., Scholes, C.P. and King, T.E. (1983) J. Biol. Chem. 258, 5348–5351.
151 Brudvig, G.W., Stevens, T.H. and Chan, S.I. (1980) Biochemistry 19, 5275–5285.
152 De Bruijn, M.H.L. (1983) Nature 304, 234–241.
153 Anderson, S., de Bruijn, M.H.L., Coulson, A.R., Eperon, I.C., Sanger, F. and Young, I.G. (1982) J. Mol. Biol. 156, 683–717.
154 Dayhoff, M.O. (1978) Atlas of Protein Sequence and Structure, Vol. 5, suppl. 3. pp. 230–233, Natl. Biomed. Res. Found., Washington D.C.
155 Wikström, M. (1977) Nature 266, 271–273.
156 Krab, K. and Wikström, M. (1978) Biochim. Biophys. Acta 504, 200–214.
157 Casey, R.P., Chappell, J.B. and Azzi, A. (1979) Biochem. J. 182, 149–156.
158 Sigel, E. and Carafoli, E. (1980) Eur. J. Biochem. 111, 299–306.
159 Casey, R.P. and Azzi, A. (1983) FEBS Lett. 154, 237–241.
160 Prochaska, L.J., Bisson, R., Capaldi, R.A., Steffens, G.C.M. and Buse, G. (1981) Biochim. Biophys. Acta 637, 360–373.
161 Chan, S.H.P. and Freedman, J.A. (1983) FEBS Lett. 162, 344–348.
162 Van Verseveld, H.W., Krab, K. and Stouthamer, A.H. (1981) Biochim. Biophys. Acta 635, 525–534.
163 Solioz, M., Carafoli, E. and Ludwig, B. (1982) J. Biol. Chem. 257, 1579–1582.

164 Sone, N. and Hinkle, P.C. (1982) J. Biol. Chem. 257, 12600–12604.
165 Hill, T.L. (1977) Trends Biochem. Sci. 2, 204–207.
166 DeVault, D. (1971) Biochim. Biophys. Acta 226, 193–199.
167 Babcock, G.T. and Callahan, P.M. (1983) Biochemistry 22, 2314–2319.
168 Babcock, G.T., Callahan, P.M., Ondrias, M.R. and Salmeen, I. (1981) Biochemistry 20, 959–966.
169 Saari, H., Penttilä, T. and Wikström, M. (1980) J. Bioenerg. Biomembr. 12, 325–338.
170 Saraste, M., Penttilä, T. and Wikström, M. (1981) Eur. J. Biochem. 115, 261–268.
171 Casey, R.P., Thelen, M. and Azzi, A. (1980) J. Biol. Chem. 255, 3994–4000.
172 Penttilä, T. and Wikström, M. (1981) In Vectorial Reactions in Electron and Ion Transport in Mitochondria and Bacteria (Palmieri, F. et al., eds.) pp. 71–80, Elsevier/North-Holland, Amsterdam.
173 Ludwig, B. (1980) Biochim. Biophys. Acta 594, 177–189.
174 Puttner, I., Solioz, M., Carafoli, E. and Ludwig, B. (1983) Eur. J. Biochem. 134, 33–37.
175 Hoppe, J. and Sebald, W. (1984) Biochim. Biophys. Acta (Reviews on Bioenergetics) 768(1), 1–27.
176 Prince, R.C., O'Keefe, D.P. and Dutton, P.L. (1982) In Electron Transport and Photophosphorylation (Barber, J., ed.) pp. 197–248, Elsevier, Amsterdam.
177 Rieske, J.S. (1976) Biochim. Biophys. Acta 456, 195–247.
178 Hauska, G., Hurt, E., Gabellini, N. and Lockau, W. (1983) Biochim. Biophys. Acta 726, 97–133.
179 Berry, E.A. and Trumpower, B.L. (1984) In Coenzyme Q (Lenaz, G. ed) John Wiley, New York, in the press.
180 Hatefi, Y. (1976) In The Enzymes of Biological Membranes, Vol. 4 (Martonosi, A.N., ed.) pp. 3–41, John Wiley, New York.
181 Wakabayashi, S., Matsubara, H., Kim, C.H. and King, T.E. (1982) J. Biol. Chem. 257, 9335–9344.
182 Anderson, S., Bankier, A.T., Barrell, B.G., de Bruijn, M.H.L., Coulsoin, A.R., Drouin, J., Eperon, I.C., Nierlich, D.P., Roe, B.A., Sanger, F., Schreier, P.H., Smith, A.J.H., Staden, R. and Young, I.G. (1981) Nature 290, 457–465.
183 Bibb, M.J., van Etten, R.A., Wright, C.T., Walberg, M.W. and Clayton, D.A. (1981) Cell 26, 167–180.
184 Nobrega, F.G. and Tzagoloff, A. (1980) J. Biol. Chem. 255, 9828–9837.
185 Waring, R.B., Davies, R.W., Lee, S., Grisi, E., McPheil Berks, M. and Scazzocchio, C. (1981) Cell 27, 4–11.
186 Citterich, M.H., Morelli, G.F., Macino, G. (1983) EMBO J. 2, 1235–1242.
187 Marres, C.A.M. and Slater, E.C. (1977) Biochim. Biophys. Acta 462, 531–548.
188 Weiss, H. and Juchs, B. (1978) Eur. J. Biochem. 88, 17–28.
189 Weiss, H. and Kolb, H.J. (1979) Eur. J. Biochem. 99, 139–149.
190 Gellerfors, P., Johansson, T. and Nelson, B.D. (1981) Eur. J. Biochem. 115, 275–278.
191 Siedow, J.N., Power, S., de la Rosa, F.F. and Palmer, G. (1978) J. Biol. Chem. 253, 2392–2399.
192 Sidhu, A. and Beattie, D.S. (1982) J. Biol. Chem. 257, 7879–7886.
193 Von Jagow, G., Schägger, H., Riccio, P. and Klingenberg, M. (1977) Biochim. Biophys. Acta 462, 549–558.
194 Wang, T.Y. and King, T.E. (1982) Biochem. Biophys. Res. Commun. 104, 591–596.
195 Yu, C.A. and Yu, L. (1981) Biochim. Biophys. Acta 639, 99–128.
196 Zhu, Q.S., Berden, J.A. and Slater, E.C. (1983) Biochim. Biophys. Acta 724, 184–190.
197 Tzagoloff, A., Yang, P.C., Wharton, D.C. and Rieske, J.S. (1965) Biochim. Biophys. Acta 96, 1–8.
198 Von Jagow, G., Engel, W.D., Schägger, H., Machleidt, W. and Machleidt, I. (1981) In Vectorial Reactions in Electron and Ion Transport in Mitochondria and Bacteria (Palmieri, F. et al., eds.) pp. 149–161, Elsevier/North-Holland, Amsterdam.
199 Slater, E.C. (1981) In Chemiosmotic Proton Circuits in Biological Membranes (Skulachev, V.P. and Hinkle, P.C., eds.) pp. 69–104, Addison-Wesley, London and Massachusetts.
200 T'sai, A. and Palmer, G. (1982) Biochim. Biophys. Acta 681, 484–495.
201 Orme-Johnson, N.R., Hansen, R.E. and Beinert, H. (1974) J. Biol. Chem. 249, 1928–1939.
202 De Vries, S., Albracht, S.P.J., Leeuwerik, F.J. (1979) Biochim. Biophys. Acta 546, 316–333.

203 Carter, K., T'sai, A. and Palmer, G. (1981) FEBS Lett. 132, 243–246.
204 Saraste, M. and Wikström, M. (1983) In Structure and Function of Membrane Proteins (Quagliariello, E. and Palmieri, F., eds.) pp. 139–144, Elsevier, Amsterdam.
204a Saraste, M. (1984) FEBS Lett. 166, 367–372.
205 Widger, W.R., Cramer, W.A., Herrmann, R. and Trebst, A. (1984) Proc. Natl. Acad. Sci. U.S.A. 81, 674–678.
206 Erecinska, M. and Wilson, D.F. (1979) Arch. Biochem. Biophys. 192, 80–85.
207 Mitchell, P. (1972) Fed. Eur. Biochem. Soc. Symp. 28, 353–370.
208 Wikström, M. (1973) Biochim. Biophys. Acta 301, 155–193.
209 T'sai, A. and Palmer, G. (1983) Biochim. Biophys. Acta 722, 349–363.
210 Wilson, D.F., Erecinska, M., Leigh, J.S., Jr. and Koppelman, M. (1972) Arch. Biochem. Biophys. 151, 112–121.
211 De Vries, S., Albracht, S.P.J., Berden, J.A. and Slater, E.C. (1982) Biochim. Biophys. Acta 681, 41–53.
212 Li, Y., Leonard, K. and Weiss, H. (1981) Eur. J. Biochem. 116, 199–205.
213 Ozols, J. and Gerard, C. (1977) Proc. Natl. Acad. Sci. U.S.A. 74, 3725–3729.
214 Dutton, P.L., Wilson, D.F. and Lee, C.P. (1970) Biochemistry 9, 5077–5082.
215 Hatefi, Y., Haavik, A.G. and Griffiths, D.E. (1962) J. Biol. Chem. 237, 1681–1685.
216 Rieske, J.S., Hansen, R.E. and Zaugg, W.S. (1964) J. Biol. Chem. 239, 3017–3021.
217 Rieske, J.S., Zaugg, W.S. and Hansen, R.E. (1964) J. Biol. Chem. 239, 3023–3030.
218 Rieske, J.S. (1967) Methods Enzymol. 10, 357–362.
219 Nishibayashi-Yamashita, H., Cunningham, C. and Racker, E. (1972) J. Biol. Chem. 247, 698–704.
220 Trumpower, B.L. and Edwards, C.A. (1979) FEBS Lett. 100, 13–16.
221 Trumpower, B.L. (1981) Biochim. Biophys. Acta 639, 129–155.
222 Li, Y., de Vries, S., Leonard, K. and Weiss, H. (1981) FEBS Lett. 135, 277–280.
223 Albracht, S.P.J. and Subramanian, J. (1977) Biochim. Biophys. Acta 462, 36–48.
224 Leigh, J.S., Jr. and Erecinska, M. (1975) Biochim. Biophys. Acta 387, 95–106.
225 Prince, R.C. and Dutton, P.L. (1976) FEBS Lett. 65, 117–119.
226 Sands, R.H. and Dunham, W.R. (1975) Quart. Rev. Biophys. 7, 443–504.
227 Marres, C.A.M., de Vries, S. and Slater, E.C. (1982) Biochim. Biophys. Acta 681, 323–326.
228 Zhu, Q.S., Berden, J.A., de Vries, S., Folkers, K., Porter, T. and Slater, E.C. (1982) Biochim. Biophys. Acta 682, 160–167.
229 De Vries, S. (1983) The Pathway of Electrons in QH_2 : Cytochrome *c* Oxidoreductase, Ph.D. Thesis, Univ. of Amsterdam.
230 Leonard, K., Wingfield, P., Arad, T. and Weiss, H. (1981) J. Mol. Biol. 149, 259–274.
231 Hovmöller, S., Leonard, K. and Weiss, H. (1981) FEBS Lett. 123, 118–122.
232 Bell, R.L., Sweetland, J., Ludwig, B. and Capaldi, R.A. (1979) Proc. Natl. Acad. Sci. U.S.A. 76, 741–745.
233 Gutweniger, H., Bisson, R., Montecucco, C. and Santato, M. (1981) J. Biol. Chem. 256, 11132–11136.
234 Ohnishi, T., Salerno, J.C. and Blum, H. (1982) In Function of Quinones in Energy Conserving Systems (Trumpower, B.L., ed.) pp. 247–261, Academic Press, New York and London.
235 Ohnishi, T. and Salerno, J.C. (1982) In Iron-Sulphur Proteins (Spiro, T.G., ed.) pp. 285–327, Vol. 4, John Wiley, New York.
236 Salerno, J.C., Blum, H. and Ohnishi, T. (1979) Biochim. Biophys. Acta 547, 270–281.
237 Harmon, H.J. and Struble, V.G. (1983) Biochemistry 22, 4394–4400.
238 Trumpower, B.L. (1981) J. Bioenerg. Biomembr. 13, 1–24.
239 De Vries, S., Albracht, S.P.J., Berden, J.A. and Slater, E.C. (1981) J. Biol. Chem. 256, 11996–11998.
240 Konstantinov, A.A. and Ruuge, E.K. (1977) FEBS Lett. 81, 137–141.
241 Ohnishi, T. and Trumpower, B.L. (1980) J. Biol. Chem. 255, 3278–3284.
242 Yu, C.A., Nagaoka, S., Yu, L. and King, T.E. (1980) Arch. Biochem. Biophys. 204, 59–70.
243 De Vries, S., Berden, J.A. and Slater, E.C. (1980) FEBS Lett. 122, 143–148.
244 Wei, Y.-H., Scholes, C.P. and King, T.E. (1981) Biochem. Biophys. Res. Commun. 99, 1411–1419.

245 Yu, C.A. and Yu, L. (1980) Biochim. Biophys. Acta 591, 409–420.
246 Mitchell, P. (1976) J. Theor. Biol. 62, 327–367.
247 Clark, W.M. (1960) Oxidation-Reduction Potentials of Organic Systems, Williams and Wilkins, Baltimore, MD.
248 Kröger, A. and Klingenberg, M. (1973) Eur. J. Biochem. 39, 313–323.
249 Matsuura, K., Packham, N.K., Mueller, P. and Dutton, P.L. (1981) FEBS Lett. 131, 17–22.
250 Kingsley, P.B. and Feigenson, G.W. (1981) Biochim. Biophys. Acta 635, 602–618.
251 Chance, B. (1972) In Biochemistry and Biophysics of Mitochondrial Membranes (Azzone, G.F. et al., eds.) pp. 85–99, Academic Press, London and New York.
252 Futami, A., Hurt, E. and Hauska, G. (1979) Biochim. Biophys. Acta 547, 583–596.
253 Quinn, P.J. and Esfahani, A. (1980) Biochem. J. 185, 715–722.
254 Katsikas, H. and Quinn, P.J. (1982) Eur. J. Biochem. 124, 165–169.
255 Baum, H., Rieske, J.S., Silman, H.I. and Lipton, S.H. (1967) Proc. Natl. Acad. Sci. U.S.A. 57, 798–805.
256 Wikström, M. and Berden, J.A. (1972) Biochim. biophys. Acta 283, 403–420.
257 Deul, D.H. and Thorn, M.B. (1962) Biochim. Biophys. Acta 59, 426–436.
258 Slater, E.C. (1973) Biochim. Biophys. Acta 301, 129–154.
259 Slater, E.C. and de Vries, S. (1980) Nature 288, 717–718.
260 Convent, B. and Briquet, M. (1978) Eur. J. Biochem. 82, 473–481.
261 Brandon, J.R., Brocklehurst, J.R. and Lee, C.P. (1972) Biochemistry 11, 1150–1154.
262 Van Ark, G. and Berden, J.A. (1977) Biochim. Biophys. Acta 459, 119–137.
263 Thierbach, G. and Reichenbach, H. (1981) Biochim. Biophys. Acta 638, 282–289.
264 Von Jagow, G. and Engel, W.D. (1981) FEBS Lett. 136, 19–24.
265 Bowyer, J.R. and Crofts, A.R. (1981) Biochim. Biophys. Acta 636, 218–233.
266 De Vries, S., Albracht, S.P.J., Berden, J.A., Marres, C.A.M. and Slater, E.C. (1983) Biochim. Biophys. Acta 723, 91–103.
267 Trumpower, B.L. and Haggerty, J.G. (1980) J. Bioenerg. Biomembr. 12, 151–164.
268 Bowyer, J.R., Edwards, C.A., Ohnishi, T. and Trumpower, B.L. (1982) J. Biol. Chem. 257, 8321–8330.
269 Briquet, M., Purnelle, B., Faber, A.-M. and Goffeau, A. (1981) Biochim. Biophys. Acta 638, 116–119.
270 Garland, P.B., Clegg, R.A., Boxer, D., Downie, J.A. and Haddock, B.A. (1975) In Electron Transfer Chains and Oxidative Phosphorylation (Quagliariello, E. et al., eds.) pp. 351–358, Elsevier, North-Holland, Amsterdam.
271 Van Ark, G. (1980) Electron Transfer Through the Ubiquinol: Ferricytochrome *c* Oxidoreductase Segment of the Mitochondrial Respiratory Chain, Ph.D. Thesis, University of Amsterdam.
272 Bowyer, J.R. and Trumpower, B.L. (1981) In Chemiosmotic Proton Circuits in Biological Membranes (Skulachev, V.P. and Hinkle, P.C., eds.) pp. 105–122, Addison-Wesley, London and Massachusetts.
273 Rich, P.R. and Bendall, D.S. (1980) Biochim. Biophys. Acta 592, 506–518.
274 Rich, P.R. (1981) FEBS Lett. 130, 173–178.
275 Rich, P.R. (1982) Biochem. Soc. Trans. 10, 482–484.
276 Van den Berg, W.H., Prince, R.C., Bashford, L., Takamiya, K., Bonner, W.D., Jr. and Dutton, P.L. (1979) J. Biol. Chem. 254, 8594–8604.
277 Matsuura, K., O'Keefe, D.P. and Dutton, P.L. (1983) Biochim. Biophys. Acta 722, 12–22.
278 Van Dam, K. and Engel, G.L. (1973) In Mechanisms in Bioenergetics (Azzone, G.F. et al., eds.) pp. 141–148, Academic Press, New York and London.
279 Leung, K.H. and Hinkle, P.C. (1975) J. Biol. Chem. 250, 8467–8471.
280 Guerrieri, F. and Nelson, B.D. (1975) FEBS Lett. 54, 339–342.
281 Suzuki, H. and King, T.E. (1983) Fed. Proc. 42, 1771.
282 Ringler, R.L., Minakami, S. and Singer, T.P. (1963) J. Biol. Chem. 238, 801–810.
283 Watari, H., Kearney, E.B. and Singer, T.P. (1963) J. Biol. Chem. 238, 4063–4073.

284 Biggs, D.R., Hauber, J. and Singer, T.P. (1963) 238, 4563–4567.
285 Hatefi, Y. and Stempel, K.E. (1969) J. Biol. Chem. 244, 2350–2357.
286 Hatefi, Y. and Rieske, J.S. (1967) Methods Enzymol. 10, 235–239.
287 Hatefi, Y. (1978) Methods Enzymol. 53, 11–14.
288 Dooijewaard, G., de Bruin, G.J.M., van Dijk, P.J. and Slater, E.C. (1978) Biochim. Biophys. Acta 501, 458–469.
289 Smith, S. and Ragan, C.I. (1980) Biochem. J. 185, 315–326.
290 Heron, C., Smith, S. and Ragan, C.I. (1979) Biochem. J. 181, 435–443.
291 Ohnishi, T. (1975) Biochim. Biophys. Acta 387, 475–490.
292 Ohnishi, T. (1979) In Membrane Proteins in Energy Transduction (Capaldi, R.A., ed.) pp. 1–87, Marcel Dekker, New York.
293 Ingledew, W.J. and Ohnishi, T. (1980) Biochem. J. 186, 111–117.
294 Beinert, H. and Albracht, S.P.J. (1982) Biochim. Biophys. Acta 683, 245–277.
295 Ragan, C.I., Galante, Y.M., Hatefi, Y. and Ohnishi, T. (1982) Biochemistry 21, 590–594.
296 Ragan, C.I., Galante, Y.M. and Hatefi, Y. (1982) Biochemistry 21, 2518–2524.
297 Earley, F.G.P. and Ragan, C.I. (1980) Biochem. J. 191, 429–436.
298 Chen, S. and Guillory, R.J. (1981) J. Biol. Chem. 256, 8318–8323.
299 von Bahr-Lindström, H., Galante, Y.M., Persson, M. and Jörnvall, H. (1983) Eur. J. Biochem. 134, 145–150.
300 Boekema, E.J., van Breemen, J.F.L., Keegstran, W., van Bryggen, E.F.J. and Albracht, S.P.J. (1982) Biochim. Biophys. Acta 679, 7–11.
301 Orme-Johnson, N.R., Hansen, R.E. and Beinert, H. (1974) J. Biol. Chem. 249, 1922–1927.
302 Ohnishi, T., Blum, H., Galante, Y.M. and Hatefi, Y. (1981) J. Biol. Chem. 256, 9216–9220.
303 Salerno, J.C., Ohnishi, T., Blum, H. and Leigh, J.S., Jr. (1977) Biochim. Biophys. Acta 494, 191–197.
304 Paech, C., Reynolds, J.G., Singer, T.P. and Holm, R.H. (1981) J. Biol. Chem. 256, 3167–3170.
305 Singer, T.P., Gutman, M. and Massey, V. (1973) In Iron-Sulphur Proteins (Lovenberg, W., ed.) pp. 225–300, Academic Press, London and New York.
306 Gutman, M., Mersmann, H., Luthy, J. and Singer, T.P. (1970) Biochemistry 9, 2678–2687.
307 Singer, T.P. and Gutman, M. (1971) Adv. Enzymol. 34, 79–153.
308 Gutman, M., Singer, T.P. and Beinert, H. (1971) Biochem. Biophys. Res. Commun. 44, 1572–1578.
309 Gutman, M. and Singer, T.P. (1970) Biochemistry 9, 4750–4758.
310 Kean, E.A., Gutman, M. and Singer, T.P. (1971) J. Biol. Chem. 246, 2346–2353.
311 Ragan, C.I. and Racker, E. (1973) J. Biol. Chem. 248, 2563–2569.
312 Gutman, M., Singer, T.P. and Beinert, H. (1972) Biochemistry 11, 556–562.
313 Ohnishi, T. (1976) Eur. J. Biochem. 64, 91–103.
314 Jones, R.W., Gray, T.A. and Garland, P. (1976) Biochem. Soc. Trans. 4, 671–673.

Ernster (ed.) Bioenergetics
© 1984 Elsevier Science Publishers B.V.

CHAPTER 4

Photosynthetic electron transfer

B. ANDREA MELANDRI and GIOVANNI VENTUROLI

Institute of Botany, University of Bologna, Via Irnerio 42, 40126 Bologna, Italy

1. Introduction

The utilization of the energy of light for driving endergonic oxidoreduction reactions is the essence of photosynthesis. This process takes place in integrated membrane systems in which biological photocatalytic events are strictly associated with esergonic oxidoreductions not requiring 'per se' light energy. Integral parts of the membrane structure, also, are the systems of accessory pigments, whose function is that of increasing up to 100-fold the cross section of interaction of the photons with the photocatalytic centers, and thereby of improving the photosynthetic efficiency.

The overall process of photosynthetic electron transfer is promoted by an array of catalytic proteins, only a few of which are real photochemical enzymes. It is now realized that these proteins form a number of well-defined complexes, partially independent from each other, but nevertheless interacting through redox carriers, freely diffusable either in the membrane lipids or at the membrane-water interface. The concept of membrane photosynthetic complex is experimentally justified by the possibility of isolating specific multiprotein associations following micellization of the membrane with mild detergents. In general these associations are characterized by well-defined catalytic activities, which are lost, however, if the complex is dissociated into the individual polypeptides by more drastic detergent treatments.

In a general sense the complexes present in photosynthetic membranes can be subdivided into photosynthetic reaction centers, usually associated with pigment antenna complexes, and in non-photosynthetic, more conventional, electron transfer complexes. The role of the former is that of photocatalyzing the donation of an electron from a more positive electron donor to a more negative acceptor. Basically four different types of reaction centers (RC) are known: two photosystems I and II (PSI and PSII) present in the membranes of cyanobacteria and all eukaryotic photosynthetic organisms, and the two types of RC present in purple and green photosynthetic bacteria, respectively. In the following sections only PSI and PSII and the RC of purple bacteria will be discussed, since too little is known on green bacteria RC to allow an extensive comparative treatment. The function of electron transfer complexes, which very often closely resemble the analogous ones active in respiratory chains of aerobic bacteria and of mitochondria, is either the redox interaction between two RC (as in the central part of the photosynthetic chain of higher plants), or to supply electrons to an RC from an exogenous donor (as for the water-splitting complex of PSII), or to accept electrons from a RC (as for the NADP-reducing enzymes), or to recycle electrons around an RC (as in bacterial photosynthesis or for cyclic electron flow around PSI). In all cases the oxidoreductions occur spontaneously and no input of energy is required for the process.

An important aspect of the function of photosynthetic complexes is their asymmetric arrangement in respect to the membrane and to the external and internal phases of the cellular compartments. This arrangement allows the catalysis of vectorial electron transfer and the performance of electrical work by promoting charge separation across the membrane dielectric barrier. It allows also in some cases the net translocation of protons across the membrane. These two processes are at the basis of the mechanism of energy conservation in photosynthesis coupled to the formation of ATP, which is added, in oxygenic photosynthesis, to the conservation of redox energy in the form of reduced pyridine nucleotide coenzymes.

In the following sections the present knowledge on the structural and functional organization of the complexes present in photosynthetic membranes will be systematically reviewed. Care has been taken in emphasizing the analogies between complexes in the different groups of organisms and the experimental uncertainties still existing in the study of these systems. This detailed description, forming the main body of this Chapter, will be concluded by a brief discussion on the integration of the function of the individual complexes into a continuous electron transfer chain and on their topologic arrangement in the membrane.

2. *Reaction centers*

2.1. *General remarks*

A photosynthetic reaction center can be defined as the minimal functional unit capable of utilizing the energy of one photon for promoting the transfer of one electron from a donor (D) to an acceptor (A) molecule. This electron transferred must be stabilized against charge recombination, a highly esergonic process, through the location of the electron on a stable reduced species and the presence of a large free energy barrier for the back transfer to the donor species. For these reasons the electron, following the primary photochemical act, is sequentially transferred to a series of electron acceptors present within the reaction center itself and characterized by increasingly positive midpoint potentials, until it is stabilized on a final acceptor at a redox potential considerably more positive than that of the primary one [1,2]. The probability of back reaction becomes smaller the larger the free energy gap to be overcome for charge recombination; this probability can be evaluated by measuring the half time of the charge-recombination reaction, which can be followed by rapid spectrophotometric methods.

The span of oxidoreduction potentials in which the different reaction centers operate varies largely according to the function of the reaction center within the photosynthetic apparatus or the type of organism considered. The evolutionary adaptation of the reaction center structure to the required function has been accomplished by varying the chemical nature of the donor and the acceptor molecules: thus, while the acceptors for medium low potential centers are quinone coenzymes, as in photosystem II (PSII) or in most of photosynthetic bacteria [3], the acceptors of reaction centers operating at more negative potentials are iron-sulphur centers, as in photosystem I (PSI) [4] or in green anaerobic bacteria (*Chlorobiaceae*) [5].

The electron transfer to the final electron acceptor occurs generally via a number of intermediary steps through intermediate acceptors. For a good efficiency of the electron trapping process it is necessary that the probability of transfer from one intermediary to the following one is higher than that of the back reaction from the intermediate acceptor to the primary electron donor (cf., Fig. 4.1), which would bring to dissipation of the excitation energy to heat. However, the probability of back reaction becomes smaller the farther the donor and the acceptor molecule are apart from each other; eventually the electron is stabilized on the final acceptor for a time long enough (10 s) to allow transfer to another electron transfer complex. A reaction center complex includes in its structure therefore several redox species, some of which have been characterized only recently due to their extremely short lifetime.

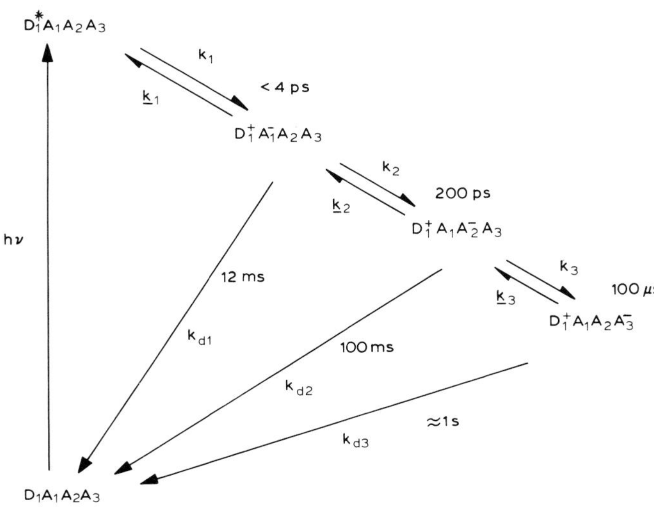

Fig. 4.1. Scheme of the electron transfer steps within the reaction center complex. Electrons from the excited primary donor are sequentially transferred to different electron acceptors characterized by increasingly positive oxidoreduction potentials. The energy barrier thus obtained prevents a rapid charge recombination process. The rate of back reaction must be small in comparison with the forward reaction for an effective electron transfer process. The rates of the forward reactions and of some back reactions in RC of *Rps. sphaeroides* are indicated as an example.

The number of pigment molecules present in a typical reaction center preparation, as isolated by a solubilization with detergents of photosynthetic membranes, is generally very small. In vivo, however, the efficiency of photon capture is enhanced by the association of the center with other accessory pigments molecules, and by a consequent energy transfer mechanism between these molecules. The accessory pigments are organized in the so-called antenna and can be present either in extrinsic proteinaceous structures, like the phycobilisome of cyanobacteria and red algae (*Rhodophyta*), or intramembrane chromoproteins. Even when located in intrinsic membrane proteins, the antenna pigment can be separated from the reaction center complexes as one or more pigment-protein structures; thus, the antenna are considered as separated complexes. In this chapter, devoted to the discussion of electron transfer mechanisms, the structure and function of the antenna complex will not be further considered. It should be mentioned, however, that whatever the nature of the antenna, intrinsic or extrinsic, a close physical association of the accessory and the photochemically active pigments must be assured (with physical distance not exceeding 20–30 Å), in order to obtain a high efficiency of energy transfer and a consequent high quantum yield of the reaction center functions.

2.2. Experimental approaches to the study of reaction centers

The most direct approach for the identification of the redox components of the reaction centers is absorption spectroscopy, whereby spectral and kinetic information on the species oxidized and reduced can be obtained. A second useful technique is ESR spectroscopy since during the photobiological reactions the production of radical species is a common event. The ability of these methods to give unambiguous information is, however, hampered by the intrinsic rates of the electron transfer, or charge recombination processes taking place within the RC, which are very high and often prevent the detection of very short-lived intermediates (especially with ESR spectroscopy [1]). These difficulties can be overcome however, if the electron transfer between two redox intermediates, e.g., between the primary and secondary acceptor, is inhibited, by specific substances, by lowering the temperature, or by simply chemically prereducing one acceptor. In all these cases the electron carrier immediately upstream from the inhibited step is reduced and, since the rate of charge recombination is usually much lower than that of electron transfer, is trapped in the reduced state. When chemical prereduction is utilized, the midpoint potential of the reduced acceptor can also be estimated even if this is spectroscopically silent. In this latter case indirect information on its properties can be obtained by studying the rate of charge recombination (determined studying the decay rate of the oxidized primary donor) or of such indirect effects as delayed fluorescence (i.e., emission of photons by the generation of the excited state of the primary donor through reverse electron transfer) or triplet state formation. However, the trapping approach cannot ever prove if the intermediate detected is a true component of the redox sequence or an artifactual side product; this can only be demonstrated by measurements of the actual rates of electron acceptance and donation, which should match kinetically those of the other species of the acceptor array [1].

The application of these general principles to the study of reaction centers of bacteria and of PSI and PSII of higher plants will be discussed in the following sections.

3. *The reaction centers of photosynthetic bacteria*

Since the first isolation of a reaction center preparation from the membrane of a facultative photosynthetic bacterium [6] our knowledge on the structure and function of these complexes has made great advances. Today the RC from purple bacteria, and particularly from the carotenoid-less strain R26 of *Rhodopseudomonas sphaeroides*, are by far the best known examples of photosynthetic complexes studied. Other RC from different bacteria species have also been studied and differences in components sometimes observed; these differences will be mentioned below, whenever necessary, while discussing the properties of the preparations from *Rp. sphaeroides* R26.

3.1. Composition and protein structure

The best preparation of the reaction center from *Rp. sphaeroides* R26 contains four molecules of bacteriochlorophyll *a*, two molecules of bacteriopheophytin *a*, two molecules of ubiquinone-10 and one atom of ferrous iron. With carotenoid-containing strains one molecule of carotenoid also is found associated with the preparation [7]. All these components are bound to a complex of three protein subunits forming a stable and compact association plugged across the membrane phospholipids bilayer. The molecular weight of the three subunits, called L, M and H for light, medium and heavy are, respectively, 21 000, 24 000 and 28 000 as judged from SDS-PAGE [7]. Based, however, on aminoacid composition the molecular weights have been estimated to be 28 000, 32 000 and 36 000. The three subunits are associated with a 1 : 1 : 1 stoichiometry and are isolated as a single complex using the detergent LDAO (lauryldimethylaminooxide). A complex containing only the L and M subunits can be, however, isolated using different procedures [8]: this preparation still contains all the above mentioned components, but exhibits a slightly different kinetic behaviour [3].

The location of the complex in respect with the membrane bilayer has been studied either with hydrophobic and hydrophilic probes, or with monospecific antibodies against single subunits. In general these studies have shown that both the M and H subunits are accessible on both sides of the membrane to hydrophylic reagents, while the L subunit can be labelled only on the periplasmic side [9]. On the other hand, labelling with the hydrophobic marker 5-iodonaphthyl-1-azide, showed that the most heavily labelled, and therefore the most extensively exposed to an hydrophobic environment, are the L and M subunits [10,11].

The exact location of the functional molecules in these subunits is still rather uncertain. The association of one ubiquinone molecule with the M subunit has been demonstrated by covalent linkage with the photoaffinity-reactive quinone analogue 2-azido-anthraquinone [12]. Similarly the association of the second quinone molecule to the same M subunit was suggested by the inhibitory effect of a specific antibody against M on the functional association of this carrier [13]. In addition, cross-linking experiments indicated a close interaction of cytochrome c_2, the direct electron donor to the RC complex, with both the L and M subunits [3]. Nothing certain is known about the location of the four bacteriochlorophylls and the two bacteriopheophytins; judging from the evaluated distance from the quinones, based on functional observations, they could also be located on the M subunit.

3.2. D_1: the bacteriochlorophyll dimer as primary electron donor

Following the scheme of Fig. 4.2, the primary photochemical reaction within the reaction center complex will promote the transfer of one electron from the primary

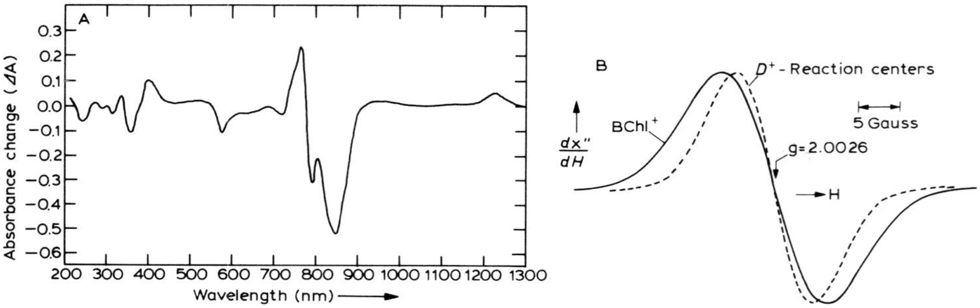

Fig. 4.2. Optical and ESR spectra of electron carriers in the RC of purple photosynthetic bacteria. (A and B) Primary electron donor (D_1): bacteriochlorophyll dimer. (A) Light-dark optical spectrum (recorded at 30 °C) and (B) ESR spectrum of D_1^+ in *Rps. sphaeroides*. The ESR spectrum is $\sqrt{2}$-times narrower than the corresponding spectrum of the Bchl cation, indicating a dimeric structure (from Ref. 3).

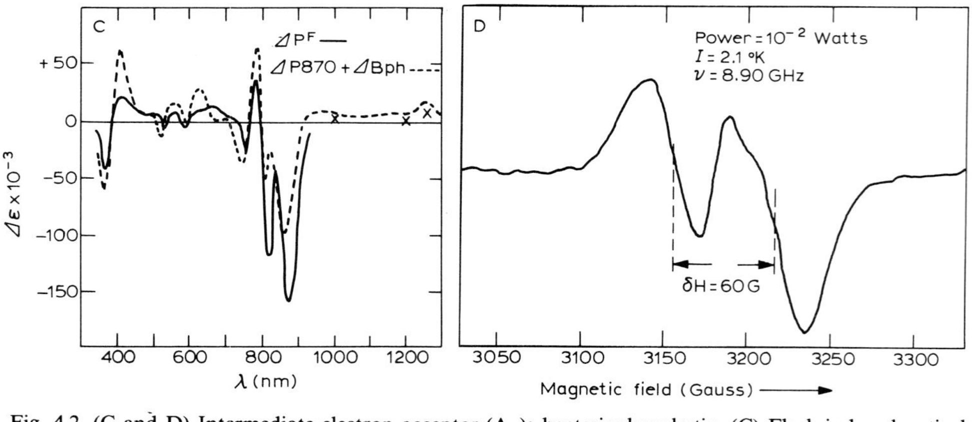

Fig. 4.2. (C and D) Intermediate electron acceptor (A_1): bacteriopheophytin. (C) Flash-induced optical changes compared to the sum of the spectra of D_1 oxidation and A_1 reduction (from Ref. 18); ESR spectrum (recorded at 2.1 °K and 10^{-2} W power) of the bacteriopheophytin anion radical trapped in RC in which ubiquinone was substituted by menaquinone. The doublet disappears if the spectrum is recorded at low power (10^{-6} W) (from Ref. 23).

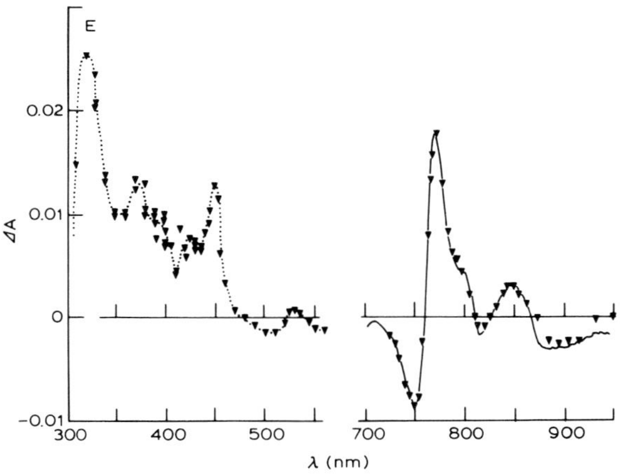

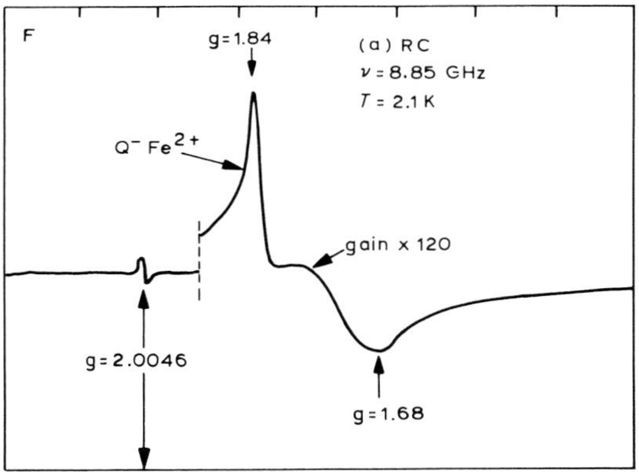

Fig. 4.2. (E and F) Primary electron acceptor (A_2): ubiquinone A. (E) Light-dark optical changes related to the production of Q_A^- radicals (maximal at 320 and 450 nm) and electrochromic effects on bacteriopheophytin (in the near infrared region) (from Ref. 288). (F) ESR spectrum of Q_A^- (obtained by chemical reduction); the large line broadening and shift is eliminated when Fe^{2+} is removed from the RC (from Ref. 3).

donor to the primary acceptor. This event can be conveniently monitored, either in reaction center preparations or in membrane fragments, following spectroscopically the bleaching of specific bands of the donor spectrum. These spectral characteristics are also the basis of the currently accepted hypothesis on the nature of the donor as a special dimeric structure formed by two bacteriochlorophyll *a* molecules. The main spectral changes observable upon illumination of the reaction center occur in the near infrared region. In *Rp. sphaeroides,* to which Fig. 4.2 refers, these correspond to a marked bleaching centered at 870 nm and a blue shift of the band at 800 nm [14]. All these spectral changes take place in less than 10 ps and are therefore monitoring

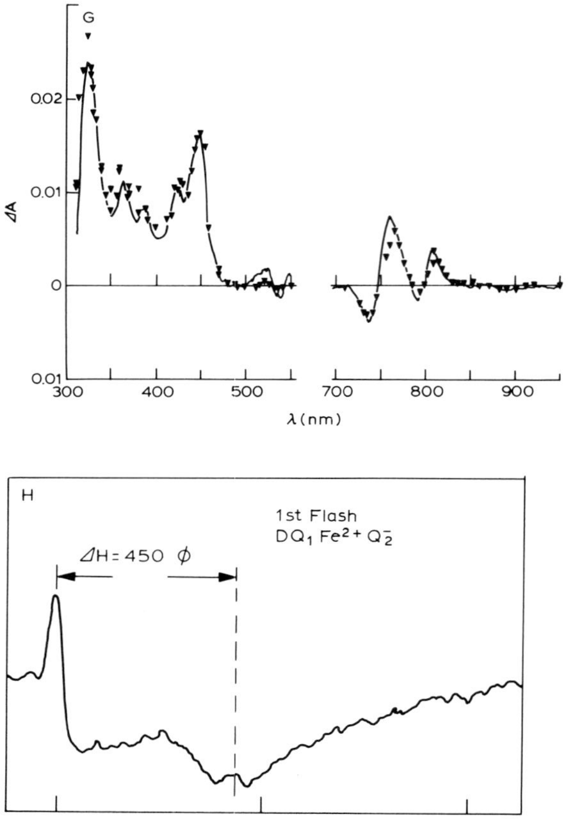

Fig. 4.2. (G and H) Secondary electron acceptor (A_3): ubiquinone B. (G) Light-dark optical spectrum of $Q_B^{\bar{\cdot}}$ and related electrochromic changes (from Ref. 288). (H) Flash-induced ESR spectrum of $Q_B^{\bar{\cdot}}$ (recorded at 2.1° K); after a second flash the signal disappears due to the formation of the Q_A Fe^{2+} $Q_B^{=}$ state, and reappears again after a third flash when the $Q_A^{\bar{\cdot}}$ Fe^{2+} $Q_B^{=}$ state is formed (from Ref. 39).

primary photochemical events [15]. These changes have been interpreted as due to excitation splitting in a bacteriochlorophyll dimer [16]. This hypothesis has been supported by the analysis of the ESR spectra appearing upon oxidation of the donor and of the ESR signal of the triplet state, which can be observed when the primary quinone acceptor (cf., below) is either prereduced or removed [17] from the complex.

Thus, most of the data collected on bacterial RC seem to agree on the presence of a bacteriochlorophyll pair which can be oxidized upon illumination. Thermodynamically this donor has been characterized by studying the dependence of photooxidation from the ambient redox potential: in *Rp. sphaeroides* the $E_{m,7}$ was found to be 0.44 V and pH independent [14].

3.3. A_1: bacteriopheophytin as an intermediate electron acceptor

The presence of a fast relaxing intermediate electron acceptor was first discovered when picosecond spectroscopy was applied to the study of photosynthetic systems. A laser flash-induced signal was detected when the primary quinone acceptor was prereduced [18,19], the lifetime of which was so short as to be observed only in the submicrosecond time range. The fast disappearance of the signal can be understood if it is considered that the rate of charge recombination in $D_1^+A_1^-$ is very fast ($t_{1/2} = 12$ ns) when the primary quinone acceptor in unavailable for reduction. A state in which the state $D_1^+A_1^-$ can be trapped in a long lifetime status can be achieved when the primary donor is rapidly reduced (e.g., by cyt. *c*) under reducing condition of the primary acceptor [18]. In this case the light-dark differential spectrum of A_1^- can be obtained; this spectrum is characteristic for bacteriopheophytin (BPh) and can be attributed to the BPh^--BPh differential spectrum [18–21]. ESR spectroscopy at liquid helium temperature (2° K), and under reducing conditions also reveals a spectrum typical of a tetrapyrrole monomer anion [20]. This spectrum presents a doublet splitting of 60 G in reaction center preparations in which menaquinone substitutes ubiquinone (either naturally as in *Chromatium vinosom* reaction centers [22], or artificially as when ubiquinone is extracted and substituted for by menaquinone [23]. This distortion of the spectrum is due to a close association of Bph with the iron-menaquinone complex. A more complicated structure of Bph with other components of the RC is also suggested by resonance Raman spectroscopy; an association with a bacteriochlorophyll molecule (other than the D donor), which could act as intermediary before Bph, has been proposed [24].

The midpoint potential of the Bph^-/Bph couple has been extimated to be −0.62 V at pH = 11 [25].

3.4. A_2: quinone as a primary electron acceptor

The identity of a ubiquinone molecule as a primary electron acceptor (A_2 also designated Q_A or Q_I) is now well established. The most conclusive evidence came following extraction of A_2 from RC of *Rp. sphaeroides* with hexane containing 0.1% methanol [26]; this procedure inhibits the occurrence of D photooxidation, when observed with millisecond spectroscopy since, as stated above, the charge recombination of $D_1^+A_1^-$ is much faster than that of the $D_1^+A_1A_2^-$ state [27]. Primary photochemistry in extracted RC could be reconstituted with pure ubiquinone-10 [26]. The electron transfer to A_2 requires 200 ps [28–30], and the transfer from A_2 to the secondary acceptor 200 μs; this last step can be inhibited by *o*-phenanthroline [31].

Again, many characteristics of A_2^- could be studied by optical and ESR spectroscopy: under suitable conditions for A_2^- trapping (i.e., rapid reduction of D_1^+ and inhibition by *o*-phenanthroline) an absorption peak is induced by a flash with a maximum at 450 nm, characteristic of a semiquinone anion [32,33]. A is therefore reduced without a simultaneous binding of a proton. The redox potential of the A_2^-/A_2 couple has been estimated to be −0.13 V [14]. ESR spectra of A_2^- are very broadened as compared to the line width expected for a free radical signal [34]; the expected narrow signal can, however, be observed if the iron atom present in RC is extracted [35]. The broadening has therefore been assigned to a magnetic interaction between the quinone molecule and the iron in an A_2FeA_3 complex. The reaction sequence $D_1 \rightarrow A_1 \rightarrow A_2$ has also been confirmed with picosecond spectroscopy by observing that the lifetime of the $D_1^+A_1^-$ state was greatly prolonged by the extraction of A_2 [17].

3.5. A_3: *quinone as secondary acceptor*

With the aid of extraction reconstitution experiments the presence of a second molecule of quinone, active as secondary electron acceptor (A_3, also designated Q_B or Q_{II}) has been established. This molecule is more loosely bound to the RC than A_2 and can be extracted from RC or from whole membranes without increasing the polarity of the extraction solvent (hexane) with methanol [36]. It has been proposed, therefore, that A_3 is not permanently bound to the RC in the phospholipid bilayer but is in rapid exchange equilibrium with the excess quinone present in the membrane [36–38]. The transfer rate of electrons from A_2 to A_3 can be estimated by double laser flash experiments, evaluating the time required after a flash to re-establish the oxidized state of A_2 and consequently the possibility of D_1 photoxidation. This time was found to be 200 μs at room temperature [31]. When the $D_1^+A_1A_2A_3^-$ state is formed the electron is finally stabilized in a very stable state, since the rate of the charge recombination from A_3^- to D_1^+ has a halftime of 1 s as compared to 100 ms for the A_2^- to D_1^+ reaction (that can be observed when A_3 is extracted or electron flow is inhibited by *o*-phenanthroline [31]). The optical spectra of A_3^- (observable 200 μs after the flash) are again those of a ubiquinone anion; this species is exceedingly stable and unprotonated [32,33]. The ESR spectrum of the $D_1^+A_1A_2A_3^-$ state shows a broad signal, again interpreted as evidence of magnetic coupling with the Fe atom in the RC [39]. Both the optical signal at 450 nm and the ESR radical signal disappear following a second flash and the production of the double reduced species $A_3^=$, which is diamagnetic and not absorbing in the blue region of the spectrum [33]. Studies of proton binding with hydrophilic pH indicator dyes have shown that one proton is bound per electron per falsh (whether even or odd number of flashes is immaterial). These results were reconciled with the spectroscopic evidence for the presence of a semiquinone unprotonated species on odd flashes, by proposing a non-chromophoric protonatable group acting as proton acceptor on odd flashes, and proton donor to $A_2^=$ on even flashes [40]. Thus, the complex behaves as a two electron gate device, capable of controlling electron transfer (through the stabilization of a semiquinone) and delivering electrons in pairs following two turnovers of the RC [40]. The reduced two electron carrier A_3H_2 (or $Q_{II}H_2$) is probably able to dissociate from the complex and transfer electrons to other carriers within the membrane [36–38]. With the production of A_3H_2 the electron donations by D_1 are almost permanently stabilized and transferred to a chemically stable reduced species.

4. Photosystem I of higher plants

The active pigment of PSI, the so-called *P*-700, active as electron donor $D_{I,1}$ in eukariotic plants and in cyanobacteria, has been the first example of photosynthetic RC identified. *P*-700 is present in very small amount in chloroplasts (about 1 per 250–300 chlorophylls), and for this reason its detection by spectroscopy, and even more its isolation, have presented many difficulties. It is now clear that *P*-700 is only part of an RC complex including many acceptors, as well as a cluster of chlorophyll *a* molecules that appear to be closely associated with the complex. In principle the operation of this RC is, therefore, very similar to that discussed for bacterial photosynthesis, in that the electron transferred to the primary acceptor is stabilized against charge recombination on secondary acceptors at more positive potential. However, the RC of PSI must operate at oxidoreduction potentials more negative than all other RCs, transferring electrons from the water-soluble cuproprotein plastocyanine ($E_{m,7} = 0.35$ V) to the water-soluble non-heme-iron protein ferredoxin ($E_{m,7} = -0.42$ V). The nature of the functional components of the RC complex is accordingly different since the acceptors are not pheophytins or ubiquinones, but chlorophyll *a* and Fe-S centers. The pathway of electron transfer within the PSI-RC goes from the primary donor *P*-700, a special pair of chlorophyll *a* molecules, to another molecule of chlorophyll *a* (or a pair), acting as intermediary acceptor, to a series of two or three Fe-S centers. The nature and the properties of all these components will be discussed in the following sections.

4.1. Polypeptide and pigment composition

The dissection of the membrane of chloroplasts in attempting to isolate the protein-pigment complex containing *P*-700 was started very early after the spectroscopic identification of this photoactive pigment in 1956 [41]. To this purpose a number of anionic (SDS) or non-ionic (Triton X-100, digitonin) detergents, associated with separative techniques such as differential centrifugation, polyacrylamide gel electrophoresis or column chromatography, have been utilized. In most of these preparations the ratio of chlorophyll *a* to *P*-700 could be decreased to 40–60 and chlorophyll *b* eliminated. In many preparations obtained with Triton X-100 cytochromes b_6 and *f* were still present; these proteins were, however, absent when utilizing SDS or digitonin (for review cf., Refs. 4, 5). Further removal of chlorophyll *a* from the PSI-RC to a level comparable to that of bacterial RC has not been achieved yet.

The functional test for PSI-RC can be either: the photobleaching of *P*-700, monitoring the photooxidation of the primary donot $D_{I,1}$ and the transfer of the electron to a stable acceptor; or the photoreduction of NADP in the overall activity of PSI. For this latter activity soluble components of the electron transfer chain must be added, such as plastocyanin (plus ascorbate) for electron donation to the RC, and ferredoxin and $NADP^+$-ferredoxin oxidoreductase as terminal electron acceptors. This activity was fully present in the PSI-RC preparation of Bengis and Nelson [42], who utilized digitonin fractionation and column chromatography on DEAE; this preparation contained six polypeptide subunits of respective molecular weights 70 000, 25 000, 20 000, 18 000, 16 000 and 8 000 (as evaluated on SDS-PAGE), 40 molecules of chlorophyll *a*, one molecule of carotenoid and several (six to ten) non-heme Fe and acid-labile S [42,43]. Further dissociation with SDS yielded an homogeneous preparation of the 70 000 Da subunit, still containing the pigment molecules and specifically *P*-700 [42]. This preparation (called *P*-700-RC) lost the ability to photoreduce NADP and oxidize plastocyanin, and also all the Fe-S center detectable by ESR spectroscopy. Following similar but milder approach, it has been suggested that subunits III (M_r 20 000) was involved in the site of oxidation of plastocyanin [43,44] and subunits IV, V and VI (M_r 18 000, 16 000 and 8 000, respectively) were related to the Fe-S centers [45,46].

The digitonin preparation of PSI-RC has also been demonstrated to be able to catalyze a light-induced proton uptake when incorporated in phospholipid liposomes and illuminated in the presence of ascorbate and phenazine methosulphate [47]; incorporation of chloroplast ATPase in the same system yielded the reconstitution of photophosphorylation in a model system. The PSI-RC preparation therefore seems to possess all the functional features of PSI for the vectorial transmembrane electron transfer [48] (see Fig. 4.7).

4.2. $D_{I,1}$: a chlorophyll a *dimer as electron donor*

The bleaching of *P*-700 upon illumination corresponds to the production of an oxidized species that has been proposed to be a chlorophyll *a* dimer. The same bleaching can be obtained by chemical oxidation and titrates with a midpoint potential of about 0.45 V [49,50]. The total differential spectrum of *P*-700 photo-oxidation presents, in addition to a major band at about 700 nm, a minor band at 685 nm and another in the Soret region at 435 nm. This spectrum has been attributed to chlorophyll *a* [51], although the presence of a new chlorophyll species in *P*-700-enriched preparations has been claimed [52].

The basis of the concept of a dimer of chlorophyll *a* in *P*-700 rests on evidence obtained with optical or ESR spectroscopy. In the *P*-700 redox difference spectrum, although very similar to that obtainable upon chemical oxidation of chlorophyll *a*, there are significant differences in the red region, which presents a splitting of the peak (at 685 and 700 nm) which is absent in chlorophyll *a* [53]. Moreover, the ESR and ENDOR spectra also present characteristics that have been interpreted as due to a dimeric arrangement [16]. Alternative interpretations have been offered suggesting that the spectral distortions are caused by a modification of the chemical environment of chlorophyll *a* in the RC complex. Definitively not in line with the dimer hypothesis is the spectrum of the light-induced triplet state of *P*-700, that can be observed when the intermediate acceptor is prereduced chemically; in this spectrum the zero field parameters are the same as those of chlorophyll *a* monomers [54]. It is not clear, however, whether the triplet state resides on *P*-700 or on other chlorophylls of the RC complex.

4.3. $A_{I,1}$: *chlorophyll* a *as intermediate acceptor*

Under the same conditions in which bacteriopheophytin was identified in bacterial RC as the intermediate A_1, i.e., under reducing conditions in which all other acceptors were prereduced (−0.75 V), an absorbance change could be detected in PSI-RC preparations [55]. This change decayed in 1 ms at 5°C, and could be observed also in RC preparations treated with SDS, that had therefore lost all the Fe-S center proteins [56]. The redox potential of this intermediate has been not determined so far.

The complete spectrum of the photoreduced intermediate is complex, presenting maxima at 480 and 672 nm, and a negative band at 445 and 700 nm; it has been shown to be very similar to that of a chlorophyll *a* anion, but shifted about 30 nm to the red. It has been therefore suggested that it is also due to a dimer of chlorophyll *a* [56,57]. ESR data are not in agreement with this interpretation, however; when the $D^{+}_{I,1}A^{-}_{I,1}$ state was trapped in *P*-700-RC preparations an ESR spectrum with $g = 2.0025$ and $\Delta H = 12$ G was observed. These values correspond to those of a chlorophyll *a* monomeric anion radical [57,58]. As for *P*-700, therefore, it is undecided whether these special chlorophyll *a* molecules present a dimeric structure or, rather, if the shifted optical spectrum is due to the special chemical environment in which a chlorophyll *a* monomer resides.

4.4. $A_{I,2}$: *the electron acceptor X*

The study of the reversible photooxidation of *P*-700 as a function of the oxidation-reduction poise has brought about the discovery of another electron acceptor ($A_{I,2}$ or X); this acceptor is reduced with a midpoint potential of −0.73 V (at pH 10), and therefore its operativity can be observed around −0.60 V, when the more positive acceptors are reduced [55,59]. The back reaction from $A^{-}_{I,2}$ to *P*-700 has a halftime of 130 ms at 5 °C, another characteristic, besides the midpoint potential differentiating X from the intermediate electron acceptor chlorophyll *a* [57].

The spectral characteristics of $A^{-}_{I,2}$ have been obtained both with optical and ESR spectroscopy. The optical spectrum exhibits band shifts near 450 and 680 nm, and a broad bleaching between 450 and 550 nm [57]. It has been suggested that the bleaching is due to the reduction of an Fe-S center and the shifts to electrochromic effects on nearby chlorophyll *a* molecules. The ESR spectrum however presents band of *g* value (2.08, 1.90 and 1.78) uncharacteristic for a normal Fe-S protein [57]. The nature of X remains therefore still indetermined.

4.5. $A_{I,3}$: *iron sulphur centers as secondary acceptors*

At more positive potentials two other secondary acceptors have been discovered. The first observation demonstrating the presence of an Fe-S center in PSI particles, reducible in light at cryogenic temperatures (10 °K) occurred simultaneously using ESR spectroscopy (center A, $g = 2.05$, 1.94 and 1.86) [60] and optical spectroscopy

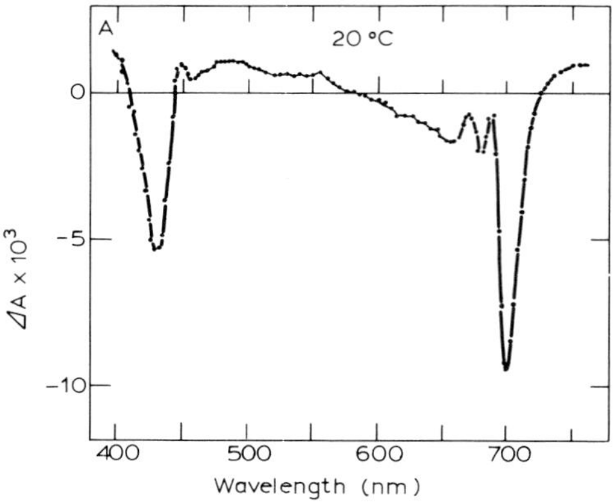

Fig. 4.3. Optical and ESR spectra of the electron carriers in the PSI-RC of higher plants and algae. (A) Primary electron donor ($D_{I,1}$): chlorophyll a dimer. (A) Light-dark difference spectrum of PSI-RC preparations (from Ref. 57).

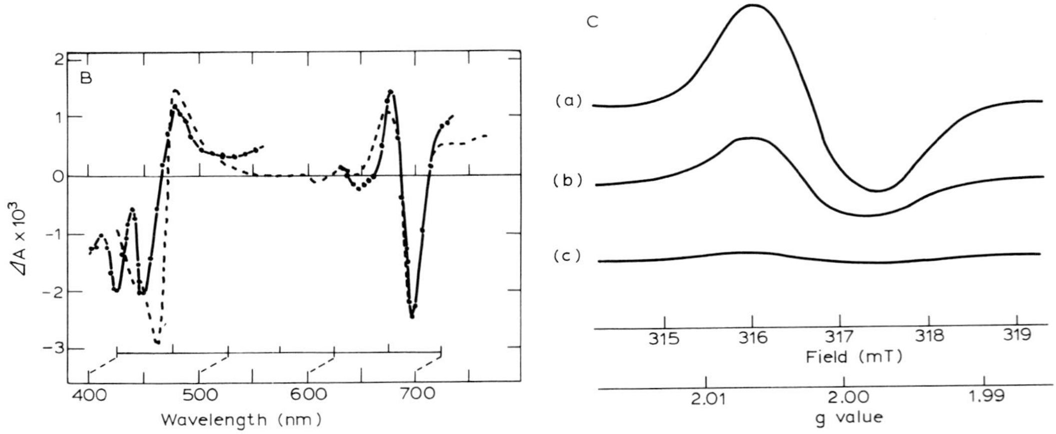

Fig. 4.3. (B and C) Intermediate electron acceptor ($A_{I,1}$): chlorophyll *a*. (B) Light-induced optical changes of $A^-_{I,1}$ obtained by subtracting the contribution of $D^+_{I,1}$. Dashed line: reduced-oxidized spectrum of isolated chlorophyll *a* anion (note the shift in the wavelength scale for this spectrum) (from Ref. 57). (C) ESR spectrum of $A^-_{I,1}$ (measured at 77 °K) and obtained by freezing dithionite-reduced samples during illumination; spectra b and c were taken in samples frozen 0 and 30 seconds after illumination (from Ref. 289).

(*P*-430) [61]. Later a second, different ESR signal (center B) could be observed only by illuminating at room, and not at cryogenic, temperature under suitable redox poise [62]. The attribution of these signals to the reduction of Fe-S centers comes from the spectral similarities with known isolated Fe-S proteins and from the

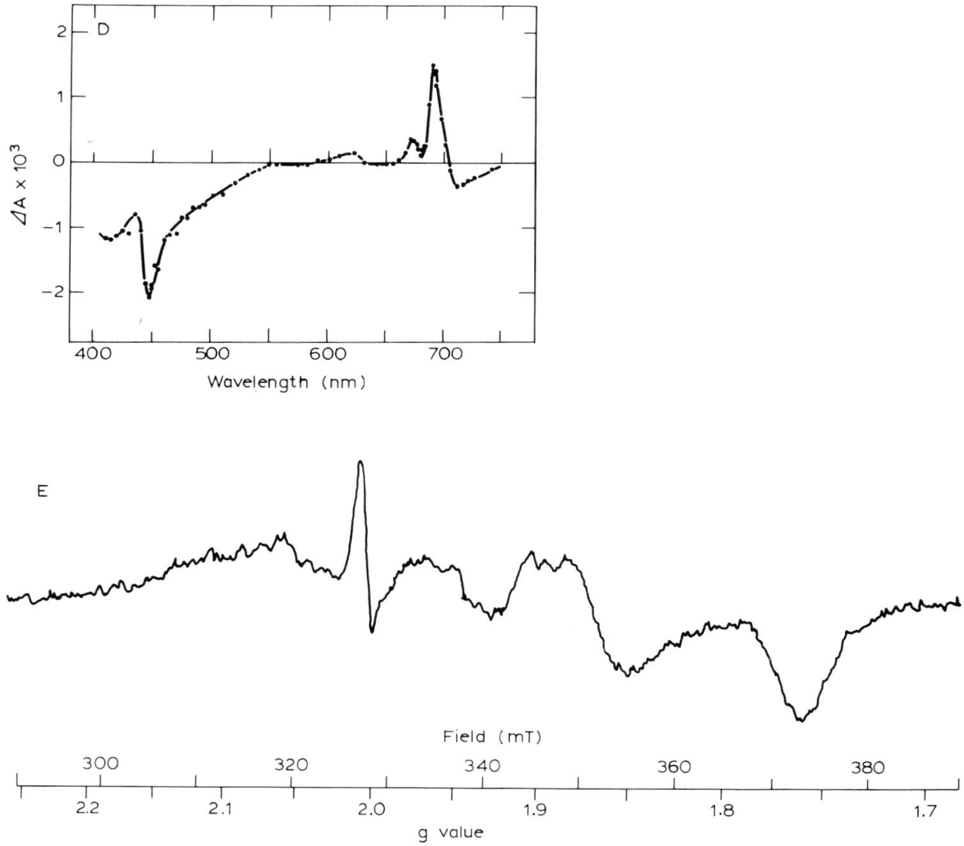

Fig. 4.3. (D and E) Primary electron acceptor ($A_{I,2}$): electron acceptor X. (D) Light-dark spectrum of the electron acceptor X in PSI-RC, recorded at 5 °C and −0.62 V — the broad band at 450–550 nm is attributed to an FeS center and the other changes to electrochromic effects (from Ref. 57). (E) ESR spectrum of X in PSI-RC, measured at 10 °K (from Ref. 290).

recognised presence in PSI-RC preparations of acid-labile Fe and S atoms [43]. The number of Fe per *P*-700 is uncertain and varies from a minimum of 4.6 to 10–12 per *P*-700 according to the preparation and the laboratories [42,63]. Their presence is clearly related, however, to the activity of PSI-RC (but not for the rapidly reversible oxidation of *P*-700 as observed in *P*-700-RC preparations), as demonstrated in studies of controlled removal of iron and sulphur from the complex.

The sequence of electron transfer steps from $A_{I,2}$ (X) to $A_{I,3}$ and $A_{I,3}$ to $A_{I,3}$ and $A_{I,3'}$ (Centers A and B respectively) is uncertain. Their midpoint potentials (−0.53 and −0.58 V for A and B, respectively [62,64]) are too close for a clear cut

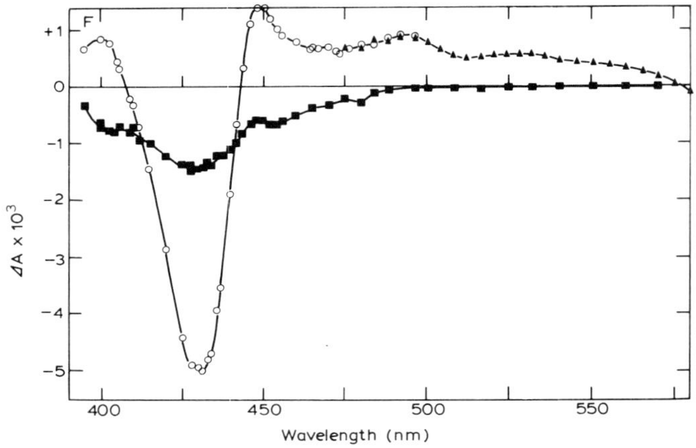

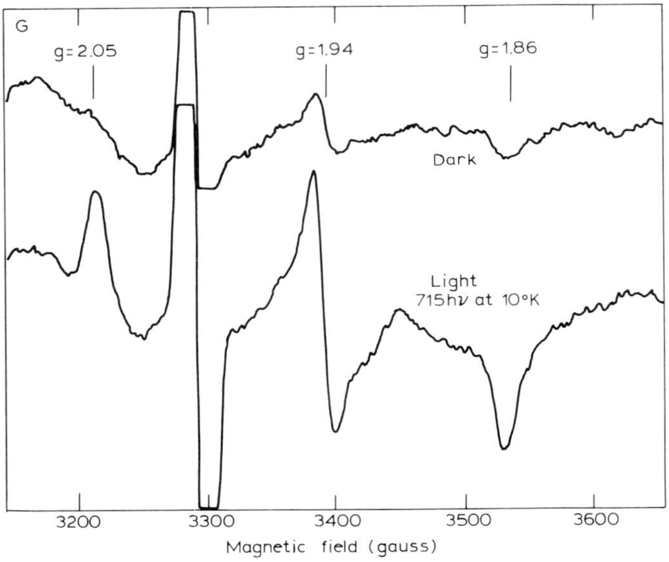

Fig. 4.3. (F and G) Secondary electron acceptor ($A_{I,3}$): FeS center A. (F) Light-dark optical spectrum of *P*-430 (■) ($A^-_{I,3}$). The second spectrum (○, ▲) is a part of the *P*-700 oxidized-reduced spectrum (from Ref. 61). (G) ESR spectrum in the dark and in the light of spinach chloroplasts. The light-induced changes are due to the reduction of FeS center A (from Ref. 58).

thermodynamic sequence, and kinetic studies on the appearance upon flash illumination, or disappearance by charge recombination, of the ESR signals are unfeasable. In fact, as mentioned above, center B, unlike center A, can be observed at cryogenic temperature but only in samples illuminated at room temperature, and freeze-quench techniques are too slow for time resolving this reduction; either centers A and B do not recombine with P-700 at cryogenic temperatures or do recombine extremely slowly [65]. According to Goldbeck and Kok [66], a progressive linear loss of activity of PSI-RC occurs concomitantly with the progressive removal of Fe-S with urea and ferricyanide (up to 10–12 Fe-S per P-700); this proportionality suggests a cooperative action of all Fe atoms in PSI-RC, perhaps organized in a single multicenter complex.

5. *Photosystem II of higher plants*

The activity of PSII can be clearly differentiated from the oxygen-evolving reaction and from the intermediary electron transfer to PSII. When the oxygen-evolving complex is inactivated by a variety of treatment, such as Tris or NH_2OH washing, alternative electron donors, such as, e.g., diphenyl carbazide, can be oxidized by chloroplasts or subchloroplast fragments; likewise electron acceptors specific for PSII have been identified. The light-induced electron transfer between a couple of such electron-donating and -accepting pairs defines the PSII activity during the attempts of fractionation of the membrane with detergents, and isolation of PSII-RC free of antenna pigments and other components of the electron transfer chain.

5.1. Polypeptide and pigment composition

In parallel with the attempts to purify PSI-RC, PSII-RC preparations were also developed: fractionations using Triton X-100 [67,68] or digitonin [69] were described, yielding preparations devoid of PSI activity. Possibly a pure PSII-RC preparation was obtained by Satoh [70] using digitonin associated with DEAE chromatography and isoelectric focusing. This preparation contained 50 molecules of chlorophyll *a* and retained a diphenyl-carbazide to dichlorophenol-indophenol electron transfer activity. As for PSI-RC and bacterial RC, a preparation of PSII-RC completely deprived of antenna pigments, has not been obtained so far. To this preparation, as well as to other less pure ones, a cytochrome of *b* type (cyt. *b*-559) [70] was found to be associated. PSII-RC contained three polypeptides, which gave three diffuse bands on SDS-PAGE, of respective molecular weights, 43 000, 27 000 and 6500; the smaller subunit is believed to be associated with cyt. *b*-559, and the larger one to the chlorophyll *a* of the PSII-RC. When urea (1–4 M) was added to the SDS-PAGE buffer the two bands of larger molecular weight could each be resolved into two components, formed by different polypeptides, as judged from fingerprint analysis [71]. One of the two intermediate polypeptides was identical to the 32 000 Da protein-binding triazine herbicides, a class of specific inhibitors of PSII activity [71].

5.2. $D_{II,I}$: chlorophyll a *as primary electron donor*

The first identification of an absorbance change associated with the primary electron donor of PSII was reported by Doring et al. [72]. A rapidly decaying bleaching at

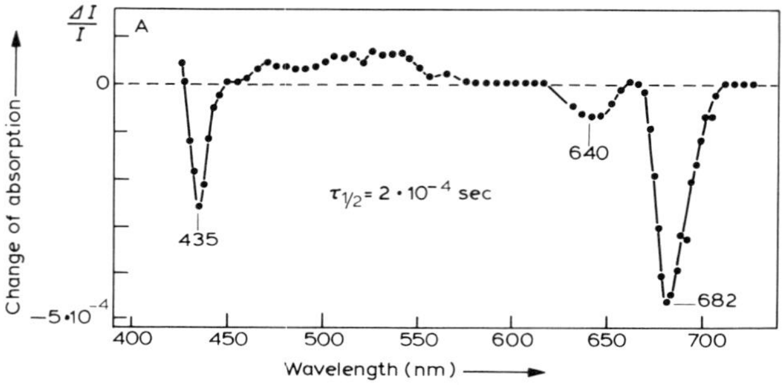

Fig. 4.4. Optical and ESR spectra of electron carriers in PSII-RC of higher plants and algae. (A) Primary electron donor ($D_{II,1}$): chlorophyll *a*. Light-dark differential spectrum of *P*-682 in chloroplasts obtained by plotting the extent of the fast-decaying component of the flash-induced optical changes including both *P*-682 and *P*-700 oxidation (from Ref. 291).

682 nm could be resolved from the overlapping *P*-700 signal by means of kinetic criteria ($t = 0.24$ ms for *P*-682 and 16 ms for *P*-700). The complete difference spectrum of *P*-682 photooxidation exhibits two negative bands at 682 and 435 nm, and a minor one at 640 nm [72]; it has been therefore attributed to a special molecule (or dimer) of chlorophyll *a*. The redox potential of *P*-682 has been determined to be 1.12 ± 0.05 V [73].

A light-induced ESR signal of *P*-682 resembling that of *P*-700 has been detected at cryogenic temperatures [74,75]; its spectral characteristics ($g = 2.002$ and line-width of 6–8 G) are similar to those observed for a ligated chlorophyll *a* cation radical in aprotic solvents [76]. It is unclear, therefore, if *P*-682 is a dimeric structure or a single chlorophyll *a* molecule ligated to a metal ion (see Fig. 4.4).

5.3. $A_{II,1}$: pheophytin a *as intermediate electron acceptor*

Similar to bacterial RC there is spectral and ESR evidence that a pheophytin *a* molecule operates as an intermediary electron acceptor in PSII-RC. Optical absorbance changes, with a spectrum similar to that of a pheophytin *a* anion radical could be detected in PSII-enriched particles illuminated at low redox potentials (-0.65 V) [57,77]. The appearance of the Ph signal could be correlated to a decrease in the extent of the rise in fluorescence of PSII of chlorophyll *a* observed upon illumination [78]. This apparent discrepancy (reduction of an electron acceptor is expected to cause an increase of fluorescence) is now explained by the fact that the fluorescence increase is in reality a delayed fluorescence emitted by the return to the ground state of P^*-682 regenerated by electron transfer from the pheophytin anion [79]. The lifetime of this transient fluorescence rise is 2–4 ns, and that of electron transfer from Ph^- to P^+-682 ≈ 4 ns, when PSII particles are poised at -0.45 V [73]. This transient fluorescence increase is, however, almost totally suppressed when $A_{II,1}$(Ph) is prereduced chemically before illumination. Using this experimental criterium the midpoint potential of the Ph^-/Ph couple has been estimated to be -0.61 V [73,80].

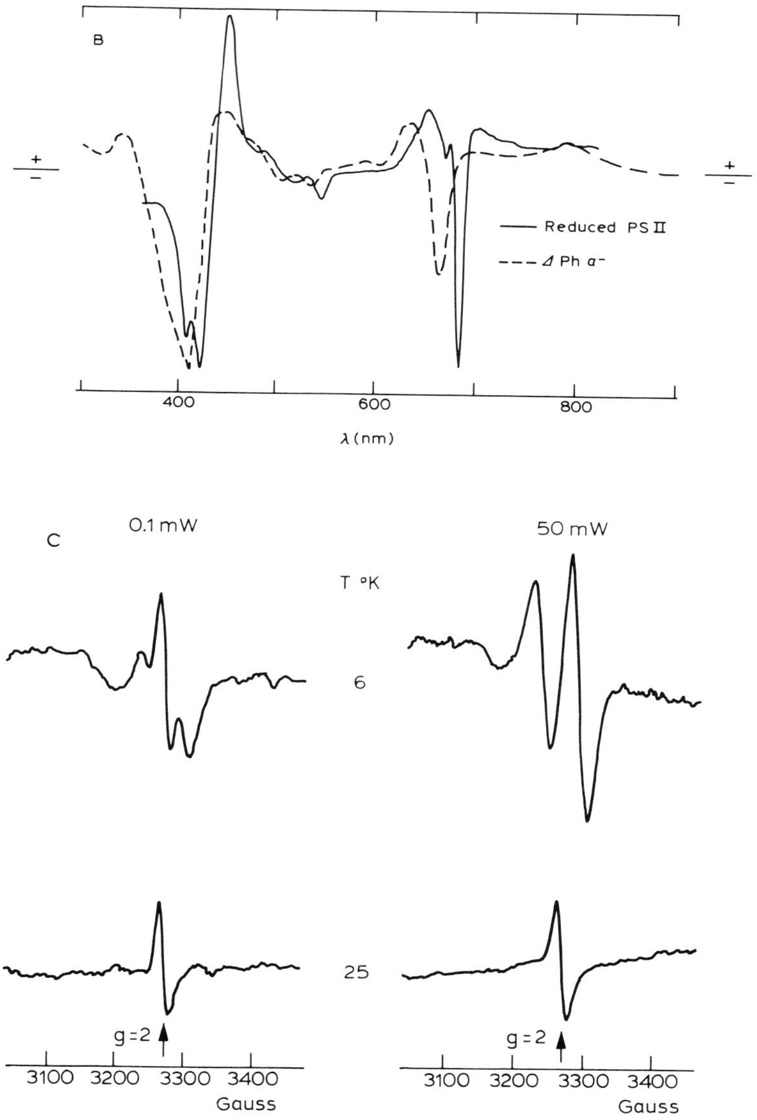

Fig. 4.4. (B and C) Intermediate electron acceptor ($A_{II,1}$): pheophytin *a*. (B) Light-induced optical changes of the intermediate electron acceptor, trapped under highly reducing conditions (−0.65 V); the signal is compared with the Ph^- −Ph differential spectrum (from Ref. 77). (C) ESR spectrum of $A_{II,1}^-$, trapped at 77 ° K and −0.45 V after preillumination at 220 ° K. The doublet appearing at high microwave power and at 6 ° K is attributed to an interaction of Ph^- with the PQ-Fe complex (cf. Fig. 4.2D, for analogy with bacterial RC) (from Ref. 82).

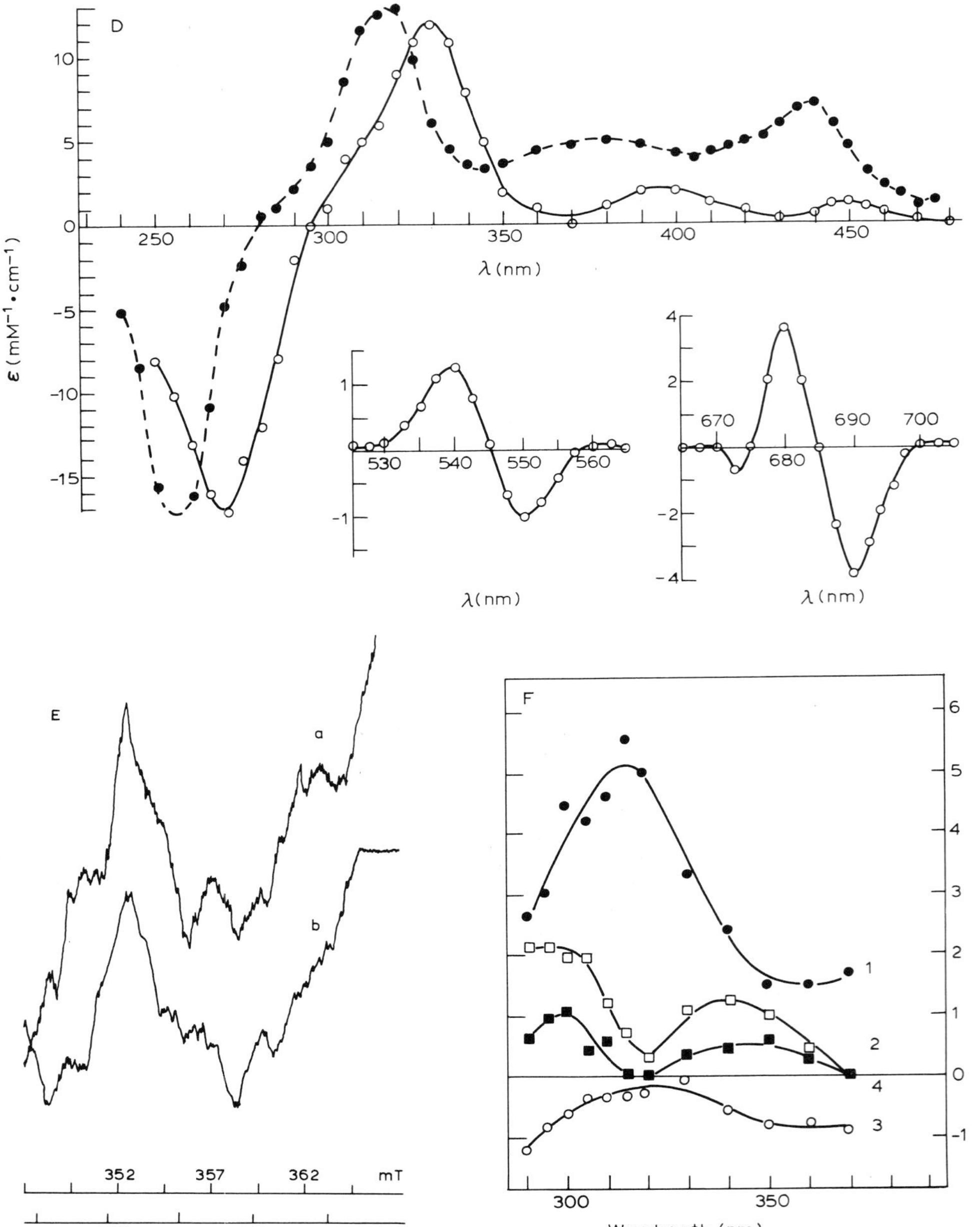

Fig. 4.4. (D and E) Primary electron acceptor ($A_{II,2}$): plastoquinone (or quencher Q). (D) Optical spectrum of the light induced plastosemiquinone anion (○), compared to the spectrum of PQ semiquinone in non-aqueous solvent (●). The additional spectral shifts at 545 and 685 nm are attributed to electrochromic effects on pheophytin (from Ref. 85). (E) ESR spectrum of $A^{-}_{II,2}$ reduced by light (a) or by dithionite (b) in *Chlamydomonas* PSII particles (from Ref. 87). (F) Secondary electron acceptor ($A_{II,3}$): plastoquinone. (F) Flash-induced optical changes due to the reduction of the secondary electron acceptor plastoquinone: the spectra oscillate in a dampened sequence following subsequent flashes indicating the production of semiquinone (1st and 3rd flash) and quinol species (2nd and 4th) (from Ref. 103).

The ESR spectrum of the Ph^- anion radical has been shown to have a $g = 2.0025$ and a linewidth of 12.5 G [57,81]. When PSII-RC were illuminated at 220 °K and ESR spectra taken at 6 °K, a split signal was observed at high microwave power (50 mW), but not at low power (0.1 mW) [82]. Similar observations, made in bacterial RC, were interpreted as due to an exchange interaction between BPh^- and the A_2FeA_3 complex [23]. A similar situation and geometry may therefore exist also in PSII-RC.

5.4. $A_{II,2}$: *plastoquinone as primary electron acceptor*

The existence of a stable electron acceptor in PSII has long been recognized indirectly by observing the fluorescence increase of chlorophyll *a* occurring upon illumination. This increase was attributed to the reduction of an acceptor and the consequent 'closing' of a PSII-RC. The acceptor was defined as Q (for quencher) and proposed to quench fluorescence when oxidized [83].

An actual absorbance signal of $A_{II,2}^-$ (previously designated X 320) was first observed by Stiehl and Witt [84]; the complete spectrum of the reduced acceptor was demonstrated by van Gorkom [85] to be very similar to that of a plastosemiquinone anion overlapping, in the green-red region of the spectrum with a band shift due to a neighbouring pheophytin *a*. This latter signal (previously designated C 550) [86] is again probably due to the interactions in close association between Ph and the plastoquinone [85].

The ESR signal of $A_{II,2}^-$ could be observed in membrane preparation from a strain of *Chlamydomonas reinhardii* lacking PSI. The broad peak at $g = 1.84$ is again similar to what observed in bacterial RC and suggests a plastoquinone-Fe complex [87]. Also similar is the inhibitory effect of *o*-phenanthroline [88].

Studies of extraction with apolar solvents and reconstitution with pure plastoquinone have offered the final demonstration of the chemical nature of $A_{II,2}$ [89]. Both *P*-682 photooxidation and the splitting of the ESR signal due to $A_{II,2}^-$ were eliminated upon extraction of the membrane with slightly polar solvents (e.g., hexane containing traces of methanol); both phenomena could be reconstituted with pure plastoquinone [89]. As a general conclusion all these studies converge to indicate a very close similarity between PSII-RC and bacterial RC in the nature and organization of the electron acceptor complex.

Redox titrations of the fluorescence rise, however, revealed a more complicated situation. The extent of fluorescence increase titrates as if two different acceptors were present: a first increase in fluorescence titrated with an $E_{m,7} \simeq 0$ V and a pH dependence of -60 mV per pH unit [90,91] indicating the presence of a $1H^+/1e^-$ carrier. This is consistent with the properties of the $PQH^{\cdot}/PQ$ couple. The pH dependency disappears above approx. pH 8.9, the functional p*K* of the $A_{II,2}$, at which the E_m was equal to -0.130 V [92]. This is believed to be the functional oxidoreduction potential of $A_{II,2}$ since, on the basis of the spectroscopic evidence for a plastosemiquinone anion radical, no protonation takes place during $A_{II,2}$ reduction due to the kinetic constraints on the accessibility of H^+ to $A_{II,2}$ [93].

A second fluorescence increase titrates with an apparent midpoint potential of -0.25 to -0.30 V [90,94] (at these potentials the acceptor has been designated Q_L, as opposed to the high potential Q_H). Also, the E_m of Q_L is pH dependent, suggesting that it could be another plastosemiquinone/plastoquinone couple. The presence of Q_L and Q_H could be related to the kinetic pattern of the fluorescence rise, which is clearly biphasic [95]. When the rate, and not the extent, of the fluorescence increase is studied as a function of the redox potential, the faster component of the rise correlates with Q_L and the slower with Q_H [96,97]. Q_L seems therefore to be the most efficient quencher; on the other hand Q_H is usually the more prominent component of the fluorescence rise and correlates with the electrochromic signal *C*-550 linked to the reduction of $A_{II,2}$ [98].

5.5. $A_{II,3}$: *plastoquinone as tertiary electron acceptor?*

An oscillatory behaviour of the electrons avsailable to PSI from PSII has been observed in chloroplasts or whole algal cells excited with short flashes of light [99]. This behavior has been interpreted, in analogy with the bacterial system, as due to the presence of a stable plastoquinone anion radical ($A_{II,3}$ or B) not easily reoxidized unless a second electron would reduce it to plastoquinol in a two electron gate mechanism [100]. This conclusion was supported by studies on PSII fluorescence, which revealed oscillations in the rate of decay of the high-fluorescence yield due to $A_{II,2}$ reduction, being the rate from $A^-_{II,2}$ to $A_{II,3}$ ($t_{1/2} = 200$–300 μs) faster than that from $A^-_{II,2}$ to $A^-_{II,3}$ ($t_{1/2} = 600$–800 μs) [101]. Spectroscopic signals at 320 nm assigned to $A^-_{II,3}$ were detected in PSII particles blocked in the donor side of PSII [102,103]; also these signals showed oscillations with maxima at odd flashes.

The overall mechanism of electron transfer proposed from PSII to the plastoquinone (PQ) pool is therefore

$$A_{II,2}A_{II,3} \xrightarrow{h\nu} A^-_{II,2}A_{II,3} \rightarrow A_{II,2}A^-_{II,3} \xrightarrow{h\nu} A^-_{II,2}A^-_{II,3} \rightleftharpoons A_{II,2}A^{=}_{II,3}$$

$$A_{II,2}A^{=}_{II,3} + 2H^+ + PQ \rightarrow A_{II,2}A_{II,3} + PQH_2$$

where PQ stands for a molecule of the plastoquinone pool. However, this scheme must still be considered a working hypothesis, which does not accomodate the low potential secondary acceptor (Q_L) and in which the pattern of protonation is still undefined.

5.6. $D_{II,2}$: *the secondary donor to PSII*

A 25 ns induction phase in the fluorescence rise observed in dark adapted *Chlorella* cells was originally interpreted as due to the reduction of *P*-682 ($D^+_{II,1}$) by a secondary electron donor to PSII ($D^-_{II,2}$ or Z) [104,105]. These observations were later substantiated in dark adapted chloroplasts (rapid transient signal $t = 25$–45 ns) and by the observation that inactivation of PSII donor site with NH_2OH resulted in a much longer lag phase ($t_{1/2} = 20$ μs) [105].

The study of the exact chemical nature of $D_{II,2}$ has been prevented so far by the extremely high redox potential of this species, which must be able to accept electrons from the water-splitting complex and donate them to P^+-682 ($E_m = 1.1$ V) [71]. An ESR signal attributed to $D_{II,2}$ has been identified [106,107] and proposed to be due to a special plastohydroquinone, with the ESR detectable species being the oxidized cation radical [108]. When the oxygen evolving complex is inactivated, the rise kinetics of the ESR signal ($t_{1/2} = 1$ ms) matches the decay of $D^+_{II,i}$ demonstrating that at least under these inhibited conditions, the immediate reductant of P^+-682 is observed [108]. The oxidoreduction potential of such a cation radical has been estimated to be sufficiently positive to oxidize the oxygen-evolving complex.

6. *The cytochrome* b/c$_1$ *complex*

6.1. *General remarks*

The intermediate electron transfer between the pool of quinones accepting electrons from the RC, and the water soluble proteins donating electrons to the RC (bacterial RC and the PSI-RC) is always promoted, at least in the systems studied so far in detail, by a multiprotein complex containing cytochromes and Fe-S proteins, the so called b/c_1 complex. The universal presence of this type of complex in many redox chains of respiration and photosynthesis has been recognized only very recently [109]. As far as photosynthesis is concerned, complexes of this kind have been characterized in facultative photosynthetic bacteria [110] in cyanobacteria [111], and in higher plant chloroplasts [112]. All these preparations share common characteristics and composition; these properties are also very similar to those of analogous complexes isolated from mitochondria of mammals and fungi [109].

The general function of this complex is that of transferring electrons from ubiquinone (or plastoquinone) to a hydrophilic protein acceptor (cytochrome *c* or plastocyanin). Therefore, in bacterial photosynthesis, it catalyzes the recycling of electrons from the secondary electron acceptor (Q_{II}) to the secondary electron donor (cyt. c_2), completing thereby the cyclic electron transfer system. In chloroplasts and cyanobacteria, an analogous system transfers the electrons from plastoquinone (the secondary acceptor of PSII, $A_{II,3}$) to plastocyanin (the secondary donor to PSI, $D_{I,2}$) and provides in this way an intersystem redox connection between PSII and PSI. The same complex is also involved in the cycling of electrons around PSI.

The general composition of the complex in all systems studied so far is also universal; they always contain two *b*-type cytochromes, one cytochrome of *c* type and a high potential Fe-S protein (the Rieske protein, so called after its discoverer in Complex III of the respiratory chain of beef heart mitochondria). In addition to these functions in electron transfer, the b/c_1 complexes also play a role in energy transduction, since they represent an essential part of the proton translocating apparatus of photosynthetic electron transfer chains.

The components of the quinol-cytochrome *c* (plastocyanin) oxidoreductase of chloroplasts, cyanobacteria and photosynthetic bacteria have been demonstrated to be very similar. This analogy proves the substantial unity of the mechanism of electron flow in all photosynthetic systems. For this reason the different components of the complexes will be discussed unitarily in the following sections in order to emphasize the functional and structural similarities between them.

6.2. Isolation procedures and properties of the complexes

6.2.1. The ubiquinol-cytochrome c *oxidoreductase of photosynthetic bacteria*

A b/c_1 complex endowed with a ubiquinol-cytochrome *c* oxidoreductase activity has been recently isolated from the membranes of *Rps. sphaeroides* GA [110,113], following micellization with octylglucoside and cholate, fractionation with $(NH_4)_2SO_4$, and separation by sucrose gradient centrifugation [110]. The catalytic activity has been estimated to be about 30–50% of that in the whole membrane, though the kinetic comparison is made uncertain due to the micellar nature of the complex preparation and of the electron donor ubiquinol; this activity is still fully sensitive to specific inhibitors of the b/c_1 segment of the chain (antimycin A, UHDBT, DBMIB).

The complex can be resolved on SDS-PAGE in three major bands of molecular weights 40 000, 34 000 and 25 000, plus a minor heterogeneous and diffuse one of 6000. Heme staining of the gels allows to clearly associate band 34 000 with a hemoprotein (cyt. c_1) and, with lower certainty, band 40 000 with cyt. *b*. The stoicheiometry in which these subunits are present in the complex is not known yet; the isolated preparation behaves on gel chromatography as a monomer of molecular weight around 100 000.

6.2.2. The b_6/f *complex of higher plant chloroplasts and cyanobacteria*

The recognition of the presence of a b/c_1 complex in higher plant chloroplasts followed immediately the attempts of fractionation of their membrane with digitonin, and the report that a complex containing cytochrome b_6 and cytochrome *f* could be obtained by sucrose density gradient fractionation. The first reported preparation of the b_6/f complex, obtained by digitonin fractionation, protamine precipitation and further purification by column chromatography, contained cytochrome b_6, cytochrome *f* (and traces of cytochrome *b*-559 as well) and an Fe-S protein [114]. The analogy of this complex with the b/c_1 complex of mitochondria was immediately recognized. This preparation was enzymatically inactive.

The isolation of an active, structurally intact complex was obtained using an association of cholate and octylglucoside and sucrose gradient centrifugation [111]. This preparation did not contain cytochrome *b*-559 and possessed a plastoquinol-plastocyanine oxidoreductase activity, inhibited by specific inhibitors (DBMIB, UHDBT). The complex was essentially free of chlorophyll and contaminations by other membrane components, specifically of the ATPase complex.

On SDS-PAGE the b_6/f preparation could be resolved into four to five subunits of intermediate molecular weight, plus one to three small components of molecular weight around 5000. The larger components could be assigned to specific electron carriers: cytochrome b_6 could be identified with a band of molecular weight 23 000; cytochrome f with a double band of 33–34 000 (reports of heterogeneity of cyt. f have been published); and the Rieske Fe-S protein with one at 20 000 [115]. An additional band at 17 000 Da is also present with an undefined assignment.

The stoicheiometry of the subunits has been estimated to be 1 : 1 : 1 : 1 on the basis of Coomassie blue staining intensities [115]; a stoicheiometry of 1 : 2 : 1 : 2 was however obtained by UV absorption scanning of the gel, indicating a possible differential staining efficiency of the bands [115]. Based on the estimated molecular weight and on the amount of cyt. f in the preparation (7.25 nmol/mg protein), a monomeric structure of the complex solubilized in Triton X-100 of molecular weight around 160–180 000 was proposed [111]; a dimeric structure within the membrane was, however, suggested by the size of the protein particles observed by freeze fracture electron microscopy of proteoliposomes containing the b_6/f complex [109].

An analogous complex could be obtained with a similar procedure from the thylakoids of the cyanobacterium *Anabaena variabilis* [112]. This complex, endowed with a similar catalytic activity, can utilize either plastocyanin or cytochrome c (purified from cyanobacteria) as electron acceptors. It is formed by four to five subunits (M_r 31–34 000 (cyt. f), 22 500 (cyt. b_6), 22 000 (Fe-S protein) and 16 000) plus a heterogeneous band around 8000 Da.

6.3. Cytochromes of b *type*

Cytochromes of b type are invariably involved in electron transfer in photosynthetic bacteria and in plant chloroplasts and cyanobacteria, and take active part in the mechanism of the quinol-cytochrome c (plastocyanin) oxidoreductase complex.

In purple photosynthetic bacteria, and specifically in *Rps. sphaeroides* and *Rps. capsulata*, three cytochromes of b type have been identified by means of redox titration, in the dark, of isolated chromatophores [116]. They are characterized by midpoint potentials at pH = 7.0 equal to 0.155, 0.050 and −0.090 V (in *Rps. sphaeroides*); the E_m of the 0.050 V species is pH dependent (−60 mV per pH unit) [116,117]. The presence of a cytochrome cc' in these organisms, interfering spectrally with cytochrome b, makes the situation unclear as far as the existence of cyt. b ($E_{m,7} = 0.155$ V) is concerned [118]. The two other cytochromes ($E_{m,7} = 0.050$ and −0.090 V) have also been resolved kinetically in studies on the photosynthetic electron transport and on the basis of their spectral characteristics (band at 561 nm and a split bands at 558 and 556 nm, respectively; these two cytochromes will be referred to as b-561 and b-566 in the following) [119].

In the isolated oxidoreductase only two of the cytochrome b species are retained, cyt. b-561 and b-566. Both their spectral and thermodynamic features are maintained. These two electron carriers have been identified with the 40 000 Da subunit of the isolated complex; it is not clear, however, whether the two hemes are carried by two different apoproteins with very similar molecular weights or rather by a single polypeptide [110]. This latter possibility is suggested by the stoicheiometry observed in other oxidoreductases (mitochondria and chloroplasts) whose structural and functional analogy is now evident.

A very similar situation exists in chloroplasts and cyanobacteria. Two (and possibly three, as will be seen below) cytochromes of *b* type have been identified [120]: a cytochrome with band at 563 nm (cyt. b_6) and one at 559 nm (cyt. *b*-559). In the active purified oxidoreductase, solubilized with cholate-octylglucoside, only cyt. b_6 is retained [111]. Cyt. *b*-559, that was found associated with this complex in other less pure preparations, is also found in many preparation of PSII-RC. Its role in the plastoquinol-plastocyanin oxidoreductase seems therefore to be excluded, and will not be further discussed in this section. Titrations of the oxidoreductase complex have indicated the thermodynamic heterogeneity of cyt. b_6, with two components of apparent $E_{m,7} = -0.050$ and -0.170 V, respectively [115]; these midpoint potentials were pH dependent between pH 6 and 9. Both forms of cyt. b_6 present a band at 563 nm. Again both cytochromes appear to be associated with the 23000 Da subunit of the chloroplast b_6/f complex (22500 in *A. variabilis*). Many observations indicate that, both in the membrane and in the isolated complex, the properties of cyt. b_6 are rather labile and can vary in response to detergents or aging [109].

6.4. Cytochromes of c *type*

Soluble forms of cytochrome of *c* type are present in bacteria and, under some conditions of growth, in cyanobacteria; these are considered to act as diffusable redox carriers between the oxidoreductase complexes and bacterial RC or PSI-RC, respectively. A similar role is assumed by plastocyanin in chloroplasts and in cyanobacteria (again dependeing on the conditions of growth plastocyanin biosynthesis can be repressed and replaced in its function by cyt. *c*). These electron carriers represent, therefore, the natural electron acceptors of the oxidoreductases and the secondary electron donors to the RC (D_2 in bacteria and $D_{I,2}$ in chloroplasts).

A second bound form of cytochrome *c* is an integral part of the oxidoreductase complexes. Cytochrome c_1 present in photosynthetic bacteria has been distinguished from cyt. c_2 (the soluble electron carrier) both thermodynamically and kinetically [121,122]. It is present in the isolated oxidoreductase with a stoicheiometry of one per two cytochromes of *b* type, and it is associated with the 34000 Da subunit. According to kinetic evidence this cytochrome acts as immediate electron donor to cyt. c_2 and electron acceptor from the high potential Fe-S protein [122]. The midpoint potential of cyt. c_1 is 0.285 V at pH 7 [121,122].

Cytochrome *f*, the first cytochrome identified in chloroplasts, is found in the purified oxidoreductase with a stoicheiometry of one per two cyt. b_6, and associated with the 33–34000 Da heterogeneous band (34000 in *A. variabilis*) [111,115]. The nature of the splitting of the band associated with the cyt. *f* subunit can be purely artifactual (e.g., due to a limited proteolysis) since a single structural gene codifies for cyt. *f* in the chloroplast genome. It is now established that cyt. *f*, analogous to cyt. *c* in bacteria, is the electron acceptor from the Fe-S protein and the electron donor to plastocyanin. Its midpoint potential has been reported to range between 0.4 and 0.37 V and is pH dependent only above pH 9 [120,123,124]. Soluble homogeneous preparations of cyt. *f* have been described; a general good agreement exists between the properties of the purified cytochrome and those detected in the purified oxidoreductase or in the intact membrane.

6.5. The high-potential Fe-S protein (Rieske protein)

A band with a molecular weight of 25 000 of the bacterial oxidoreductase has been identified with the high-potential Fe-S protein, by means of cross reaction with a monospecific antibody against the analogous electron carrier from *Neurospora crassa* mitochondria. The existence of this type of Fe-S center in photosynthetic bacteria was first discovered by ESR spectroscopy [125] and its involvement in photosynthetic electron transport was demonstrated. The midpoint potential in *Rps. sphaeroides* is 0.285 V, and is pH dependent above pH 8, with a decrease of 60 mV per pH unit [125].

The *g* values of the ESR spectral lines ($g = 2.02$, 1.89 and 1.81 in mitochondria) have been ascribed to an Fe_2S_2 (S Cys)$_4$ center, i.e., containing two tetracoordinated iron atoms bridged by two sulphurs [126]. Similar spectra can be observed in chromatophores or in the isolated complex. The high-potential Fe-S protein appears to be the carrier involved in the binding of specific inhibitors of the oxidoreductase, such as UHDBT and DBMIB, which are structurally analogues of ubiquinone [127].

An Fe-S protein has been found to be present also in the b_6/f complex of chloroplasts and cyanobacteria. This protein has been dissociated from the complex with Triton X-100 and hydroxyapatite column chromatography, and was shown to be associated with the 20 000 Da subunit in chloroplasts [111] and the 22 000 Da one in *A. variabilis* [128]. In all cases the dissociation resulted in an irreversible loss of activity; the involvement of the Fe-S protein in electron transport was also proved by the inhibition by an antibody raised against the Triton isolated protein (but not by one against the SDS denatured subunit) [129,130]. An oxidoreduction potential of 0.290 V was measured in intact chloroplast membranes and in the complex or in the isolated homogeneous preparation; the potential was pH independent below pH 8 [111].

ESR spectra of chloroplast of the isolated complex or subunit are consistent with the presence of a Fe_2S_2 center [129]. The Rieske proteins appear to be the target of many specific inhibitors of the complex: the plastoquinone analogues UHDBT, DBMIB and DNP-INT, all potent inhibitors of the plastoquinol-plastocyanin oxidoreductase activity, alter the ESR spectrum [131, 132]. Moreover, UHDBT shifts the midpoint potential of the FeS center to higher values, indicating its higher affinity for the reduced form. In the case of DBMIB the change in the ESR spectrum of the chloroplast protein is reversed by plastoquinone [132], suggesting a role of the FeS protein in the plastoquinol oxidizing site.

6.6. The mechanism of electron transfer within the b/c_1 *complex*

The key observation at the basis of the current interpretations of the mechanism of electron transfer in the b/c_1 complex was made in 1966 by Baum and Rieske [133]; these authors observed that addition of an oxidant (ferricyanide) to antimycin A-inhibited mitochondria brought about an unexpected reduction of cytochromes of *b* type. This apparently paradoxical behaviour led Mitchell to propose (further elaborating ideas of Wikström and Berden [134]) the scheme of the so called 'Q cycle' [135] and, subsequently, Wikström and colleagues that of the 'b cycle' [136]. In both schemes the reduction of *b* cytochromes by oxidant addition is recognised as the consequence of the generation of a strongly reducing species, a semiquinone, upon subtraction of one electron from the mild reductant fully reduced ubiquinol. The theoretical difference in midpoint potential between the $QH_2/QH^{\overline{\cdot}}$ and the $QH^{\overline{\cdot}}/Q$ couples is related to the equilibrium constant of the semiquinone dismutation reaction and is larger the more stable is the semiquinone radical [93]; with this mechanism, the delivery of one electron to the oxidizing branch of the complex (the sequence of the FeS protein and of the two cytochromes *c*, or plastocyanin) will result in the possibility of reducing more negative electron acceptors, cytochromes *b* [135]. The fate of this second electron is different according to the Q cycle model as compared to the *b* cycle scheme: in the former, oxidized ubiquinone is reduced to semiquinone in a separate site of the complex and the semiquinone is returned to the pool for further reduction by independent electron donors (e.g., photosynthetic RC); in the latter, another semiquinone molecule is reduced back to quinol on the same site of quinol oxidation, thus returning again one half of the equivalents to the quinone pool. Specific inhibitors of the oxidoreductase are thought to act at different steps of this reaction sequence. Thus, antimycin A is considered to block the reduction of quinone by cytochrome *b*; UHDBT and DBMIB the oxidation of the Fe S center by cytochrome $c_1(f)$; and myxothiazol the reduction of cytochrome *b* by the FeS protein (see Ref. 109 for review).

Several observations made in whole membranes or in the isolated complexes are in line with these concepts: the shifts induced by antimycin A [110,137] and myxothiazol on the absorption spectra of cytochromes *b* and the alterations of the ESR spectrum of the FeS protein by UHDBT or DBMIB [131]. Moreover, the oxidant-induced reduction of cytochromes *b*, the key observation for accepting these electron transfer schemes, has been demonstrated in all b/c_1 complexes isolated so far from mitochondria [134], chloroplasts [111], cyanobacteria [112] and photosynthetic bacteria [110]. In the chloroplast b_6/f complex this reaction has been demonstrated also in the absence of any exogenously added quinol, indicating that possibly a structurally bound quinone (quinone is always present in the isolated complexes with a stoicheiometry of about 0.5–0.7 mol/mol of cyt. c_1 [110,111]) is sufficient to drive the reduction of cytochromes b_6 [138]. Since a detailed treatment of the general mechanism, as well as of the more specific problems of the mitochondrial respiratory chain, are reported in Chapter 3 of this volume, the following discussion will deal only with the specific features of the electron transfer chains in photosynthetic membranes.

In photosynthetic bacteria, and particularly in chromatophores of *Rps. sphaeroides*, the kinetics of electron transfer in the b/c_1 segment of the photosynthetic cycle have been studied in extreme details [116,139,140]. These studies have been made technically easier by the special characteristics of bacterial photosynthesis: (a) the photochemical mechanism of the primary redox reactions, which allows the activation of RC in single turnover flashes ($t_{1/2} < 10$ μs), and consequently the control of the electron transfer through the complex; (b) the cyclic nature of chain, which permits on one hand the performance of experiment in anaerobiosis and under perfectly controlled redox poise, and on the other the delivery of one electron from the RC secondary acceptor (Q_{II}) and the simultaneous withdrawal of one electron by the secondary donor to the RC (cyt. c_2) with the promotion of a single turnover of the oxidoreductase. In this way the redox kinetics within the b/c_1 complex can be studied in single turnover with an excellent time resolution. Early studies of this sort have indicated the central role of an electron carrier in controlling the kinetics of reduction of cytochromes *b* and *c*. This carrier was characterized by an apparent midpoint potential of 0.155 V at pH 7, $n = 2$ and a pH dependence of -60 mV per pH unit [137,141]; all properties consistent with those of a ubiquinone molecule, except that the midpoint potential was displaced to more positive values as compared to that of the large pool present in the membrane ($E_{m,7} = 0.090$ V at pH 7, $n = 2$ and -60 mV per pH unit [142]). The existence of a special, bound molecule of ubiquinone (Q_z), acting as reductant both of the two cytochrome *b* sequences (*b*-566 → *b*-561) [143] and of the of cyt. c_2 (in a carrier sequence including the FeS center and cyt. c_1), was postulated [122,127].

The nature of Q_z was confirmed by extraction reconstitution experiments [143,145]. Q_z would correspond, therefore, to a special Q molecule bound to the site of electron donation to the b/c_1 complex. The proposed electron transfer schemes following these observations can be considered as the bacterial photosynthesis counterparts of the 'Q cycle' [139] or of the 'b cycle' [146], although linear arrangements have also been considered.

More recently a modified Q cycle scheme has been proposed by Crofts and colleagues [38], in which the actual existence of Q was invalidated. The apparent displacement of the midpoint potential of Q_z as compared to the Q pool was interpreted as due to the large size of the pool, so that a sufficient supply of electron to the oxidoreductase could be provided by a small reduced fraction of the pool. In this scheme the RCs are proposed to function in pairs as compared to the b/c_1 complexes, so that per flash a doubly reduced quinol is available to the oxidoreductase; the quinol is then oxidized at the Q_z site (now visualized as a site in rapid exchange equilibrium with the pool) in a concerted reaction on the FeS center, reducing the *b* cytochrome sequence and cytochrome c_1. After the oxidation of two quinols two electrons are delivered to the secondary electron donors of two RCs. The other two electrons are utilized to reduce a molecule of oxidized Q in a second reducing site of the oxidoreductase, either following two turnovers of cyt. *b* reduction, or via the concerted action of a dimer of the b/c_1 complex (evidence for a dimeric structure of the complex is available for *N. crassa* mitochondria and for chloroplasts). In this way one half of the equivalents are returned to pool according to a Q cycle scheme. This model, which has been quantitatively supported by many measurements of the kinetics of electron transfer, is discussed 'in extenso' in Ref. 38; it represents today the most detailed kinetic study of the mechanism of electron transfer within a b/c_1 complex.

The rate of electron transfer within the b/c_1 complex matches kinetically the appearance of a large electrochromic signal of endogenous carotenoids, monitoring an electrogenic step in the electron pathway [147]. The amplitude of this slow signal can be equal, under appropriate redox conditions, to that of the signal associated with the primary photosynthetic reactions, believed to occur across the entire thickness of the membrane. For this reason the b/c_1 complex has been proposed to promote an electron transfer through the membrane during one step of its catalytic mechanism, possibly coincident with the oxidation of cytochrome *b*-561. A proton translocating activity has also been associated with the oxidoreductase [148]; evidence for the translocation of two protons per electron has been obtained experimentally in proteoliposomes containing all known preparations of b/c_1 complexes [149]. In intact chromatophores, proton binding of slow risetime inhibited by antimycin A (H^+_{II}) and therefore associable with the b/c_1 complex can be demonstrated, but only in the presence of valinomycin and K^+ collapsing the transmembrane electrostatic potential [148].

Evidence for the existence of a 'Q cycle' mechanism also in the intersystem chain of higher plant chloroplasts has been obtained; this evidence rests, however, on more indirect experimental approaches, such as a slow phase of the electrochromic shift of carotenoids and an H^+/e^- stoicheiometry of proton translocation higher than one, as would be postulated by a linear loop including plastoquinone as a transmembrane proton translocator.

A slow electrochromic signal, resembling the one associated with the b/c_1 complex in photosynthetic bacteria, has also been observed in whole *Chlorella* cells [150] and in chloroplasts. The signal could be associated on the basis of its response to specific inhibitors of cyclic (antimycin A) [150,151] and non-cyclic (DCMU) electron transfer, or on the basis of action spectra [152], both to the action of PSII or of PSI. The appearance of the slow carotenoid signal is, however, dependent on the light intensity (being absent at high light intensities when the transmembrane proton gradient is large) [153] and on the appropriate redox poise of the assay medium [154]. The signal was also inhibited by DBMIB [155]. By studying the response to the redox ambient potential, Bouges-Bocquet [154] identified an electron carrier (U), with an $E_{m,7} = 120–160$ mV, whose prereduction was necessary for observing a slow electrochromic signal. U appears to share many properties with Q_z of photosynthetic bacteria, and has been proposed to be a special form of quinone bound to the b_6/f complex. Under the same experimental conditions for observing the slow electrochromic signal, an extra proton is translocated through the membrane per electron transferred between PSII and PSI [156–158], again in line with the operativity of a Q cycle mechanism. Moreover, direct measurements of the reduction of cyt. b_6 indicated that this carrier can be reduced either by PSI, via ferredoxin [159], or by PSII in a DCMU-sensitive pathway.

It appears, therefore, that the electrons delivered either by PSII or cyclically by PSI can enter in the mechanism of the Q cycle and generate additional charge translocation across the membrane and additional proton translocation. This conclusion is supported by many experimental analogies in the intact membrane with other electron transfer chains and by the extreme similarity of the b_6/f complex with the oxidoreductase of the other systems in which the Q cycle could be more convincingly demonstrated. The utilization of the Q cycle pathway appears, however, to be facultative in chloroplasts and algae and to depend on the presence of a low transmembrane $\Delta\tilde{\mu}_{H^+}$ and a correct redox poise, possibly of the plastoquinone pool. The mechanism by which the bypass of the Q cycle pathway can be regulated is still obscure. Equally undefined is the pathway to the plastoquinol-oxidizing site of the oxidoreductase for the electrons derived from PSI. This point will be further discussed when considering the interaction between complexes.

7. *Oxygen-evolving complex*

7.1. *General remarks*

Among the partial reactions of photosynthetic electron transport, the redox coupling between the reaction center of PSII and the oxygen-evolving complex is the most labile. A variety of treatments that do not impair electron flow from added electron donors to PSII and leave intact the associated activities on the reducing side of PSII, completely destroy oxygen evolution. As a consequence of the fragility of the oxygen-evolving site, the isolation of a specific complex responsible for water-splitting has not been possible so far. Despite this difficulty there are a number of observations, mainly based on extraction and reinsertion procedures, supporting the central role of a manganese protein in the water-splitting process. Most likely the O_2-evolving component involves changes in the Mn oxidation state and in the oxidation level of bound water.

In recent years there have been many attempts to isolate PSII proteinaceous components containing Mn and possibly restoring the O_2 evolving ability. It has been reported on proteins [162,163] partially able to restore O_2 evolution in protein-depleted vesicles; the Mn associated to these preparations was however insufficient to account for the Mn content correlated to O_2 evolution in the native membranes. By washing inside-out thylakoids with 250 mM NaCl, Åkerlund [164] has extracted a 23 kDa protein free of Mn but reconstituting O_2 activity. Several proteinaceous components, not all associated with Mn, could be involved in the oxidizing side of PSII.

7.2. *Involvement of manganese and other cofactors*

It is well known from nutritional deficiency studies that manganese is required for the O_2-evolving activity [165], and generally involved with the activity of PSII [166].

The Mn content of well washed chloroplast preparations from plant and algal sources is approximately 5–8 atoms per 400 Chl molecules [167]. These bound atoms of Mn are not readily released from chloroplasts upon washing with metal complexing agents. Treatments that inactivate, primarily, the O_2-evolving site, such as gentle heating [168], washing with NH_2OH or chaotropic agents [169], and with alkaline Tris [170] have been shown to result in Mn depletion of various degrees. Cheniae and Martin [169] established that two-thirds of the Mn bound to PSII centers (4–5 Mn atoms per 400 Chl) could be removed by these treatments and found a concomitant linear decrease in oxygen evolution. The remaining third finally became tightly bound to the centers and did not appear to be involved with electron transfer. Manganese ESR signals, indicative of $Mn^{2+}(H_2O)_6$ and not present in untreated chloroplasts, can be observed after Tris treatment [171]. Centrifugation indicates that the Mn responsible for the ESR signal is released within the inner space of intact thylakoids and diffuses across the thylakoid membrane with $t_{1/2}$ of a few hours. Utilizing this ESR signal to monitor the debinding of Mn, Blankenship et al. [172] found a complete block of O_2 evolution when 60% of the total chloroplast-bound Mn was removed. Their observation could also explain, in terms of the permeability properties of different preparations, some discrepancies in the evaluation of the Mn pool required for O_2 evolution [173,174].

The most convincing arguments supporting the idea that Mn acts at the O_2 evolving site come from reconstitution experiments. In algae grown in Mn-deficient media [175,176] or extracted with NH_2OH [177] the oxygen evolving activity can be restored through a light-dependent incorporation of Mn. The reconstitution process includes (a) an accumulation of Mn^{2+} within the thylakoids which can occur in the dark but which is stimulated by light and inhibited by uncoupling agents [178], (b) a photoreactivation catalyzed by PSII-RC which promotes the binding of Mn to the site responsible for O_2 evolution. The rate of photoreinsertion, which is a function of Mn concentration, displays a first-order dependence on the number of inactive O_2-evolving complexes [177]. Reactivation by the above-mentioned procedure is paralleled by the progressive disappearance of the ESR signal attributed to free Mn.

Recently Yochum et al. [179] were able to obtain thylakoid membrane preparations fully active in oxygen evolution and retaining four Mn atoms per PSII unit. Subsequent Mn-removing treatments, resulting in different degrees of further Mn extraction depending upon the specific treatment employed, caused a complete inhibition of oxygen evolution. As a whole, the results of both extraction and reinsertion experiments seem to indicate the functional association of four Mn atoms to each O_2-evolving site.

Chloride has been recognized as a cofactor required for efficient transfer of electrons from water to *P*-682, but not for its reduction when artificial electron donors are used [180,181]. The exact mechanism of Cl^- action in O_2 evolution still remains unclear. Direct interactions between Mn and Cl^- ions have been hypothesized by Gol'dfel'd et al. [182]. Indications in this sense come from the effects of Cl^- in relation to the inhibition of O_2 evolution by NH_2OH [181] and by exogenous Mn [183]. Moreover, by utilizing $^{35}Cl^-$-NMR, a correlation between oxygen evolution and $^{35}Cl^-$ line broadening has been obtained [184], possibly indicating the binding of Cl^- to active sites. A Ca^{2+} requirement for O_2 evolution has been claimed [185,186], but the role of calcium ions in the O_2-evolving side of PSII is completely unknown.

7.3. Kinetic studies

The evolution of an O_2 molecule from water requires the removal of four electrons, according to the formal reaction

$$2H_2O \rightarrow 4e^- + O_2 + 4H^+ \quad (E'_o, \text{pH } 7 = 0.82 \text{ V})$$

Since only one electron is transferred following the excitation of the PSII reaction centers, the question of whether four RCs cooperate together to promote the evolution of one O_2 molecule, or whether charge accumulation occurs within individual PSII systems, arise. The kinetic characterization of oxygen evolution has provided an answer to this question, and has given a considerable insight into the chemical mechanisms involved with the transfer of the oxidizing equivalents to the O_2-evolving sites.

Allen and Franck [187] were the first to show that no oxygen evolved in dark-adapted anaerobic algae following a single ms flash of light, while O_2 could be produced following two closely spaced flashes or when a single flash was preceded by a weak continuous illumination. The availability of μs duration flash sources, and the development of fast and sensitive polarographic techniques for O_2 detection, enabled Joliot et al. [188] to demonstrate the occurrence of a characteristic pattern of oxygen evolution following a sequence of short saturating light flashes. The maximum O_2 yield per flash occurs after the third flash; subsequent yields show oscillation characterized by a periodicity of 4 which damps out to a steady state value. Several models have been put forward to explain these results, all implying a mechanism of positive charge accumulation on the water-splitting enzyme. The reaction scheme proposed by Kok et al. [189] has been widely accepted as the minimum model which can account for the existing kinetic data. In this hypothesis the O_2-evolving system cycles through four photoactivated charged states (S) following excitation by a sequence of flashes (Fig. 4.5). The suffix of S represents the number of oxidized equivalents stored on the O_2-evolving complex; each $S_n^* \rightarrow S_{n+1}$ transition, which follows light activation ($S_n \rightarrow S_n^*$), corresponds to the dark relaxation the system must undergo before a new photon can be utilized by S_{n+1}. Thus, in terms of a specific charge accumulating intermediate M and of the electron donor $D_{II,2}$ to *P*-682, the transitions $S_1 \xrightarrow{h\nu} S_1^* \rightarrow S_2$ would be equivalent to the conversions

$$M^+D_{II,2}D_{II,1}A_{II,1}A_{II,2}A_{II,3} \xrightarrow{h\nu} M^+D_{II,2}D_{II,1}^+A_{II,1}A_{II,2}^-A_{II,3}$$

$$\rightarrow M^{2+}D_{II,2}D_{II,1}A_{II,1}A_{II,2}A_{II,3}^-$$

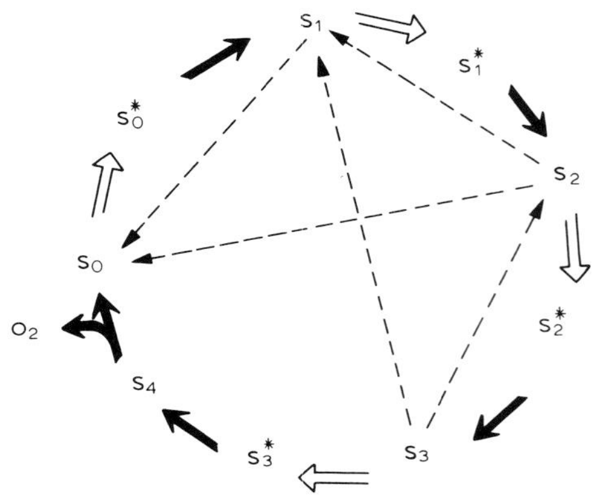

Fig. 4.5. Scheme of the transition states (S states) of the oxygen-evolving complex. Light-dependent and dark reactions are indicated with open and filled arrows, respectively. Dashed arrows indicate proposed dark deactivation processes.

A satisfactory agreement can be obtained between the kinetic features observed and the quantitative predictions of the model when the following assumptions are made.

(**a**) Electron transfer chains, each comprising a reaction center and an associated water-splitting complex, act as individual units independent of each other. Different electron transfer chains do not cooperate in the charge accumulation process.

(**b**) All light-induced transitions $S_n \rightarrow S_n^*$ have the same quantum efficiency and involve a one-quantum process.

(**c**) The S_1 state is stable in the dark, while states S_2 and S_3 decay to S_1; the S_0 state, stable in the dark, can only be generated by the $S_4 \rightarrow S_0$ transition.

(**d**) Upon each single excitation, a small fraction of the O_2-evolving units undergo two transitions ($S_n \rightarrow S_{n+2}$) (double hits), while another fraction of the centers fail the transition $S_n \rightarrow S_{n+1}$ (misses).

Assumption (**a**), i.e., the concept of non-cooperativity, gives rise to the oscillatory pattern observed. If a complete independency of the electron transfer chains is postulated with respect to accumulation of oxidizing equivalents, partial inhibition which blocks a fraction of O_2-evolving centers should decrease the flash-induced O_2 yield without affecting the pattern of oscillation. The same degree of inhibition was indeed observed for all flash yields in a sequence, when the number of active centers was decreased as much as 30-fold by DCMU, UV irradiation or Mn depletion [189].

Oscillation was not essentially observed in the oxygen yields when a flash sequence was preceeded by weak continuous illumination [188,190]. This result has been taken as evidence that the steady-state concentrations of the S states are very similar, and consequently the quantum yields of the photosteps must be equal. If this conclusion is correct, one quantum is involved in each transition, since in non-saturating flashing light the O_2 yield, monitoring the final transition $S_4 \rightarrow S_0$, depends linearly upon the flash intensity [189].

In the framework of the S-state model, the occurrence of a maximum O_2 yield on the third (and not the fourth) flash, can be explained in principle assuming that either one of the first three flashes generates two equivalents instead of one, or S_1 and not S_0 is the predominant state in the dark. Although the former possibility, implying that all the centers deactivate completely to the ground state S_0 seems the simpler one, it is not supported experimentally [191–193]. Strong arguments in favour of a relative stability of the S_1 state in the dark came from the preillumination experiments by Forbush et al. [194]. Assuming that the S_2 and S_3 states can deactivate to S_1, and that S_0 remains constant during the decay period, the uniform distribution of the S states reached under steady-state illumination should decay in the dark to the distribution $1S_0 : 3S_1 : OS_2 : OS_3$. When this status is perturbed by one or three flashes, different dark distributions are attained following relaxation; it is evident from the reaction scheme that in the decativation period following a single flash all centers should convert to S_1, while following three flashes the S_1/S_0 ratio should decrease. These dark distributions, in turn determine the ratio of O_2 yield on the third (Y_3) to the fourth flash (Y_4) in a subsequent train of light pulses. Using this rationale, and neglecting in a first approximation the effect of misses and double hits, Forbush et al. obtained results in agreement with the predictions implied by assumption (**c**). Utilizing the same method, Bouges [195] found that when a longer dark time was allowed between perturbing and measuring flashes, the Y_3/Y_4 ratio was rather independent of the preillumination regime. This finding could reflect a slow dark equilibration between S_0 and S_1 states to the distribution observed after decay from the steady state. In line with this interpretation the S_1/S_0 equilibrium ratio, as judged from the analysis of the Y_3/Y_4 ratio, is changed by chemical oxidation and reduction with ferricyanide and DCPIP-ascorbate [196]. Kok et al. [197] confirmed the redox results of Bouges-Boquet and proposed a model for $S_0 \rightarrow S_1$ interconversions in which the S_1/S_0 ratio is determined by the balance between oxidation of S_0 to S_1 by oxygen and the reduction by a pool of endogenous (and eventually exogenous) reductants.

The deactivation rates of S_2 and S_3 states have been measured in algae and chloroplasts [194,198] monitoring the O_2 yield of one and two flashes fired after an appropriate preillumination. Forbush et al. [194] found a second order kinetics for both S_2 and S_3 decays. The S_2 recovery kinetics following a preillumination regime that generates a mixture of S_2 and S_3 states, showed a marked transient reflecting most likely a single-step deactivation process $S_3 \rightarrow S_2 \rightarrow S_1$. However, on the basis of similar results, it has been suggested [198] that a fraction of the S_3 or of the S_2 states can also deactivate directly to S_0 states.

Deactivation of S_2 and S_3 states could proceed either via a redox component which replaces water as the electron donor or through a back reaction within PSII. A class of reagents, including CCCP and substituted thiophenes (the so-called 'ADRY' compounds) has been demonstrated to accelerate up to 50-fold the decay of S_2 and S_3 states. In the presence of these compounds deactivation probably occurs through a cyclic electron transfer around PSII which involves an endogenous reductant [199]. S_2 and S_3 states are on the contrary stabilized by ammonia, an inhibitor of O_2

evolution. Velthuys [200] has shown that this inhibitor interacts with the water-splitting enzyme in states S_2 and S_3 (but not in S_0 and S_1) and that light activation of a center in state $S_3(NH_3)$ generates an $S_4(NH_3)$ state which deactivates to $S_3(NH_3)$ through charge recombination.

The occurrence of double hits and misses (assumption **d**) was introduced to account for the increasing degree of disorder in S-state distribution as revealed by the damping of O_2-yield oscillation in a sequence of flashes. This assumption can equally explain why the maximum yield observed on the third flash is not quite three times the yield obtained in steady state. The extent of double hitting is expected to be related to the duration of the light pulse as compared to the relaxation time of the transitions $S_n^* \rightarrow S_{n+1}$, the limiting step in the PSII turnover rate. Experiments with saturating xenon flashes [194,198,201] have shown that the probability of double hits as estimated by the O_2 yield after the second flash and by the damping of the oscillatory pattern is decreased by decreasing the duration of microsecond light activations. Using ns laser pulses, Weise and co-workers [202,203] concluded that no double hits occurred following very brief energizations; more recent investigations, however, have revealed that measurable amounts of O_2 can be produced following two closely spaced ns laser pulses by freshly prepared thylakoid vesicles, and that this ability depends on the degree of physiological intactness of the thylakoid preparations [204]. These results suggest that non-photochemical (e.g., cooperative) double hits also determine the oscillatory pattern of O_2 yield. Under conditions that minimize double-hitting the flash yield oscillation is still damped and a slight phase retardation can be detected. To explain this kind of behaviour in terms of the S-state model significant frequencies of misses (ranging from about 10% in chloroplasts to about 20% in algal cells) must be assumed; moreover the estimated frequencies are independent of the flash intensity well above the saturation level. Misses seem, therefore, to be the predominant intrinsic perturbation in determining the damping of the oscillatory pattern. Both the failure of the transition $S_n \xrightarrow{h\nu} S_n^*$ (i.e., primary photochemistry) and a back reaction which competes with the S-state forward transitions by dissipating the positive charge could cause the postulated misses. A charge recombination at the level of the secondary (or tertiary) electron donors involving the primary or secondary acceptors seems to be irreconcilable with the parallel damping of the O_2 yields and $A_{II,3}^-$ (plastoquinone B) oscillations (see Ref. 205). The back reaction between P^+-682 and $A_{II,2}^-$ could be more reasonably invoked as a source of misses. This reaction, as monitored by the Chl luminescence, has been reported to undergo oscillation with the same periodicity of the O_2 yield [206]. Any speculation on the mechanisms which causes misses in terms of redox photochemical events is further complicated by the fact that the quantum yield in steady state and the calculated frequency of misses are both higher in intact cells than in isolated chloroplasts. On the contrary, by assuming photochemical misses, one predicts a decreasing quantum yield at increasing probability of failures. To account for this anomaly Lavorel and Lemasson [207] assumed that a lateral carrier could exchange charges with the oxygen-evolving intermediate. Since the damping of the O_2 oscillation is determined, in this scheme, by the relative concentrations of the carrier and of the charge-storing intermediate M, the anomaly is easily explained, but the model contradicts the concept of non-cooperativity of the water-splitting enzymes (assumption **a**).

As to the chemical interpretation of the S-state component M, which plays the role of charge accumulator, manganese is the most attractive candidate in view of its involvement in oxygen activity and of its multiple oxidation states. The occurrence of light-induced redox changes of chloroplast-bound Mn is strongly supported by the ESR measurements of Widrzynski and Sauer [202]. These authors, using a temperature shock to rapidly remove Mn after a variable number of short light pulses, were able to observe an Mn ESR signal oscillating with a period of 4. Since only free Mn give rise to an ESR signal in aqueous solution at room temperature, the light-induced changes in ESR signal were interpreted as reflecting changes in the oxidation states of Mn functionally associated to O_2 evolution. Following this interpretation the charge stored on Mn increased after the first and second flash and decreased after the third activation. This oscillatory pattern was found consistent with a simple model based on the S-state scheme and involving an Mn dimer.

8. Cytochrome b*-559*

The presence in chloroplasts and algae of a cytochrome *b* with the band at 559 nm, and therefore distinct from cyt. b_6, was described in the mid 1970s [120]. However, in spite of this early discovery and of the efforts of many laboratories, the actual role of this carrier in the photosynthetic chain is still obscure.

The midpoint potential of this cytochrome is heterogeneous. When titrated in slow equilibrium conditions, two forms can be detected, a high potential (E_m = 0.35–0.40 V at pH 7) and a low potential form (E_m = 0.05–0.10 V) [120]; the two forms appear to be interconvertible, at least partially, with the high potential species being transformed into the low potential one by denaturing treatments, such as NH_2OH-Tris washing, heating, detergents or proteolysis. Cytochrome *b*-559 is present in freshly prepared chloroplasts with a stoicheiometry of about two per reaction center, with the high potential form being the predominant one. With more rapid evaluations of the redox properties by the addition of various reducing or oxidizing agents, the presence of other forms at potentials intermediate between the ones cited above have been assessed [98,208,209]. At the moment, therefore, it is not clear whether the variously detected cytochromes *b*-559 represent various forms of the same redox carrier, capable of changing its redox characteristics following regulatory or functional modulation, or rather really different unrelated proteins with distinct structural and functional properties.

The association of cyt. *b*-559 with intramembrane complexes is also obscure; this cytochrome has been found associated with the b_6/f complex and with PSII-RC preparations [70]. After the purification of an intact active b_6/f complex, however, the lack of involvement of cyt. *b*-559 in the plastoquinol-plastocyanine oxidoreductase activity has become evident [111], since the catalytic activity and the usual presence of two cytochromes of *b* type in this complex seems to involve only cyt. b_6. Cyt. *b*-559 is, however, always found associated with even the purest preparations of PSII-RC, and more precisely with its 6000 Da subunit [70,210]. But the size of the apoprotein seems too small to enclose a heme group; it is likely therefore that this subunit is formed as a proteolysis product. A molecular weight of 37000 for cyt. *b*-559 has been indeed claimed [211].

Many suggestions have been put forward for the role of cyt. *b*-559 in electron transfer. These include: (a) electron transfer between PSII and PSI in series or in parallel with the b_6/f complex [212,213]; (b) cycling of electrons around PSII [214]; (c) electron carrier involved in the water-splitting complex [215]. Role (a) for a parallel pathway seems unlikely in view of the observation that non-cyclic electron transfer is fully sensitive to specific inhibitors of the plastoquinol-plastocyanine oxidoreductase, and that cyt. *b*-559 is not associated with the active isolated b_6/f complex [111]. However, a side branch could be involved in the bypass of the electrogenic [153,154] and proton-translocating [154–157] steps of the oxidoreductase, and seems to be operative at high light intensities (or at high proton motive force). Roles (b) or (c) are consistent with the redox properties of the high potential form [215], and with its association with PSII-RC [71]. Recycling of electrons around PSII-RC is also supported by the stimulation that ADRY reagents exert on the photooxidation of cyt. *b*-559 [214–217]. This carrier could therefore be involved in the deactivation at low light intensities of the oxygen-evolving complex [218].

9. The redox interaction between complexes

9.1. The secondary electron donors to bacterial and PSI reaction centers

As stated on several occasions in the previous sections, electrons are delivered to bacterial and PSI-RC by electron carriers which can be isolated as water soluble homogeneous proteins, cytochromes of *c* type or plastocyanine. These carriers represent also the physiological electron acceptors for the b/c_1 complexes. It has been conceived, therefore, that these proteins can act as diffusable redox mediators between the different complexes, which in turn are thought to be laterally and independently mobile in the membrane lipid bilayer [219]. The location of these carriers would be the interface on one side of the asymmetrically arranged coupling membrane, namely towards the periplasmic space in bacteria (corresponding to the internal volume of chromatophores) or the inner lumen of thylakoids.

The location of cytochrome c_2 in the periplasmic space of purple photosynthetic bacteria has been demonstrated directly by its prompt release following the preparation of sphaeroplasts, and by its accessibility to antibodies in these preparations [220]. Cytochromes *c* are oxidized in single turnover experiments with a biphasic kinetics ($t_{1/2} \approx 5$ and 200–400 μs); this pattern has been interpreted as due to the presence in chromatophores of both cyt. c_1 and c_2, which are oxidized in series [122].

On the other hand a biphasic behaviour was also observed with isolated RC and, in this case, the interpretation considered the oxidation of a cyt. c_2 complexed to the RC, followed by a slower diffusion-controlled step [221,222]. Specific studies on the mobility of cyt. c_2 between complexes are very limited. Mobility of cyt. c_2 in respect both to the RC and the oxidoreductase was inferred from the observations that: (a) cyt. c_2 is reduced in the millisecond time range by the oxidoreductase with a second order kinetics, although the Fe-S protein is substoicheiometric to cyt. c_2 [223]; (b) cyt. c_2 can act as electron donor, not only to the RC but also to the respiratory cyt. c_2 oxidase, with a kinetic behaviour indicating a single homogeneous pool [224]. On the other hand experiments utilizing non-saturating flashes, oxidizing only a fraction of the RC have indicated that the oxidized cyt. c_2 dissociates very slowly from the RC, suggesting a lower mobility [225]. The question whether cyt. c_2 shuttles between the membrane complexes or is part of a supramolecular structure is still uncertain.

A similar situation exists for plastocyanin. This carrier is present in chloroplasts or algae in an excess of 2- [226] to 5- [227] fold with respect to *P*-700, as compared to cyt. *f* which is present with a stoicheiometry 1 per *P*-700 [228]. The rate of electron transfer to *P*-700 is also biphasic with haltimes of 20 and 200 μs; this biphasicity has been interpreted as being due to the formation of a complex with PSI-RC and to a diffusion-controlled phase [229]. The rate of *P*-700 reduction is accelerated at high osmolarities, and this has been explained with an increase of the concentration of plastocyanin in the inner lumen of thylakoids. The question as to whether plastocyanin is actually shuttling between cyt. *f* and *P*-700, or rather is part of a supramolecular structure between the two complexes, has been complicated by the observation that in granal chloroplasts PSII is prevalently, if not exclusively, present in the appressed regions, while PSI is located in the stroma lamellae [219]; the possibility that plastocyanin could act as a diffusable carrier for the entire length of the grana, were the oxidoreductases also present in the grana stacks, seems to be rule out by the estimated diffusion rate of this protein, as compared to the velocity of electron transfer. In very recent experiments Haenel [230] has studied the effect of partial inactivation of plastocyanin with KCN or $HgCl_2$; he observed effects on both the rate and the extent of *P*-700 reduction and interpreted the results as kinetic evidence of a supramolecular organization between plastocyanin and the b_6/f complex. Studies with antibodies [231] as well as with hydrophilic inhibitors [232] or electron donors [233] to plastocyanin indicate that this carrier is located in a phase inaccessible to water soluble molecules, i.e., the inner lumen of the thylakoids; on the other hand many experimental results opposing this conclusion have been presented [93], so that no agreement yet exists, on the precise compartmentation of the Cu prosthetic group of this carrier.

9.2. The role of quïnones in the interaction between complexes

Invariably quinones are present in large excess over the other electron carriers, in chloroplasts about 40-fold and in bacteria between 25- and 35-fold depending on the bacterial strain and the growth conditions [235]. Mainly for this reason quinones, in addition to playing a specific role in a bound form as electron acceptors in some RC (and perhaps in the b/c_1 complex), are thought to act as a freely diffusable species, in a way similar to cyt. c_2 or plastocyanin but in a lipid phase. In this way quinone could mediate the redox interaction between PSII-RC and the b_6/f complex, also if the two complexes are located in two different regions of the grana [219,236], or complete the redox cycle in bacterial photosynthesis between the RC and the b/c_1 complex [38]. Plastoquinone could also mediate the cyclic electron flow around PSI promoted by ferredoxin [159,161,237].

The mobility of quinone in lipid bilayers is fast enough to assure a high rate of electron transfer (the diffusion coefficient has been estimated to be 10^{-8}–10^{-9} $cm^2 \cdot s^{-1}$) by a diffusion-controlled process. This point has been recently confirmed experimentally in chromatophores by studying the effect on the rate of electron transfer of phospholipid enrichment of the membrane (obtained either by fusion with liposomes [238,239] or by exploiting the naturally occurring change in the RC-phospholipid ratio during synchronous cell growth [239]); it was shown that the rate constant of electron transfer is linearly related to the average area of the membrane per RC, indicating a diffusion-controlled process in two dimensions [238]. In line with these ideas are the conclusions by Siggel et al. [240], based on partial inhibition of PSII-RC with DCMU, that one PSI can exchange electrons with about ten PSII. At odds with these conclusions, however, is the demonstration that a large amount of ubiquinone, up to 20–22 per RC can be extracted from chromatophores with apolar solvents, without delaying the rate of electron flow in the cycle; in this case the lowering of the quinone concentration in the membrane lipid phase does not affect the rate of electron exchange (observations of this sort are at the basis of the concept of UQ_z) [145,241]. It can be tentatively proposed that when the phospholipid-protein ratio is unchanged with respect to the naturally occurring one (as is the case with solvent extraction), supramolecular structures between complexes can be formed, stable enough in time to allow electron transfer in the absence of long-range diffusional processes of ubiquinone. This proposal is supported by the observed spontaneous formation in an aqueous buffer of functional associations between RCs of *Rps. sphaeroides* and the mitochondrial b/c_1 complex in the absence of added ubiquinone [242].

The delivery of electrons from PSII-RC to the plastoquinone pool was originally proposed to occur via the formation of free semiquinone and the spontaneous dismutation to quinol [236]. It is clear now that, both in PSII and in bacterial RC, this electron transfer is controlled by a two electron gate mechanism involving bound forms of quinones, in which the semiquinone anion is stabilized for a time long enough to allow two turnovers of the same reaction center [40]. If diffusion of the quinone indeed happens, this occurs after complete reduction to quinol at the Q_B site of bacterial RC or the B site of PSII-RC and dissociation from these sites [37,38].

Less clear is the pathway of cyclic electron transfer around PSI. This pathway involves cyt. b_6 and cyt. f, and an energy conserving site (demonstrated by the formation of ATP, proton pumping and a slow rising electronchromic signal, sensitive to DBMIB and UHDBT): the role of the b_6/f complex is therefore beyond doubt. The pathway requires the intactness of the outer membrane and the maintainanace in an undiluted state of the stromal proteins, or alternatively high concentrations of externally added ferredoxin. The involvement of ferredoxin has been also confirmed by immunological methods [159]. The exact mechanism of reduction of plastoquinone by PSI is however obscure and might involve other electron carriers at present still unidentified. Ferredoxin-dependent cyclic electron transfer is in fact inhibited by antimycin A [109,243], an inhibitor inactive on the non-cyclic reaction and on the plastoquinol-plastocyanin oxidoreductase activity; this suggests a binding site for antimycin A outside the b_6/f complex or at least unrelated to the above activity. Antimycin A must, however, be used at high concentrations (20 mol/mol f [109.243]), much higher than those inhibiting the b/c_1 complex of bacteria or mitochondria. The pathway of cyclic electron flow is still an open problem and its regulation with respect to the non-cyclic transfer of electrons to $NADP^+$ is interconnected with that of the possible bypass of the Q cycle, and probably with the redox state of the plastoquinone pool.

9.3. The reduction of $NADP^+$ by photosystem I

The final destination of the electrons transferred to low redox potentials by PSI-RC is $NADP^+$. The electron transfer system to this pyridine nucleotide coenzyme is formed by a sequence of two enzymes, ferredoxin and ferredoxin-$NADP^+$ reductase.

Chloroplast ferredoxin is a small water soluble protein (M_r 11 000) containing an Fe-S center [245]. Its midpoint potential (−0.42 V [246]) is suitable for acting as an electron acceptor from the PSI Fe-S secondary acceptors (Centers A and B) and as a donor for a variety of functions on the thylakoid membrane surface and in the stroma. Due to its hydrophylicity and its abundance in the stromal space, ferredoxin is generally considered as a diffusable reductant not only for photosynthetic non-cyclic and cyclic electron flow, but also for such processes as nitrite and sulphite reduction, fatty acid desaturation, N_2 assimilation and regulation of the Calvin cycle enzyme through the thioredoxin system [245]. Its possible role in cyclic electron flow around PSI has already been discussed. The mobility of ferredoxin along the membrane plane could be an essential feature of this electron transfer process; the actual electron acceptor for this function and the pathway of electron to plastoquinone is, however, still undefined.

Ferredoxin-$NADP^+$ oxidoreductase is a flavoprotein bound to the membrane that can nevertheless be isolated as a water soluble homogeneous preparation [247]. The isolated enzyme has a molecular weight of 40 000 and contains one mol of FAD per mol of apoprotein [248]; its midpoint potential is −0.38 V at pH 7 [249]. The enzyme can also accept NAD^+ as electron acceptor, but is very specific for ferredoxin as donor; the formation of a 1 : 1 complex with this latter protein could be demonstrated by differential spectrophotometry [250]. Immunological studies have revealed the location of this flavoprotein on the stromal surface of the membrane, in the vicinity of the ATPase complexes and therefore predominantly on the stroma lamellae [251]. Consistent with this location ferredoxin is bound to the membrane because of the amount of flavoprotein present.

The reduction of $NADP^+$ by PSI is therefore catalyzed by an association of a flavoprotein and an Fe-S protein, somewhat in analogy with the dehydrogenases operating in the same redox span in respiratory chains; the dissociability of ferredoxin from the complex, however, can assure a large degree of flexibility in the utilization of the reducing power generated by photosynthesis.

10. Membrane topology and proton translocation

Although this topic is not extensively dealt with here, this chapter cannot be ended without including a brief discussion on the orientation of the complexes within the membrane in relation to proton translocation. The formation of a transmembrane protonic gradient, coupled to the electron transfer reactions, is, according to Mitchell's proposal [252], one of the main functions of the photosynthetic chain. In Mitchell's model protons are translocated through a direct electroneutral mechanism, by which they cross the membrane under the form of reduced and protonated carriers. The charge-separating steps necessary to complete a net electrogenic proton movement is obtained by an electron transfer in the opposite direction, in reactions catalyzed by carriers different, to the proton translocating ones. These electroneutral and electrogenic segments of the chain should be alternating and form one or more protonic loops, giving origin to a mechanism compatible only with a stoicheiometry of one proton per electron per loop [93,139]. As an alternative to this arrangement, in the concept of the proton pump, protons are translocated through a mechanism depending on conformational changes of proteins, in turn coupled to the redox reactions; with this mechanism stoicheiometry of more than one proton per electron per pump can be envisaged [253]. Reaction centers, oxidoreductive complexes and diffusable redox carriers are catalysts of these activities, and their structural arrangement must be compatible with this function.

In bacterial chromatophores the RC and the b/c_1 complex are arranged to form a cyclic electron transfer system possibly mediated by the diffusion of ubiquinone and cyt. c_2; these carriers are, however, also coupled to other multienzyme complexes forming the respiratory chain and perform the aerobic metabolism of these facultative photosynthetic organisms [254]. The electrogenic steps of the photosynthetic cycle take place both within the RC and the b/c_1 complexes and can be monitored by the electrochromic spectral shift of endogenous carotenoids and on the basis of their response to specific inhibitors and kinetics. When induced by a short laser flash the carotenoid signal displays three distinct kinetic phases ($t_{1/2} < 10$ ns, $t_{1/2} \approx 5$ μs and $t_{1/2} \approx 2$ ms); under optimal ambient redox conditions ($E_h = 100$ mV) the amplitudes of these phases are 30, 20 and 50% of the total, respectively. The first two, more rapid phases have been attributed to an electron transfer process taking place within the RC [255]: the faster to the primary photochemical process and the other to the reduction of D_1^+ by cyt. c_2 [256]. Since no electrogenic event can be observed during electron transfer from A_2^- to A_3 it is believed that the combination of these two charge-separating processes spans the whole membrane thickness, with the primary donor situated about 40% inside the membrane on the periplasmic side [139]. The orientation of the RC is therefore such as to catalyze electron transfer from the periplasmic to the cytoplasmic face of the membrane, being the periplasmic location of cyt. c_2 (a well established fact; see Ref. 220). The transmembrane arrangement of the RC has also been demonstrated with direct electric measurements after inserting isolated RC into artificial planar membranes [257,258] or liposomes [259].

The third kinetic phase of the carotenoid signal is related to the function of the b/c_1 complex as suggested by its sensitivity to antimycin A and UHDBT [139]. Its amplitude and risetime are dependent on the ambient redox potential in a way consistent with the presence of a control by reduced Q_z; it is believed that this third electrogenic phase is connected with the electron flow through cytochromes *b* and possibly with the oxidation of cyt. *b*-561 [147]. Also, in the case of the b/c_1 complex, the shift of the carotenoid indicates a transmembrane orientation of the electrogenic step, being the amplitude and the direction of the shift comparable to that coupled to the RC. The transmembrane arrangement of the b/c_1 complex is also supported by the generation of a membrane potential coupled to the ubiquinol-cyt. *c* reductase reaction in proteoliposomes containing the complex, as monitored with the exogenous $\Delta\psi$ indicator carbocyanin [149].

Proton translocation has been documented in bacterial chromatophores since the 1970s [140,260]; the time resolution of this phenomenon has also been attempted using fast spectrophotometry and sensitive hydrophilic pH-indicating dyes [148]. Due to intrinsic experimental limitation these studies have been so far restricted to the external face of the chromatophores (cytoplasmic side), and only proton-binding phenomena have been studied, i.e., the utilization of protons for oxidoreduction reactions, probably for the reductive protonation of electron carriers (e.g., ubiquinone, according to the loop model [140]). The experimental results have shown that one proton (H_I^+) per turnover of the RC is bound with a kinetics matching the reduction of A_3 (Q_B) [261]. A second proton is bound much more slowly — with a halftime consistent with that of the b/c_1 complex; this latter proton (H_{II}^+) is inhibited by antimycin A but can be observed only after addition of K^+ and valinomycin, i.e., possibly preventing the formation of a membrane potential [148,262]. According to the modified Q cycle model [38] presented in Fig. 4.6, H_I^+ is the result of the direct photoreduction of ubiquinone by the RC, while the slower H_{II}^+ is derived from the ubiquinone-reducing activity at the 'Q_c' site of the b/c_1 complex. In order to obtain a net translocation of protons across the membrane utilizing ubiquinol as a proton carrier, it is sufficient to assume that the Q_z and 'Q_c' (ubiquinone reductase) sites of the oxidoreductase are located near the opposite faces of the membrane [135]. No direct information on the location of these sites, or even on the actual existence of the reductase site, is yet available, and other possibilities exist (see, e.g., the 'b cycle' scheme [136]). This model requires that ubiquinol crosses the lipid bilayer with a rate kinetically competent for cyclic electron flow. Thermodynamic considerations seem to exclude this possibility; recent NMR data on ubiquinol mobility in artificial dimyristoyl lecithin appear however to support a sufficient transmembrane mobility [263]. Moreover, ubiquinone has been demonstrated to be able to mediate the transmembrane reduction of ferricyanide by dithionite, a reaction requiring the transmembrane crossing of reduced and oxidized ubiquinone [264,265].

The topologic scheme of the photosynthetic chain of chloroplasts and cyanobacteria is depicted in Fig. 4.7; in this figure the complexes are arranged in series in the familiar Z scheme, and their main properties in relation to proton translocation are indicated [93]. The function performed by plastoquinone is completely analogous to the one, already discussed, of ubiquinone in bacterial photosynthesis. The large pool of plastoquinone should catalyze proton translocation by means of its reductive protonation on the stromal face of the thylakoids by PSII-RC (or by PSI-RC in the cyclic electron transfer pathway) and by the Q_c site of the b_6/f complex; analogously protons should be released following plastoquinol oxidation at the U site (equivalent to Q_z) at the lumen face. The assumption of the existence of these sites in the b_6/f complex is still fully hypothetical, and based only on the well documented analogies between all the b/c_1 complexes isolated from photosynthetic and respiratory systems, and on the proposition of a common mechanism of electron transfer. In addition, evidence for a 'Q cycle' behaviour, such as oxidant-induced reduction of cytochromes *b* [109], or an H^+/e^- stoicheiometry higher than one in proton translocation by proteoliposomes coupled to the ubiquinol-cyt. *c* reductase reaction [138], has been obtained for all the b/c complexes so far isolated, including the b_6/f complex from chloroplasts and cyanobacteria. There are, however, other possibilities different from the Q cycle, and the *b* cycle (assuming the existence of a common alternating site for plastoquinol oxidation and plastoquinone reduction, and of a protonic pump) is an example of an alternative mechanism. In chloroplasts and algae the problem is further complicated by the lex apparent possibility of excluding the translocation of protons with a stoicheiometry higher than $1H^+/e^-$ in the intersystem chain (i.e., avoiding the Q cycle) [156–158]; this alternative route seems operative at high protonic forces and/or under specific redox conditions of the chain, as suggested by the absence of a slow electrochromic response (and of extra protons) at high light intensities [153,154].

In addition to the possible release of protons upon plastoquinol oxidation at the U site of the b_6/f complex, a second mechanism of H^+ release is present in chloroplasts and algae at the oxygen evolving site. Water oxidation implies the liberation of four protons per O_2 and it is obvious that this phenomenon must be related to the S states of the oxygen-evolving complex. The topology and kinetics of H^+ release has been followed, either with specially arranged glass electrodes, or spectroscopically with the membrane-permeable pH-indicating dye neutral red. Oscillation in the proton release as a function of the activation of different S states by flash sequences has been observed with glass electrodes [266]; these measurements required the addition of protonophores in the assay medium, demonstrating that H^+ were released in the inner lumen of the thylakoids, or at least beyond a

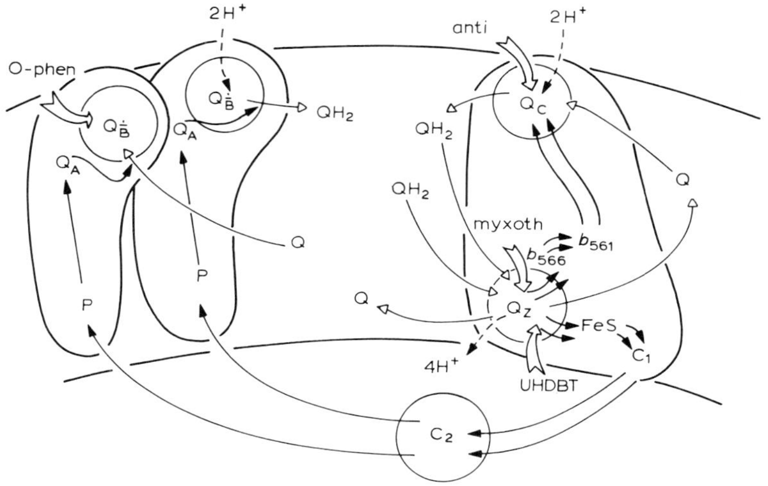

Fig. 4.6. The proposed arrangement of the RC and b/c_1 complex in the photosynthetic chain of *Rps. sphaeroides*. The scheme indicates the reduction of Q to QH_2 by a pair of RC complexes and the net oxidation of QH_2 by two turnovers of the oxidoreductase, as a balance of the oxidation of two quinols and the reduction of one quinone at the Q_c site. The proposed sites of proteolytic reactions are also indicated (from Ref. 93).

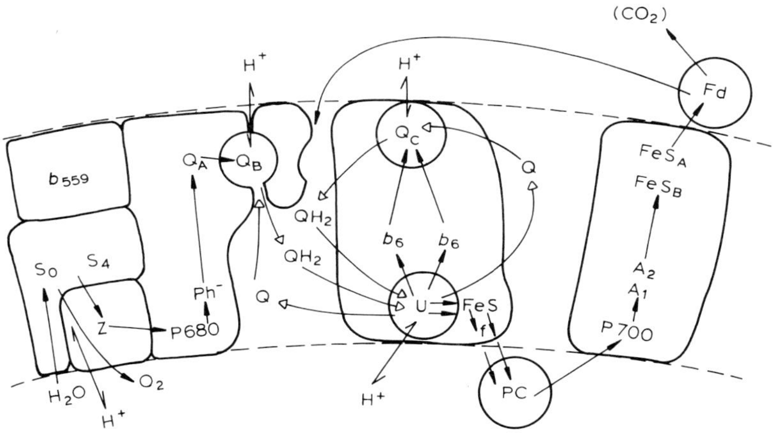

Fig. 4.7. The Z scheme of the higher plant photosynthetic chain visualized as an arrangement of multiprotein complexes. The two RC of PSII and PSI are arranged in parallel across the membrane and are interconnected by the b_6/f complex. The electron transfer pathway within this complex follows a modified Q cycle scheme analogous to that proposed for bacterial photosynthesis. The oxygen-evolving complex is proposed to face the inner thylakoid lumen and to release protons in this compartment. The association of cytochrome *b*-559 with PSII-RC and the cyclic role of ferredoxin are also depicted. Proton-binding and proton-releasing sites are illustrated (from Ref. 93).

diffusion barrier. The internal location of the releasing site is also demonstrated by the impossibility of buffering out the pH changes monitored by neutral red with bovine serum albumin acting as a membrane-impermeable buffer [267].

The correlation of proton liberation with the $S_0 \rightarrow S_1$, $S_1 \rightarrow S_2$, $S_2 \rightarrow S_3$, $S_3 \rightarrow S_4/S_0$ transitions is still a matter of debate; same authors advocate a sequence of release of the type 1 : 0 : 1 : 2 [267–269], other one of 0 : 1 : 1 : 2 [270] or even of 1 : 1 : 1 : 1 [271]. The difference in the experimental results, obtained generally with neutral red, are probably related to the dark adaptation of chloroplasts before flashing: exposition to light even tens of minutes before the experiment seems sufficient for observing proton release after the first flash, i.e., following the $S_1 \rightarrow S_2$ transition [272]. Whatever the reason of this discrepancy, all authors agree on the conclusion that protons are not released in bursts of four in phase with O_2 evolution, but separately during the charge accumulation process within the oxygen-evolving complex. It is possible, therefore, that water oxidation takes place out of phase with H^+ release and that OH^- groups act as counterions to the Mn^{3+} accumulating in the complex [273,274].

The orientation of the charge-separating processes in the thylakoid membrane has generally been followed by studying the electrochromic signal of endogenous pigments [236,275]; the polarity, positive inside, of the membrane potential generated by photosynthesis has been confirmed also by ion-distribution methods [276] or by inserting microelectrodes directly into the chloroplast [277]. Analogously to bacteria the two reaction centers promote the translocation of one electron each from the inner lumen of the thylakoids to the external face. This topology is consistent with the known location of plastocyanin and of the O_2-evolving site and with the outer location of ferredoxin. That the parallel orientation of PSII-RC- and PSI-RC-mediated charge separation, approximately equivalent as far as the thickness of the membrane dielectric is concerned, has been supported by the observation that, following a single turnover flash, about one half of the carotenoid signal is generated by PSII and one half by PSI [236].

The orientation of the electrogenic step within the b_6/f complex has a large component parallel to the RCs, as judged from the direction and the amplitude of the electrochromic signal (in the conditions in which it is observable experimentally [151,154]). The nature of the electron carriers promoting this latter charge separation is, however, unknown; it can be suggested, in analogy with bacterial photosynthesis, that it coincides with the oxidation of cyt. b_6 at the Q_c site.

The location of the H^+-accepting and H^+-releasing sites on the two opposite surfaces of the membranes, and in equilibrium with the external aqueous phases, is still a matter a controversy and experimentation. In bacterial chromatophores the kinetics of proton binding has been followed with a group of anionic pH-indicating dyes [148], which are believed to report on pH changes occurring in very hydrophylic regions in equilibrium with the bulk external water phase. In these studies the kinetics of H_I^+ binding (antimycin A insensitive) has been found to match Q_B reduction. The same situation does not hold for H_{II}^+ (antimycin A sensitive) that can be observed only in the presence of K^+ and valinomycin [262]. This ionophore, in addition to preventing the formation of a membrane potential, could perturb possible diffusion barriers for protons existing at the Q_c site of the b/c_1 complex [278]. No indication so far is available on the kinetics of proton release at the inside of chromatophores.

The situation is more complex in chloroplasts: even when the hydrophylic pH indicator, cresol red, was utilized the H^+-binding kinetics at the outer surface of the thylakoids did not correspond to the reduction of $A_{II,3}$ (B), but was approximately 30-fold slower [279]. A faster equilibration of protons could be obtained only after harsh surface treatments such as mechanical disruption by sand grinding or detergents. Attempts in the evaluation of the kinetics of proton release in the inner thylakoid lumen have been made with the permeant pH indicator, neutral red, buffering the outer phase with BSA [272,280]. Initially the two kinetic phases observed seemed to match kinetically the rate of water and plastoquinol oxidation [279]. Later, it was discovered that this kinetic consistency could be observed only in chloroplasts stored frozen in the presence of dimethylsulphoxide [281]. In freshly prepared chloroplasts (or stored frozen in the presence of ethyleneglycol), the fast phase could be eliminated by concentration of gramicidin (10–15 nM) affecting neither the plastoquinol-derived protons, nor the transmembrane proton conductivity [281]. The effect of gramicidin was, however, transient and disappeared after five to eight flashes were fired to dark-adapted chloroplasts. Thus, it was proposed that water-derived protons are initially deposited in a domain, not in equilibrium with the bulk water phase and with the plastoquinol oxidizing site. This domain would be preferentially hit by gramicidin at low concentration, and therefore would be deprived of protons in the dark; only after the domain has been replenished by firing closely spaced flashes, would protons be released into the aqueous phase. This proposal is consistent with the observations by Prochaska and Dilley [282] and Theg and Homann [283], suggesting the presence of localized proton gradients in close connection with the water-oxidizing complex. Moreover, it was recognized that neutral red is partitioned between the membrane and the water phase, and this consequently reports on changes in pH at the interface; in freshly prepared chloroplasts, at variance with freeze-stored ones, the water oxidation-related pH changes were unaffected by changes in the ionic strength, demonstrating that the region monitored by the dye was not in equilibrium with the ionic double layer at the interface [284]. This conclusion is also supported by studies with inverted chloroplasts [285], in which, with the hydrophylic pH indicator, bromocresol purple, the kinetics of proton release was considerably slower than plastoquinol oxidation.

There is good experimental basis therefore to believe that, in chloroplasts and possibly in the b/c_1 complex of bacteria, protons are not directly released or taken up from the water bulk phase but must overcome a diffusion barrier of unknown nature. The presence of these barriers could help in interpreting the severe kinetic discrepancies in the rate of ATP synthesis and its relation with the rate of electron transfer and the extent of the bulk-to-bulk proton gradient, for which a model including H^+ diffusion barriers has been recently proposed [286] (for reviews see Refs. 278, 287).

References *

1 Parson, W.W. and Ke, B. (1982) In Photosynthesis. Energy Conversion by Plants and Bacteria. (Govindjee, ed.) pp. 331–385, Academic Press, New York.
2 Sauer, K. (1979) Annu. Rev. Phys. Chem. 30, 155–178.
3 Okamura, M.Y., Feher, G. and Nelson, N. (1983) In Photosynthesis. Energy Conversion in Plants and Bacteria (Govindjee, ed.) pp. 195–264, Academic Press, New York.
4 Ke, B. (1978) Curr. Top. Bioenerg. 7, 1–33.
5 Olson, J.M. and Thornber, J.P. (1979) In Membrane Proteins in Energy Transduction (Capaldi, R.A., ed.) pp. 279–340, Dekker, New York.
6 Reed, D.K. and Clayton, R.K. (1968) Biochem. Biophys. Res. Commun. 30, 471–475.
7 Feher, G. and Okamura, M.Y. (1978) In Photosynthetic Bacteria (Clayton, R.K. and Sistrom, W.R., eds.) pp. 349–386, Plenum, New York.
8 Broglie, R.M., Hunter, C.N., Delepelaire, P., Niederman, R.A., Chua, N.-H and Clayton, R.K. (1980) Proc. Natl. Acad. Sci. U.S.A. 77, 87–91.
9 Valkirs, G. and Feher, G. (1982) J. Cell. Biol. 95, 179–188.
10 Odermatt, E., Snozzi, M. and Bachofen, R. (1980) Biochim. Biophys. Acta 591, 372–381.
11 Zurrer, H., Snozzi, M., Hanselmann, K. and Bachofen, R. (1977) Biochim. Biophys. Acta 460, 273–279.
12 Marinetti, T.D., Okamura, M.J. and Feher, G. (1979) Biochemistry 18, 3126–3133.
13 Okamura, M.Y., Debus, R.J., Kleinfeld, D. and Feher, G. (1982) In Function of Quinones in Energy Conserving Systems (Trumpover, B., ed.) pp. 299–317, Academic Press, New York.
14 Prince, R.C. and Dutton, P.L. (1978) In The Photosynthetic Bacteria (Clayton, R.K. and Sistrom, W.R., eds.) pp. 439–453, Plenum, New York.
15 Moskowitz, E. and Malley, M.M. (1978) Photochem, Photobiol. 27, 55–59.
16 Norris, J.R., Scheer, H. and Katz, J.J. (1975) Ann. N.Y. Acad. Sci. 244, 261–280.
17 Okamura, M.Y., Isaacson, R.A. and Feher, G. (1975) Proc. Natl. Acad. Sci. U.S.A. 72, 3491–3495.
18 Fajer, J., Brune, D.C., Davis, M.S., Forman, A. and Spaulding, L.D. (1975) Proc. Natl. Acad. Sci. U.S.A. 72, 4956–4960.
19 Parson, W.W., Clayton, R.K. and Cogdell, R.J. (1975) Biochim. Biophys. Acta 387, 265–278.
20 Shuvalov, V.A. and Klimov, V.V. (1976) Biochim. Biophys. Acta 440, 587–599.
21 van Grondelle, R., Romijn, J.C. and Holmes, N.G. (1976) FEBS Lett. 72, 187–192.,
22 Tiede, D.M., Prince, R.C. and Dutton, P.L. (1976) Biochim. Biophys. Acta 449, 447–467.
23 Okamura, M.Y., Isaacson, R.A. and Feher, G. (1979) Biochim. Biophys. Acta 546, 394–417.
24 Lutz, M., Kleo, J. and Reiss-Housson, F. (1976) Biochem. Biophys. Res. Commun. 69, 711–717.
25 Klimov, V.V., Shuvalov, V.A., Krakhmaleva, I.N., Klevanik, A.V. and Krasnovsky, A.A. (1977) Biokhimiya 42, 519–530.
26 Cogdell, R.J., Brune, D.C. and Clayton, R.K. (1974) FEBS Lett. 45, 344–347.
27 McElroy, J.D., Maurezall, D.C. and Feher, G. (1974) Biochim. Biophys. Acta 333, 261–278.
28 Kaufmann, K.J., Dutton, P.L., Netzel, T.L. and Rentzepis, P.M. (1975) Science 188, 1201–1304.
29 Rockley, M.G., Windsor, M.W., Cogdell, R.J. and Parson, W.W. (1975) Proc. Natl. Acad. Sci. U.S.A. 72, 2251–2255.
30 Schenck, C.C., Parson, W.W., Holten, D. and Windsor, M.W. (1981) Biochim. Biophys. Acta 635, 383–392.
31 Blankenship, R.E. and Parson, W.W. (1979) Biochim. Biophys. Acta 545, 429–444.

---

* Due to the extension of the topics covered in this chapter, the reference list has been kept to a minimum and only a few representative research papers and reviews are mentioned per section. We wish to apologize for this to all our colleagues. For more detailed information the reader should refer to the following reviews: For bacterial RC, Refs. 1, 2, 3, 5, 7, 292–294; for PSI-RC and PSII-RC, Refs. 1, 3–5, 295; for ubiquinol-cyt. *c* (plastocyanin) oxidoreductase, Refs. 93, 109, 116, 139, 140, 296; for the oxygen-evolving complex, 93, 297–301.

32 Vermeglio, A. (1977) Biochim. Biophys. Acta 459, 516–524.
33 Wraight, C.A. (1977) Biochim. Biophys. Acta 459, 525–531.
34 Dutton, P.L., Leigh, J.S. and Reed, D.W. (1973) Biochim. Biophys. Acta 292, 654–664.
35 Feher, G., Okamura, M.Y. and McElroy, J.D. (1972) Biochim. Biophys. Acta 267, 222–226.
36 Baccarini-Melandri, A., Gabellini, N., Melandri, B.A., Jones, K.R., Rutherford, A.W., Crofts, A.R. and Hurt, E. (1982) Arch. Biochem. Biophys. 216, 566–580.
37 O'Keefe, D., Prince, R.C. and Dutton, P.L. (1981) In Function of Quinones in Energy Conserving Systems (Trumpover, B., ed.) pp. 271–276, Academic Press, New York.
38 Crofts, A.R., Meinhardt, S.W., Jones, K.R. and Snozzi, M. (1983) Biochim. Biophys. Acta 723, 202–218.
39 Okamura, M.Y., Isaacson, R.A. and Feher, G. (1978) Biophys. J. 21, 8a.
40 Wraight, C.A. (1981) In Function of Quinones in Energy Conserving Systems (Trumpover, B., ed.) pp. 181–188, Academic Press, New York.
41 Kok, B. (1956) Biochim. Biophys. Acta 22, 399–401.
42 Bengis, C. and Nelson, N. (1975) J. Biol. Chem. 250, 2783–2788.
43 Bengis, C. and Nelson, N. (1977) J. Biol. Chem. 252, 4564–4569.
44 Haenel, W., Hesse, V. and Propper, A. (1980) FEBS Lett. 111, 79–82.
45 Nelson, N. and Notsani, B. (1977) Rev. Bioenerg. Biomembr. 1, 233–244.
46 Nelson, N., Bengis, C., Silver, B.L., Getz, D. and Evans, M.C.W. (1975) FEBS Lett. 58, 363–365.
47 Orlich, G. and Hauska, G. (1980) Eur. J. Biochem. 111, 525–533.
48 Hauska, G., Samoray, D., Orlich, G. and Nelson, N. (1980) Eur. J. Biochem. 111, 535–543.
49 Kok, B. (1961) Biochim. Biophys. Acta 48, 527–533.
50 Knaff, D.B. and Malkin, R. (1973) Arch. Biochem. Biophys. 159, 555–562.
51 Philipson, K.D., Sato, V.L. and Sauer, K. (1972) Biochemistry 11, 4591–4594.
52 Donerman, D. and Senge, H. (1980) In Photosynthesis (Akoyonoglou, E., ed.) Vol. 5, pp. 223–231, Vol. 5, Balaban Int. Services. Philadelphia.
53 Fuhrhop, J.H. and Maurezall, D. (1969) J. Am. Chem. Soc. 91, 4174–4181.
54 Frank, H.A., McLean, M.B. and Sauer, K. (1979) Proc. Natl. Acad. Sci. U.S.A. 76, 5124–5128.
55 Sauer, K., Mathis, P., Acker, S. and van Best, J.A. (1978) Biochim. Biophys. Acta 503, 120–134.
56 Mathis, P., Sauer, K. and Remy, R. (1978) FEBS Lett. 88, 275–278.
57 Shuvalov, V.A., Dolan, E. and Ke, B. (1979) Proc. Natl. Acad. Sci. U.S.A. 76, 770–773.
58 Baltimore, B.G. and Malkin, R. (1980) Photochem. Photobiol. 31, 485–490.
59 Ke, B., Dolan, E., Sugahara, K., Hawkridge, F.M., Demeter, S. and Shaw, E.R. (1977) Plant Cell Physiol., Spec. Issue 3, 187–199.
60 Malkin, R. and Bearden, A.J. (1971) Proc. Natl. Acad. Sci. U.S.A. 68, 16–19.
61 Hiyama, T. and Ke, B. (1971) Arch. Biochem. Biophys. 147, 99–108.
62 Evans, M.C.W., Reeves, S.G. and Cammack, R. (1974) FEBS Lett. 49, 111–114.
63 Goldbeck, J.H., Lien, S. and San Pietro, A. (1977) Arch. Biochem. Biophys. 178, 140–150.
64 Ke, B., Hansen, R.E. and Beinert, H. (1973) Proc. Natl. Acad. Sci. U.S.A. 70, 2941–2945.
65 Ke, B., Demeter, S., Zamaraev, K.I. and Kairutdinov, R.F. (1979) Biochim. Biophys. Acta 545, 265–284.
66 Goldbeck, J.H. and Kok, B. (1978) Arch. Biochem. Biophys. 188, 233–242.
67 Vernon, L.P. and Klein, S.M. (1975) Ann. N.Y. Acad. Sci. 244, 281–296.
68 Ke, B., Sugahara, K. and Shaw, E.R. (1975) Biochim. Biophys. Acta 408, 12–25.
69 Wessels, J.S.C., Alphen-van Waveren, O. and Voorn, G. (1973) Biochim. Biophys. Acta 292, 741–752.
70 Satoh, K. (1979) Biochim. Biophys. Acta 546, 84–92.
71 Satoh, K., Nakasani, H.Y., Steinback, K.E., Watson, J. and Arntzen, C.J. (1983) Proc. Int. Congr. Photosynth., 6th (Abst.) p. 173.
72 Doring, G., Renger, G., Vater, J. and Witt, H.T. (1969) Z. Naturforsch. Teil B. 48B, 1139–1143.
73 Klimov, V.V. (1984) In Advances in Photosynthetic Research (Sybesma, C., ed.) Vol. I, pp. 131–138, Martinus Nijhoff, The Hague.
74 Malkin, R. and Bearden, A.J. (1975) Biochim. Biophys. Acta 396, 250–251.

75 Visser, J.W.M., Rijgersberg, C.P. and Gast, P. (1977) Biochim. Biophys. Acta 460, 36–46.
76 Davis, M.S., Forman, A. and Fajer, J. (1979) Proc. Natl. Acad. Sci. U.S.A. 76, 4170–4174.
77 Klimov, V.V., Klevanik, A.V., Shuvalov, V.A. and Krasnovsky, A.A. (1977) FEBS Lett. 82, 183–186.
78 Klimov, V.V., Allackhverdiev., S.I., Shutilvova, N.I. and Krasnovsky, A.A. (1980) Sov. Plant. Physiol. (Eng. Transl.) 27, 315–326.
79 Klimov, V.V., Klevanik, A.V., Shuvalov, V.A. and Krasnovsky, A.A. (1977) FEBS Lett. 82, 183–186.
80 Klimov, V.V., Allackhverdiev, S.I., Demeter, S. and Krasnovsky, A.A. (1979) Dokl. Akad. Nauk. SSSR 249, 227–230.
81 Klimov, V.V., Allackhverdiev, S.I. and Krasnovsky, A.A. (1979) Dokl. Akad. Nauk. SSSR 249, 485–488.
82 Klimov, V.V., Dolan, E. and Ke, B. (1980) Proc. Natl. Acad. Sci. U.S.A. 77, 7227–7231.
83 Duysens, L.N.M. and Sweers, H.E. (1963) In Studies on Microalgae and Photosynthetic Bacteria (Jpn. Soc. Plant Physiol., ed.) pp. 353–372, Univ. of Tokyo Press, Tokyo.
84 Stiehl, H.H. and Witt, H.T. (1969) Z. Naturforsch. Teil B, 24B, 1588–1598.
85 van Gorkom, H.J. (1974) Biochim. Biophys. Acta 347, 439–442.
86 Knaff, D.B. and Arnon, D.I. (1969) Proc. Natl. Acad. Sci. U.S.A. 63, 956–962.
87 Nugent, J.H.A., Diner, B.A. and Evans, M.C.W. (1981) FEBS Lett. 124, 241–244.
88 Izawa, S. (1977) Encycl. Plant Physiol., New Ser. 5, 266–282.
89 Klimov, V.V., Dolan, E., Shaw, E.R. and Ke, B. (1980) Proc. Natl. Acad. Sci. U.S.A. 77, 7227–7231.
90 Cramer, W.A. and Butler, W.L. (1969) Biochim. Biophys. Acta 172, 503–510.
91 Ke, B., Hawkridge, F.M. and Sahu, S. (1976) Proc. Natl. Acad. Sci. U.S.A. 73, 2211–2215.
92 Knaff, D. (1975) FEBS Lett. 60, 331–335.
93 Cramer, W.A. and Crofts, A.R. (1982) In Photosynthesis. Energy Conversion by Plant and Bacteria (Govindjee, ed.) pp. 387–467, Academic Press, New York.
94 Goldbeck, J.H. and Kok, B. (1979) Biochim. Biophys. Acta 547, 347–360.
95 Joliot, P. and Joliot, A. (1973) Biochim. Biophys. Acta 305, 302–316.
96 Melis, A. (1978) FEBS Lett. 95, 202–206.
97 Horton, P. and Croze, E. (1979) Biochim. Biophys. Acta 545, 188–201.
98 Erixon, K. and Butler, W.L. (1971) Biochim. Biophys. Acta 234, 381–389.
99 Bouge-Bocquet, B. (1973) Biochim. Biophys. Acta 314, 250–256.
100 Velthuys, B.R. and Amesz, J. (1974) Biochim. Biophys. Acta 333, 85–94.
101 Bowes, J.M., Crofts, A.R. and Arntzen, C.J. (1980) Arch. Biochem. Biophys. 200, 303–308.
102 Mathis, P. and Haveman, J. (1976) Biochim. Biophys. Acta 461, 167–181.
103 Pulles, M.P.J., van Gorkom, H.J. and Willemsen, J.G. (1976) Biochim. Biophys. Acta 449, 536–540.
104 Butler, W.L. (1972) Proc. Natl. Acad. Sci. U.S.A. 69, 3420–3422.
105 Duysens, L.N.M., Den Haan, G.A. and van Vest, J.A. (1975) Proc. Int. Congr. Photosynth., 3rd, pp. 1–12.
106 Blankenship, R.E., Babcock, G.T., Warden, J.T. and Sauer, K. (1975) FEBS Lett. 51, 287–293.
107 Warden, J.T., Blankenship, R.E. and Sauer, K. (1976) Biochim. Biophys. Acta 423, 462–478.
108 Babcock, J.T., Buttner, W.J. Ghanotakis, D.F., O'Malley, P.J., Yerkes, C.T. and Yocum, C.F. (1984) In Advances in Photosynthesis Research (Sybesma, C., ed.) Vol. I, pp. 243–252, Martinus Nijhoff, The Hague.
109 Hauska, G., Hurt, E., Gabellini, N. and Lockau, W. (1983) Biochim. Biophys. Acta 726, 97–133.
110 Gabellini, N., Bowyer, J.R., Hurt, E., Melandri, B.A. and Huska, G. (1982) Eur. J. Biochem. 126, 105–111.
111 Hurt, E. and Hauska, G. (1981) Eur. J. Biochem. 117, 591–599.
112 Krinner, M., Hauska, G., Hurt, E. and Lockau, W. (1982) Biochim. Biophys. Acta 681, 110–117.
113 Yu, C.-A., Yu, L. and King, T.E. (1975) Biochem. Biophys. Res. Commun. 66, 1194–1200.
114 Nelson, N. and Neumann, J. (1972) J. Biol. Chem. 247, 1917–1924.
115 Hurt, E. and Hauska, G. (1982) J. Bioenerg. Biomembr. 14, 119–138.
116 Dutton, P.L. and Prince, R.C. (1978) In The Photosynthetic Bacteria (Clayton, R.K. and Sistrom, W.R., eds.) pp. 525–570, Plenum, New York.
117 Bowyer, J.R., Meinhardt, S.W., Tierney, G.V. and Crofts, A.R. (1981) Biochim. Biophys. Acta 635, 167–186.

118 Petty, K.M. and Dutton, P.L. (1976) Arch. Biochem. Biophys. 172, 346–353.
119 Crofts, A.R. and Meinhardt, S.W. (1982) Biochem. Soc. Trans. 10, 201–203.
120 Cramer, W.A. and Whitmarsh, J. (1975) Ann. Rev. Plant Physiol. 28, 133–17.
121 Wood, P.M. (1980) Biochem. J. 189, 385–391.
122 Meinhardt, S.W. and Crofts, A.R. (1983) FEBS Lett. 149, 223–227.
123 Boehme, H., Bruetsch, S., Weithmann, G. and Boeger, P. (1980) Biochim. Biophys. Acta 590, 248–260.
124 Bendall, D.S., Davenport, H.E. and Hill, R. (1971) Methods Enzymol. 23A, 327–344.
125 Prince, R.C., Lindsay, J.G. and Dutton, P.L. (1975) FEBS Lett. 51, 108–111.
126 Beinert, H. (1977) In Iron-Sulfur Proteins (Lovenberg, W., ed.) pp. 61–100, Academic Press, New York.
127 Bowyer, J.R., Dutton, P.L., Prince, R.C. and Crofts, A.R. (1980) Biochim. Biophys. Acta 592, 445–460.
128 Krinner, M. (1981) Thesis, University of Regensburg.
129 Hurt, E., Hauska, G. and Malkin, R. (1981) FEBS Lett. 134, 1–5.
130 Hurt, E. and Hauska, G. (1982) Photobiochem. Photobiophys. 4, 9–15.
131 Malkin, R. (1981) FEBS Lett. 131, 169–172.
132 Malkin, R. (1982) Biochemistry 21, 2945–2949.
133 Baum, H., Rieskie, J.S., Silman, H.J. and Lipton, S.H. (1967) Proc. Natl. Acad. Sci. U.S.A. 57, 798–805.
134 Wikström, M. and Berden, J. (1972) Biochim. Biophys. Acta 283, 403–420.
135 Mitchell, P. (1976) J. Theor. Biol. 62, 327–367.
136 Wikström, M. and Krab, K. (1980) Curr. Top. Bioenerg. 10, 52–103.
137 Takamiya, K.I., Bonner, W.D. and Dutton, P.L. (1979) J. Biol. Chem. 254, 8594–8604.
138 Hurt, E. and Hauska, G. (1982) Biochim. Biophys. Acta 682, 466–473.
139 Crofts, A.R. and Wood, P.M. (1978) Curr. Top. Bioenerg. 7, 175–244.
140 Baccarini-Melandri, A., Casadio, R. and Melandri, B.A. (1981) Curr. Top. Bioenerg. 12, 197–258.
141 Evans, E.H. and Crofts, A.R. (1974) Biochim. Biophys. Acta 357, 89–102.
142 Takamiya, K.I. and Dutton, P.L. (1979) Biochim. Biophys. Acta 546, 1–16.
143 Meinhardt, S.W. and Crofts, A.R. (1983) Biochim. Biophys. Acta 638, 219–230.
144 Baccarini-Melandri, A. and Melandri, B.A. (1977) FEBS Lett. 80, 459–464.
145 Takamiya, K.I., Prince, R.C. and Dutton, P.L. (1979) J. Biol. Chem. 254, 11307–11311.
146 Matsuura, K., Packam, N.K., Tiede, D.M., Mueller, P. and Dutton, P.L. (1982) In Function of Quinones in Energy Conserving Systems (Trumpover, B., ed.) pp. 277–283, Academic Press, New York.
147 Bashford, C.L., Prince, R.C., Takamiya, K.I. and Dutton, P.L. (1979) xx 545, 223–235.
148 Petty, K.M., Jackson, J.B. and Dutton, P.L. (1979) Biochim. Biophys. Acta 546, 17–42.
149 Hurt, E., Gabellini, N., Shahak, Y., Lockau, W. and Hauska, G. (1983) Arch. Biochem. Biophys. 225, 879–885.
150 Joliot, P. and Delosme, R. (1973) Biochim. Biophys. Acta 357, 267–284.
151 Hind, G., Crowther, D., Shahak, Y. and Slovacek, R.E. (1981) Proc. Int. Congr. Photosynth., 5th. Vol. II 87, 97.
152 Govindjee and Govindjee, R. (1965) Photochem. Photobiol. 4, 675–683.
153 Bouges-Bocquet, B. (1977) Biochim. Biophys. Acta 462, 371–379.
154 Bouges-Bocquet, B. (1981) Biochim. Biophys. Acta 635, 327–340.
155 Malkin, R. (1981) Proc. Int. Congr. Photosynth., 5th. Vol. II 643–653.
156 Velthuys, B.R. (1978) Proc. Natl. Acad. Sci. U.S.A. 75, 6031–6034.
157 Velthuys, B.R. (1980) Annu. Rev. Plant. Physiol. 31, 545–567.
158 Olsen, L.F. and Cox, R.P. (1979) Eur. J. Biochem. 95, 427–432.
159 Böhme, H. (1976) Z. Naturforsch. 31C, 68–77.
160 Böhme, H. (1979) Eur. J. Biochem. 93, 287–293.
161 Velthuys, B.R. (1979) Proc. Natl. Acad. Sci. U.S.A. 76, 2765–2769.
162 Nakatani, H.Y. and Barber, J. (1981) Photobiochem. Photobiophys. 2, 69–78.

163 Sayre, R.T. and Cheniae, G.M. (1982) Plant Physiol. 69, 1084–1095.
164 Åkerlund, H.-E., Jansson, C. and Andersson, B. (1982) Biochim. Biophys. Acta 681, 1–10.
165 Pirson, A., Tichy, C. and Wilhelmi, G. (1952) Planta 40, 199–253.
166 Kessler, E., Arthur, W. and Brugger, J.E.(1957) Arch. Biochem. Biophys. 71, 326–335.
167 Cheniae, G.M. and Martin, I.F. (1971) Plant Physiol. 47, 568–575.
168 Katoh, S. and San Pietro, A. (1967) Arch. Biochem. Biophys. 122, 144–152.
169 Cheniae, G.M. and Martin, I.F. (1970) Biochim. Biophys. Acta 197, 219–239.
170 Yamashita, T. and Horio, T. (1968) Plant Cell. Physiol. 9, 267–284.
171 Blankenship, R.E. and Sauer, K., (1974) Biochim. Biophys. Acta 357, 252–266.
172 Blankenship, R.E., Babcock, G.T. and Sauer, K. (1975) Biochim. Biophys. Acta 387, 165–175.
173 Itoh, M., Yamashita, K., Nishi, T., Konishi, K. and Shibata, K. (1969) Biochim. Biophys. Acta 180, 509–519.
174 Selman, B.R., Bannister, T.T. and Dilley, R.A. (1973) Biochim. Biophys. Acta 292, 566–581.
175 Cheniae, G.M. and Martin, I.F. (1966) Brookhaven Symp. Biol. 19, 409–417.
176 Gerhardt, B. and Wiessner, W. (1967) Biochem. Biophys. Res. Commun. 28, 958–964.
177 Cheniae, G.M. and Martin, I.F. (1971) Biochim. Biophys. Acta 253, 167–181.
178 Homann, P.H. (1967) Plant Physiol. 42, 997–1006.
179 Yocum, C.F., Yerches, C.T., Blankenship, R.E., Sharp, R.R. and Babcock, G.T. (1981) Proc. Natl. Acad. Sci. U.S.A. 78, 7507–7511.
180 Izawa, S., Heath, R.L. and Hind, G. (1969) Biochim. Biophys. Acta 180, 388–398.
181 Kelley, P.M. and Izawa, S. (1978) Biochim. Biophys. Acta 502, 198–210.
182 Gol'dfel'd, M., Khaolova, I.I. and Vanin, A.F. (1979) Biofizika 24, 550–551.
183 Muallem, A. and Izawa, S. (1980) FEBS Lett. 115, 49–53.
184 Critchley, C., Baianu, I.C., Govindjee, R. and Gutowsky, H.S. (1982) Biochim. Biophys. Acta 682, 436–445.
185 Piccioni, R.G. and Mauzerall, D.C. (1976) Biochim. Biophys. Acta 423, 605–609.
186 Barr, R., Troxel, K.S. and Crane, S.L. (1982) Biochem. Biophys. Res. Commun. 104, 1182–1188.
187 Allen, F. and Franck, J. (1955) Arch. Biochem. Biophys. 58, 124–143.
188 Joliot, P., Barbieri, G. and Chabaud, R. (1969) Photochem. Photobiol. 10, 309–329.
189 Kok, B., Forbush, B. and McGloin, M. (1970) Photochem. Photobiol. 11, 457–475.
190 Bouges-Bocquet, B., Bennoun, P. and Taboury, J. (1973) Biochim. Biophys. Acta 325, 247–254.
191 Doshek, W.W. and Kok, B. (1972) Biophys. J. 12, 832–838.
192 Diner, B. (1977) In Photosynthesis '77 (Hall, D.O., Coombs, J. and Goodwin, T.W., eds.) pp. 359–372, Biochem. Soc., London.
193 Eckert, H.J. and Renger, G. (1980) Photochem. Photobiol. 31, 501–511.
194 Forbush, B., Kok, B. and McGloin, M. (1971) Photochem. Photobiol. 14, 307–321.
195 Bouges, B. (1971) These de Troisieme Cycle, Paris, France.
196 Bouges-Bocquet, B. (1973) Biochim. Biophys. Acta 314, 250–256.
197 Kok, B., Radmer, R. and Fowler, C.F. (1975) Proc. Int. Congr. Photosynth., 3rd, 1974, pp. 485–496.
198 Joliot, P., Joliot, A., Bouges, B. and Barbieri, G. (1971) Photochem. Photobiol. 12, 287–305.
199 Renger, G., Bouges-Bocquet, B. and Deslome, R. (1973) Biochim. Biophys. Acta 292, 796–802.
200 Velthuys, B. (1975) Biochim. Biophys. Acta 396, 392–401.
201 Wydrzynsky, T. and Sauer, K. (1980) Biochim. Biophys. Acta 589, 56–70.
202 Weise, C., Kenneth, J., Solnit, T. and Von Gutfeld, R.J. (1971) Biochim. Biophys. Acta 253, 298–301.
203 Weise, C. and Sauer, K. (1970) Photochem. Photobiol. 11, 495–501.
204 Jursinic, P. (1981) Biochim. Biophys. Acta 635, 38–52.
205 Bouges-Bocquet, B. (1980) Biochim. Biophys. Acta 594, 85–104.
206 Zankel, K. (1971) Biochim. Biophys. Acta 245, 373–385.
207 Lavorel, J. and Lemasson, C. (1976) Biochim. Biophys. Acta 430, 501–516.
208 Horton, P. and Croze, E. (1977) Biochim. Biophys. Acta 462, 86–101.
209 Rich, P. and Bendall, D.S. (1980) Biochim. Biophys. Acta 591, 506–518.
210 Satoh, K. (1981) Proc. Int. Congr. Photosynth., 5th III, 607–616.
211 Lach, H.-J. and Boeger, P. (1977) Z. Naturforsch, C, 32C, 877–879.

212 Whitmarsh, J. and Cramer, W.A. (1977) Biochim. Biophys. Acta 460, 280–289.
213 Whitmarsh, J. and Cramer, W.A. (1978) Biochim. Biophys. Acta 501, 83–93.
214 Heber, U., Kirk, M.R. and Boardman, N.K. (1979) Biochim. Biophys. Acta 546, 292–306.
215 Butler, W.L. (1978) FEBS Lett. 95, 19–25.
216 Ben-Hayyim, G. (1974) FEBS Lett. 41, 191–196.
217 Velthuys, B.R. (1981) Proc. Int. Cong. Photosynth., 5th, II, 75–85.
218 Cramer, W.A., Whitmarsh, J. and Horton, P. (1979) In The Porphyrins (Dolphin, D., ed.) Vol. 7, pp. 71–106, Academic Press, New York.
219 Kaplan, S. and Arntzen, C.J. (1982) In Photosynthesis. Energy Conversion in Plants and Bacteria (Govindjee, ed.) pp. 67–151, Academic Press, New York.
220 Prince, R.C., Baccarini-Melandri, A., Hauska, G., Melandri, B.A. and Crofts, A.R. (1975) Biochim. Biophys. Acta 387, 212–227.
221 Overfield, R.E. and Wraight, C.A. (1980) Biochemistry 19, 3322–3327.
222 Overfield, R.E. and Wraight, C.A. (1980) Biochemistry 19, 3328–3334.
223 Bowyer, J.R. (1979) Ph.D. Thesis, Univ. of Bristol, p. 222.
224 Baccarini-Melandri, A., Jones, O.T.G. and Hauska, G. (1978) FEBS Lett. 86, 151–154.
225 Prince, R.C., Bashford, L., Takamiya, K., van den Berg, W.H. and Dutton, P.L. (1978) J. Biol. Chem. 253, 4137–4142.
226 Katoh, S., Shiratori, I. and Takamiya, A. (1962) J. Biochem. (Tokyo) 51, 32–40.
227 Plesnicar, M. and Bendall, D.S. (1970) Biochim. Biophys. Acta 216, 192–199.
228 Boardman, N.K., Björkman, O., Anderson, J.M., Goodchild, D.J. and Thorne, S.W. (1972) Proc. Int. Congr. Photosynth., 3rd, pp. 1809–1827.
229 Haenel, W., Proepper, A. and Krause, H. (1980) Biochim. Biophys. Acta 593, 384–399.
230 Haenel, W. and Krause, H. (1982) Second Eur. Conf. Bioenerg., Short Rep., pp. 217–218.
231 Hauska, G.A., McCarty, R.E., Berzborn, R.J. and Racker, E.F. (1971) J. Biol. Chem. 346, 3524–3531.
232 Plesnicar, M. and Bendall, D.S. (1973) Eur. J. Biochem. 34, 483–488.
233 Whitmarsh, J. and Cramer, W.A. (1979) Proc. Natl. Acad. Sci. U.S.A. 76, 4417–4420.
234 Crane, F.L. (1965) In The Biochemistry of Quinones (Morton, R.A., ed.) pp. 183–206, Academic Press, New York.
235 Hauska, G. and Hurt, E. (1981) In Function of Quinones in Energy Conserving Systems (Trumpover, B., ed.) pp. 87–110, Academic Press, New York.
236 Witt, H.T. (1979) Biochim. Biophys. Acta 505, 355–427.
237 Slovacek, R.E., Crowther, D. and Hind, G. (1979) Biochim. Biophys. Acta 547, 138–148.
238 Casadio, R., Venturoli, G. and Melandri, B.A. (1982) Second Eur. Conf. Bioenerg., Short Rep., pp. 185–186.
239 Snozzi, M. and Crofts, A.R. (1983) Proc. Int. Congr. Photosynth., 6th, (Abstr.) p. 92.
240 Siggel, U., Renger, G., Stiehl, H.H. and Rumberg, B. (1972) Biochim. Biophys. Acta 256, 328–335.
241 Takamiya, K. and Dutton, P.L. (1979) Biochim. Biophys. Acta 546, 1–16.
242 Packham, N.K., Trede, D.M., Mueller, P. and Dutton, P.L. (1980) Proc. Natl. Acad. Sci. U.S.A. 77, 6339–6343.
243 Huber, S.C. and Edwards, G.E. (1977) FEBS Lett. 79, 207–211.
244 Crowther, D., Mills, J.D. and Hind, G. (1979) FEBS Lett. 98, 386–390.
245 Hall, D.O. and Rao, K.K. (1977) Encycl. Plant Physiol., New Ser. 5, 206–216.
246 Evans, M.C.W., Hall, D.O., Bothe, H. and Whatley, F.R. (1968) Biochem. J. 110, 485–489.
247 San Pietro, A. and Keister, D.L. (1962) Arch. Biochem. Biophys. 98, 235–243.
248 Zanetti, G. and Forti, G. (1966) J. Biol. Chem. 241, 279–285.
249 Forti, G., Encycl. Plant Physiol., New Ser. 5, 222–226.
250 Foust, G.P., Mayhew, S.G. and Massey, V. (1969) J. Biol. Chem. 244, 964–970.
251 Berzborn, R.J. (1969) Z. Naturforsch., Teil B, 24B, 436–446.
252 Mitchell, P. (1966) Biol. Rev. Cambridge Phil. Soc. 41, 445–502.
253 Papa, S. (1976) Biochim. Biophys. Acta 456, 39–84.
254 Baccarini-Melandri, A. and Zannoni, D. (1978) J. Bioenerg. Biomembr. 10, 109–139.
255 Jackson, J.B. and Crofts, A.R. (1971) Eur. J. Biochem. 18, 120–130.
256 Jackson, J.B. and Dutton, P.L. (1973) Biochim. Biophys. Acta 325, 102–115.

257 Schoenfeld, M., Montal, M. and Feher, G. (1979) Proc. Natl. Acad. Sci. U.S.A. 76, 6351–6355.
258 Packham, N.K., Packham, C., Mueller, P., Tiede, D.M. and Dutton, P.L. (1980) FEBS Lett. 110, 101–106.
259 Crofts, A.R., Crowther, D., Celis, H., Calis, S. and Tierney, G.V. (1977) Biochem. Soc. Trans. 5, 491–495.
260 Wraight, C.A., Cogdell, R.J. and Chance, B. (1978) In The Photosynthetic Bacteria (Clayton, R.K. and Sistrom, W.R., eds.) pp. 471–511, Plenum, New York.
261 Patty, K.M. and Dutton, P.L. (1976) Arch. Biochem. Biophys. 172, 335–345.
262 Cogdell, R.J., Jackson, J.B. and Crofts, A.R. (1973) J. Bioenerg. 4, 211–227.
263 Kingsley, P.B. and Feigenson, G.W. (1981) Biochim. Biophys. Acta 635, 602–618.
264 Futami, A. and Hauska, G. (1979) Biochim. Biophys. Acta 547, 597–608.
265 Futami, A., Hurt, E. and Hauska, G. (1979) Biochim. Biophys. Acta 547, 583–596.
266 Fowler, C.F. and Kok, B. (1974) Biochim. Biophys. Acta 357, 299–307.
267 Fowler, C.F. (1977) Biochim. Biophys. Acta 459, 351–363.
268 Saphon, S. and Crofts, A.R. (1977) Z. Naturforsch., Teil B, 32B, 617–626.
269 Velthuys, B.R. (1980) FEBS Lett. 115, 167–170.
270 Junge, W., Renger, G. and Auslaender, W. (1977) FEBS Lett. 79, 155–159.
271 Hope, A.B. and Moreland, A. (1979) Aust. J. Plant. Physiol. 6, 1–6.
272 Foerster, V., Hong, Y.-Q. and Junge, W. (1981) Biochim. Biophys. Acta 638, 141–152.
273 Govindjee, Wydrzynski, T. and Marks, S.B. (1977) In Bioenergetics of Membranes (Packer, L., Papgeorgiou, G. and Trebst, A., eds.) pp. 305–316, Elsevier, Amsterdam.
274 Sauer, K. (1980) Acc. Chem. Res. 13, 246–256.
275 Junge, W. and Jackson, J.B. (1982) In Photosynthesis. Energy Conversion by Plants and Bacteria (Govindjee, ed.) pp. 589–646, Academic Press, New York.
276 Rottenberg, H., Gruenwald, T. and Avron, M. (1972) Eur. J. Biochem. 25, 54–63.
277 Bulychev, A.A. and Vredenberg, W.J. (1976) Biochim. Biophys. Acta 423, 548–556.
278 Kell, D.B. (1979) Biochim. Biophys. Acta 549, 55–99.
279 Ausländer, W. and Junge, W. (1975) FEBS Lett. 59, 310–315.
280 Junge, W., Ausländer, W., McGeer, A. and Runge, R. (1979) Biochim. Biophys. Acta 546, 121–141.
281 Theg, S.M. and Junge, W. (1983) Biochim. Biophys. Acta 723, 294–307.
282 Prochaska, L.J. and Dilley, R.A. (1978) Arch. Biochem. Biophys. 187, 61–71.
283 Theg, S.M., Johson, J.D. and Homann, P.H. (1982) FEBS Lett. 145, 25–29.
284 Hong, Y.-Q. and Junge, W. (1983) Biochim. Biophys. Acta 722, 187–208.
285 Renger, G. and Voelker, M. (1982) FEBS Lett. 149, 203–207.
286 Westerhoff, H., Venturoli, G., Melandri, B.A., Azzone, G.F. and Kell, D.B. (1983) FEBS Lett. 165, 1–5.
287 Ort, D.R. and Melandri, B.A. (1982) In Photosynthesis. Energy conversion by Plants and Bacteria (Govindjee, ed.) pp. 537–587, Academic Press, New York.
288 Vermeglio, A. and Clayton, R.K. (1977) Biochim. Biophys. Acta 461, 159–165.
289 Heathcote, P., Timofeev, K.N. and Evans, M.C.W. (1979) FEBS Lett. 101, 105–109.
290 Evans, M.C.W., Shira, C.K. and Slabas, A.R. (1977) Biochem. J. 162, 75–85.
291 Doering, G., Stiehl, H.H. and Witt, H.T. (1967) Z. Naturforsch., Teil B, 22B, 639–644.
292 Gingras, G. (1978) In The Photosynthetic Bacteria (Clayton, R.K. and Sistrom, W.R., eds.) pp. 119–131, Plenum, New York.
293 Blankenship, R.E. and Parson, W.W. (1978) Annu. Rev. Biochem. 47, 635–653.
294 Hoff, A.J. (1979) Phys. Rev. 54, 75–200.
295 Boardman, N.I., Anderson, J.M. and Goodchild, D.J. (1978) Curr. Top. Bioenerg. 7, 75–138.
296 Baccarini-Melandri, A. and Zannoni, D. (1978) J. Bioenerg. Biomembr. 10, 109–138.
297 Cheniae, G.M. (1970) Annu. Rev. Plant Physiol. 21, 467–498.
298 Redmer, R. and Kok, B. (1975) Annu. Rev. Biochem. 44, 409–433.
299 Diner, B. and Joliot, P. (1977) Encycl. Plant Physiol., New Ser. 5, 187–205.
300 Velthuys, B. (1980) Annu. Rev. Plant Physiol., 31, 545–567.
301 Wydrzynsky, T. (1982) In Photosynthesis. Energy Conversion by Plants and Bacteria (Govindjee, ed.) pp. 469–506, Academic Press, New York.

Ernster (ed.) Bioenergetics
© 1984 Elsevier Science Publishers B.V.

CHAPTER 5

Proton motive ATP synthesis

YASUO KAGAWA

Department of Biochemistry, Jichi Medical School, Minamikawachimachi, Tochigi-ken, Japan 329-04

1. Introduction

Two major ATP synthesizing reactions in living organisms are oxidative phosphorylation and photophosphorylation. Both reactions take place in H^+-ATPase (F_0F_1), which is driven by an electrochemical potential difference of protons across the biomembrane, as predicted by Mitchell [1]. In Racker's laboratory, ATPases related to oxidative phosphorylation were prepared, but their relationship to Mitchell's chemiosmotic hypothesis [1] was not described [2]. Later, an insoluble ATPase (H^+-ATPase) was shown to translocate protons across the membrane when it was reconstituted into liposomes [3]. H^+-ATPase was shown to be composed of a catalytic moiety called F_1 (coupling factor 1) [4], and a membrane moiety called F_0 [5], which confers inhibitor sensitivity to F_1. F_0 was shown to be a proton channel, which translocates H^+ down an electrochemical potential gradient across the membrane when F_0 is reconstituted into liposomes (Fig. 5.1) [6]. Thus, H^+-ATPase was called F_0F_1 or ATP synthetase.

For definition of H^+-ATPase, the following questions must be answered: I. What is it? II. What does it do? III. How does it do it? The next three sections report studies on these questions, i.e., the structure, function and mechanism of H^+-ATPase. Excellent reviews are available on this enzyme from mitochondria [7,8], chloroplasts [9,10] and bacteria [11,12], and also on the mechanism of its reaction [13–17].

The H^+-ATPase described here, called F_0F_1, differs greatly from another proton translocating ATPase discovered in fungal plasma membranes, etc. [18]. The latter,

Abbreviations: H^+ ATPase, entire proton translocating ATPase (F_0F_1); F_0, integral membrane portion of the proton translocating ATPase; F_1, catalytic portion of the proton translocating ATPase; F_0F_1, entire proton translocating ATPase; CF_1, CF_0, chloroplast F_1 and F_0; EF_1, EF_0, *Escherichia coli* F_1 and F_0; MF_1, MF_0, mitochondrial F_1 and F_0; TF_1, TF_0, thermophilic F_1 and F_0 obtained from thermophilic bacteria; 9AA, 9-aminoacridine; AMP-PNP, adenyl-5′-yl imidophosphate; ANS, 8-anilinonaphthalene-1-sulfonate; ATPS, thiophosphate analogue of ATP; DCCD, *N,N′*-dicyclohexylcarbodiimide; FCCP, carbonylcyanide *p*-trifluoromethoxyphenylhydrazone; FSBA, *p*-fluorosulfonyl-benzoyl-5′-adenosine; NBDCl, 4-chloro-7-nitro-2-oxal-1,3-diazole; P_i, inorganic orthophosphate.

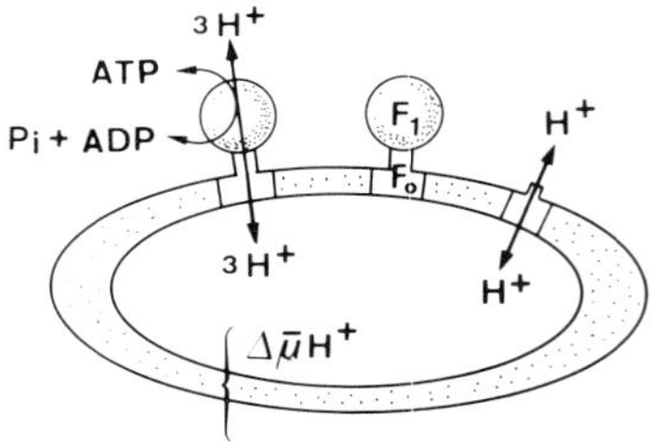

Fig. 5.1. H^+ ATPase (F_0F_1) and proton motive ATP synthesis. F_0F_1 reconstituted into liposomes transform energy of proton efflux driven by an electrochemical potential difference of protons across the membrane ($\Delta\tilde{\mu}_{H^+} = F\Delta\psi - 2.3RT \ln \Delta pH$) [3,38]. F_0 (right hand side of the liposome) without F_1 is an H^+ channel that allows free passage of H^+ down the $\Delta\tilde{\mu}_{H^+}$ [6,77].

like Ca^{2+}-ATPase and Na^+,K^+-ATPase, belongs to E_1E_2 type ATPase, which assumes two distinct conformations during ion transport. In contrast to the latter ATPase, F_0F_1 synthesizes ATP in physiological conditions without forming a phosphorylated intermediate, and has a complicated subunit organization.

In this review, the chemiosmotic hypothesis [1] at the physiological level, i.e., ATP synthesis in F_0F_1 driven by an electrochemical potential difference of protons (Fig. 5.1) is supported, while the hypothesis [1] at the level of molecular mechanism, i.e., the direct participation of the translocated protons in the dehydration of phosphate during ATP synthesis in F_0F_1 is excluded. The solid chemical and physical experiments on the purified F_1, F_0 and F_0F_1, and genetic analysis of the F_0F_1 established a new concept on the proton motive ATP synthesis.

2. *Structure of H^+ ATPase (F_0F_1)*

2.1. Subunits of F_0F_1 and its reconstitution

2.1.1. Subunits of F_1 and F_0

Purified prokaryotic F_0F_1 was shown to be composed of eight distinct subunits, five in F_1 and three in F_0 [19,20]. These biochemical observations were confirmed by genetic analysis of the F_0F_1 operon of *Escherichia coli* [11,21,22]. Similarly, chloroplast F_0F_1 is composed of eight subunits [23]. However, mitochondrial F_0F_1 is more complex: about 16 subunits are found in beef heart F_0F_1 [24,25], and 10 subunits in yeast F_0F_1 [26]. The nomenclature and molecular weights of these subunits are summarized in Table 5.1. In contrast to the gene of *Escherichia coli* F_0F_1, the gene of mitochondrial F_0F_1 has only partially been sequenced [27,28]. Moreover, mitochondrial F_0F_1 has several coupling factors such as F_B [29], which are not shown in Table 5.1. The three largest subunits of F_1 are common to all prokaryotic and eukaryotic F_1's [11,30,31], but the prokaryotic δ subunit corresponds to oligomycin sensitivity-conferring protein (OSCP) of prokaryotic F_0 [32]. The ϵ subunit and ATPase inhibitor [33,34] of eukaryotic F_0F_1 have no counterparts in prokaryotic

TABLE 5.1

Molecular weights of subunits of F_0F_1 [a]

Source	*Escherichia coli* [11]	*Saccharomyces cervisiae* [26]	Beef heart mitochondria [24]
Subunits			
α	55 264	56 000	56 000
β	50 316	52 000	51 500
γ	31 387	32 000	33 500
δ (OSCP)	19 558 (δ)	21 900	(22 500) (OSCP)
ϵ	14 914	13 700	14 000 δ
ϵ (animal)	–	–	6 000 ϵ
ATPase inhibitor	–	7 383 [34]	9 578 [33]
a	30 258	25 300	25 000, 21 000
b	17 233	23 700, 21 900	11 500, 10 500 10 000, 4 200
c (DCCD-binding p)	8 246	8 000 [67]	7 800 [67], 4 500

[a] Factor B, (14 600) [29] is not included.

F_0F_1's. The β and ϵ subunits of chloroplast F_1 are similar to those of prokaryotic F_1 [35,36]. There are three subunits (a, b and c) in prokaryotic F_0, but only subunit c was shown to be common to eight species, including plants and animals [37]. Subunit 6 (226 residues) of human mitochondrial F_0 has been sequenced [28], and shows some homology with the subunit b of *E. coli* F_0 [11,21].

For establishment of the functions of these subunits in proton motive ATP synthesis, it is essential to reconstitute F_0F_1 from its subunits. However, only thermophilic F_0F_1 can be reconstituted after complete destruction of its higher structure [38,39]. EF_1 was reconstituted after mild treatment for dissociation of its subunits without impairing their tertiary structure [40,41]. Genetic analysis [42–47], chemical modifications of subunits [48–50], and immunological inactivations [30] are also useful for elucidating the characters of F_0F_1.

2.1.2. Organization of subunits in F_0F_1

The subunit organization of prokaryotic F_0F_1 is shown in Fig. 5.2. This figure is based on the subunit stoichiometry ($\alpha:\beta:\gamma:\delta:\epsilon:a:b:c = 3:3:1:1:1:1:2:6$) of TF_1 [51], which is similar to that reported for F_0F_1's of other prokaryotes [15], mitochondria [26,52] and chloroplasts [23]. However, this stoicheiometry is still controversial [53], since 3 : 3 : 1 [54] and 2 : 2 : 2 [55] models of the major F_1 subunits have been proposed. The molecular weight supporting the 2 : 2 : 2 model is 342 898 (EF_1, Table 5.1), which is close to the value obtained by small angle X-ray [56] and neutron scattering [57], while that supporting the 3 : 3 : 1 model is 382 629, which is close to the values obtained by low and high speed ultracentrifugation (TF_1, EF_1, MF_1 and CF_1, 380 000) [30], X-ray crystallography of MF_1 (380 000) [58] and chemical analysis [15,23,26,52,53]. According to the latter model, the total molecular weight of F_0F_1 should be 496 829.

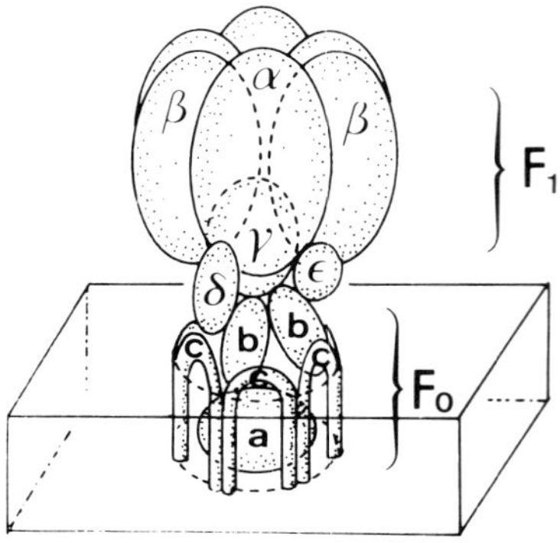

Fig. 5.2. Subunit organization of prokaryotic H^+ ATPase. The subunit stoicheiometry is based on that of thermophilic F_0F_1 [51]. $\alpha:\beta:\gamma:\delta:\epsilon:a:b:c = 3:3:1:1:1:1:2:6$. The F_0 portion forms a plug through the phospholipid bilayer. The c subunit assumes a hairpin shape [67].

As shown in Fig. 5.2, the major portion of F_1 is composed of α, β and γ subunits, and this portion shows ATPase activity. Thus, it is the core of F_1. The δ and ε subunits of F_1 constitute the portion connecting the F_1 core to F_0 in collaboration with the b subunit of F_0. In fact, the δ subunit of prokaryotic F_1 is OSCP of prokaryotic F_0 [32]. The major portion of F_0 is composed of a and c subunits. There are 6–10 c subunits, and each assumes hairpin shape. Evidence in support of this model will be discussed in the following sections.

2.2. *Primary structure and gene analysis*

2.2.1. *F_0F_1 gene*

The gene for F_0F_1 of *Escherichia coli* is a single operon called *unc* [42], *pap* [11] or *atp* [59]. The operon (Fig. 5.3) is approximately 7 kilo basepairs (Kb) in size and is located at 83 min on the chromosome [11,42]. The order of structural genes is a, c and b subunits of EF_0, followed by those for the δ, α, γ, β and ε subunits of EF_1 [11,21,59]. There is a reading frame of 130 amino acids preceeding the gene of the subunit a, but the gene product called 14K may not be essential for F_0F_1, since it is not found in F_0F_1, and transposon Tn_{10} insertion in this gene does not stop the synthesis of F_0F_1 [60]. There is a promotor sequence (P1) before the 14K gene with a typical Pribnow box, but two other promotors (P2, P3) are found in the 14K gene [11]. P1 with a complete structure may be active in vivo, though a plasmid carrying only P3 and the structural gene for F_0F_1 still produces F_0F_1 [61]. A characteristic sequence for termination of transcription is found in the region downstream of the gene for the ε subunit [11]. A single operon structure of the genes for F_0F_1 may not be universal even in prokaryotic cells.

The subunits of F_0F_1 in eukaryotic cells are coded partly by nuclear and partly by organellar genes. Mitochondrial DNA of human cells contains only the gene for subunit 6 [28], but that of yeast cells contains genes for both subunits 6 and 9 of MF_0 [27,62,63], corresponding to subunits a and c of EF_0, respectively. Chloroplast

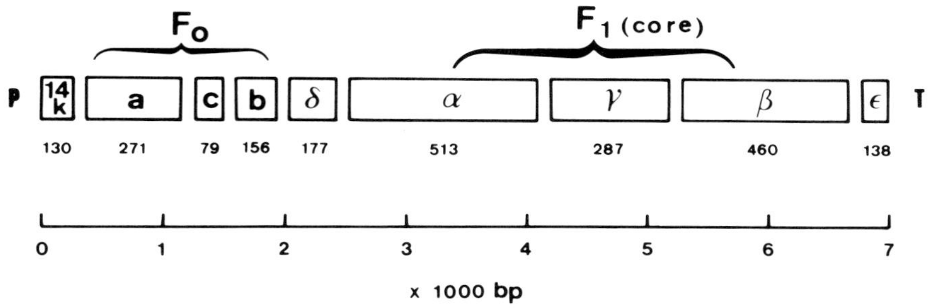

Fig. 5.3. The structure of the gene cluster coding for H^+ ATPase (F_0F_1). The F_0F_1 operon of *Escherichia coli* is shown [11,21,59]. Reading frames are shown in squares, and numbers below frames are numbers of amino acid residues. The reading frames are arranged from upstream (left) to downstream (right). A Shine-Dalgarno sequence is found before each reading frame. F_1 (core), the α, β and γ subunit complex that show ATPase activity. F_0, H^+ channel portion. In eukaryotic F_0, the δ subunit (OSCP) is included in F_0. F_1, prokaryotic F_1 is composed of the α, β, γ, δ and ϵ subunits. P, the promotor region. Of the three promoter regions (P1, P2 and P3), only P1 is complete with a Pribnow box. P2 and P3 are located in the coding frame of 14K. T, terminator region of transcription.

DNA contains genes for α, β and ϵ of CF_1 [35,36,44], and a [64,44] and c [35,36,44] of CF_0, and the β and ϵ genes are connected [35,36,44]. The other nuclear genes of F_0F_1 have been extensively studied by the cell fusion technique [65] and gene cloning [66]. The in vivo assembly of subunits of F_0F_1 is described in a later chapter of this volume.

2.2.2. Homologies in primary structure of F_0F_1

The gene for F_0F_1 has been conserved during evolution, and homologies in the sequence are found among (1) F_0F_1's of different species, (2) related subunits of F_0F_1 and (3) related enzymes and proteins, especially nucleotide binding or metabolizing proteins. With regard to subunits of F_0F_1, the α and β subunits show strong homologies of categories (1), (2) and (3), perhaps because of their central catalytic function in ATP synthesis. Detailed nucleotide and amino acid sequences of these subunits are shown in Fig. 5.4.

(*1*) There is a close homology in the β subunits of EF_1, CF_1 (66% of EF_1 in maize, 67% in spinach) and MF_1 (72% of EF_1 in beef heart, 70% in *S. cerevisiae*). Chloroplast ϵDNA shows some homology with the gene for ϵ subunit of EF_1 (23% for maize [36] and 26% in spinach [35]). EF_0 shows some homology with MF_0 of *S. cerevisiae* in both the a and c subunit [66]. All the c subunits of nine species examined (animal, plants, and mesophilic and thermophilic prokaryotes) have two hydrophobic sequences connected by a central polar sequence [67]. DCCD binds to a Glu (or Asp) residue in the center of the second hydrophobic sequence from the amino terminus [67]. ATPase inhibitory peptides of beef heart [33] and yeast [34] also show some homology.

(*2*) As shown in Fig. 5.4, there is homology between the α and β subunits, since both are nucleotide binding sites of F_0F_1. The nucleotide sequences of these subunits

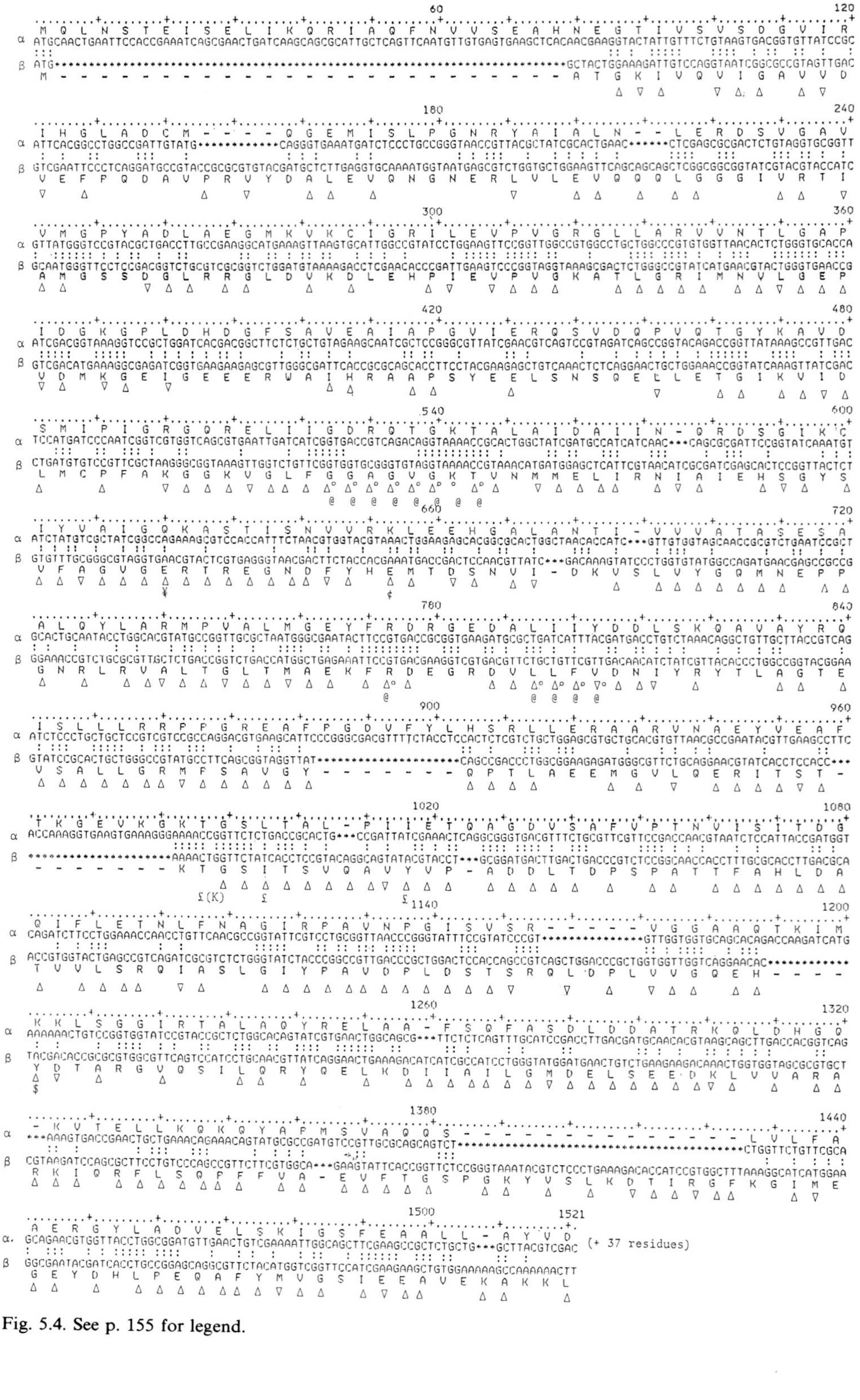

Fig. 5.4. See p. 155 for legend.

are more relevant to this homology. Homology between the prokaryotic δ subunit and eukaryotic OSCP has also been reported [32]. The prokaryotic ϵ subunit is also equivalent to the eukaryotic δ subunit [32]. No equivalent subunit was found for the eukaryotic ϵ subunit. There is some homology between subunit 6 of MF_0 and subunit a of EF_0.

(*3*) The α and β subunits show homologous sequences with several nucleotide binding proteins. There is a common sequence of Glu-Arg-Gly-Leu-Ala in the α subunit of EF_1, alanyl- and tyrosyl tRNA synthase, and the β subunit of RNA polymerase [11]. Significant homology is also found between the β subunit of EF_1 and Ca^{2+}-ATPase, ATP/ADP exchange protein, adenylate kinase and phosphofructokinase [21]. Residues 150–156 in the sequence of the β subunit are similar to those of adenylate kinase, and residues 186–202 to those of recA protein of *E. coli* [68]. The subunit is also homologous to myosin (rabbit and nematode) [69], and oncogene product p21 [70].

2.2.3. Chemical modification of the primary structure

In addition to the homologies described above, chemical modifications of specific amino acid residues provide clues to the meaning of the primary structure. As shown in Fig. 5.4, when the tyrosine residue (Y-355 of EF_1 or Y-368 of MF_1) in the β subunit was specifically modified by *p*-fluorosulfonyl-benzoyl-5′-adenosine, the ATPase activity of F_1 was completely lost [51]. Dicyclohexylcarbodiimide (DCCD) inactivates TF_1, MF_1 and EF_1 by modifying a specific glutamic acid residue in the β subunit, but it reacts with a different residue (11 residues from the glutamic acid residues of MF_1 and EF_1) of TF_1 [72] (Fig. 5.4). 8-Azido-ATP also modifies three residues in the Rossmann fold (see the next section) of the β subunit of MF_1 [49]. The crosslinking of the α and β subunits by the divalent azide nucleotide analogue [73] (Fig. 5.5) strongly suggests that the catalytic site of F_1 is located at the interface of the two subunits. Chemical modifications of F_1 suggested that Tyr, Lys, Asp or Glu, and Arg are functional residues at the catalytic site [74–76]. These experiments were supported by the effect of ATP, ADP, Mg or P_i added during the modification. However, the positions of residues modified have not been identified. It should be emphasized that modification of any residue essential for maintaining the structure of F_1 will reduce the ATPase activity, and the addition of those ligands of F_1 are known to stabilize the higher structure of F_1. Genetic approach to delete or replace

Fig. 5.4. Primary structures of the α and β subunits of EF_1 and the genes coding for them. (:) Nucleotide sequence homology between the two genes. (Δ) Homology common to the β subunits of CF_1 of maize [36] and spinach [35], MF_1 of beef heart [31], and EF_1 [11,21]. (∇) Similar amino acid residues common to the above mentioned subunits of two CF_1's, MF_1 and EF_1. ($) Binding site of *p*-fluorosulfonyl-benzoyl-5′-adenosine (Y-355 of EF_1, Y-368 of MF_1) [52]. (¢) Binding site of DCCD to EF_1 (E-193) and MF_1 (E-199) [72]. (Y) Binding site of DCCD to TF_1 (corresponding to E-182 of EF_1 and E-188 of MF_1) [50]. (£) Binding site of 8-azido-ATP to MF_1 (K-301, I-304 and Y-311) [49]. (@) Region weakly homologous to other enzymes employing ATP in catalysis. (Computer analysis was kindly performed by Dr. T. Miyata of Kyushu University.)

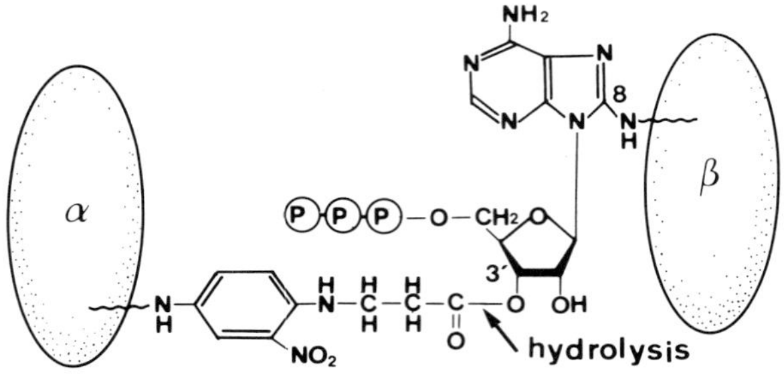

Fig. 5.5. Crosslinking of α and β subunits of F_1 at the catalytic site with 3′-*O*-{3-[*N*-(4-azido-2-nitrophenyl)amino]propionyl}8-azido-adenosine 5′-triphosphate. The crosslinked product (molecular weight, 120000) was isolated by sodium dodecyl sulfate polyacrylamide gel electrophoresis, and then hydrolyzed at the position shown by an arrow. Analysis of each fraction confirmed the structure [73].

any residue of F_0F_1 also results in nonspecific inactivation of F_0F_2. For example, a temperature sensitive mutant does not always reveal the functional group. Modification of Glu (or Asp of EF_0) of the c subunit with DCCD was found to block proton flow [6,12,15]. Iodination of the only Tyr of subunit c of TF_0 was found to slow down proton flow [77], though there is no homologous Tyr in other F_0's.

2.3. Secondary structure of the subunits of F_0F_1

Owing to the instability of the isolated subunits of F_0F_1, direct measurement of the secondary structure by circular dichroic spectroscopy or infrared spectroscopy has been limited to the subunits of TF_1 [30,78]. For example, the secondary structures of TF_1 and its subunits are lost in the presence of 8 M urea, but restored on its removal [30]. The secondary structure has also been estimated from the amino acid sequence of EF_1 [11,21] by the method of Chou and Fassman [79]. Table 5.2 summarizes the contents of α-helices and β-sheets of subunits of TF_1 [30] and EF_1 [11]. Similar values were obtained for TF_1 and EF_1. The CO-stretching vibrations of the α and β subunits of TF_1 were 1648 cm^{-1} and 1640 cm^{-1}, respectively, indicating that the subunit has a greater amount of antiparallel β-sheet structure than the subunit [80].

The α-helix contents of the δ [81] and ϵ [82] subunits of EF_1 were also measured directly. The high α-helix content of the subunit is consistent with the long Stokes' radius of the subunit [81], which might be a stalk connecting the αβγ complex with F_0.

The validity of applying the method of Chou and Fassman [79] to the membrane proteins is still controversial, but the estimated values of the a, b and c subunits of EF_0 are also shown in Table 5.2. The b subunit is also very rich in α-helices, and for the most part hydrophilic residues, except for the 22 residues at the N terminus [11,21]. The distribution of the secondary structure in these F_0 subunits has been described in detail [67]. As shown in Fig. 5.6, a Rossmann fold [83], an alternating structure of α-helices and β-sheets, is found in the subunit β around residues

TABLE 5.2

EF_1 and TF_1 subunit contents of α-helices and β-sheets

Subunit	α-helix		β-sheet	
	EF_1 [a]	TF_1 [b]	EF_1 [a]	TF_1 [b]
α	46.4	31	19.1	19
β	27.2	34	13.7	23
γ	35.0	49	15.0	4
δ	61.0	65	19.2	15
ε	43.4	33	12.3	24

[a] Calculated from DNA sequence [11].
[b] Directly measured on isolated subunits by circular dichroic spectroscopy [30].

237–330 [11,21], which is the highly conserved region in this subunit. A Rossmann fold is known to be the nucleotide binding site of dehydrogenases, kinases, etc. [83] and, in fact, 8-azido-ATP binds to this region [49]. However, the contents of α-helix and β-sheets in the α and β subunits of TF_1 were not affected by the addition of AT(D)P [80]. The tertiary structures of the above subunits and the quarternary structure of F_1 are strongly affected by nucleotides [11,80,84], as will be discussed in the following sections. Although the α subunit binds AT(D)P, no Rossmann fold was found in it.

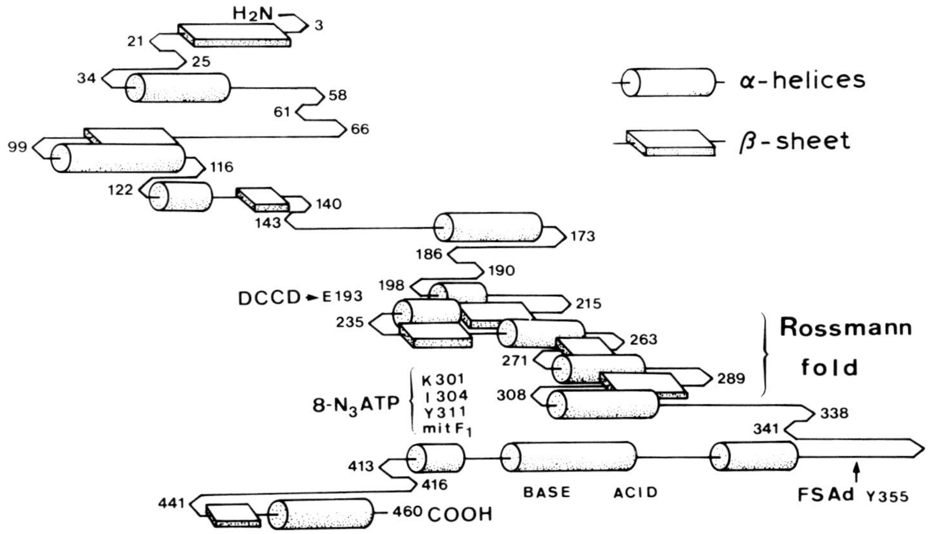

Fig. 5.6. Secondary structure of the β subunit of *E. coli* F_1. The primary structure of the protein was deduced from the DNA sequence [11]. The binding residues and numbers of binding sites for covalent inhibitors are also shown.

2.4. *Tertiary and quaternary structure of F_0F_1*

2.4.1. *Stepwise reconstitution of F_0F_1*

The proton motive ATPase activity of F_0F_1 was restored only after its quaternary structure was reconstituted [38]. A tripartite structure of F_0F_1 was found electron microscopically [51,85,86] (Fig. 5.7). The spheric portion (cf., Fig. 5.2) of this structure was lost on removal of F_1 from F_0F_1 by urea [51,58, and was restored on adsorption of [^{3}H]acetyl F_1 to F_0 [85]. At the same time, the proton channel activity of F_0 was converted into the proton motive ATPase activity [3]. Thus, the structure of F_0F_1 is interpreted as that shown in Figs. 5.1 and 5.2. The F_0 seems smaller than F_1 (Fig. 5.7) [3,51], but in some preparations, F_0 is much larger than F_1 [86]. Perhaps owing to the hydrophobicity of F_0, some lipid components may be adsorbed on F_0. In fact, the aggregation of F_0F_1 always takes place at the F_0 portion (Fig. 5.7).

Further reconstitution of F_1 from its subunit was possible only in TF_1 [38,39] and EF_1 [40,41]. The $\alpha\beta\gamma$ or $\alpha\beta\delta$ complexes are the essential portion of the ATPase activity [40,41,87]. The recovery of the activity was 46% of the original TF_1 [87]. The formation of hybrid F_1's from subunits of EF_1 and TF_1 [88] suggests that subunits of the different bacteria have similar roles and structural homologies. The stability of TF_1 was shared in the following hybrid F_1's: (a) complex of α and β of TF_1 and γ of

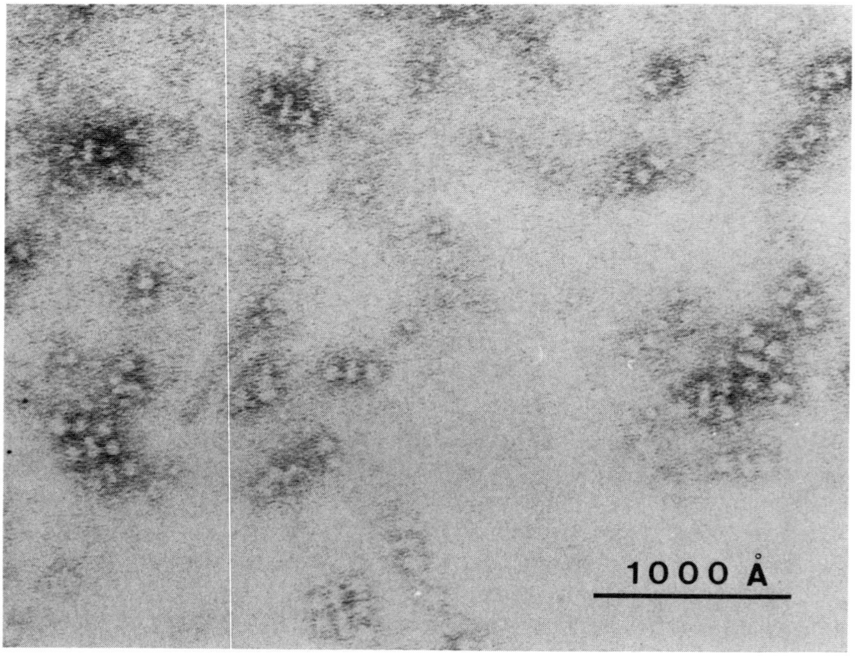

Fig. 5.7. Electron micrograph of F_0F_1. Beef heart mitochondrial F_0F_1 was stained negatively. Many dimers and trimers connected at the F_0 portion are seen.

EF_1 was resistant to both methanol and a detergent, and activated with Cd^{2+} and SO_3^-; (b) a complex of α and β of EF_1 and γ of TF_1 was halophilic and relatively thermophilic; (c) a complex of α and γ of EF_1 and β of TF_1 was similar to EF_1 [88]. Extensive tryptic digestion of EF_1 results in an ATPase containing both α and β subunits and fragments of the γ subunit [89]. Both the δ and ϵ subunits are required to bind TF_1 to TF_0 [39,90]. However, the role of δ and ϵ subunits in the binding of the $\alpha\beta\gamma$ complex to EF_0 is still controversial [91]. In mitochondrial F_0F_1, OSCP which is homologous to the δ subunit [32] is shown to bind both MF_1 and MF_0 [92]. The major evidence of the direct binding of both α and β subunits to the a and b subunits is the in vivo assembly experiments of F_0F_1 by using mutants [93]. However, reconstitution experiments show that three major subunits of F_1 do not interact directly with subunits of F_0 [38]. Although some point mutations in certain subunits may cause drastic reduction of activity or reconstitutability of F_0F_1 [91–93], the active hybrid F_1 from subunits of very different F_1's was easily reconstituted [88].

2.4.2. *Crystallographic analysis of F_1*

The molecular mechanism of ATPase may be elucidated if the detailed tertiary and quaternary structure of F_0F_1 is shown by crystallographic analysis. To date, MF_1 [58,94–97], CF_1 [98] and TF_1 [51,99,100] have been crystallized. Fourier image reconstruction of two dimensional crystal of TF_1 (spacings, $a = b = 90$ Å) demonstrated a molecule with pseudohexagonal symmetry and a central hollow region [99]. A similar structure was found in thin crystalline plates of beef heart MF_1 [97]. The form I crystals were space group $P2_12_12$ with unit cell parameters of $a = 164$ Å, $b = 324$ Å and $c = 118$ Å [97]. The form II crystals of beef MF_1 have unit cell parameter of $a = 156$ Å, $c = 162$ Å, $\beta = 90$ Å and are either space group $P2_12_12$ or $P222_1$ [97]. However, the cube shaped crystals of spinach CF_1 have a tetragonal lattice, $a = b = 135$ Å, $c = 280$ Å with eight molecules per unit cell [98]. The three dimensional crystal of rat liver MF_1 diffracts up to a resolution of 3.5 Å on X-ray diffraction [94], and belong to space group R32 and have hexagonal cell dimension of $a = 146$ Å and $c = 367$ Å [94]. The molecular weight of the asymmetric unit is 180 000–190 000, or approximately one-half of the molecular weight of the MF_1. Since the most probable subunit stoicheiometry ($\alpha_3\beta_3\gamma_1\delta_1\epsilon_1$) is not dimeric, some adjustments must be needed to reconcile both experimental observations [94]. The asymmetric unit of the crystal was found to be divided into three regions of approximately equal size ($40 \times 50 \times 60$ Å) [94], and one of them may contain some of the minor subunits. Thus, the three pairs of α and β subunits are not equal. In fact, the aurovertin was shown to bind to the β subunit at a multiplicity of two aurovertins per MF_1 molecule [101]. Moreover, NBDCl binds to the β subunit with a multiplicity of one NBDCl per MF_1 molecule with complete inhibition of ATPase activity [13]. Further details of the crystallographic analysis have not been obtained. The hexagonal molecular shape with a core structure was observed by small angle X-ray scattering of TF_1 [102]. This structure may be interpreted as three α and three β subunits surrounding a complex composed of the $\gamma\delta\epsilon$ subunits [102].

2.4.3. Dynamic conformational change of F_1 and F_0

The importance of conformational changes of F_1 during ATP synthesis [8,14,16,17] (cf., Section 4) stimulated the direct measurement of the conformational change [80,84,103]. The rapid incorporation of $^3H^+$ into CF_1 molecules (slowly exchanging H^+) during photophosphorylation provided the first solid evidence of the conformational change of F_1 [103]. The isolated CF_1 after the illumination of chloroplasts contained 90 $^3H/CF_1$ molecules which could be exchanged by loosening the quaternary structure with urea.

The conformational change of the subunits of TF_1 was estimated from the $^1H-^2H$ exchange rate measured in a Fourier transform infrared spectrometer [80]. The relaxation spectrum (Abscissa: $\log_{10}(ke \times t)$, where $ke = 386 \times 10^{[6+pH-(3850/T)]} \cdot min^{-1}$, T is the absolute temperature, and t is the time in min. The ordinate (percentage of undeuterated NH groups in the proteins) of the α and β subunits showed that the α subunit has greater structural fluctuation than the β subunit. Experiments with predeuterated α and β subunits (α^* and β^*) on formation of hybrid complexes ($\alpha^*\beta$ and $\alpha\beta^*$) clearly showed that the β subunit stabilized the α subunit [80]. The addition of AT(D)P to intact TF_1 decreased its $^1H-^2H$ exchange rate, and this stabilization is mainly caused by tightening of the β subunit by the predeuterated ATP-α-subunit complex [80]. Stabilization of mesophilic F_1's, and the α and β subunits with ATP has been reviewed [11]. The stabilization was measured by the degree of proteolysis, denaturation and reaction with a fluorescent probe. The nucleotide binding and the conformational change will be discussed in detail in Section 4.

The conformational change of F_0 may also be important for the proton flux, since the rate of flux is very slow compared with the extremely rapid rate of $^1H-^2H$ exchange of free NH_2 or COOH groups on the surface of F_0. NMR spectroscopy of subunit c of TF_0 revealed no hydrogen bonding around Tyr^{69}, and free movements of C-2/6 and C-3/5 of the residue in organic solvent (K. Nagayama and Y. Kagawa, unpublished), though Tyr^{69} is needed for H^+-flux [77]. The flexible sequences, Gly at every second position (Gly^{13} to Gly^{24}) which allow for a large range of dihedral angles at C-2 atoms, will help to transfer H^+ through Arg^{41}, Glu^{56} and Tyr^{69}, which are surrounded by very hydrophobic regions, yet essential for H^+-flux in TF_0 [77].

3. Function of F_0F_1

3.1. Phosphorylation in biomembranes

3.1.1. ATPase and H^+ transport in intact membranes

ATPase activity and exchange reactions, such as P_i-ATP exchange, are partial reactions of oxidative and photosynthetic phosphorylation. These reactions have been described in detail and have been considered to consist of a series of reversible chemical reactions forming high energy intermediates (X-Y and X-P) [104–106]. In the previous series of Comprehensive Biochemistry, immediately after the isolation and reconstitution of F_0F_1 [2,5,107], Mitchell predicted the roles of F_0 and F_1 in proton translocation [108]. His chemiosmotic theory simplified many complicated observations on phosphorylation. For example, he concluded that stimulation of ATPase of intact mitochondria by dinitrophenol is not the result of hydrolysis of X-Y, but of the increase in permeability of protons by this uncoupler. ATPase is tightly coupled to proton translocation, and the electrochemical potential difference ($\Delta\tilde{\mu}_{H^+}$) of about 200 mV established across the mitochondrial inner membrane will convert ATPase into ATP synthetase, which results in P_i-ATP exchange. Therefore, by reducing $\Delta\tilde{\mu}_{H^+}$ with an uncoupler, ATPase is activated and oxidative phosphorylation and P_i-ATP exchange are inhibited. ATPase, phosphorylation and P_i-ATP exchange are all sensitive to oligomycin and other energy transfer inhibitors, because proton translocation through F_0 is blocked by these inhibitors [6]. When F_1 is detached from F_0, it is no longer sensitive to these inhibitors [5].

The driving force of ATP synthesis in F_0F_1 is the $\Delta\tilde{\mu}_{H^+}$ [1]. The $\Delta\tilde{\mu}_{H^+}$ is usually supplied from the electron transport system, but in *Halobacterium halobium*, it is supplied by bacteriorhodopsin [109]. Anaerobic bacteria, such as *Streptococcus faecalis*, have no proton translocating electron transport system, but they contain F_1-like EF_1 [110]. When a $\Delta\tilde{\mu}_{H^+}/F$ is applied to the membrane of anaerobic bacteria, ATP is synthesized [111]. But the physiological role of F_0F_1 in these bacteria is to translocate protons from the cytosol to the environment [112]. In mitochondria, ATPase driven proton extrusion [113] and reverse electron flow [114] are also observed, but their physiological role is unknown.

The phenomena not expected by the chemiosmotic theory were the synthesis of ATP (F_1-bound form) and P_i-^{18}O exchange reaction without the direct participation of the $\Delta\tilde{\mu}_{H^+}$. These are observed in soluble system containing isolated F_1. These will be discussed in Section 3.4.

3.1.2. Electrochemical potential of H^+, localized and delocalized

Transport of protons across the membrane produces an electrochemical potential difference of protons ($\Delta\tilde{\mu}_{H^+}$), which is related to the membrane potential ($\Delta\psi$) and pH difference (ΔpH) as follows:

$$\Delta^{-}\tilde{\mu}_{H^+} = F\Delta\psi - 2.3RT\,\Delta\text{pH}$$

where R and T are the gas constant and absolute temperature. The $\Delta\psi$ and ΔpH have different properties as summarized in Table 5.3. Because of these differences, ATP synthesis in F_0F_1 was studied either with ΔpH or $\Delta\psi$ as will be discussed in Section 3.3. Owing to the ion permeability of the membrane, ΔpH is the major driving force of F_0F_1 in chloroplasts.

The route of transfer of protons and electrons from the electron transport system to F_0F_1 has been a matter of controversy [1,115], because these highly charged particles are generated and consumed at the specific atoms, such as Fe or COO^-,

TABLE 5.3

Comparison of membrane potential and pH difference

	Membrane potential ($\Delta\psi$)	pH difference (ΔpH)
Capacity	Low (1 μF·cm^2); depends on the thickness of membrane (about 95 Å)	High depends on the internal volume of the vesicle (0.75·mg^{-1} protein)
Intensity	High	Low
Velocity of generation	Rapid	Slow
Localization	Non-diffusible, localized by external electric field	Buffering, diffuse
Detection	Permeant ions (fluorescent probes or ion electrode); carotenoid shift; anilinonaphthalen-sulfonate	pH meter (inside after detergent); ^{31}P-NMR; flow dialysis; aminoacridine
Removal	Permeant ions; K^+ valinomycin	Permeant weak base or acid

which are localized in very small catalytic sites. The route can be either delocalized in the bulk phase [1], or localized in a certain part of the membrane [115]. The direct measurement of medium pH during energization of mitochondria [113] and the reconstituted F_0F_1 liposomes [3], and ATP synthesis driven by an artificially imposed ΔpH in the F_0F_1 liposomes (see Section 3.3), support the delocalized pathway. However, owing to the very low concentration of H^+ (10^{-7} M) in water, the velocity of movement of H^+ through the bulk phase, even by the jumping through H-O-H chains, is much slower than that through the surface of lipid bilayer [115]. The electron transport complexes are moving laterally along the bilayer by thermal agitation with diffusion coefficients of about 4×10^{-10} cm. F_0F_1 reaction is much slower than the time required for the proton movement, but if three protons are used to drive one ATP synthesis, the reaction will be a 4th order reaction. Thus, in physiological conditions, protons may circulate from electron transport chain to the nearest F_0F_1 through the membrane surface. The localized $\Delta\tilde{\mu}_{H^+}$ in the membrane equilibrates with the bulk phase with some time lag, and this phenomenon will reduce the proton leakage and free energy loss by diffusing out the localized energy.

3.1.3. H^+/ATP ratio

The reaction of F_0F_1 in membranes is formulated as

$$ATP + H_2O + xH_o^+ = ADP + P_i + xH_i^+ + yH^+$$

where H_o^+ and H_i^+ are protons on the F_1 side and F_0 side, respectively, of the membrane. In intact mitochondria and bacteria, F_1 is located on the inner surface of the membranes, while in chloroplasts and submitochondrial particles, it is located on the outer surface of the membranes. Thus, protons are extruded from mitochondria, but accumulated in submitochondrial particles when ATP is hydrolyzed by F_0F_1. x in the equation is the number of protons translocated per mole of ATP hydrolyzed, and y is the number of protons liberated by the difference in pK_a of ATP and ADP + P_i. In the early stage of the chemiosmotic theory, y was confused with x, but y is the scalar component and x is the vectorial component of this membrane reaction. The value of y is about one at pH 8, but it becomes 0 at pH 6.25 in the presence of Mg^{2+}. At the latter pH, net proton uptake (xH^+) driven by ATP hydrolysis was observed [3]. On the other hand, if the membrane structure is destroyed or an uncoupler is added, x becomes zero. The experimental determination of x, or H^+/ATP stoicheiometry, is disturbed by proton leakage through the membrane [13,15]. The value of $\Delta\tilde{\mu}_{H^+}$ should exceed that of the phosphorylation potential (ΔG_p) during ATP synthesis. The value of ΔG_p is related to the standard free energy change of ATP hydrolysis ($\Delta G_o'$) and the concentrations of ATP, ADP and P_i as follows:

$$\Delta G_p = \Delta G_o' + RT \ln[\mathrm{ATP}]/[\mathrm{ADP}][\mathrm{P_i}]$$

Thus, the stoicheiometry x (= H^+/ATP) at equilibrium can be expressed simply as

$$x = \Delta G_p/\Delta\tilde{\mu}_{H^+}$$

The value of $x = 2$ was proposed by Mitchell [113], but recently the values have been estimated as $x = 3$ [13,15] in mitochondria [116], submitochondrial particles [117], *E. coli* [118,119] and chloroplasts [120]. Details of the thermodynamics of this reaction will be discussed in another chapter. The possibility of loose coupling of H^+-flux and ATP synthesis will be discussed in the last part of Section 4.

3.2. F_0F_1-Proteoliposomes

Biomembranes containing an electron transport system and ion translocating proteins, such as ATP/ADP exchanger, are not suitable for studying the molecular mechanism of energy conversion, since any manipulation of these complex systems may have, in addition to direct effects, indirect effects through electron transport or ion translocation. As reviewed by Boyer, conformational changes in some proteins may be transferred to F_0F_1 by direct contact [121]. As discussed in Section 3.1.2, local electrochemical changes near electron transport components may be transmitted directly to F_0F_1 [115], but not through the $\Delta\tilde{\mu}_{H^+}$ delocalized in the two bulk phases across the membrane [1]. It is thus necessary to reconstitute the pure F_0F_1 into liposomes to measure its function [3,122].

F_0F_1 proteoliposomes can be reconstituted by the cholate dialysis method [123] or another method [124]. The resultant reconstituted F_0F_1 may have two different orientations in liposomes, but since the membranes are not permeable to ATP, protons are translocated from the outside to the inside when ATP is hydrolyzed [3,123]. The F_0 portion of F_0F_1 plugs the lipid bilayer [5,85], and a crosslinking experiment with lipid confirmed the interaction of F_0 with phospholipids [125]. Further analysis showed that Cys-21 of the b subunit of EF_0 was specifically labeled with membrane permeating photoreactive trifluoromethyl-iodo-diazirine [67]. F_0 proteoliposomes have membranes that allow passive proton flux down the $\Delta\tilde{\mu}_{H^+}$ [6,15]. To some extent, the flow of protons is proportional to the $\Delta\tilde{\mu}_{H^+}$ (Ohm's law) [6]. As shown in Fig. 5.1, when F_1 is attached to the F_0 proteoliposomes, active proton translocation is restored [3,38]. The passive proton flux through F_0 is blocked by F_1, since some gating activity is present in the F_1 molecule [126].

The $\Delta\tilde{\mu}_{H^+}$ formed in F_0F_1 proteoliposomes by ATP hydrolysis could be determined by measuring 9-aminoacridine accumulation (for ΔpH) and anilinonaphthalene sulfonate adsorption (for $\Delta\psi$) by following the fluorescence changes of these reagents [127]. Permeant anions, such as NO_3^-, decreased $\Delta\psi$, while permeant buffers decreased the ΔpH of $\Delta\tilde{\mu}_{H^+}$ formed in F_0F_1 proteoliposomes. Uncouplers released protons accumulated in the F_0F_1 proteoliposomes, while energy transfer inhibitors blocked the accumulation of protons [3]. Thus, typical inhibitors of oxidative phosphorylation behaved in F_0F_1 proteoliposomes as in intact mitochondria.

Proteoliposomes are useful for elucidating the function of a single molecule. F_0F_1 proteoliposomes after sonication become microliposomes, while those prepared by evaporation of an organic solvent are macroliposomes of as much as 20 μm in diameter [124]. The particle weight of a unilamellar microliposome is 2.1×10^6. Therefore, the addition of 0.04% (w/w) of an F_0F_1 molecule (5.0×10^5 Da) should result in a mixture of liposomes and proteo-liposomes mainly containing one molecule of F_0F_1 per vesicle. The number of remaining unloaded liposomes can be estimated from the Poisson distribution by the formula $P = me^{-ax}$, where m is the total number of liposomes in the absence of F_0F_1, a is the efficiency of incorporation of F_0F_1 into liposomes, and x is the molar ratio of F_0F_1. Thus, it is improbable that any direct contact with other proteins or small molecules contaminating the preparation affects F_0F_1 molecules isolated in the liposomes. The single channel recording method of an ion translocator, such as an acetylcholine receptor in a lipid bilayer, actually measures the ion transport activity of the molecule [128].

The roles of lipids in proton translocation can also be studied with F_0F_1 proteoliposomes. Fluidity and acidity of the phospholipids are required for proton translocation [129,130].

3.3. ATP synthesis driven by $\Delta\tilde{\mu}_{H^+}$

3.3.1. Ion gradient applied to F_0F_1 proteoliposomes

Direct evidence for the chemiosmotic theory [1] is proton motive ATP synthesis in purified F_0F_1 proteoliposomes on applying H^+ [38,131]. Jagendorf and Uribe [132] first demonstrated ATP synthesis in subchloroplast particles that had been loaded with a weak acid (pH 4.5) and were then transferred to alkaline media at pH values greater than 7.7 in the presence of an inhibitor of electron transport. However, the energy might have been stored in other components in subchloroplast particles during the acid-base transition. Thus, experiments with F_0F_1 proteoliposomes were needed. Most liposome preparations become too leaky to synthesize ATP under these drastic conditions unless additional energy is continuously supplied, as from ATP hydrolysis (P_i-ATP exchange) [3,123], electron transport [133], or illumination of added bacteriorhodopsin [134]. This difficulty was overcome by using thermophilic F_0F_1 (which is resistant to acid-base change and high salt [87]) and thermophilic phospholipids (which are composed of branched saturated phospholipids [135] and do not show proton leakage) for reconstitution of F_0F_1 [131]. Experiments were carried out by first incubating the F_0F_1 proteoliposomes with acidic malonate buffer and valinomycin, and then making the solution alkaline with glycylglycine buffer which is impermeable [131,136]. Proton motive ATP synthesis is summarized in Table 5.4. This instantaneous transition should create a $\Delta\tilde{\mu}_{H^+}$ equal to 275 mV, composed of ΔpH (2.38 units acidic inside) and $\Delta\psi$ (125 mV positive inside) across the liposome membrane. ATP synthesis occurs at a velocity of 650 nmol $\cdot$ mg^{-1} $F_0F_1 \cdot min^{-1}$, which is faster than substrate oxidation in mitochondria. Thus, the primary role of $\Delta\tilde{\mu}_{H^+}$ translocation in oxidative phosphorylation has been confirmed [137]. The maximal level of ATP synthesis is about 100 nmol $\cdot$ mg^{-1} F_0F_1 in F_0F_1 proteoliposomes, whereas it is less than 2.5 nmol $\cdot$ mg^{-1} protein in submitochondrial particles [138], bacterial membranes [139] and intact mitochondria [140]. The $\Delta\psi$ and ΔpH of equal $\Delta\tilde{\mu}_{H^+}$ value elicit identical rates of ATP formation [131,136]. Thus, there may be only one rate-limiting step responsive to the $\Delta\psi$ and ΔpH. The rate-limiting step is the ATP-release from F_1-ATP complex as will be discussed in Section 4.3.

3.3.2. Electric field applied to F_0F_1 proteoliposomes

Analysis of ATP synthesis driven by an ion gradient, as described in the previous section, has the disadvantage that time resolution is poor and the energy components

TABLE 5.4

The reconstituted liposomes (0.055 mg of F_0F_1 and 4 mg of thermophilic phospholipids) were first incubated in acidic medium (pH 5.5, final volume 0.25 ml) containing 10 μmol malonate, 1 μmol ADP, and 0.1 μg valinomycin at 40 °C for 10 min. Then 0.25 ml of an alkaline medium consisting of glycylglycine (40 μmol, pH 8.5), 75 μmol KCl, 0.5 μmol $MgSO_4$, 2 μmol $^{32}P_i$ (6×10 cpm, sodium salt), 25 μmol glucose, and 10 units of hexokinase was added. The final pH was 8.3. The reaction was carried out at 40 °C and terminated after 5 min.

Conditions	ATP formed (nmol $\cdot$ mg^{-1} of F_0F_1 protein)
Complete system (pH 5.5–8.3)	53.3
Complete system (pH 8.0–8.3)	5.4
Without ADP	3.9
With DCCD (an energy transfer inhibitor)	3.6
With FCCP (an H^+-carrier, or uncoupler)	1.9

are complex: i.e., both $\Delta\psi$ and ΔpH. Moreover, the ion gradient applied is continuously decreasing during ATP synthesis, and to maintain a certain steady state level of the $\Delta\tilde{\mu}_{H^+}$ is difficult even when the activity of an electron transport system or bacteriorhodopsin present is regulated by substrate concentration or light intensity. For overcoming these difficulties, ATP synthesis by an external electric field on thyakoid membranes was studied in Witt's laboratory [141,142]. This method has the following advantages; (a) it gives high time resolution; (b) the membrane potential is proportional to the electric field applied; (c) the pH is maintained relatively constant. (d) energy can be applied easily and repeatedly to one sample. This method was first applied to intact mitochondria, and net ATP synthesis of 0.05 nmol · mg^{-1} protein/rectangular pulse of 760 V · cm^{-1}, 30 ms, was observed [143]. Considering the electrical capacity (0.5 μF · cm) [144] and the rotational relaxation time (0.9 s) [145], the electric pulse could energize the mitochondrial membrane. The role of the outer membrane as an electrically resistant barrier was disproved by purification of a nonspecific diffusion channel [146]. An interesting finding in electric field-driven ATP synthesis is that a very short pulse duration (less than 50 μs) is enough to synthesize ATP [143]. This was confirmed in submitochondrial particles [143,147]. Perhaps the very slow step in ATP synthesis, the ATP release from F_1 (see Section 4.2.3), is accelerated by the electric pulse. The mechanism of energization may involve a conformational change by the membrane potential, because tightly bound nucleotide in CF_1 is released by application of an electric pulse [142]. The ATP synthesis was sensitive to uncouplers and energy-transfer inhibitors [143,147]. A considerable time-lag after irradiation with an electric pulse before synthesis of ATP was reported [148], but its meaning is still unknown.

Finally, net ATP synthesis in reconstituted F_0F_1 proteoliposomes was demonstrated in F_0F_1 from a thermophile [137, 149], a chloroplast [150] and mitochondria [151]. The membrane potential produced by an external electric field is proportional to the diameter of the proteoliposomes, so in macroliposomes [137,149] about 1000 V · cm^{-1} is enough, but in microliposomes of about 50 nm diameter [150,151] 15 000 V · cm^{-1} was the threshold for the ATP synthesis [151]. The yield with a 30 000 V · cm^{-1}, 10 μs pulse was 80 pmol · mg^{-1} per pulse [151]. The maximum $\Delta\psi$ is 114 mV, which is consistent with the value measured on electrochromy of thylakoids under an electric field [152,153]. Since the experiments with purified F_0F_1 liposomes could synthesize ATP, neither a special protein nor a special structural arrangement should be prerequisites for ATP synthesis.

3.4. Formation of F_1-bound ATP without $\Delta\tilde{\mu}_{H^+}$

Mitchell advocated the hypothesis that $2H^+$ driven by $\Delta\tilde{\mu}_{H^+}$ directly attack the O^- of PO^{4-} in a complex with $ADPO^-$ and Mg^{2+} at the active site of F_1 [113]. However, ATP synthesis takes place in the absence of proton flux down the $\Delta\tilde{\mu}_{H^+}$ [154–156]. Feldman and Sigman found that CF_1 synthesized enzyme-bound ATP from P_i [154]. The reaction did not depend on ADP in the medium indicating that the ADP substrate was tightly bound to CF_1. The addition of hexokinase and glucose did not reduce the yield of CF_1-ATP, showing that the ATP is not released from the enzyme. Since ADP in the medium is not utilized in this reaction, the reaction may not be a partial reaction of photophosphorylation. However, Sakamoto and Tonomura succeeded in synthesizing ATP with MF_1 from ADP and P_i in the medium in the presence of dimethylsulfoxide [155], which is known to promote phosphorylation of Ca^{2+}-ATPase by P_i [157]. The pH optimum for ATP synthesis in the presence of 35% dimethylsulfoxde was 6.7 at 30 °C. The rate and extent of [α-^{32}P]ATP synthesis from [α-^{32}P]ADP and P_i were equal to those of [γ-^{32}P]ATP synthesis from ADP and $^{32}P_i$. No nucleoside-5′-triphosphate was synthesized by MF_1, when GDP, IDP or UDP was used as substrate. The following relationship was found:

$$[\text{ATP formed}] = [\text{maximum ATP formed}]/(1 + K_{ADP}/[\text{ADP}] \text{ or } K_P/[P_i])$$

where $K_{adp} = 3$ μM, $K_p = 0.5$ mM, and [maximum ATP formed] = 0.4–0.6 mol/mol MF_1 [155]. There are several kinds of tightly bound nucleotides in MF_1 or CF_1 [158], but only non-tightly bound nucleotides were found in purified TF_1 [84]. Thus, TF_1-ATP synthesis was confirmed in 50% dimethylsulfoxide solution [156]. The TF_1-ATP formed contained 0.8 mol ATP/TF_1 molecule [156], and did not release ATP unless it is denatured or treated with ammonium sulfate. Instead of using dimethylsulfoxide, high concentrations of P_i (50–200 mM) stimulated the formation of CF_1-ATP [154], MF_1-ATP [155] and TF_1-ATP. As will be discussed in Section 4.2.2, F_1-ATP is in equilibrium with F_1-ADP-P_i. For the formation of F_1-ADP-P_i from F_1-ADP, a high concentration of P_i is necessary, because the binding sites for ADP and P_i partially overlap to synthesize ATP. It is interesting that DCCD-treated MF_1 still synthesizes MF_1-ATP by this method, DCCD treated TF_1 does not synthesize TF_1-ATP, perhaps because of the difference in the Glu residues modified (Fig. 5.4). The synthesis of ATP by this method takes place at the usual catalytic site of F_1, because [^{3}H]ADP formed on F_1 by hydrolyzing newly added [^{3}H]ATP with F_1 (and removing the remaining ATP) was converted into [γ-^{32}P, ^{3}H]ATP (0.23 mol/mol F_1) in the presence of 100 mM $^{32}P_i$ [158].

The energy of ATP formation from P_i and ADP may be derived from the energy of conformational change and nucleotide binding to F_1 [8,16,155], which will be discussed in Section 4.3.

4. *Mechanism of the H^+-ATPase reaction*

4.1. Stereochemistry of the ATPase reaction

4.1.1. Stereochemical course

H^+-ATPase catalyzes terminal transphosphorylation in ATP synthesis. The presence of phosphorylated intermediate in this reaction was suggested by kinetic analysis [103–105], and isolation of an acylphosphate phosphoenzyme in ion transporting ATPases, including Ca^{2+}-ATPase [159], Na^+,K^+-ATPase [160] and plasma membrane-type H^+-ATPase [18]. However, no phosphoenzyme intermediate of F_0F_1 has been isolated. Even if a phosphorylated intermediate was too labile to be isolated, the stereochemical course should reveal its presence. As shown in Fig. 5.8, a phosphoryl acceptor approaches the P atom at an angle of 180° to the leaving group. This in-line displacement pathway, via a pentavalent phosphorus transition state, results in an inversion of the configuration of the P atom [161]. Therefore, ATPγS labeled stereospecifically with ^{18}O in the position will produce chiral (^{16}O, ^{17}O, ^{18}O) thiophosphate when it is hydrolyzed with ATPase in ^{17}O enriched water (Fig. 5.8). The configuration of this chiral thiophosphate was determined with ^{31}P-NMR [162] after its stereospecific conversion to ATPβS with glycolytic enzymes and adenylate kinase [163]. The γP atom of ATPγS undergoes inversion during hydrolysis with beef heart MF_1 [163] and TF_1 [164]. These results excluded (1) formation of phosphoenzyme (retention), (2) pseudorotation (retention) and (3) formation of metaphosphate (racemization) during the reaction. In contrast to the experiments with myosin [165] which does not contract with ATPγS, TF_0F_1-liposomes translocated protons on hydrolysis of ATPγS. Thus, the in-line S_{N2} reaction was established in F_1.

4.1.2. Cation-dependent diastereoisomer preference

The true substrate of F_0F_1 is MgATP, rather than the free nucleotide. There is a strict stereochemical requirement for the structure of metal-nucleotide complexes in the phosphorylation reaction [161]. In the case of myosin ATPase, for example, the Sp diastereoisomer (A in Fig. 5.9) of ATPβS is more rapidly hydrolyzed than its Rp diastereoisomer (B in Fig. 5.9) with Mg with an Sp/Rp ratio of > 3000, while with

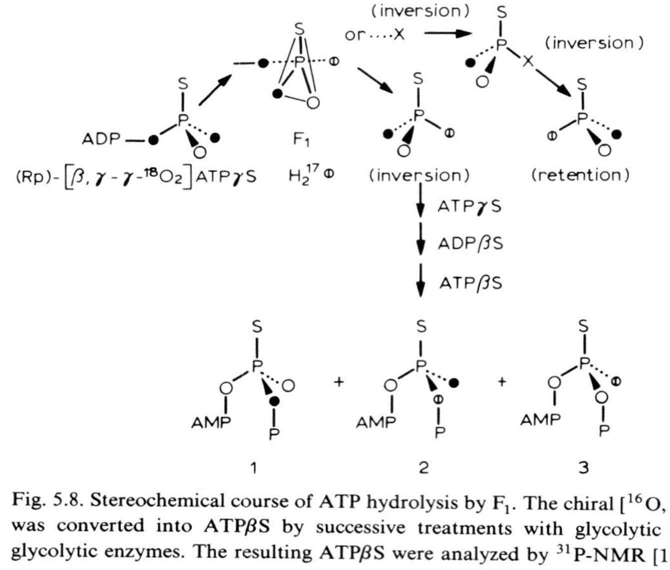

Fig. 5.8. Stereochemical course of ATP hydrolysis by F_1. The chiral [^{16}O, ^{17}O, ^{18}O]-thiophosphate librated was converted into ATPβS by successive treatments with glycolytic enzymes, adenylate kinase and glycolytic enzymes. The resulting ATPβS were analyzed by ^{31}P-NMR [164].

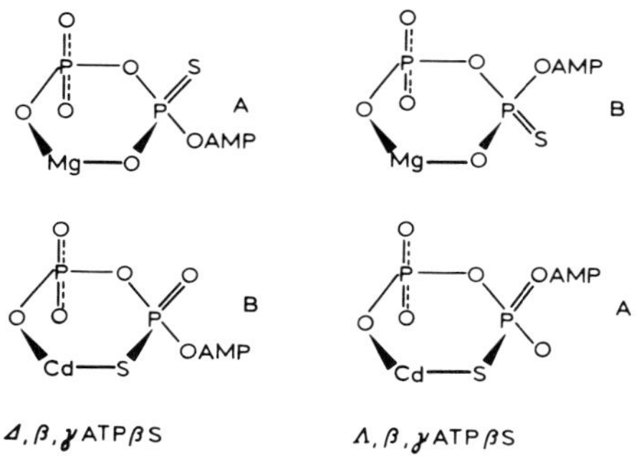

Fig. 5.9. Configuration of the metal ATPβS chelate. A, Sp diastereoisomer; B, Rp diastereoisomer [164].

Cd this ratio is only 0.2 [166]. On the other hand, the Rp diastereoisomer of ATPβS is preferred by hexokinase when Mg is the cation (Sp/Rp ratio 0.002), and the ratio is reversed with Cd (Sp/Rp ratio, 30) [167]. As shown in Fig. 5.9, Mg prefers the O atom and Cd prefers the S atom as a ligand. Thus, it is reasonable that an enzyme should prefer only one geometrical arrangement. TF_1 is several times more active than EF_1 [88] and MF_1 [87] when Cd is used as the divalent cation. Therefore, the cation-dependent diastereoisomer preference of TF_1 was determined [164]. Mg(Sp)-ATPβS and Cd(Rp)-ATPβS, which have the Δ-configuration, were better substrates than Mg(Rp)-ATPβS and Cd(Sp)-ATPβS, both of which have the Λ-chelate configuration (Fig. 5.9). The Sp/Rp ratios with Mg and Cd were 575 and 0.5, respectively. Both diastereoisomers of ATPαS were substrates for TF_1 and no metal-dependent diastereomeric selectivity was observed. These results suggest that TF_1 uses the Δ, β, γ-bidentate nucleotide chelate as substrate [164]. This kind of experiment on other F_1's was disturbed by the presence of tightly bound Mg and nucleotides.

The addition of Mg^{2+} strongly influences the effects of chemical labeling agents, such as DCCD [50]. An interesting finding is that the photoactivatable ATP analog, 3′-benzoylbenzoic ATP, is covalently incorporated into both the α and β subunits in the presence of Mg^{2+}, but only into the β subunit in the absence of Mg^{2+} [168]. It was postulated that MgATP binds to the β subunit and undergoes hydrolysis, with subsequent transfer of ADP to the α subunit [168], because the catalytic site may be at the interface of the two subunits (cf., Fig. 5.5).

Since the Mg exchanges oxygen ligands such as ATP very rapidly, Cr^{3+}, which forms a stable nucleotide chelate, was used to analyze the transition state on F_1 [169]. Both CrATP and CrADP interact with F_1 very much like the corresponding Mg nucleotide chelates. Incubation of F_1 with monodentate CrADP and $^{32}P_i$ resulted in the formation of F_1-P_iCrADP, from which P_iCrADP was isolated. Incubation of F_1 with bidentate CrATP also resulted in the same product [169]. This suggests that the synthesis and hydrolysis of ATP-metal chelate takes place via the same transition state on F_1. The 1 : 1 : 1 ADP-Mg-TF_1 complex [170] forms ATP-Mg-TF_1 P_i solution.

4.2. *Energetics of the F_1-bound nucleotides*

4.2.1. *Binding sites of nucleotides and inhibitors in F_1*

There are many reports that less than five nucleotides are bound per F_1 molecule (see review in Ref. 158). However, the total number of adenine nucleotide binding sites on beef heart MF_1 was finally established as six of which three are noncatalytic (AMP-PNP-binding) sites and three exchangeable catalytic sites [171]. The binding of nucleotides to noncatalytic sites is tight, but the bound nucleotides were removed with 50% glycerol, and the replaced radioactive nucleotides were not lost on further addition of nonradioactive nucleotides [171]. The isolation of subunits in the active state [8,9,68], and the finding that the subunit stoicheiometry is $\alpha : \beta = 3:3$ also supported the existence of six binding sites/F_1. Circular dichroic spectrometry of TF_1 indicated that the total number of AT(D)P bound to TF_1 is six : three to the α subunit and three to the β subunit [84]. The α subunit shows an ellipticity change corresponding to the absorption curve of the *anti*-form of nucleotides tested, while the β subunit shows negative ellipticity at 275 nm corresponding to a tyrosyl residue on binding nucleotides [84]. Very tight binding of ATP to the α subunit of EF_1 was

TABLE 5.5

Inhibitors of H^+-ATPase (F_0F_1)

Inhibitors	Affected sites	References
ATPase inhibitor protein	$F_1\beta$ noncovalent [a]	33, 34, 186–189
AMPPNP	$F_1\alpha\beta$ noncovalent [a]	171, 190
Aurovertin	$F_1\beta$ noncovalent [a]	5, 101, 187
Bathophenanthroline iron complex	F_1 noncovalent [a] (released by uncouplers)	188
Citreoviridin	$F_1\beta$ noncovalent [a]	193, 199
Quercetin	F_1 noncovalent	173, 194
Efrapeptin	$F_1\beta$ noncovalent [a]	192
Tentoxin	F_1 noncovalent	148
Azide	F_1(γ?) noncovalent	4, 39
3′-Benzoylbenzoic ATP	$F_1\alpha\beta$ covalent	168
DCCD	$F_1\beta$ covalent (Fig. 5.4)	48, 50, 72, 170
FSBA	$F_1\beta$ covalent (Fig. 5.4)	52
NBDCl	$F_1\beta$ covalent (Figs. 5.4, 5.6)	191
8-Azido-ATP	$F_1\beta$ covalent (Figs. 5.4, 5.6)	49
Phenylglyoxal	F_1 covalent	74, 196
Divalent azide ATP analogue	$F_1\alpha\beta$ covalent (Fig. 5.5)	75
Oligomycin, rutamycin	MF_0 noncovalent	5
Ventruicidin A	MF_0 noncovalent	37
Tributyltin Cl	F_0 noncovalent	19, 127
DCCD	F_0 subunit c covalent	6, 19, 37, 127
Phenylglyoxal	F_0 covalent	198

[a] The inhibitor interacts with F_1 in some energized conformation described in Section 4.3.2.

also reported, with binding of one ATP/α subunit [41]. The experiments with photoactivatable ATP analogues, described in Sections 2.2.3 (Fig. 5.5) [7] and 4.1.1 (3′-benzoylbenzoic ATP) [168], showed covalent incorporation at the interface of the α and β subunits, which is the putative catalytic site.

There are many reports on the ligands of F_1 [1,16,91]. P_i is loosely bound to F_1 [172], perhaps to the β subunit judging from the experiments with the isolated subunit (Kagawa, unpublished). This result also suggests that at least part of the β subunit is the catalytic site. Inhibitors of F_0F_1, which are called energy transfer inhibitors, are listed in Table 5.5, because they are useful in elucidating proton motive ATP synthesis. In contrast to mesophilic F_1's, TF_1 is resistant to aurovertin [87], efrapeptin, quercetin and local anesthetics [173]. Thus, the noncovalent inhibitor binding site may not have a direct relationship to the structure of the catalytic site [173]. The presence of three catalytic sites, three noncatalytic sites, and perhaps many different inhibitor binding sites complicates the kinetic analysis of F_0F_1.

4.2.2. Energy requiring step in ATP synthesis in F_0F_1

Although the formation of an anhydride bond between ADP and P_i requires a large free energy change in aqueous solution, this catalytic step may not be the energy requiring step in ATP synthesis on F_0F_1. As described in Section 3.4, net synthesis of F_1-bound ATP was observed without adding energy of proton flux [154–156,158]. It was proposed that enzymes, including F_1, utilize the binding energy of the specific substrate to increase the reaction rate (Ref. 174, see energy diagram).

Direct measurement of enzyme-bound nucleotides with ^{31}P-NMR confirmed this idea [175]. At a catalytic concentration of pyruvate kinase (or 3-phosphoglycerate kinase), the equilibrium constant of the reaction, K_{eq} is as low as 3×10^{-4}, whereas the equilibrium constant for enzyme-bound ligands for both enzymes is 1.

$$K_{eq} = \frac{[\text{MgADP}][P\text{-enolpyruvate (or 1,3-bis-}P\text{-glycerate)}]}{[\text{MgATP}][\text{pyruvate (or 3-}P\text{-glycerate)}]} = 3 \times 10^{-4}$$

$$K_{eq'} = \frac{[\text{Enzyme-MgADP-}P\text{-enolpyruvate (or 1,3-}P\text{-glycerate)}]}{[\text{Enzyme-MgATP-pyruvate (or 3-}P\text{-glycerate)}]}$$

$$= 1.0\text{–}2.0 \text{ (or 1)}$$

Likewise, F_1-MgATP is easily synthesized from F_1-MgADP-P_i ($K_{eq'} = 0.5$, see Section 4.3 [176]). The energy requiring step might be the release of MgATP from the F_1 molecule, as first proposed by Boyer [177] and Slater [178]. From the presence of tightly bound AT(D)P on F_1, Harris and Slater supported this idea [158]. However, as described in the previous section, there are three tightly bound nucleotides at the noncatalytic sites [171]. These are the stabilizing or allosteric ligands, removal of which results in destabilization of MF_1. TF_1 does not contain tightly bound AT(D)P at noncatalytic sites [84], but TF_1-bound ATP (0.8 mol/TF_1) has been synthesized [156], either from the 3 : 3 : 1 ADP-Mg-TF_1 [156] or 1 : 1 : 1 ADP-Mg-TF_1 [170].

To avoid the complications of tightly bound noncatalytic nucleotides, the transition state on the enzyme at the equilibrium ($K_{eq'}$) of the ^{18}O-P_i exchange reaction was measured [179,180]. It was pointed out earlier that any enzyme binds tightly to substrates in the transition state and that the binding energy affects the rate in the transition state, not in the enzyme-substrate complex [174]. $H_2^{18}O$-P_i exchange of submitochondrial particles was only partially inhibited by an uncoupler [180]. If the formation of the anhydride bond of ATP requires $\Delta\tilde{\mu}_{H^+}$, it should be inhibited by an uncoupler [180]. Moreover, $H_2^{18}O$ exchange during [γ-^{32}P, ^{18}O]ATP hydrolysis by soluble F_1 resulted in the incorporation of about 3 oxygen atoms per P_i at very low concentration (1 μM) of ATP, with or without added ADP (100 μM).

Together with results on the synthesis of F_1-bound ATP [154–156], this result indicates that formation of a covalent structure of ATP at the catalytic site is not directly coupled to proton translocation.

4.3. Uni-site and multi-site kinetics of F_1

4.3.1. Positive cooperativity in V_{max} and negative cooperativity in K_d

The velocity of hydrolysis of [γ-^{32}P]ATP [181] or 2′,3′-*O*-(2,4,6-trinitrophenyl)-ATP [182] at a single site of MF_1 is very low when measured at very low substrate concentration. However, it is accelerated more than 10^6-fold when non-labelled ATP is added to allow binding at multiple catalytic sites [181,182] (Table 5.6). The existence of this strong cooperative interaction is supported by the following observations.

1. One mole of AMP-PNP/F_1 or NBDCl/F_1 is sufficient to inhibit catalysis despite the fact that there are three catalytic sites per F_1 [16].

2. The $H_2^{18}O$-P_i exchange activity of isolated F_1 is lost at high ATP concentrations, because the rate of ADP-release from F_1 is accelerated by the binding of ATP to the second catalytic site [180].

3. A lag time of 2 s is observed in TF_1-ATPase which does not contain any tightly bound noncatalytic nucleotides [183].

As shown in Table 5.5, the K_d (or K_m) for bi- and tri-site catalysis is 10^7-fold that of the K_d for uni-site catalysis. This value represents the negative cooperativity of nucleotide binding, whereas the difference in the V_{max} values represents the positive cooperativity of catalysis.

TABLE 5.6

Negative cooperativity of nucleotide binding (K_d) and positive cooperativity of catalysis (V_{max}) between catalytic sites of F_1 (kinetic constants for ATP hydrolysis at multiple catalytic sites of beef heart MF_1)

ATP-promoted hydrolysis of [γ-^{32}P]ATP bound in a single site (uni-site) on F_1 was measured on MF_1 with a quenched flow apparatus [181]. The rates were analyzed using Lineweaver-Burk plots.

Mode of catalysis	K_d or K_m (M)	V_{max} (s^{-1})
Uni-site	10^{-12}	10^{-4}
Bi-site	3×10^{-5}	300
Tri-site	1.5×10^{-4}	600

The kinetic constants for the uni-site catalysis of MF_1 are summarized in Fig. 5.10 [176]. The findings relevant to ATP synthesis are the very tight binding of ATP to MF_1 ($K_1 = 10^{12}$ M^{-1}), and the equilibration of $F_1 \cdot$ ATP and $F_1 \cdot$ ADP $\cdot$ P_i ($K_2 = K_{eq'}$ = 0.5) [176]. Moreover, since the simultaneous binding of P_i and ADP to MF_1 is very unfavorable ($K_3 > 6 \times 10^{-4}$ M), perhaps P_i and ADP may be forced to be in close spatial proximity to synthesize ATP on the catalytic site. However, ADP binds tightly to MF_1 ($K_4 = 3 \times 10^{-7}$ M). The product of the equilibrium constants for the individual steps shown in Fig. 5.10 ($K_1 \times K_2 \times K_3 \times K_4$) has a minimum value of 10^2. This value is four orders of magnitude less than the overall equilibrium constant for ATP hydrolysis ($K_{eq} = 10^6$). The difference may be largely due to the low value for K_3 estimated (the actual value may be 10^{-1} M) [176]. Extremely slow release of ATP from F_1-ATP ($K_{-1} = 7 \times 10^{-6}$ s^{-1}) is also important, and this step may be facilitated by applying an electric field pulse of 10–50 μs [143,147] as described in Section 3.3.2.

The kinetics described in Fig. 5.10 does not represent the very rapid turnover of F_1 or F_0F_1 under normal assay conditions (MgATP 10^{-4}–10^{-3} M). The positive cooperativity of catalysis is caused by enhancement of the rate of release of P_i (10^5-fold) and ADP (10^6-fold) [181]. The release of F_1-bound ADP by ATP in the other site was also confirmed with pyruvate kinase which does not react with bound ADP [179]. On the other hand, ADP binding is necessary for ATP release during ATP synthesis [179]. The 18 K values of the F_1-ATPase reaction during uni- and multisite catalysis were determined [179]. Although there is a large difference in the value of K_1 (10^{12} M^{-1} in Fig. 5.10 and 10^6 M^{-1} in Ref. 179), the presence of both positive cooperativity in the V_{max} and negative cooperativity in the K_d is well established. Based on these observations, an alternating three-site model was proposed by Boyer's group. However, it is still not certain that three catalytic sites are really alternating, because several nucleotide analogues that covalently occupy one or two catalytic sites of F_1, did not inhibit ATPase activity of F_1 down to the uni-site catalytic velocity [184].

Tight nucleotide binding in negative cooperativity is essential for the rapid equilibrium of $F_1 \cdot$ ATP and $F_1 \cdot$ ADP $\cdot$ P_i, while positive cooperativity of catalysis is

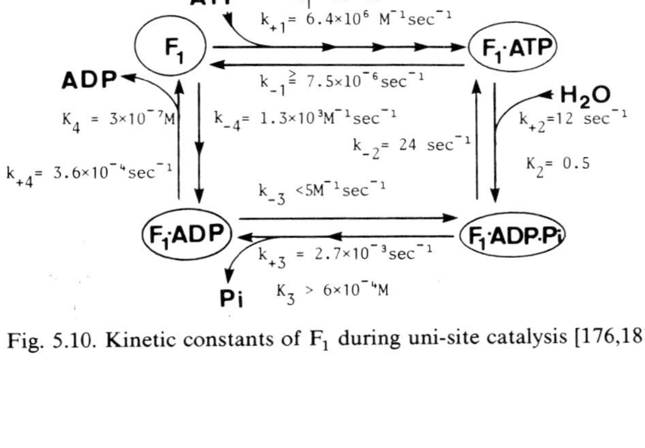

Fig. 5.10. Kinetic constants of F_1 during uni-site catalysis [176,181].

needed for the rapid overall reactions. Use of a fluorescent ATP analogue showed that the kinetics of F_1-ATPase was similar to that of myosin [185].

4.3.2. Control of ATPase activity

The cooperative interaction between catalytic sites may be modified by many factors. ATPase inhibitor protein is a well defined factor from mitochondria [33,34,186] (Table 5.1). Its physiological regulatory function may prevent unnecessary ATP hydrolysis [186–189]. However, there is no corresponding protein inhibitor of prokaryotic F_0F_1. One mole of inhibitor is enough to inhibit MF_1. The inhibitors listed in Table 5.5 interact with F_1 in some energized conformation. For example, the rate of combination of the protein inhibitor decreases by 2 orders of magnitude ($t_{1/2}$ decreases from 30 s to 90 min) in the presence of oligomycin which blocks energization of MF_1 by F_0 [187]. The energized state of F_1 during the multi-site catalysis may be simplified as follows:

$$F_1 \rightleftarrows F_1ATP \rightleftarrows F_1^*ATP \rightleftarrows F_1^*ADP \cdot P_i \rightleftarrows F_1^*ADP \rightleftarrows F_1ADP \rightleftarrows F_1$$

where F_1^* = energized F_1. ATPase inhibitor is bound to the F_1^*ADP [180], while AMP-PNP is bound to F_1^*ATP, because hydrolysis of ATP is impossible [190]. The NBDCl-F_1 complex hydrolyzes ATP, but the products are not released, so $F_1^*ADP \cdot P_i$ is attained [191]. NBDCl [188], efrapeptin [187] and bathophenanthroline-Fe [187] prevent release of ATPase inhibitor from F_1 even under energized conditions.

Many inhibitors of F_0F_1 are known [192–199] (Table 5.5), but the details of their mechanisms of inhibition during energization are unknown. In simplified F_0F_1-liposomes, even Mg^{2+} concentration had profound effects on P_i-ATP exchange (7-fold) and inhibition by oligomycin and DCCD [197].

4.4. Coupling of proton flux and ATP synthesis

4.4.1. Mechanism of proton translocation

The proton translocating activity of F_0 reconstituted into liposomes was measured [6,77] (Fig. 5.1). The H^+ channel activity of F_0 is specific for H^+. The velocity of H^+ translocation (V_{max} = 31 μg ion per min per mg TF_0, K_m = 0.095 μg ion per liter) is somewhat proportional to $\Delta\tilde{\mu}_{H^+}$ (Ohm's law). This velocity is much slower than that of Na^+ translocation through Na^+ channels [77].

There are several hypotheses on H^+ translocation through F_0. One is based on an analogy with H^+ conduction in solid metal hydroxides, such as $Mg(OH)_2$ [200]. Another is the H^+ jumping through a network of H^+-bonded groups in F_0 or H_2O, much like H^+ conductance through ice [201,202]. However, these hypotheses are too simple to explain the specific function of F_0, and the distance between protonating groups in subunit c of F_0 is too long for formation of a chain of hydrogen bonds [37]. Purified subunit c of F_0 was reported to translocate H when it was reconstituted into liposomes [203] or planar lipid bilayer [204]. The author could not reproduce these liposomes [77]. In fact, genetic analyses clearly demonstrated that all three subunits of F_0 were required for H^+ conduction [205]. Most of the other F_1 binding subunits could be removed by proteolysis without impairing H^+ translocation [198]. Subunits a and b may be necessary for the assembly of subunit c into an oligomer. Subunits a and b, individually as well as together, were found to bind F_1 [205].

The amino acid sequence of subunit c of EF_0 suggested a hairpin model containing two hydrophobic, transmembrane α-helices (no. 15–40, and no. 51–76 residues, each 43 Å long) are connected with a hydrophilic random coil (no. 41–50 residues) (Fig. 5.2) [37,67,206]. Asp-61, which is buried 15 Å from the surface of the membrane, plays a central role in H^+ conduction. Modification of Asp-61 by DCCD or mutation completely abolished the H^+-translocating activity of EF_0 [37,206]. Genetic replacement of Leu-31 by Phe-31, which is situated close to Asp-61 in the hairpin, resulted in slower H^+ translocation, because Phe is 2 Å longer than Leu [206]. Replacement of Ile-28 by Val-28 resulted in DCCD resistant EF_0 [37,67]. The Gly-rich segment containing Leu-31 and Ile-28 is essential for H^+ translocation, because of its high flexibility as discussed earlier [207,208]. There are only three dissociating residues (Glu, Arg and Tyr) in subunit c of TF_0. Moreover, Tyr is not found in c subunits of five other species [37]. Thus, since there is no corresponding functional group in the area of Tyr, only Glu (or Asp-61 or in *E. coli*) and homologous Arg-41 could be the proton translocating residues. The oligomeric structure of subunit c is essential for H^+ translocation [12] (Fig. 5.2), and only one or two DCCD per 6–15 c subunits in F_0 are enough to block H^+ translocation [12,37,67]. The F_0 molecule binds exactly one F_1, and passive H^+ translocation through F_0 is blocked by the gating activity of F_1 [6,77].

4.4.2. Release of ATP from F_0F_1-ATP by $\Delta\tilde{\mu}_{H^+}$

As described in Section 4.3, ATP-release from F_1-ATP is the energy requiring step. The energy is indirectly supplied to F_1 from H^+-flux driven by $\Delta\tilde{\mu}_{H^+}$, via an interesting unknown pathway. An electric pulse releases nucleotides from F_0F_1 [142], and the electric potential seems to lower the enthalpy of the activated F_0F_1 complex during ATP synthesis. However, Arrhenius plots of the initial rates of ATP synthesis at different temperatures and $\Delta\psi$ values, revealed the lack of such an effect [209]. The $\Delta\psi$ acts not by changing the structure of F_1-ATP but rather by changing the concentration of H^+ (mass action). F_0 acts as a 'proton well' that converts $\Delta\psi$ to ΔpH, not as a 'proton sink' that converts ΔpH to $\Delta\psi$. On the other hand, protonation-deprotonation steps in reactions of H^+-driven ATP synthesis are not rate limiting, because the initial velocity of ATP synthesis is not decreased in ^2H_2O [209].

The conformational change observed in energized CF_1 [103] and TF_1 [80,84] may be related to release of nucleotides from F_1. In F_0F_1, it is proposed that a proton gradient along a membrane-spanning α-helix is coupled to small changes in the torsional angles which finally accumulate to cause a large conformational change [210]. The energy transfer through conformational change in F_0F_1 may be compared with that of H^+ flux-driven flagellar motor rotation [211] and ATP-driven bending of myosin head [212]. Both the latter reactions are loosely coupled, since their torque or tension is generated even under fixed or isometric conditions. In the motor, the minimum energy that can cause rotation is less than kT (30 meV, thermal fluctuation) and the rotation speed decreases inversely with the viscosity; i.e., the torque generated by $\Delta\tilde{\mu}_{H^+}$ is independent of rotation [211]. This loose coupling is convenient for regulation of the efficiency of energy transfer in response to wide ranges of resistance. The State 4 respiration of mitochondria observed in the absence of ADP + P_i, is similar to this situation. Although most of the State 4 respiration is explained by H^+ leakage through the lipid bilayer, DCCD-resistant ATPase activity of F_0F_1 is much higher than the uni-site ATPase activity. DCCD-treated cytochrome oxidase shows loose H^+ pumping activity, yet its cytochrome *c* oxidation is intact [213]. Similar molecular slipping or loose coupling is also proposed for F_0F_1 [214]. An interesting possibility is that alternating catalytic sites on $\alpha_3\beta_3$ [179] are revolved by $\Delta\tilde{\mu}_{H^+}$ through the F_0 complex [215]. However, it is still uncertain whether F_1 rotation on F_0 during ATP synthesis is necessary and, for example, when a part of the nucleotide binding sites of F_1 is blocked by a covalent nucleotide analogue, the resulting modified F_1 is active [184]. It is still difficult to detect this rapid change even by nanosecond fluorometry [216] and small angle X-ray scattering [102], though conformational change is detected by hydrogen exchange in peptide bonds [80,84,103]. The flagellar motor is not specific for H^+, and in alkalophilic bacillus it is driven by Na^+ [217]. Purple membrane and other H^+ pumps do not rotate.

4.4.3. A new model: acid-base cluster hypothesis

Finally, a new model consistent with the experimental results described in this chapter, is shown in Fig. 5.11 and Table 5.7. The acid-base cluster device (the upper portion of Fig. 5.11) converts energy of H^+ flux into a conformational change. Conformational change caused by protonation-deprotonation is well known in the pH-dependent helical change of polyglutamate [218] and Bohr effect of hemoglobin. Proton transport through a single salt bridge relay composed of proton donating and accepting amino acid residues has been proposed on bacteriorhodopsin [219] and cytochromes [220]. However, in order to satisfy the stoicheiometry of ATP synthesis ($3H^+$/ATP), more than three basic and three acidic groups are needed in one device. The clustering of these groups is also necessary to release or adsorb protons by mutual electrostatic interactions, at a narrow area of F_1 to receive H^+ from one γ connected to F_0. A mixture of poly-Glu and poly-Lys forms a pH-sensitive structure.

This device should be common to all F_1's, and in fact these clusters are found in the homologous sequences in the subunits of MF_1, CF_1 and EF_1. The acidic cluster is composed of Asp-Glu-Leu-Ser-Glu-Glu-Asp (residues no. 381–387 of EF_1) and the basic cluster of Arg-Ala-Lys-Ile-X-Arg (no. 393–399) (Fig. 5.4) [31]. There are other small clusters which are also homologous in F_1's, but the clusters mentioned above are located very close to the Rossmann fold, as shown in Fig. 5.6. There are a few clusters in subunits a and b, which may transfer energy of H^+ flux. The invariant Arg-41 of subunit c may become a candidate of the basic cluster if c subunits form an oligomer. Intracistronic mapping of the defective site of the 12β subunit mutants in the domain II (from residue no. 288 to the C-terminus) revealed that many of

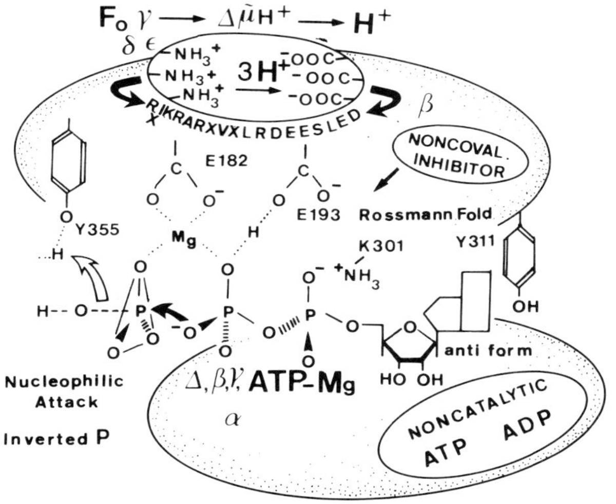

Fig. 5.11. Acid-base cluster hypothesis. The cluster of acidic residues and basic residue is indicated on the upper portion of the figure. F_0 is located on the left side where H^+ flux is transferred to the cluster. The movement of three protons from three NH_3^+ to three COO^- results in the charge neutralization, which causes conformational change to release tightly bound Δ, β, γ bidentate ATP-Mg complexes. After the energy is used up, three COOH's are dissociated and protons are released into the matrix side because of their pK_a. Three NH_2 are protonated from the F_0 side. The formation of ATP takes place without added energy, via in-line nucleophilic substitution (SN_2) of O^- of ADP to the P of inorganic phosphate. For details, refer to the text.

them lost H^+-translocating and reconstitutive activities [47]. A mixture of poly-acid and poly-base is a barrier for any ions except for H^+, and in fact, H^+ channel (F_0) is inhibited by modification of either COOH or Arg [198]. On removal of acid-base cluster in the b subunit, active H^+ transport was lost without impairment of F_1-binding or passive H^+ transport [221]. When $\Delta\tilde{\mu}_{H^+}$ is applied to this device through F_0-$\gamma\delta\epsilon$, three protons are transported from NH_3^+ to COO^- to form neutral residues (NH_2 and COOH). The charge neutralization in the device causes conformational change in the Rossmann fold which now releases tightly bound ATP. The acid-base device undergoes a cyclic protonation-deprotonation so that COOH's eject protons to F_1 side bulk water after the conformational change and may then be protonated when $\Delta\tilde{\mu}_{H^+}$ is applied to the cluster after ATP is synthesized at the catalytic site, and NH_2 in the cluster is protonated from the F_0 side. The role of localized $\Delta\tilde{\mu}_{H^+}$ outside the F_0F_1 [222] (Section 3.1.2) and a protein labeled with an anisotropic inhibitor by $\Delta\psi$ [223] have been discussed, but the localized H^+ in the cluster is important. Kozlov [224] suggested that the positively charged complex of ATP at the catalytic site is translocated directly down the $\Delta\psi$. However, the bidentate MgATP complex [164] is negatively charged (Figs. 5.9 and 11). Only one tightly bound ATP was found in TF_1 after the synthesis [156] from the 1 : 1 : 1 ADP-Mg-TF_1 [170]. In the reverse reaction, conformational change causes the

TABLE 5.7

Acid base cluster hypothesis in proton motive ATP synthesis

Requirements	Experimental facts
Structure (genetics and physical properties)	
Primary: Homology in the acid base cluster. Mutation in the area should cause inactivation.	Homology in β of CF_1, MF_1 and EF_1. Invariant acidic (381–387) and basic (393–399) residues.
Secondary: Rossmann fold should be affected.	The cluster is located near the fold.
Tertiary: ATP-binding should cause stabilization of F_1 and β because of 3 ionic bonds.	Slower ^1H-^2H exchange of F_1 and β in the presence of ATP than in its absence.
Quaternary: Narrow area of contact in β to receive H^+ flux from F_0-γ.	Narrow cluster area. Crosslinking of $\alpha\beta$, $\beta\gamma$, $\gamma\alpha$, etc.
Energy-dependent conformation change: Effect of $\Delta\tilde{\mu}_{H^+}$ $\Delta\psi$ and ΔpH.	^3H incorporation into F_1 during phosphorylation.
^{32}P-NMR of TF_1 nucleotide complex.	ATP-release by an electric pulse.
Energetics	
Stoicheiometry: $3H^+$/ATP bound at one subunit should be explained.	$3H^+$ transfer from $3NH_4^+$ to $3COO^-$ in the cluster is possible at once.
ΔG of ATP synthesis: 7–10 kcal/ATP, and ionic bond energy in proteins.	The energy of an ionic bond is larger than 2–3 kcal. Three bonds.
Activation enthalpy (ΔH_a) should explain the mass action of [H^+] during ATP synthesis.	Arrhenius plot of the initial velocity of ATP synthesis showed that ΔH_a is constant at different $\Delta\psi$.
Equalization of the internal thermodynamics is required (K_{eq} of near unity).	The 1:1:1 ADP-Mg-TF_1 is converted into ATP-TF_1 in P_i.
Kinetics	
Rapid protonation-deprotonation in the cluster, compared with the rate limiting conformational change.	There is no isotopic effect of ^2H_2O during ATP synthesis. Relaxation spectrum of TF_1, $\alpha\beta$.
Nucleotide effect on H^+ flux	ATP-stimulated H^+ uptake (gate).
Conformation-dependent kinetics during ATP synthesis or hydrolysis.	Positive cooperativity in V_{max} and negative cooperativity in K_d for AT(D)P.
Chemical modification	
The following should inactivate.	Loss of active H^+ transport
Removal of essential acid-base clusters in b.	without impairment of passive H^+ transport and F_1 binding.
Energy dependent amphipatic cation inhibitor.	Inhibition by octylguanidin.
pH dependent polyanion-polycation CD change.	CD change of poly-Glu–poly-Lys complex
Energy dependent change in gate.	Crosslinking of γ with divalent agent.
Essential amino and carboxyl groups.	Inactivation by DCCD etc.

release of protons from NH_3^+ to the direction of F_0, and adsorption of protons to COO^-.

This protonation and deprotonation is not rate limiting as discussed by Khan and

Berg [209], because our direct determination revealed that the ^1H-^2H exchange of COOH and NH_3^+ is much more rapid than the conformational change (hydrogen-bonded protons) [80,84]. A hydrogen-bonded chain mechanism for H^+ conduction and H^+ pumping was proposed, but the exchange velocity of the H^+ was not discussed [225]. This device is consistent with the fact that $\Delta\psi$ and ΔpH of equal $\Delta\tilde{\mu}_{H^+}$ value elicit identical rates of ATP formation in purified F_0F_1-liposomes [131,136]. The mass action of H^+ concentration, rather than lowering the activation enthalpy (ΔH_a) [209], drives this device. This indirect coupling mechanism allows some loose coupling, depending on the conditions which facilitates conformational change. The very rapid ATP synthesis in an electric pulse experiment [143,148] is also consistent with this model [14].

A chemical approach to study of this device has been tried. Both hydrophobic anions and cations (octylguanidine) [226] are inhibitory in ATP synthesis. Acetylation of MF_1 and TF_1 [87] with ^3H-labeled acetic anhydride resulted in loss of activity, though the first few highly accessible Lys residues were modified without inactivation. The modification of COOH and Arg also inactivated TF_1 [196], but the details are unknown. The H^+-gating function of the γ subunit of TF_1 [126] was confirmed in CF_1 [227]; a number of bifunctional maleimides form crosslinks within the γ subunit, and the resulting CF_1 becomes permeable to protons.

The nucleotide binding site of this model has been discussed in the previous sections. In short, this model is also consistent with the stereochemical course during the nucleophilic attack of a P atom by the terminal oxygen of ADP (Figs. 5.8, 5.9 and 5.11), which takes place at the interface of the α and β subunits (Fig. 5.5) on the Rossmann fold (Fig. 5.6). The effects of Mg and ADP during the modification with DCCD [50,72] identify the two Glu residues in Fig. 5.11. E182 is more directly related to the site of phosphorylation, since modification of this residue inhibited ATP synthesis, but that of E193 did not inhibit synthesis of ATP in dimethylsulfoxide [15]. The tyrosine residue (Y355 in Fig. 5.11) functions as a general acid catalyst during ATP synthesis which donates a proton to the oxygen of P_i to facilitate the removal of OH^- from P_i when it is attacked by oxygen from ADP. As shown by circular dichroic spectrometry the anti-form of ATP interacts with protonated tyrosine at the catalytic site [84]. The tight binding of one ADP molecule on a hard portion of TF_1 and the bending of the P-O bond at the β position is confirmed by ^{31}P-NMR (50 °C, 80 mg/ml) [228]. This hypothesis will be tested by many sophisticated methods.

Epilogue

Proton motive ATP synthesis in F_0F_1 liposomes (Fig. 5.1, Table 5.4) and $\Delta\tilde{\mu}_{H^+}$-driven H^+ translocation through F_0 [6] strongly supported Mitchell's chemiosmotic theory described in the previous series of Comprehensive Biochemistry [108]. The molecular mechanism of function of F_0F_1 was not predicted by this theory, but it has been elucidated by new methods. For example, the presence of X-P was demonstrated in

H^+ ATPase of fungal plasma membranes [18], while existence of X-P in F_0F_1 was disproved by studies using [^{16}O, ^{17}O, ^{18}O]-thiophosphate (Figs. 5.8, 5.11) [163,164]. The important methods used are reconstitution of F_0F_1 from its subunits (Fig. 5.2) [38,40], physicochemical analyses [7–10,12–17] (Figs. 5.2 and 5.7) and genetic studies (Figs. 5.3, 5.4 and 5.6) [11,21,67,208] on F_0F_1. The interaction of nucleotides with F_1 was studied by crosslinking methods (Fig. 5.5), stereochemical analyses (Figs. 5.8, 5.9, 5.11) [161–166] and enzyme kinetics (Fig. 5.10) [176–182].

The direct participation of translocated protons in anhydride bond formation during ATP synthesis [113] was disproved by the net synthesis of F_1-bound ATP from ADP, P_i and F_1 [154–156] and the kinetics of uni-site and multi-site catalysis of F_1 [16,176–182] (Table 5.6, Fig. 5.10). The conformational change of F_0F_1 resulting in release of ATP from the F_1-ATP complex is the energy-requiring step for free ATP synthesis [154–156,176–182]. This releasing step is not the detachment of ATP-complex from F_1 by $\Delta\psi$, because the true substrate is Δ,β,γ-bidentate Mg-ATP [164]. The mechanism of proton translocation will probably be elucidated first in bacteriothodopsin [229], because of its high reconstitutability, simple structure and easy genetic manipulation of the reacting groups. How is the energy of proton flux converted into conformational energy of F_1 to release ATP? Cross [230] and Boyer [231] proposed that the movement of three protons through F_0F_1 drives the trimer of F_1 through a 120° rotation such that asymmetric interactions with other subunits cause the catalytic sites to change their affinities and product. This mechanism is similar to that of flagellar motion driven by proton flux [211,215]. Since only one γ subunit is present per three $\alpha\beta$ subunits, $3H^+/ATP$ stoicheiometry and conformational change of one $\alpha\beta$-ATP complex must be explained at the molecular level. The acid-base cluster hypothesis [232] is thus proposed (Fig. 5.11 and Table 5.7).

Acknowledgement

The author expresses his thanks to Prof. E. Racker, Dr. P. Mitchell and many other biochemists for helpful discussion at the International Congress of Biochemistry (1979 and 1982), Gordon Research Conferences (1977, 1979, 1981 and 1983), etc.

The encouragement of pioneer Japanese biochemists in this field is also deeply appreciated: namely, Prof. S. Kakiuchi, who reconstituted oxidative activity by adding back phospholipids to acetone extracted mitochondria in 1927 (J. Biochem. 7, 263); Prof. K. Makino, who first reported the triphosphate structure of ATP in 1935 (Biochem. Z. 278, 161); and Prof. K. Okunuki, who used bile acids in the extraction of mitochondrial components in 1940 (Proc. Imper. Acad. 16, 140). Thanks are also due to my colleagues Drs. N. Sone, H. Hirata, M. Yoshida, H. Okamoto, S. Ohta, T. Hamamoto, Y. Yanagita, and Mrs. N. Nukiwa. This work was supported by a grant from the Ministry of Education, Science and Culture of Japan (No. 58114006) and from the Naito Foundation.

References

1 Mitchell, P. (1961) Nature 191, 144–148.
2 Racker, E. (1965) Mechanisms in Bioenergetics. Academic Press, New York.
3 Kagawa, Y. (1972) Biochim. Biophys. Acta 265, 297–338.
4 Pullman, M.E., Penefsky, H.S., Datta, A. and Racker, E. (1960) J. Biol. Chem. 235, 3322–3329.
5 Kagawa, Y. and Racker, E. (1966) J. Biol. Chem. 241, 2461–2466.
6 Okamoto, H., Sone, N., Hirata, H., Yoshida, M. and Kagawa, Y. (1977) J. Biol. Chem. 252, 6125–6131.
7 Penefsky, H.S. (1979) Adv. Enzymol. 49, 223–280.
8 Amzel, L.M. and Pedersen, P.L. (1983) Ann. Rev. Biochem. 52, 801–824.
9 Shavit, N. (1980) Ann. Rev. Biochem. 49, 111–138.
10 McCarty, R.E. (1979) Ann. Rev. Plant Physiol. 30, 79–104.
11 Kanazawa, H. and Futai, M. (1982) Ann. N.Y. Acad. Sci. 402, 45–64.
12 Fillingame, R.H. (1981) Curr. Topics Bioenerg. 11, 35–106.
13 Ferguson, S.J. and Sorgato, M.C. (1982) Ann. Rev. Biochem. 51, 185–217.
14 Hamamoto, T. and Kagawa, Y. (1984) In The Enzymes of Biological Membranes, Vol. 4 (Martonosi, A., ed.) pp. 149–176, Plenum Press, New York.
15 Fillingame, R.H. (1980) Ann. Rev. Biochem. 49, 1079–1113.
16 Cross, R.L. (1981) Ann. Rev. Biochem. 50, 681–714.
17 Skulachev, V.P. and Hinkle, P.C. (eds.) (1981) Chemiosmotic Proton Circuits in Biological Membranes, pp. 3–633, Addison-Wesley, Massachusetts.
18 Goffeau, A. and Slayman, C.W. (1981) Biochim. Biophys. Acta 639, 197–223.
19 Sone, N., Yoshida, M., Hirata, H. and Kagawa, Y. (1975) J. Biol. Chem. 250, 7917–7923.
20 Foster, D.L. and Fillingame, R.H. (1979) J. Biol. Chem. 254, 8230–8236.
21 Walker, J.E., Eberle, A., Gay, N.J., Runswick, M.J. and Saraste, M. (1982) Biochem. Soc. Trans. 10, 203–206.
22 Hansen, F.G., Nielsen, J., Riise, E. and von Meyenburg, K. (1981) Mol. Gen. Genet. 183, 463–472.
23 Süss, K-H and Schmidt, O. (1982) FEBS Lett. 144, 213–221.
24 Montecucco, C., Dabbini-Sala, F., Friedel, P. and Galante, Y.M. (1983) Eur. J. Biochem. 132, 189–194.
25 Galante, Y.M., Wong, S-Y. and Hatefi, Y. (1979) J. Biol. Chem. 254, 12372–12378.
26 Todd, R.D., Griesenbeck, T.A. and Douglas, M.G. (1980) J. Biol. Chem. 255, 5461–5467.
27 Macino, G. and Tzagoloff, A. (1980) Cell 20, 507–517.
28 Anderson, S., Bankier, A.T., Barrell, B.G., de Bruijn, M.H.L., Coulson, A.R., Drouin, J., Eperon, I.C., Nierlich, D.P., Roe, B.A., Sanger, F., Schreier, P.H., Smith, A.J.H., Staden, R. and Young, I.G. (1981) Nature 290, 457–465.
29 Sanadi, D.R. (1982) Biochim. Biophys. Acta 683, 39–56.
30 Yoshida, M., Sone, N., Hirata, H., Kagawa, Y. and Ui, N. (1979) J. Biol. Chem. 254, 9525–9533.
31 Runswick, M.J. and Walker, J.E. (1983) J. Biol. Chem. 258, 3081–3089.
32 Walker, J.E., Runswick, M.J. and Saraste, M. (1982) FEBS Lett. 146, 393–396.
33 Frangione, B., Rosenwasser, E., Penefsky, H.S. and Pullman, M.E. (1981) Proc. Natl. Acad. Sci. U.S.A. 78, 7403–7407.
34 Matsubara, H., Hase, T., Hashimoto, T. and Tagawa, K. (1981) J. Biochem. 90, 1159–1165.
35 Zurawski, G., Bottomley, W. and Whitfield, P.R. (1982) Proc. Natl. Acad. Sci. U.S.A. 79, 6260–6264.
36 Krebbers, E.T., Larrinua, I.M., McIntosh, L. and Bogorad, L. (1982) Nucleic Acid Res. 10, 4985–5002.
37 Sebald, W. and Hoppe, J. (1981) Curr. Top. Bioenerg. 12, 1–64.
38 Kagawa, Y. (1978) Biochim. Biophys. Acta 505, 45–93.
39 Yoshida, M., Sone, N., Hirata, H. and Kagawa, Y. (1977) J. Biol. Chem. 252, 3480–3485.
40 Futai, M. (1977) Biochem. Biophys. Res. Commun. 79, 1231–1237.
41 Dunn, S.D. and Futai, M. (1980) J. Biol. Chem. 255, 113–118.

42 Downie, J.A., Gibson, G. and Cox, G.B. (1979) Ann. Rev. Biochem. 48, 103–131.
43 Gibson, F. (1983) Biochem. Soc. Trans. 11, 229–240.
44 Deno, H., Shinozaki, K. and Sugiura, M. (1983) Nucleic Acid Res. 11, 2185–2185.
45 Bragg, P.D., Stan-Lotter, H. and Hou, C. (1982) Arch. Biochem. Biophys. 213, 669–679.
46 Hoppe, J., Schaier, H.U. and Sebald, W. (1980) FEBS Lett. 109, 107–111.
47 Kanazawa, H., Noumi, T., Oka, N. and Futai, M. (1983) Arch. Biochem. Biophys. 227, 596–608.
48 Lunardi, J. and Vignais, P.V. (1982) Biochim. Biophys. Acta 682, 124–134.
49 Hollemans, M., Runswick, M.J., Fearnley, I.M. and Walker, J.E. (1983) J. Biol. Chem. 258, 9307–9313.
50 Yoshida, M., Poser, J.W., Allison, W.S. and Esch, F.S. (1981) J. Biol. Chem. 256, 148–153.
51 Kagawa, Y., Sone, N., Yoshida, M., Hirata, H. and Okamoto, H. (1976) J. Biochem. 80, 141–151.
52 Esch, F.S. and Allison, W.S. (1979) J. Biol. Chem. 254, 10740–10746.
53 Kagawa, Y., Sone, N., Hirata, H. and Yoshida, M. (1979) J. Bioenerg. Biomembr. 11, 39–78.
54 Catterall, W.A. and Pedersen, P.L. (1971) J. Biol. Chem. 246, 4987–4994.
55 Senior, A.E. (1975) Biochemistry 14, 660–664.
56 Paradies, H.H. and Schmidt, U.D. (1979) J. Biol. Chem. 254, 5257–5263.
57 Satre, M. and Zaccai, G. (1979) FEBS Lett. 102, 244–248.
58 Amzel, L.M. and Pedersen, P.L. (1978) J. Biol. Chem. 253, 2067–2069.
59 Nielsen, J., Hansen, F.G., Hoppe, J., Friedel, P. and von Meyenburg, K. (1981) Mol. Gen. Genet. 184, 33–39.
60 von Meyenburg, K., Jørgensen, B.B., Nielsen, J. and Hansen, F.G. (1982) Mol. Gen. Genet. 188, 240–248.
61 Kanazawa, H., Tamura, F., Mabuchi, K., Miki, T. and Futai, M. (1980) Proc. Natl. Acad. Sci. U.S.A. 77, 7005–7009.
62 Hensgens, L.A.M., Grivell, L.A., Bor, P. and Bos, J.L. (1979) Proc. Natl. Acad. Sci. U.S.A. 76, 1663–1667.
63 Macino, G. and Tzagoloff, A. (1979) J. Biol. Chem. 254, 4617–4623.
64 Nelson, N., Nelson, H. and Schatz, G. (1980) Proc. Natl. Acad. Sci. U.S.A. 77, 1361–1364.
65 Webster, K.A., Oliver, N.A. and Wallace, D.C. (1982) Somatic Cell Genet. 8, 223–244.
66 Gellerfors, P., Takeda, M., Szekely, E. and Douglas, M.G. (1983) Fed. Proc. 42, 1938.
67 Sebald, W., Friedl, P., Schairer, H.U. and Hoppe, J. (1982) Ann. N.Y. Acad. Sci. 402, 28–44.
68 Kanazawa, H., Kayano, T., Mabuchi, K. and Futai, M. (1981) Biochem. Biophys. Res. Commun. 103, 604–612.
69 Walker, J.E., Saraste, M., Runswick, M.J. and Gay, N.J. (1982) EMBO J. 1, 945–951.
70 Gay, N.J. and Walker, J.E. (1982) Nature 301, 262–264.
71 Esch, F.S. and Allison, W.S. (1978) J. Biol. Chem. 253, 6100–6106.
72 Yoshida, M., Allison, W.S., Esch, F.S. and Futai, M. (1982) J. Biol. Chem. 257, 10033–10037.
73 Schäfer, H-J., Scheurich, P., Rathgeber, G., Dose, K., Mayer, A. and Klingenberg, M. (1980) Biochem. Biophys. Res. Commun. 95, 562–568. cf., Hoppe-Seylers Z. Physiol. Chem. 365, 267 (1984).
74 Ting, L.P. and Wang, J.H. (1982) Biochemistry 21, 269–275.
75 Schäfer, H.J., Scheurich, P., Rathgeber, G. and Dose, K. (1980) Anal. Biochem. 104, 106–111.
76 Cantley, L.C. Jr. and Hammes, G.G. (1975) Biochemistry, 14, 2968–2975.
77 Sone, N., Hamamoto, T. and Kagawa, Y. (1981) J. Biol. Chem. 256, 2873–2877.
78 Kagawa, Y., Ohta, S., Yoshida, M. and Sone, N. (1980) Ann. N.Y. Acad. Sci. 358, 103–117.
79 Chou, P.Y. and Fasman, G.D. (1978) Adv. Enzymol. 47, 45–148.
80 Ohta, S., Tsuboi, M., Yoshida, M. and Kagawa, Y. (1980) Biochemistry 19, 2160–2165.
81 Sternweis, P.C. and Smith, J.B. (1977) Biochemistry 16, 4020–4025.
82 Sternweis, P.C. (1978) J. Biol. Chem. 253, 3123–3128.
83 Rossmann, M.G. and Argos, P. (1981) Ann. Rev. Biochem. 50, 497–532.
84 Ohta, S., Tsuboi, M., Oshima, T., Yoshida, M. and Kagawa, Y. (1980) J. Biochem. (Tokyo) 87, 1609–1617.
85 Kagawa, Y. and Racker, E. (1966) J. Biol. Chem. 241, 2475–2482.

86 Soper, J.W., Decker, G.L. and Pedersen, P.L. (1979) J. Biol. Chem. 254, 11170–11176.
87 Kagawa, Y. and Nukiwa, N. (1981) Biochem. Biophys. Res. Commun. 100, 1370–1376.
88 Takeda, K., Hirano, M., Kanazawa, H., Nukiwa, N., Kagawa, Y. and Futai, M. (1982) J. Biochem. 91, 695–701.
89 Smith, J.B. and Sternweis, P.C. (1982) Arch. Biochem. Biophys. 217,. 376–387.
90 Dunn, S.D. and Heppel, L.A. (1981) Arch. Biochem. Biophys. 210, 421–436.
91 Senior, A.E. and Weise, J.G. (1983) J. Membr. Biol. 73, 105–124.
92 Senior, A.E. (1979) In Membrane Proteins in Energy Transduction (Capaldi, R.A., ed.) pp. 233–278, Marcel Dekker, New York.
93 Cox, G.B., Downie, J.A., Langman, L., Senior, A.E., Ash, G., Fayle, D.R.H. and Gibson, F. (1981) J. Bacteriol. 148, 30–42.
94 Amzel, L.M., Narayanan, P. and Pedersen, P.L. (1982) Ann. N.Y. Acad. Sci. 402, 21–27.
95 Paradies, H.H. (1980) Biochem. Biophys. Res. Commun. 92, 1076–1082.
96 Spitsberg, V. and Haworth, R. (1977) Biochim. Biophys. Acta 492, 237–240.
97 Akey, C.W., Spitzberg, V. and Edelstein, S.J. (1983) J. Biol. Chem. 258, 3222–3229.
98 Paradies, H.H. (1979) Biochem. Biophys. Res. Commun. 91, 685–692.
99 Wakabayashi, T., Kubota, M., Yoshida, M. and Kagawa, Y. (1977) J. Mol. Biol. 117, 515–519.
100 Kagawa, Y. (1979) Methods Enzymol. 5, 372–377.
101 Chang, T. and Penefsky, H.S. (1973) J. Biol. Chem. 248, 2746–2754.
102 Furuno, T., Ikegami, A., Kihara, H., Yoshida, M. and Kagawa, Y. (1983) J. Mol. Biol. 170, 137–153.
103 Ryrie, I.J. and Jagendorf, A. (1972) J. Biol. Chem. 247, 4453–4459.
104 Slater, E.C. (1971) Q. Rev. Biophys. 4, 35–71.
105 Lehninger, A.L. and Wadkins, C.L. (1962) Ann. Rev. Biochem. 31, 47–78.
106 Ernster, L. and Lee, C.P. (1964) Ann. Rev. Biochem. 33, 729–788.
107 Kagawa, Y. and Racker, E. (1966) J. Biol. Chem. 241, 2467–2474.
108 Mitchell, P. (1967) In Comprehensive Biochemistry (Florkin, M. and Stotz, E.H., eds.) Vol. 22, pp. 167–197, Elsevier, Amsterdam.
109 Stoeckenius, W., Lozier, R.H. and Bogomolni, R.A. (1979) J. Biol. Chem. 50, 215–278.
110 Leimgruber, R.M., Jensen, C. and Abrams, A. (1981) J. Bacteriol. 147, 363–372.
111 Clarke, D.J. and Morris, J.G. (1979) Eur. J. Biochem. 98, 613–620.
112 Kobayashi, H., Murakami, N. and Unemoto, T. (1982) J. Biol. Chem. 257, 13246–13252.
113 Mitchell, P. and Moyle, J. (1968) Eur. J. Biochem. 4, 530–539.
114 Chance, B. (1961) J. Biol. Chem. 236, 1544–1554.
115 Williams, R.J.P. (1961) J. Theor. Biol. 1, 1–17.
116 Alexandre, A., Reynafarje, B. and Lehninger, A.L. (1978) Proc. Natl. Acad. Sci. U.S.A. 75, 5296–5300.
117 Berry, E.A. and Hinkle, P.C. (1983) J. Biol. Chem. 258, 1474–1486.
118 Cirillo, V.P. and Gromet-Elhanan, Z. (1981) Biochim. Biophys. Acta 636, 244–253.
119 Kashket, E.R. (1982) Biochemistry 21, 534–538.
120 Avron, M. (1978) FEBS Lett. 96, 225–232.
121 Boyer, P.D. (1968) In Biological Oxidation (Singer, T.P., ed.) p. 193, Wiley, New York.
122 Eytan, G.D. (1982) Biochim. Biophys. Acta 649, 185–202.
123 Kagawa, Y. and Racker, E. (1971) J. Biol. Chem. 246, 5477–5487.
124 Kagawa, Y., Ide, C., Hamamoto, T., Roegner, M. and Sone, N. (1982) In Membrane Reconstitution (Poste, G. and Nicolson, G.L., eds.) pp. 137–160, Elsevier, Amsterdam.
125 Montecucco, C., Bisson, R., Dabbeni-Sala, F., Pittoti, A. and Gutweniger, H. (1980) J. Biol. Chem. 255, 10040–10043.
126 Yoshida, M., Okamoto, H., Sone, N., Hirata, H. and Kagawa, Y. (1977) Proc. Natl. Acad. Sci. U.S.A. 74, 936–940.
127 Sone, N., Yoshida, M., Hirata, H. and Kagawa, Y. (1976) J. Membr. Biol. 30, 121–134.
128 Suarez-Isla, B.A., Wan, K., Lindstrom, J. and Montal, M. (1983) Biochemistry 22, 2319–2323.
129 Brown, R.E. and Cunningham, C.C. (1982) Biophys. J. 37, 91–93.

130 Kagawa, Y., Kandrach, A. and Racker, E. (1973) J. Biol. Chem. 248, 676–684.
131 Sone, N., Yoshida, M., Hirata, H. and Kagawa, Y. (1977) J. Biol. Chem. 252, 2956–2960.
132 Jagendorf, A.T. and Uribe, E. (1966) Proc. Natl. Acad. Sci. U.S.A. 5, 170–177.
133 Racker, E. (1976) A New Look at Mechanisms in Bioenergetics, pp. 1–97, Academic Press, New York.
134 Racker, E. and Stoeckenius, W. (1974) J. Biol. Chem. 249, 662–663.
135 Kagawa, Y. and Ariga, T. (1977) J. Biochem. 81, 1161–1165.
136 Kagawa, Y., Ohno, K., Yoshida, M., Takeuchi, Y. and Sone, N. (1977) Fed. Proc. 36, 1815–1818.
137 Kagawa, Y. (1982) Curr. Top. Membr. Transplant 16, 195–213.
138 Thayer, W.S. and Hinkle, P.C. (1975) J. Biol. Chem. 250, 5336–5342.
139 Tsuchiya, T. and Rosen, B.P. (1976) J. Bacteriol. 127, 154–161.
140 Reid, R.A., Moyle, J. and Mitchell, P. (1966) Nature 212, 257–258.
141 Witt, H.T., Schlodder, E. and Gräber, P. (1976) FEBS Lett. 69, 272–276.
142 Witt, H.T. (1979) Biochim. Biophys. Acta 505, 355–427.
143 Hamamoto, T., Ohno, K. and Kagawa, Y. (1982) J. Biochem. 91, 1759–1766.
144 Pauly, H., Packer, L. and Schwann, P.H. (1960) J. Biophys. Biochem. Cytol. 7, 589–601.
145 Tao, T. (1969) Biopolymers 8, 609–632.
146 Zalman, L.S., Nikaido, H. and Kagawa, Y. (1980) J. Biol. Chem. 255, 1771–1774.
147 Teissie, J., Knox, B.E., Tsong, T.Y. and Wehrle, J. (1981) Proc. Natl. Acad. Sci. U.S.A. 78, 7473–7477.
148 Vinkler, C. and Korenstein, R. (1982) Proc. Natl. Acad. Sci. U.S.A. 79, 3183–3187.
149 Rögner, M., Ohno, K., Hamamoto, T., Sone, N. and Kagawa, Y. (1979) Biochem. Biophys. Res. Commun. 91, 362–367.
150 Gräber, P., Rögner, M., Buchwald, H.-E., Samoray, D. and Hauska, G. (1982) FEBS Lett. 145, 35–40.
151 Knox, B.E. and Tsong, T.Y. (1983) J. Biol. Chem 259, 4757–4763.
152 Schlodder, E. and Witt, H.T. (1980) FEBS Lett. 112, 105–113.
153 Schlodder, E. and Witt, H.T. (1981) Biochim. Biophys. Acta 635, 571–584.
154 Feldman, R.I. and Sigman, D.S. (1982) J. Biol. Chem. 257, 1676–1683.
155 Sakamoto, J. and Tonomoura, Y. (1983) J. Biochem. 93, 1601–1614.
156 Yoshida, M. (1983) Biochem. Biophys. Res. Commun. 114, 907–912.
157 de Meis, L., Martins, O.B. and Alves, E.W. (1980) Biochemistry 19, 4252–4261.
158 Harris, D.A. and Slater, E.C. (1975) Biochim. Biophys. Acta 387, 335–348.
159 Hasselbach, W. and Waas, W. (1982) Ann. N.Y. Acad. Sci. 402, 459–469.
160 Skou, J.C. (1982) Ann. N.Y. Acad. Sci. 402, 169–184.
161 Knowles, J.R. (1980) Ann. Rev. Biochem. 49, 877–919.
162 Cohn, M. (1982) Ann. Rev. Biophys. Bioeng. 11, 23–42.
163 Webb, M.R., Grubmeyer, C., Penefsky, H.S. and Trentham, D.R. (1980) J. Biol. Chem. 255, 11637–11639.
164 Senter, P., Eckstein, F. and Kagawa, Y. (1983) Biochemistry 22, 5514–5518.
165 Webb, M.R. and Trentham, D.R. (1981) J. Biol. Chem. 256, 4884–4887.
166 Connolly, B.A. and Eckstein, F. (1981) J. Biol. Chem. 256, 9450–9456.
167 Jaffe, E.K. and Cohn, M. (1979) J. Biol. Chem. 254, 10839–10845.
168 Williams, N. and Colemans, P.S. (1982) J. Biol. Chem. 257, 2834–2841.
169 Bossard, M.J., Vik, T.A. and Schuster, S.M. (1980) J. Biol. Chem. 255, 5342–5346.
170 Yoshida, M. and Allison, W.S. (1983) J. Biol. Chem. 258, 14407–14412.
171 Cross, R.L. and Nalin, C.M. (1982) J. Biol. Chem. 257, 2874–2881.
172 Kasahara, M. and Penefsky, H.S. (1978) J. Biol. Chem. 253, 4180–4187.
173 Saishu, T., Kagawa, Y. and Shimizu, R. (1983) Biochem. Biophys. Res. Commun. 112, 822–826.
174 Jencks, W.P. (1980) Adv. Enzymol. 51, 75–106.
175 Nageswara Rao, B.D., Kayne, F.K. and Cohn, M. (1979) J. Biol. Chem. 254, 2689–2696.
176 Grubmeyer, C., Cross, R.L. and Penefsky, H.S. (1982) J. Biol. Chem. 257, 12092–12100.

177 Kayalar, C., Rosing, J. and Boyer, P.D. (1976) Biochem. Biophys. Res. Commun. 72, 1153–1159.
178 Harris, D.A., Rosing, J., van de Stadt, R.J. and Slater, E.C. (1973) Biochim. Biophys. Acta 314, 149–153.
179 Gresser, M.J., Myers, J.A. and Boyer, P.D. (1982) J. Biol. Chem. 257, 12030–12038.
180 Boyer, P.D., Cross, R.L. and Monsen, W. (1973) Proc. Natl. Acad. Sci. U.S.A. 70, 2837–2839.
181 Cross, R.L., Grubmeyer, C. and Penefsky, H.S. (1982) J. Biol. Chem. 257, 12101–12105.
182 Grubmeyer, C. and Penefsky, H.S. (1981) J. Biol. Chem. 256, 3728–3734.
183 Rectenwald, D. and Hess, B. (1979) FEBS Lett. 108, 257–260.
184 Schäfer, G. (1982) FEBS Lett. 139, 271–275.
185 Matsuoka, I., Watanabe, T. and Tonomura, Y. (1981) J. Biochem. 90, 967–989.
186 Pullman, M.E. and Monroy, G.C. (1963) J. Biol. Chem. 238, 3762–3769.
187 Power, J., Cross, R.L. and Harris, D.A. (1983) Biochim. Biophys. Acta 724, 128–141.
188 Carlsson, C. and Ernster, L. (1981) Biochim. Biophys. Acta 638, 345–357.
189 De Pierre, J.W. and Ernster, L. (1977) Ann. Rev. Biochem. 46, 201–262.
190 Philo, R.D. and Sewlyn, M.J. (1974) Biochem. J. 143, 745–749.
191 Ferguson, S.J., Lloyd, W.J., Radda, G.K. and Slater, E.C. (1976) Biochim. Biophys. Acta 430, 189–193.
192 Cross, R.L. and Kohlbrenner, W.E. (1978) J. Biol. Chem. 253, 4865–4873.
193 Linnet, P.E., Mitchell, A.D., Osselton, M.D., Mulheirn, L.J. and Beechey, R.B. (1978) Biochem. J. 170, 503–510.
194 Lang, D.R. and Racker, E. (1974) Biochim. Biophys. Acta 333, 180–186.
195 Gould, J.M. (1978) FEBS Lett. 94, 90–94.
196 Arana, J.L., Yoshida, M., Kagawa, Y. and Vallejos, R.H. (1980) Biochim. Biophys. Acta 593, 11–16.
197 Fuyu, Y., Beiqi, G. and Yuguo, H. (1983) Biochim. Biophys. Acta 724, 104–110.
198 Sone, N., Ikeba, K. and Kagawa, Y. (1979) FEBS Lett. 97, 61–64.
199 Satre, M., Bof, M. and Vignais, P.V. (1980) J. Bacteriol. 142, 768–776.
200 Freund, F. (1981) Trends Biochem. Sci. 6, 142–145.
201 Williams, R.J.P. (1978) Biochim. Biophys. Acta 50, 1–44.
202 Nagle, J.F., Mille, M. and Morowitz, H.J. (1980) J. Chem. Phys. 72, 3959–3971.
203 Sigrist-Nelson, K. and Azzi, A. (1980) J. Biol. Chem. 25, 10638–10643.
204 Sindler, H. and Nelson, N. (1982) Biochemistry 21, 5787–5794.
205 Friedle, P., Hoppe, J., Gunsalus, R.P., Michelsen, O., von Meyenburg, K. and Schairer, H.U. (1983) EMBO J. 2, 99–103.
206 Senior, A.E. (1983) Biochim. Biophys. Acta 726, 81–95.
207 Kagawa, Y. (1981) In Chemiosmotic Proton Circuits in Biological Membranes (Skulachev, V.P. and Hinkle, P.C., eds.) pp. 425–433, Addison-Wesley, Massachusetts.
208 Futai, M. and Kanazawa, H. (1983) Microbiol. Rev. 47, 285–312.
209 Khan, S. and Berg, H.C. (1983) J. Biol. Chem. 258, 6709–6712.
210 Marvin, D.A. (1983) FEBS Lett. 156, 1–5.
211 Oosawa, F. and Masai, J. (1982) J. Phys. Soc. Japan 51, 631–641.
212 Huxley, A.F. and Simmons, R. (1971) Nature 233, 533–540.
213 Sone, N. and Hinkle, P.C. (1983) J. Biol. Chem. 257, 12600–12604.
214 Pietrobon, D., Zoratti, M. and Azzone, G.F. (1983) J. Biol. Chem. 723, 317–321.
215 Oosawa, F. and Hayashi, S. (1983) J. Phys. Soc. Japan 52, 1575–1581.
216 Kinoshita, K. Jr., Ikegami, A., Yoshida, M. and Kagawa, Y. (1982) J. Biochem. 92. 2043–2046.
217 Hirota, N. and Imae, Y. (1983) J. Biol. Chem. 253, 10577–10581.
218 Adler, A.L., Greenfield, N.J. and Fasman, G.D. (1973) Methods Enzymol. 27, 675–735.
219 Fischer, U.C. and Oesterhelt, D. (1980) Biophys. J. 31, 139–146.
220 Papa, S. (1982) J. Bioenerg. Biomembr. 14, 69–86.
221 Hoppe, J., Friedl, P., Schairer, H.V., Sebald, W., von Meyenburg, K and Jorgensen, B.B. (1983) EMBOJ 2, 105–110.
222 Skulachev, V.P. (1982) FEBS Lett. 146, 1–4.

223 Higuchi, T., Ohe, T., Arakaki, N. and Kotera, Y. (1981) J. Biol. Chem. 256, 985–9860.
224 Kozlov, I.A. (1981) In Chemiosmotic Proton Circuits in Biological Membranes (Skulachev, V.P. and Hinkle, P.C., eds.) pp. 407–420, Addison-Wesley, Massachusetts.
225 Nagle, J.F. and Tristram-Nagle, S. (1983) J. Membr. Biol. 74, 1–14.
226 Pansini, A., Guerrieri, F. and Papa, S. (1978) Eur. J. Biochem. 92, 541–551.
227 McCarty, R.E. (1982) Ann. New York Acad. Sci. 402, 84–90.
228 Yokoyama, S., Miyazawa, T. and Kagawa, Y. (1983) Jpn. J. Biochem. 55, 949 (abstr.).
229 Stoeckenius, W. and Bogomolni, R.A. (1982) Ann. Rev. Biochem. 52, 587–616.
230 Cross, R.L., Cunningham, D. and Tamura, J.K. (1984) Curr. Top. Cell. Regul. in press.
231 Boyer, P.D., Kohlbrenner, W.E., McIntosh, D.B., Smith, L.T. and O'Neal, C.C. (1982) Ann. N.Y. Acad. Sci. 402, 65–83.
232 Kagawa, Y. (1983) J. Biochem. 95, 295–298.

Ernster (ed.) Bioenergetics
© 1984 Elsevier Science Publishers B.V.

CHAPTER 6

The synthesis and utilization of inorganic pyrophosphate

MARGARETA BALTSCHEFFSKY and PÅL NYRÉN

Department of Biochemistry, Arrhenius Laboratory, University of Stockholm, S-106 91 Stockholm, Sweden

1. Introduction

The pyrophosphate structure $-\overset{\|}{\underset{|}{P}}-O-\overset{\|}{\underset{|}{P}}-$ plays an important role in biology. It is the main chemical form in which energy is transmitted in living cells. The simplest compound containing this entity is inorganic pyrophosphate (PP_i) which increasingly appears to play a central role in the bioenergetic processes. PP_i is produced in a variety of biosynthetic reactions such as the synthesis of polysaccharides, proteins, nucleic acids and lipids. Many of these reactions are reversible and must in some way be pulled in the direction of biosynthesis. This can be done by enzymatic hydrolysis of PP_i catalyzed by inorganic pyrophosphatase (PPase), which also replenishes P_i to the adenine nucleotide energy conversion systems. If the PP_i is hydrolyzed by a soluble PPase, the energy liberated upon hydrolysis of the anhydride is lost as heat. Much of this energy can, however, be conserved if PP_i is hydrolyzed by the action of a membrane-bound energy-linked PPase as in the photosynthetic bacterium *Rhodospirillum rubrum* [1]. This enzyme allows PP_i to support a number of energy-dependent processes, such as reversed electron transport [2–4], carotenoid band shift [5], ion transport [6] transhydrogenation [7,8], NAD^+ reduction [9,10] or ATP synthesis [11,12].

The membrane-bound PPase does not only hydrolyze PP_i for the maintenance of a pool of high energy, it can also form PP_i at the expense of the energy liberated in the electron transport chain [13,14]. This enzyme has been found in both purple nonsulfur [13–17] and sulfur photosynthetic bacteria [18], and in mitochondria of lower [19] and higher [20] heterotrophic organisms, and also in chloroplasts from algae and higher plants [21].

Although it is generally accepted that the PP_i formed in synthetic reactions is hydrolyzed, and that this provides a thermodynamic driving force for these reactions, the same driving force would be provided no matter how the concentration of PP_i was lowered. For instance, the PP_i concentration can be decreased by the reactions that use PP_i as an energy and phosphate donor, instead of ATP, to yield energy-rich phosphates. The direct use of PP_i as an energy source in metabolic reactions has been demonstrated not only in the chromatophores of photosynthetic bacteria, but also in several prokaryotes and the eukaryote *Entamoeba histolytica*, which contains two enzymes that utlize PP_i as the phosphate donor [22–24].

2. *Properties of inorganic pyrophosphate*

It is well known that the affinity in biochemical reactions, which proceed in an aqueous medium at rather closely controlled pH and ionic composition, and at constant pressure and temperature, will be determined by the change in the free energy (ΔG) of the system undergoing a transformation. Attempts to understand the role of PP_i are facilitated by accurate thermodynamic data for its hydrolysis, particularly at pH values and ionic strengths prevailing in the cell. In intact cells PP_i is largely present as the 1 : 1 $MgPP_i$ complex, because of the high affinity of the pyrophosphate group for divalent cations and the relatively high concentrations of Mg^{2+} in the intracellular fluid. In the absence of divalent cations the free energy of hydrolysis of PP_i is rather close to that at the γ-bond of ATP, but in the presence of magnesium ions $\Delta G^{0'}$ for PP_i becomes relatively less negative because, unlike the case of ATP, the product of its hydrolysis does not strongly chelate with Mg^{2+}. Due to the higher stability of the $MgPP_i$ complex, as compared to the corresponding ATP complex, the PP_i-forming reaction of ATP cleavage is about 2.6 $kcal \cdot mol^{-1}$ more negative than the P-forming reaction of ATP cleavage at pH 7.5, 37 °C, in the presence of excess Mg^{2+} [25]. Very accurate values for $\Delta G^{0'}$ for the hydrolysis of PP_i to P_i at pH 7.4, 25 °C, might well be those of Flodgaard and Fleron [26]. They used an isotope derivative method for determination of microquantities of PP_i to measure final concentrations of PP_i after equilibrium was established from both directions in the reaction: $PP_i \leftrightarrows P_i + P_i$. In the presence of 1 mM Mg^{2+}, 150 mM K^+ and 0.25 M ionic strength, pH 7.4, the $\Delta G^{0'}$ was calculated to be -4.0 $kcal \cdot mol^{-1}$. In the absence of Mg^{2+} the $\Delta G^{0'}$ was estimated to be -5.7 $kcal \cdot mol^{-1}$. The last value is in agreement with results reported by Stiller et al. [27]. Janson, Degani and Boyer [28] used a direct assay for measurement of PP_i and they obtained somewhat lower values for the equilibrium concentration of PP_i than Flodgaard and Fleron. Otherwise the results obtained by Flodgaard and Fleron for ΔG are considerably less negative than previously and later reported values [25,29–33]. These authors have calculated the equilibrium constant for the PPase reaction by algebraic combination of several different reactions. However, great uncertainties exist about the equilibrium constants for many of the reactions, and different values for the stability constants of magnesium complexes reported in the literature have been used. From experimentally determined values for the concentration of P_i (2.4 mM), PP_i (6 μM), Mg^{2+} and K^+ in rat liver the actual free energy of hydrolysis of PP_i in this tissue was estimated to be -4.0 $kcal \cdot mol^{-1}$ [26]. In the parasitic amoeba, *E. histolytica*, which contain 200 μM PP_i, the $\Delta G'$ was calculated to be -6 $kcal \cdot mol^{-1}$ [22]. The predominant free-energy contribution to the large negative free energy of PP_i hydrolysis is suggested to be the differences in solvation energy of reactants and products, whereas intramolecular effects such as opposing resonance and electrostatic repulsion are of secondary importance [31,32].

3. Formation of inorganic pyrophosphate

PP_i is produced in a variety of processes and is a common reactant in many of the major metabolic pathways in all parts of the cell: fat metabolism through the conversion of acetate into acetyl-CoA *via* the acetyl-CoA synthetase and by enzymes which activate both medium-chain and long-chain fatty acids; amino acid activation through the action of a number of specific aminoacyl-tRNA synthetases forming aminoacyl-tRNA; DNA synthesis through the action of DNA polymerase; RNA synthesis through the action of RNA polymerase; and nucleoside diphosphate sugar synthesis through glycosyl-1-phosphate nucleotidyltransferases. In addition PP_i is produced during biosynthesis of nucleotides, lipids and urea, and in other NTP dependent reactions. PP_i is also formed (like ATP) by photosynthetic [12,13] and oxidative phosphorylation [20,34–36]. In the photosynthetic bacterium *Rhodospirilium rubrum* the membrane-bound coupling factor PPase system can couple light induced electron transport to PP_i synthesis. PP_i can be formed in the bacterial chromatophores in parallel with, and independently of, ATP. During oxidative phosphorylation PP_i is produced by an analogous system, which has been found in both animal [20] and plant mitochondria [21]. Another example of PP_i production is the sulphate-reducing bacteria of the genus *Desulfovibrio* [37,38]. In these organisms the respiration is based on the reduction of sulphate to sulphide, in place of the reduction of oxygen to water, fundamental to aerobes. PP_i is produced in the initial step of this reaction, the 'activation' of sulphate with ATP to form adenosine phosphosulphate.

4. Inorganic pyrophosphate as phosphate and energy donor in soluble systems

Several enzymes using PP_i as phosphoryl donor have been described in bacteria and parasitic amoebae [22–24]. In both *Propionibacterium shermanii* and the eukaryote *Entamoeba histolytica*, the enzyme PP_i-phosphofructose dikinase has been found [39,40]. This enzyme uses PP_i as the phosphoryl donor to fructose-6-phosphate instead of ATP:

$$\text{fructose-6-P} + PP_i \rightleftarrows \text{fructose-1,6-diP} + P_i \qquad \text{(reaction 1)}$$

The rate of this reversible reaction is virtually identical in both directions. The enzyme appears to play a major role in the glycolytic pathway of both *P. shermanii* and *E. histolytica* [41,40]. This enzyme has later been found in both bacterial and plant species, i.e., in *Pseudomonas marina* [42], a marine *Alcaligenes* species [42], *Bacteroides fragilis* [43], *R. rubrum* [44] in pineapple leaves [45] and in extracts of mung bean sprouts [46]. The PP_i-phosphofructose dikinase from mung bean sprouts is activated about 20-fold by the product fructose-1,6-diP [46].

Another enzyme that uses the energy of the PP_i is the pyruvate, phosphate dikinase. This enzyme converts P-enolpyruvate to pyruvate, but unlike pyruvate kinase, which uses ADP as the phosphate acceptor yielding ATP and pyruvate. This enzyme uses AMP and in combination with PP_i yields ATP, P_i and pyruvate:

$$\text{pyruvate} + \text{ATP} + P_i \rightleftarrows \text{P-enolpyruvate} + \text{AMP} + PP_i \qquad \text{(reaction 2)}$$

The enzyme has been discovered in the leaves of topical grasses such as sugar cane [47], in the amoeba *E. histolytica* [48], in *Bacteroides symbiosus* [49], in *P. shermanii* [50], in *Acetobacter xylinum* [51] and in several photosynthetic bacteria including *R. rubrum* [52]. The dikinase appears to function in different directions in different organisms. For example in *A. xylinum, P. shermanii,* and in the C4 plants it functions in the direction of P-enolpyruvate formation. On the other hand, in *E. histolytica* and *B. symbiosus,* this enzyme provides a means of formation of pyruvate and ATP.

PP_i can also serve as a source of energy for growth in the presence of fixed carbon (acetate and yeast extract) of anaerobic sulfate-reducing bacteria belonging to the genus *Desulfotomaculum* [53,54] as well as other anaerobic microorganisms like the thermophilic fermentative anaerobe *Thermoanaerobacter ethanolicus* [55] and a new species of *Clostridium* [56].

Desulfotomaculum is able to conserve the energy of the PP_i produced by ATP-sulfurylase (reaction 3) by means of the enzyme acetate: PP_i phosphotransferase (reaction 4). ATP can then be produced from acetylphosphate and ADP by acetate kinase (reaction 5).

$$\text{ATP} + SO_4^{2-} \rightleftarrows \text{APS} + PP_i \qquad \text{(reaction 3)}$$

$$\text{acetate} + PP_i \rightleftarrows \text{acetyl phosphate} + P_i \qquad \text{(reaction 4)}$$

$$\text{ADP} + \text{acetyl phosphate} \rightleftarrows \text{acetate} + \text{ATP} \qquad \text{(reaction 5)}$$

Thus, it is not necessary for this microorganism to carry out electron transport-coupled phosphorylation during growth with lactate and sulfate. This is in contrast to the sulfate-reducing bacteria belonging to the genus *Desulfovibrio,* which do not have the acetate, PP_i phosphotransferase, and therefore appear to have to carry out electron transport-coupled phosphorylation in order to obtain a net yield of ATP.

Other enzymes with the ability to use PP_i as an energy source are certain acetyl-CoA synthetases and PP_i-serine phosphotransferase. Acetyl-CoA synthetases catalyze a reversible reaction with enzyme kinetic properties generally appropriate for reaction in either direction:

$$\text{ATP} + \text{acylate} + \text{CoA} \rightleftarrows \text{AMP} + \text{PP}_i + \text{acyl-CoA} \qquad \text{(reaction 6)}$$

In the many eukaryotic organisms in which acetate, propionate and butyrate are metabolic end products, the net flux is right to left [22].

The PP_i-serine phosphotransferase catalyzes the following reaction:

$$\text{PP}_i + \text{L-serine} \rightleftarrows \text{L-serine-P} + \text{P}_i \qquad \text{(reaction 7)}$$

The $\Delta G^{0'}$ for this reaction is $-6.7\ \text{kcal} \cdot \text{mol}^{-1}$ for the hydrolysis of PP_i [57]. This reaction is not part of a major pathway. The enzyme has been found in *P. shermanii* [58].

Reports that the concentration of PP_i in some cells was found to be high and inconsistent with the PPase equilibrium [33,59–67] indicate that PP_i conserving reactions might be more widespread than to just the above mentioned organisms.

5. Membrane bound pyrophosphatases

Since the first report by one of us [1] on the existence of membrane-bound pyrophosphatases in mitochondria and in *R. rubrum* chromatophores there have been a number of indications of a more frequent occurrence. In this section we will try to point to the available evidence for the participation of PP_i in membrane bioenergetics. The most extensively studied membrane PPases are those in rat liver and beef heart mitochondria and in *R. rubrum* chromatophores. They are treated separately in Sections 6 and 7, respectively, in this review.

In photosynthetic bacteria at least four other species than *R. rubrum* have well-documented PP_i-linked energy metabolism. In *Rhodopseudomonas viridis* Jones and Saunders [17] reported that succinate-linked NAD^+ reduction occurs in the dark with PP_i as energy source at about one third of the rate when ATP or light are energy sources. In *Rhodopseudomonas sphaeroides* Ga chromatophores Sherman and Clayton [16] obtained large carotenoid band shifts induced by PP_i. The spectrum of the PP_i-induced band shift was spectrally identical to the light-induced one and the extent similar. Carotenoid band shift induced by PP_i as well as an active PPase, slightly stimulated by uncouplers, has also been found in *Chromatium* D chromatophores by Knaff and Carr [18]. Knobloch [15] has shown that energy-linked transhydrogenase activity is obtained with PP_i as energy donor in the membrane fraction of *Rhodopseudomonas palustris.*

In all these examples, the reaction could not have been obtained unless the hydrolysis of PP_i would lead to the establishment of an electrochemical gradient of protons. This implies the existence of an H^+ translocating PPase in these species of photosynthetic bacteria.

Chloroplasts both from pea leaves and the alga *Acetabularia mediterranea* show light-induced PP_i synthesis which is stimulated if ADP is omitted from the reaction medium. The pea chloroplasts also show increased rates of PP_i synthesis in CF_1-depleted chloroplasts. Furthermore, inhibitors of electron transport and energy transduction inhibit the PP_i synthesis [21].

Yeast mitochondria depleted of ADP and ATP also have PP_i synthesis coupled to electron transport as shown by Mansurova in preparations from *Endomyces magnusii* [19]. Also, recent work in this laboratory (unpublished) with mitochondria from *Saccharomyces cerevisiae* shows the existence of a membrane-bound PPase, the activity of which is stimulated by uncouplers and inhibited by fluoride. Earlier, we showed [4] PP_i-induced cytochrome redox changes, indicative of reversed electron flow in mitochondria from *Saccharomyces cerevisiae.*

Taken all together, these scattered examples show a wide-spread occurrence of PP_i-linked energy metabolism in both prokaryotes and lower as well as higher eukaryotes. It is very likely that more extensive investigations will show that the PP_i system is almost as common as the ATP system, providing cells with an alternative pathway for energy conversion operating at a lower level of energy.

6. The mitochondrial membrane-bound PP_iase

6.1. Electron transport-coupled synthesis of PP_i

Early reports indicated the formation of PP_i in connection with oxidative phosphorylation in animal tissues [68,69], but it was not until it had become firmly established in a bacterial system, that PP_i can serve as a competent energy converter alternative to ATP [2,13,8], that the mammalian systems were consistently investigated for this capacity. Some indications that PP_i might be involved also in mammalian energy transduction were that mitochondria contain a membrane-bound pyrophosphatase, the activity of which is stimulated under uncoupling conditions and that reversed electron transport may be induced by PP_i in isolated yeast and rat liver mitochondria [1,70]. Mansurova et al. [20] have shown that isolated rat liver mitochondria indeed are able to synthesize PP_i at 10–20% of the rate of ATP synthesis, coupled to the oxidation of succinate. The reaction is insensitive to oligomycin, inhibited by uncoupling concentrations of 2,4-dinitrophenol and thus appears to be intimately coupled to the electron transport system of the mitochondria. Inhibitors of electron transport, such as antimycin and NaCN, as well as malonate, also inhibit the PP_i synthesis. It is noteworthy, however, that the inhibition caused by all these components never is as extensive as it is on ATP synthesis. Fluoride, a well known inhibitor of soluble PPases, is very efficient also with the mitochondrial membrane-bound pyrophosphatase, without affecting the ATPase to any significant degree.

Subsequently, the same investigators showed that beef heart mitochondria and submitochondrial particles also contain a membrane-bound, as well as a soluble, pyrophosphatase [71]. The membrane-bound activity proved to be solubilized rather easily.

6.2. PP_i-synthesis in relation to ATP synthesis

The PP_i synthesis in mitochondria is consistently much lower than the ATP synthesis (only 10–20%) under normal conditions [20]. The same is true in chromatophores from *R. rubrum* [14,72]. On the other hand, when the energy input from electron transport is limited, either by partial inhibition or by partial uncoupling, the PP_i synthesis in rat liver mitochondria reaches the level of the ATP synthesis or even exceeds it [74]. Under such conditions, where the $\Delta\mu_{H^+}$ is low, the synthesis of ATP is strongly inhibited (Fig. 6.1). A similar change in the relationship between PP_i synthesis and ATP synthesis has been shown in *R. rubrum* chromatophores with low light intensities [72]. An interesting feature of the mitochondrial system is that at moderate inhibition (about 25%) of electron transport, either by the addition of KCN or by malonate, the PP_i synthesis is arrested, but returns at higher (50%) inhibition, and then is at least as high as in the uninhibited system. Pullaiya et al. [73] relate this switch to the inhibition of ATP synthesis. This in turn may well be a consequence of a greatly lowered $\Delta\mu_{H^+}$, to a value where the available energy is more compatible with the lower $\Delta G^{0'}$ required for PP_i synthesis. This may constitute an emergency system for living cells to conserve energy even under energetically unfavourable conditions. The PP_i formed may even be used to support ATP synthesis [20].

6.3. Solubilization and purification

In contrast to the case with the H^+-translocating PPase of *R. rubrum* chromatophores, the membrane-bound mitochondrial PPase activity is more easily removed from the membrane than the ATPase. Two min sonication of beef heart sub-

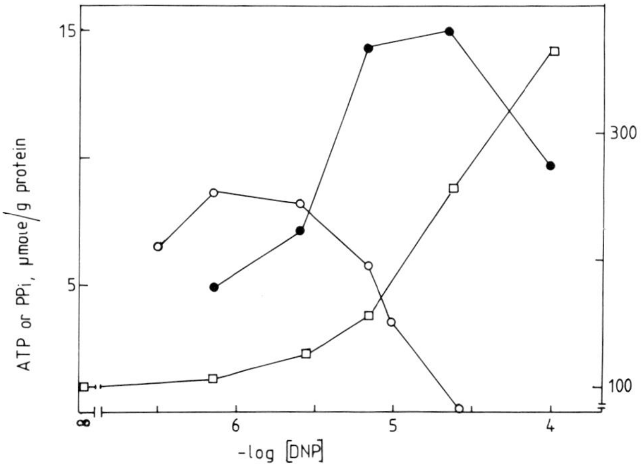

Fig. 6.1. Influence of uncoupling conditions on the ATP and PP_i content of rat liver mitochondria: ○, ATP content; ●, PP_i content; □, rate of respiration (redrawn from Ref. 73).

mitochondrial particles removes almost 80% of the PPase activity, while having only marginal effects on the coupling factor ATPase.

In the first reports [71,74], the two forms of mitochondrial PPase, membrane bound and free, were claimed to be isoenzymes, both having a molecular weight of 75 000. The membrane-bound form, termed pyrophosphatase II in accordance with a suggestion by Iria et al. [75], is a lipoprotein, containing phosphatidylcholine. Moreover, it was shown that the activity of the soluble enzyme, pyrophosphatase I, was greatly enhanced by the addition of mitochondrial lipids [74] and that addition of phosphatidylcholine to PPase I would convert it to PPase II. In this form it could be incorporated into PPase-depleted mitochondrial membranes and catalyze electron transport-linked PP_i synthesis [76].

In contrast to these results, Volk et al. in a very recent publication [77] using a detergent based solubilization described in [78], found very distinct differences between pyrophosphatases I and II. PPase I has a molecular weight of 60 000 and consists of two different subunits, α and β. PPase II has a molecular weight of 185 000 and is composed of four subunits, α, β, γ and δ. Because of the similarity in mass between subunits α and β in the two enzymes, Volk et al. propose that these two subunits are the same in both enzymes and form the catalytic part of PPase II. PPase I from mitochondria thus seems to differ from other soluble pyrophosphatases in that the two subunits are nonidentical.

It is interesting to note that Volk et al. are able to differentiate between the two enzyme activities by the lack of fluoride inhibition of PPase II, which is in contradiction with the results in [20]. Volk et al. also obtained, though careful differentiation of mitochondrial membranes, a localization of the PPases. They came to the conclusion that PPase I is localized in the mitochondrial matrix. PPase II is situated in the inner mitochondrial membrane, a localization in line with its proposed role as a coupling factor [71].

It may appear difficult to reconcile the differences in data from the two Russian groups. But one possible explanation is that Mansurova et al. [71,74] have extracted only part of PPase II, possibly subunits α and β. The isolation method they actually have used is somewhat undetailed, but if it only consists of a buffer extraction from an acetone powder of mitochondria, without added detergents, one may well envisage that only part of the membrane-bound enzyme is liberated. The same may be true for their sucrose washing of submitochondrial particles with or without sonication.

6.4. Resolution and reconstitution

The PPase activity may be removed from beef heart submitochondrial particles by washing with 0.25 M sucrose. The removal is parallel with loss of PP synthesis [71].

The PPase protein isolated from an acetone powder of mitochondrial 'PPase II', probably consisting of only the α and β subunits of the total membrane bound PPase, can reconstitute both PPase and PP_i synthesis activities, the latter to above the original rate [68]. The main difference between 'PPase II' and 'PPase I' appears to be the presence of phosphatidylcholine in 'PPase II', and it has also been shown by Shakhov et al. [74] that this lipid will convert PPase I into a form in which it may substitute for 'PPase II' in reconstituting PP_i synthesis to depleted submitochondrial particles. Addition of this lipid, as well as some other lipids, to PPase I also greatly enhances the PPase activity of the enzyme [73].

Since the ATP synthesis is highly active in the resolved particles it seems unlikely that the removal of 'PPase II' causes much damage to the membrane structure. It also seems unlikely that the whole membranous protein has been removed, and our guess is that the interconvertibility between the two enzymes described [73] is limited to the two smallest subunits of the preparation described in Ref. 71.

No H^+-pumping activity has yet been described for either preparation and this aspect of the coupling between electron transport and PP_i synthesis in the mitochondrial system is thus still an unsolved question.

7. *The H^+-PPase from* Rhodospirillum rubrum

7.1. *Electron transport-coupled synthesis of PP_i*

The membrane of chromatophores of *R. rubrum* contains a PPase or PP_i synthase, system for energy-dependent synthesis of PP_i. The synthesis of PP_i is coupled to light-driven electron transport and is catalyzed by a reversible membrane-bound proton-translocating PPase. This enzyme is functionally similar to the ATP synthase but may well have a simpler structure; it has a structurally simpler substrate and is oligomycin-insensitive. The light-induced formation of PP_i by chromatophores from *R. rubrum* was first shown by Baltscheffsky et al. and Baltscheffsky and von Stedingk [12,13] partly based on their earlier results obtained in collaboration with Horio [79]. During illumination, in the presence of P_i, but in the absence of ADP, they found a net synthesis of PP_i, which was inhibited by electron transport inhibitors and uncouplers. The reaction requires Mg^{2+} and light but is not inhibited by oligomycin, which inhibits photophosphorylation to ATP and the membrane-bound ATPase [79,80]. Oligomycin rather slightly stimulates the PP_i synthesis. Only very limited, if any, PP_i synthesis occurs in the dark.

Guillory and Fisher [72] studied the light-dependent synthesis of PP_i by a trapping system, which allowed simultaneous investigation of the rates of formation of both PP_i and ATP. They found that maximum activation of PP_i synthesis occurred at a lower light intensity than that required for ATP synthesis. The ratio of rates of PP_i and ATP synthesis ranged from 1.0 at low light intensities to 0.25 at saturating light intensities. PP_i synthesis was inhibited by as much as 50% if ATP synthesis took place simultaneously [82]. Maximal rates of PP_i and ATP synthesis occurred at rather similar Mg^{2+} concentrations (3.3 and 3.0 mM, respectively) and pH values (7.5 and 7.0, respectively) [72].

The PP_i synthase is a different enzyme from the ATP synthase. This conclusion comes from the findings that LiCl-treated particles neither catalyse the photophos-

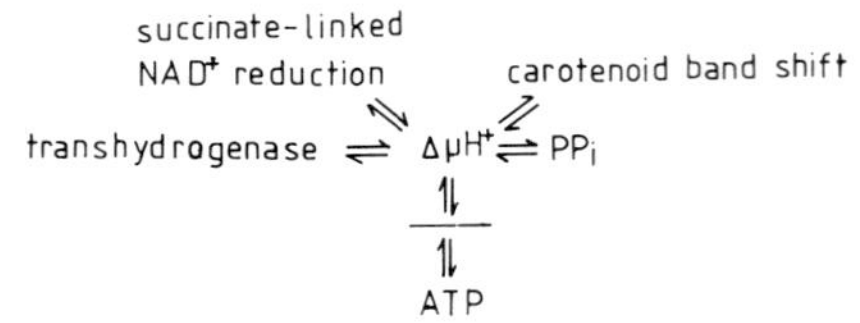

Fig. 6.2. Scheme for some PP_i- and ATP-induced reactions in *R. rubrum* chromatophores.

phorylation of ADP nor the hydrolysis of ATP, but retain the light-induced PP_i synthesis activity [72]. In LiCl-extracted particles oligomycin promotes PP_i synthesis by 3-fold, up to that found for untreated chromatophores [72]. Furthermore, an antibody against the purified ATPase, which completely inhibits ATP synthesis as well as other energy-linked reactions dependent on ATP, does not affect the PP_i-linked reactions to any significant degree [83]. Results from sonication experiments, which selectively solubilize the ATPase, confirm the stronger binding of the PPase, than of the ATPase, to the chromatophore membrane [84].

7.2. *PP_i-driven energy-requiring reactions*

Chromatophores from *R. rubrum* can utilize the energy of PP_i to sustain several energy-dependent reactions as has been stated above (Fig. 6.2). This feature is conferred by the membrane-bound proton-translocating PPase, which translocates protons across the membrane, coupled to PP_i hydrolysis. According to Moyle and Mitchell [85] the stoicheiometry is one proton translocated per two PP_i hydrolyzed. The utilization of PP_i involves the formation of an energized state, a specific electrochemical potential of H^+ ions on the chromatophore membrane. The electrochemical potential of H^+ involves the electric component ($\Delta\Phi$) and the osmotic component given as the transmembrane difference of H^+ activity (ΔpH). This electrochemical potential, generated by the proton-translocating PPase, can be utilized as energy source for several energy-requiring processes.

7.2.1. *PP_i-induced changes in the redox state of cytochromes*

PP_i or ATP, when added to coupled chromatophores of *R. rubrum*, induces oxidation-reduction reactions which can be ascribed to energy-requiring reversed electron transport [2–4,7–12,90] (Fig. 6.3).

Addition of PP_i in the dark to a suspension of chromatophores, supplied with Mg^{2+} ions, causes a reduction of endogenous *b*-type cytochrome and a simultaneous

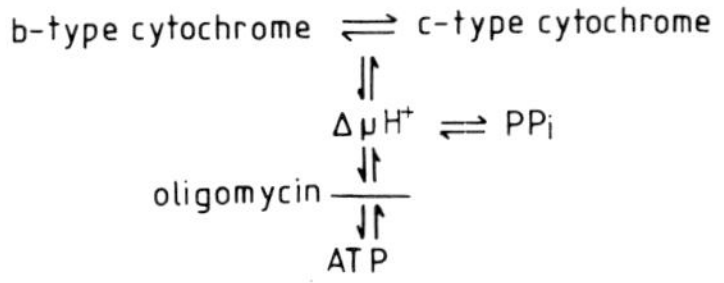

Fig. 6.3. Scheme for PP_i- and ATP-induced oxidation reduction reactions in cytochromes in *R. rubrum* chromatophores.

oxidation of endogenous *c*-type cytochrome. The facts that the initial rates of oxidation of cytochrome c_2 and reduction of cytochrome *b* are the same, and that the criterion of the cross-over theorem is satisfied show that a phosphorylation site is localized between these two cytochromes.

The halftime for the PP_i-induced changes is around 0.4 s which is one-tenth of the half-time when ATP is used as the energy donor [4]. The extent of the changes obtained with PP_i is usually about two to three times that obtained with ATP. The effects of PP_i and ATP are additive [2]. The reoxidation of *b*-type cytochrome starts when the concentration of PP_i falls below 25 μM [4]. Uncouplers of photophosphorylation and electron transport inhibitors abolish both the PP_i and the ATP-induced reactions. In line with the results with PP_i synthesis and hydrolysis, the PP_i-induced reactions are not inhibited by oligomycin.

7.2.2. *PP_i-induced carotenoid absorbance change*

The coloured carotenoids of wild type *R. rubrum* respond in the dark to ATP or PP_i with a reversible shift in their absorption spectrum [3,5]. This shift is qualitatively identical to the light-induced shift which, however, occurs at a more than 10^5-times faster rate. Quantitatively, the extent of the PP_i-induced shift approaches the light-induced one, whereas the ATP-induced one usually is only half as great [5]. The half-time is 300 ms for the PP_i-induced shift and 3 s for the ATP-induced shift [5]. The PP_i-induced, as well as the ATP-induced, carotenoid change is abolished by uncouplers, but not by electron-transport inhibitors. Oligomycin prevents the ATP-induced change, but has no, or a slightly, stimulatory effect on the PP_i-induced change [2]. On the other hand, we found that Dio-9 inhibits both the ATP-induced and the PP_i-induced change (unpublished).

It was shown in *Rhodopseudomonas sphaeroides* chromatophores that the carotenoid shift could be obtained in response to an induced diffusion potential [86] and this finding initiated a widespread use of this uniquely endogenous energy probe as potential indicator. We suggested that at least in *R. rubrum* chromatophores the carotenoid shift is not entirely due to a membrane potential, but in part also a reflection of conformational changes of the ATPase or PPase complexes [87]. This suggestion was partly based on the fact that PP_i consistently induces a much larger shift than ATP, something which is difficult to explain with the membrane potential indicator hypothesis.

7.2.3. *PP_i-driven energy-linked transhydrogenase*

Chromatophore preparations of *R. rubrum* contain a particle-bound transhydrogenase which catalyzes the energy-dependent reduction of $NADP^+$ by NADH. The following equations describe the reaction:

$$PP_i \rightarrow 2P_i + \Delta\mu_{H^+} \qquad \text{(reaction 8)}$$

$$NADH + NADP^+ + \Delta\mu_{H^+} \rightarrow NAD^+ + NADPH \qquad \text{(reaction 9)}$$

The $\Delta\mu_{H^+}$ can be generated by light-induced or dark electron transport, as well as by hydrolysis of ATP, GTP or PP_i [7,8]. The rates of the ATP- and PP_i-driven reactions are about 70 and 56% of the light-driven reaction, respectively. The addition of both ATP and PP_i stimulates the rate more than either alone, and gives a rate almost as high as light. The PP_i-driven reaction is inhibited by uncouplers of phosphorylation but not by electron-transport inhibitors or oligomycin. The stoicheiometries of ATP and PP_i hydrolyzed to $NADP^+$ reduced are 1 and 10, respectively.

7.2.4. PP_i-driven succinate-linked NAD^+ reduction

Succinate-linked NAD^+ reduction by *R. rubrum* chromatophores can be driven by PP_i [9,10,88] according to the following equations:

$$PP_i \rightarrow 2P_i + \Delta\mu_{H^+} \qquad \text{(reaction 8)}$$

$$\text{Succinate} + NAD^+ + \Delta\mu_{H^+} \rightarrow \text{Fumarate} + NADH \qquad \text{(reaction 10)}$$

This reaction can also be driven by light [9,10,88] or ATP [9,10,89]. The rates of the ATP and PP_i-driven NAD^+ reduction are about 20–30 and 6–12% of the light-driven reduction, respectively. The addition of both PP_i and ATP causes a synergistic stimulation. Oligomycin inhibits the ATP-driven reaction but stimulates the PP_i- as well as the light-driven reactions with 60–70 and 20–80%, respectively. The PP_i-driven reaction is inhibited by uncouplers but not by electron-transport inhibitors.

7.2.5. PP_i-driven ATP synthesis

Chromatophores from *R. rubrum* can utilize the energy generated by the hydrolysis of PP_i to drive ATP synthesis in the presence of P and ADP [11,12] according to the following reactions:

$$PP_i \rightarrow 2P_i + \Delta\mu_{H^+} \qquad \text{(reaction 8)}$$

$$ADP + P_i + \Delta\mu_{H^+} \rightarrow ATP \qquad \text{(reaction 11)}$$

Phosphorylation and exchange reactions involving the direct transfer of P_i from PP_i to ADP or ATP are very low or nonexistent. The rate of the PP_i-driven ATP synthesis is sbout 5% of the light-driven phosphorylation of ADP. The reaction is inhibited by oligomycin and by uncouplers, whereas electron transport inhibitors have no effect. Inhibitors of the PPase, fluoride and methylene diphosphate, inhibit at concentrations similar to those which inhibit other PP_i-linked reactions (10 and 1 mM, respectively). The stoicheiometry of ATP formation to PP_i hydrolysis is about 8 PP_i hydrolyzed per ATP synthesized, which is comparable with that of the PP_i-driven transhydrogenase. The transhydrogenase competes somewhat (24%) with PP_i-driven ATP formation, but the ATP formation has little effect on the transhydrogenase. Thus, surprisingly, when both reactions are progressing simultaneously, the efficiency of energy conversion with PP_i as the energy donor is increased by almost 60%.

The opposite effects are observed with the succinate-linked NAD^+ reduction. The NAD^+ reduction is inhibited over 50% by PP_i-driven ATP formation, but ATP formation is not inhibited at all by NAD^+ reduction. NAD^+ reduction apparently requires a higher level of $\Delta\mu_{H^+}$ than does the transhydrogenase. Thus, competing reactions appear to inhibit this reaction more than the transhydrogenase.

7.3. Mechanistic aspects

Chromatophores from *R. rubrum* catalyze both oxygen exchange between water and inorganic phosphate [82,91] and $PP_i : P_i$ equilibration [92], in addition to their synthesis and hydrolysis of PP_i. The P_i-PP_i exchange, in the dark, is inhibited by uncouplers of phosphorylation and ADP, and is stimulated by oligomycin (27% stimulation) [91]. The ADP inhibition is eliminated by oligomycin, indicating that this inhibition is due to ATP synthesis. Methylene diphosphonate and fluoride, inhibitors of the membrane-bound PPase, inhibit the reaction. The mechanism of the exchange is postulated to be due to the reversal of the energy-linked hydrolytic reaction:

$$PP_i \rightleftarrows 2P_i + \Delta\mu_{H^+} \qquad \text{(reaction 8)}$$

The electron transport inhibitor, antimycin, has no inhibitory effect. Free Mg^{2+} appears to be required for the PP_i-P_i exchange reaction in that 2–3 mM Mg^{2+} is required for optimal activity of less than 1.5 mM PP_i. The P_i concentration has little effect on the Mg^{2+} requirement. The P_i requirement is high compared to the light-driven synthesis of PP_i (one-half maximal activity at 48 and 3 mM, respectively). The V_{max} for the exchange reaction, at saturating P_i concentration, is about the same as for the light-induced PP_i synthesis. About 12 mol of PP_i are hydrolyzed per mol of P_i exchange. This stoicheiometry is comparable to that found for the PP_i driven transhydrogenase reaction and for the PP_i-driven synthesis of ATP.

Chromatophores from *R. rubrum* also catalyze a rapid exchange of oxygen atoms between P_i and water [92]. The reaction is strongly inhibited by inhibitors of the membrane-bound PPase, fluoride and methylene diphosphonate, but is not inhibited by oligomycin. The $P_i \rightleftarrows$ HOH oxygen exchange is almost entirely due to the PPase. In the presence of ADP, the exchange reaction is stimulated about 40% and this portion of the exchange is sensitive to oligomycin, but not to fluoride or methylene diphosphonate. Thus, this portion of the exchange can be attributed to the ATP-synthase complex. Little oxygen exchange is observed in the absence of added Mg^{2+} and a specific Mg^{2+} to P_i ratio of about 0.8 is required for optimal activity. The exchange activity increases rapidly with PP_i concentration up to 7 mM P_i (at constant Mg^{2+} to P_i ratio), and then increases slowly and almost linearly at higher P_i concentrations. The concentration for half-maximal activity for the fast phase is 3 mM, which is the same as for the light-induced synthesis of PP_i. The uncoupler Cl-CCP, has only a small effect on the exchange (23% inhibition) at concentrations which would almost completely inhibit energy-linked reactions in the chromatophore. This result is consistent with the effects of uncouplers on the $P_i \leftrightarrows HOH$ oxygen exchange in submitochondrial particles [93,94]. It appears that oxygen exchange is not strongly dependent upon an energized membrane, as light does not stimulate the reaction. The $P_i \rightleftarrows HOH$ exchange represents a partial reaction of the net reaction catalyzed by the membrane-bound PPase. The $P_i \rightleftarrows HOH$ exchange and PP_i hydrolysis proceed at about the same rate, whereas the $PP_i \rightleftarrows P_i$ exchange rate is about 10-fold lower or about the same as the PP_i-synthesis rate. So the rate of formation of medium PP_i is too slow to account for the oxygen exchange. However, enzyme-bound PP_i might be formed from, and cleaved to, P_i sufficiently rapidly to account for most of the observed oxygen exchange, according to the mechanism for the yeast PPase [95]:

$$EP_i + P_i \rightleftarrows E_2P_i \overset{H_2O \atop \downarrow}{\rightleftarrows} EPP_i \rightleftarrows E + PP_i \qquad \text{(reaction 12)}$$

Once EPP_i is formed, it is more likely to be cleaved than to appear as medium PP_i. If this reaction mechanism is true also for the membrane-bound PPase from *R. rubrum* the contribution of a membrane potential for the synthesis of PP_i should be on the release of the product PP_i from the enzyme.

7.4. Solubilization and purification

The studies of the PPase from *R. rubrum* chromatophore membranes may provide a further insight into the mechanism of energy transduction in membranous systems. Such studies have, however, been hampered by the difficulties in the isolation of the enzyme. The membrane-bound PPase was first solubilized and partly purified from the carotenoid deficient mutant of *R. rubrum* (strain G-9) in 1978 by Rao and Keister [96]. The enzyme was solubilized by using cholate in the presence of $MgCl_2$. The solubilized enzyme was further partially purified using ammonium sulfate fractionation and gel chromatography. After fractionation the enzyme required phospholipid for activity. Storage of the enzyme in the cold, with phospholipids added, yielded an inactive preparation, indicating cold lability when phospholipid was present. The method gave only 10-fold purification from the chromatophores.

Recently, we have developed a method for obtaining a solubilized and highly purified membrane-bound PPase from *R. rubrum* (wild type strain S-1) chromatophores [96a]. The enzyme is solublized with good yield using Triton X-100 in the presence of high concentration of $MgCl_2$ and ethylene glycol. The solubilized enzyme is fairly stable for at least 5–10 days at 0 °C. The stability is due to the high concentration of ethylene glycol present. A high level of purification (80-fold) is obtained with hydroxylapatite chromatography of the Triton extract. The enzyme appears to be a very hydrophobic, integrally bound membrane protein. Triton X-100 alone reconstitutes the enzyme activity to 70%. The addition of phospholipids to the enzyme preparation increases the activity up to 40%.

TABLE 6.1

Effect of various compounds on the PPase activity before and after its incorporation into liposomes

Activity	Control	Addition							
		FCCP (1.5 μM)	Valinomycin (10 μg/ml)	Nigericin (10 μg/ml)	Valinomycin + nigericin	Oligomycin (10 μg/mg prot)	DCCD (100 μM)	NaF (10 mM)	IDP (1 mM)
Incorporated PPase nmol $P_i \cdot min^{-1}$ (%)	37 (100)	50 (135)	46 (124)	54 (146)	55 (149)	36 (97)	36 (97)	6 (16)	21 (57)
Solubilized PPase nmol $P_i \cdot min^{-1}$ (%)	40 (100)	39 (98)	40 (100)	40 (100)	39 (98)	38 (95)	39 (98)	4 (10)	18 (45)

PPase activity was measured in the same mixture with phospholipids added before (solubilized PPase) and after (incorporated PPase) freezing-thawing. The assay medium contained $5 \times 10^{-}2$ M KCl, and 50 μl of Ppase preparation was added. For details see text.

7.5. Reconstitution

Both highly and partially purified membrane-bound PPase from *R. rubrum* have been subjected to reconstitution experiments. A highly purified PPase has been reconstituted into liposomes by the freeze-thaw technique [97]. The addition of the uncoupler FCCP, which collapsed the electrochemical potential difference of H^+ across the membrane, led to stimulation of PPase activity in liposomes to a similar extent as for the chromatophore PPase activity. Likewise, the activity increased when the membrane was made permeable to K^+ by adding valinomycin (to collapse the membrane potential). The stimulatory effect was also achieved when nigericin, the H^+/K^+ exchanger, was used. The compounds used did not affect PPase activity before incorporation of the enzyme into liposomes (Table 6.1). The results strongly support the concept that this PPase is a proton-pump, and thus functions in a manner analogous to that of the coupling factor ATPase.

The highly purified PPase, functioning as a proton-pump, has been used for the energization of a liposomal membrane and the energy liberated by the hydrolysis of PP_i was coupled to the synthesis of ATP from ADP and P_i [98] (Fig. 6.4). This energy converting system was put together by mixing the highly purified membrane-bound PPase and the coupling factor F_0F_1 complex with an aqueous suspension of sonicated soybean phospholipids, and subjecting the mixture to the freeze-thaw procedure. In the presence of ADP, Mg^{2+}, P_i and PP_i the resulting liposome system catalyzed phosphorylation of ADP to ATP (Fig. 6.5).

The phosphorylation was sensitive to uncouplers and inhibitors of phosphorylation, such as oligomycin, efrapeptin and DCCD (*N,N'*-dicyclohexylcarbodiimide). The phosphorylation reaction was also sensitive to fluoride, imidodiphosphate and methylenediphosphonate.

A partly purified membrane-bound PPase, incorporated in a phospholipid membrane, has been shown to act as a PP_i-dependent electric generator [99]. Proteoliposomes reconstituted from soybean phospholipids and the PPase, were incorporated, in the presence of Mg^{2+} ions, into a phospholipid-impregnated Teflon filter separating two solutions of an identical electrolyte content. Addition of PP_i to the same compartment as proteoliposomes, induced generation of an electric potential difference between the two filter-separated compartments, the proteoliposomes-con-

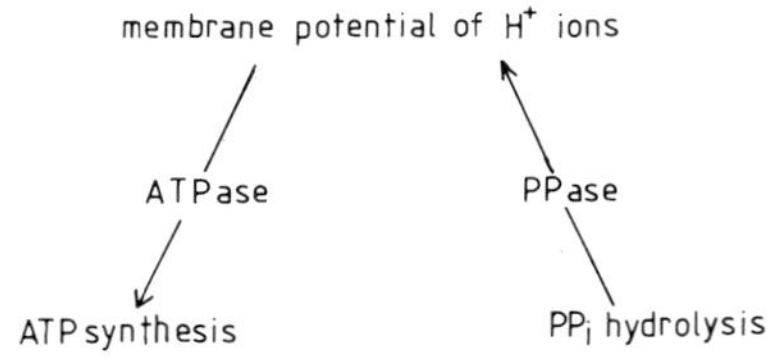

Fig. 6.4. Scheme for PP_i-driven ATP synthesis in liposomes containing membrane-bound PPase and F_0-F_1 complex from *R. rubrum*.

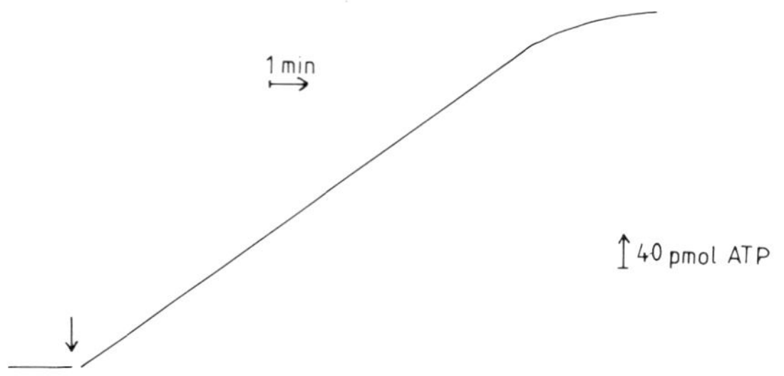

Fig. 6.5. Time-course of ATP synthesis in a suspension of PPase, ATPase liposomes. 25 μl liposomes (about 0.07 mg protein per ml) were suspended in a medium containing 1 ml 0.2 M glycylglycine (pH 7.8), 0.2 ml luciferin/luciferase assay, 20 μl 100 mM sodium phosphate, 10 μl 10 mM ADP and 50 μl 10 mM sodium pyrophosphate. The final concentration of $MgCl_2$ was about 10 mM. At the arrow, 25 μl liposomes were added. The resulting luminescence was measured in an LKB luminometer 1250. The light output was calibrated by addition of a known amount of ATP. (From Ref. 98.)

taining compartment being negatively charged. An electric potential difference of 15 mV and a current of 20 pA were observed. The electrogenic effect required Mg^{2+}, and proved to be sensitive to fluoride but not to DCCD.

8. *Outlook*

In the last decade or two, the energy metabolism of PP_i has emerged as a prominent alternative to that of ATP. The increasing number of experiments involving various aspects of PP_i in bioenergetics are notable. The relative simplicity of both the substrate and the enzymes involved, may ease our way to a more complete understanding of how the living cell converts energy.

A very unusual example of PP_i synthesis in a soluble system with clear implications also for membrane bioenergetics has recently been published. By changing the water activity and ionic environment around soluble yeast pyrophosphatase, the equilibrium of the enzyme-catalyzed reaction could be pushed towards the direction of PP_i synthesis instead of the usual hydrolysis [100].

Our new ability to obtain a highly purified preparation of a bacterial coupling factor PPase, in combination with our recently obtained capacity to clone any part of the *R. rubrum* genome, may open the road to a detailed study of the molecular properties of the enzyme. The techniques now available for directed mutations of bacterial genomes may give us possibilities to alter the dynamic properties of the enzyme in ways that should provide added insight into the mechanism of formation and utilization of inorganic pyrophosphate.

In a longer perspective, the new protein and gene technology should permit detailed molecular engineering of the subunits of the coupling factor PPase, as well as the ATPase, and also, at the three dimensional level, a fuller understanding of the minimum structural requirements for, and the dynamics and evolution of, biological coupling between electron transport and the formation of energy-rich phosphate compounds.

References

1 Baltscheffsky, M. (1964) In Abstracts of the First Meeting of the Federation of the European Biochemical Society, p. 67, London.
2 Baltscheffsky, M. (1967) Nature 216, 241–243.
3 Baltscheffsky, M. (1967) Biochem. Biophys. Res. Commun. 28, 270–276.
4 Baltscheffsky, M. (1969) Arch. Biochem. Biophys. 133, 46–53.
5 Baltscheffsky, M. (1969) Arch. Biochem. Biophys. 130, 646–652.
6 Baltscheffsky, H., Baltscheffsky, M. and von Stedingk, L.-V. (1969) In Progress in Photosynthesis Research (Metzner, H., ed.) Vol. III, pp. 1313–1318, Tübingen.
7 Keister, D.L. and Yike, N.J. (1966) Biochem. Biophys. Res. Commun. 24, 519–626.
8 Keister, D.L. and Yike, N.J. (1967) Biochemistry 6, 3847–3857.
9 Keister, D.L. and Yike, N.J. (1967) Arch. Biochem. Biophys. 121, 415–422.
10 Keister, D.L. and Minton, N.J. (1969) Biochemistry 8, 167.
11 Keister, D.L. and Minton, N.J. (1971) Biochem. Biophys. Res. Commun. 43, 932–939.
12 Keister, K.L. and Minton, N.J. (1971) Arch. Biochem. Biophys. 147, 330–338.
13 Baltscheffsky, H., von Stedingk, L.-V., Heldt, H.-W. and Klingenberg, M. (1966) Science 153, 1120.
14 Baltscheffsky, H. and von Stedingk, L.-V. (1966) Biochem. Biophys. Res. Commun. 22, 722–728.
15 Knobloch, K. (1975) Z. Naturforsch. 30C, 771.
16 Sherman, L.A. and Clayton, R.K. (1972) FEBS Lett. 22, 127.
17 Jones, O.T.G. and Saunders, V.A. (1972) Biochim. Biophys. Acta 275, 427.
18 Knaff, D.B. and Carr, J.W. (1979) Arch. Biochem. Biophys. 193, 379–384.
19 Mansurova, S.E., Ermakova, S.A., Zvyagilskaya, R.A. and Kulaev, I.S. (1975) Mikrobiologiya 44, 874–879.
20 Mansurova, S.E., Shakhov, Yu.A., Belyakova, T.N. and Kulaev, I.S. (1975) FEBS Lett. 55, 94–98.
21 Rubtsov, P.M., Efemovich, N.V. and Kulaev, I.S. (1976) Dokl. Akad. Nauk. Uzb. SSR 230, 1236–1237.
22 Reeves, R.E. (1976) Trends Biochem. Sci. 1, 53–55.
23 Wood, H.G., O'Brien, W.E. and Michaels, G. (1977) Adv. Enzymol. 45, 85–155.
24 Wood, H.G. (1977) Fed. Proc. 36, 2197–2205.
25 Schuegraf, A., Ratner, S. and Warner, R.C. (1960) J. Biol. Chem. 235, 3597.
26 Flodgaard, H. and Fleron, P. (1974) J. Biol. Chem. 249, 3465–3474.
27 Stiller, M., Diamondstone, T., Witonsky, R., Baltimore, D., Ruthman, R.J. and George, P. (1965) Fed. Proc. 24, 363.
28 Janson, C.A., Degani, C. and Boyer, P.D. (1979) J. Biol. Chem. 254, 3743–3749.
29 Alberty, R.A. (1968) J. Biol. Chem. 243, 1337.
30 Wood, H.G., Davis, J.J. and Lochmüller, H. (1966) J. Biol. Chem. 241, 5692.
31 George, P., Witonsky, R.J., Trachtman, M., Wu, C., Dorwart, W., Richman, L., Richman, W., Shurayh, F. and Lentz, B. (1970) Biochim. Biophys. Acta 223, 1–15.
32 Hayes, D.M., Kenyon, G.L. and Kollman, P.A. (1978) J. Am. Chem. Soc. 100, 4331–4340.
33 Lawson, J.W.R., Guynn, R.W., Cornell, N. and Veech, R.L. (1976) In Gluconeogenesis (Mehlman, M.A. and Hanson, R., eds.) p. 481, Wiley, New York.
34 Klungsöyr, L., King, T.E. and Cheldelin, V.H. (1957) J. Biol. Chem. 227, 135.
35 Klungsöyr, L. (1959) Biochim. Biophys. Acta 34, 586.
36 Barltrop, J.A., Grubb, P.W. and Hesp, B. (1963) Nature 199, 759.
37 Peck, H.D. (1962) Bacteriol. Rev. 26, 67.
38 Ware, D. and Postgate, J.R. (1970) Nature 226, 1250.
39 O'Brien, W.E. and Bowien, S. (1975) Fed. Proc. 34, 641.
40 Reeves, R.E., South, D.J., Blytt, H.J. and Warren, L.G. (1974) J. Biol. Chem. 249, 7737–7741.
41 O'Brien, W.E., Bowien, S. and Wood, H.G. (1975) J. Biol. Chem. 250, 8690–8695.
42 Sawyere, M.H., Baumann, P. and Baumann, L. (1977) Arch. Microbiol. 112, 169–172.
43 Macy, J.M., Ljungdahl, L.G. and Gottschalk, G. (1978) J. Bacteriol. 134, 84–91.
44 Pleidere, C. and Klemme, J.-H. (1980) Z. Naturforsch. 35C, 229–238.

45 Carnal, N.W. and Black, C.C. (1979) Biochem. Biophys. Res. Commun. 86, 20–26.
46 Sabularse, D.C. and Anderson, R.L. (1981) Biochem. Biophys. Res. Commun. 100, 1423–1429.
47 Hatch, M.D. and Slack, C.R. (1968) Biochem. J. 106, 141.
48 Reeves, R.E. (1968) J. Biol. Chem. 243, 3203.
49 Reeves, R.E., Munzies, R.A. and Hsu, D.S. (1968) J. Biol. Chem. 243, 5468.
50 Evans, H.J. and Wood, H.G. (1968) Proc. Natl. Acad. Sci. U.S.A. 61, 1448.
51 Benziman, M. and Palgi, A. (1970) J. Bacteriol. 104, 24.
52 Buchanan, B.B. (1974) J. Bacteriol. 119, 1066–1068.
53 Liu, C.L. and Peck, H.D. Jr. (1981) J. Bacteriol. 145, 966.
54 Liu, C.L., Hart, N. and Peck, H.D. Jr. (1982) Science 217, 363–364.
55 Peck, H.D. Jr, Liu, Chi-Li, Varma, A.K., Ljungdahl, L.G., Szulzynski, M., Bryant, F. and Carreira, L. (1982) In Basic Biology of New Developments in Biotechnology (Holleander, A., Laskin, A.I. and Rogers, P., eds) pp. 317–348, Plenum Press, New York and London.
56 Varma, A.K., Rigsby, W. and Jordan, D.C. (1983) Can. J. Microbiol. 29, 1470–1474.
57 Cagen, L.M. and Friedmann, H.C. (1972) J. Biol. Chem. 247, 3382–3392.
58 Cagen, L.M. and Friedmann, H.C. (1968) Biochem. Biophys. Res. Commun. 33, 528–533.
59 Silcox, D.C., Jacobelli, S. and McCarty, D.J. (1973) J. Clin. Invest. 52, 1595–1600.
60 Heinonen, J. (1974) Anal. Biochem. 59, 366–374.
61 Cook, G.A., O'Brien, W.K., Wood, H.G., King, H.T. and Veech, R.L. (1978) Anal. Biochem. 91, 557–565.
62 Ermakova, S.A., Mansurova, S.E., Kalebina, T.S., Lobakova, E.S., Selyach, I.A. and Kulaev, I.S. (1981) Arch. Microbiol. 128, 394–397.
63 Lust, G. and Seegmiller, J.E. (1976) Clin. Chim. Acta 66, 241–249.
64 Reich, J.G., Fill, U., Gunther, J., Zahn, D., Tschisgale, M. and Frunder, H. (1968) Eur. J. Biochem. 6, 384–394.
65 Guyn, R.W., Veloso, D., Lawson, J.W.R. and Veech, R.L. (1974) Biochem. J. 140, 369–375.
66 Kukko, E. and Heinonen, J. (1982) Eur. J. Biochem. 127, 347–349.
67 Bornefeldt, T. (1981) Arch. Microbiol. 129, 371–373.
68 Cori, C.F. (1942) In Symposium on Respiratory Enzymes. Madison, Wisconsin.
69 Cross, R.J., Taggart, J.V., Covo, G.A. and Green, P.E. (1949) J. Biol. Chem. 177, 655–678.
70 Baltscheffsky, M. (1968) In Regulatory Functions of Biological Membranes (Järnefelt, J., ed.) Biochim. Biophys. Acta Library 11, 277–286.
71 Mansurova, S.E., Shakhov, Yu.A. and Kulaev, J.S. (1977) FEBS Lett. 74, 31–34.
72 Guillory, R.J. and Fisher, R.R. (1972) Biochem. J. 129, 471–481.
73 Pullaiya, T., Shakhov, Yu.A., Mansurova, S.E. and Kulaev, J.S. (1980) Biokhimiya 45, 1093–1097.
74 Shakhov, Yu.A., Mansurova, S.E. and Kulaev, J.S. (1981) Biochem. Int. 3, 139–145.
75 Iria, M., Yabuto, A., Kimura, K., Shindo, Y. and Tomita, K. (1970) Biochem. J. (Tokyo) 67, 47–58.
76 Shakhov, Yu.A., Dukhovich, V.A., Velandia, A., Spiridonova, V.A., Mansurova, S.E. and Kulaev, J.S. (1982) Biokhimiya (USSR) 47, 601–607.
77 Volk, S.E., Baykov, A.A., Kostenko, E.B. and Avaeva, S.M. (1983) Biochim. Biophys. Acta 744, 127–134.
78 Efremovich, N.V., Volk, E.S., Baykov, A.A. and Shakhov, Yu.A. (1980) Biokhimiya 45, 1033–1040.
79 Horio, T., von Stedingk, L.-V. and Baltscheffsky, H. (1966) Acta Chem. Scand. 20, 1–10.
80 Baltscheffsky, H. and Baltscheffsky, M. (1960) Acta Chem. Scand. 14, 257.
81 Baltscheffsky, M., Baltscheffsky, H. and von Stedingk, L.-V. (1966) Brookhaven Symp. Biol. 19, 246–257.
82 Horio, T., Nishikawa, K., Horiuti, Y. and Kakuno, T. (1968) In Comparative Biochemistry and Biophysics of Photosynthesis (Shibata, K., Takamiya, A., Jagendorf, A.T. and Fuller, R.C., eds.) pp. 408–424, Tokyo University Press, Tokyo.
83 Johansson, B.C. (1975) Doctorial thesis, University of Stockholm, Sweden.
84 Johansson, B.C., Baltscheffsky, M. and Baltscheffsky, H. (1972) In Proc. 2nd International Photosynthesis Res. (Forti, G., Avron, M. and Melandri, B.A., eds.) Vol. 2, pp. 1203–1209, Junk, The Hague.

85 Moyle, J., Mitchell, R. and Mitchell, P. (1972) FEBS Lett. 23, 233–236.
86 Jackson, J.B. and Crofts, A.R. (1969) FEBS Lett. 4, 185–189.
87 Baltscheffsky, M. (1977) In Structure of Biological Membranes (Abrahamsson, S. and Pascher, J., eds.) pp. 41–62, Plenum, New York.
88 Vernon, L.P. and Ash, O.K. (1960) J. Biol. Chem. 235, 2721–2727.
89 Löw, H. and Alm, B. (1964) Abstr. of First FEBS Meeting, p. 68.
90 Dutton, P.L. and Baltscheffsky, M. (1972) Biochim. Biophys. Acta 267, 172–178.
91 Keister, D.L. and Raveed, N.J. (1974) J. Biol. Chem. 249, 6454–6458.
92 Harvey, G.W. and Keister, D.L. (1981) Arch. Biochem. Biophys. 208, 426–430.
93 Mitchell, R.A., Hill, R.D. and Boyer, P.D. (1967) J. Biol. Chem. 242, 1793–1801.
94 Boyer, P.D., Cross, R.L. and Momsen, W. (1973) Proc. Natl. Acad. Sci. U.S.A. 70, 2837–2839.
95 Hackney, D.D. and Boyer, P.D. (1978) Proc. Natl. Acad. Sci. U.S.A. 75, 3133–3137.
96 Rao, P.V. and Keister, D.L. (1978) Biochem. Biophys. Res. Commun. 84, 465–473.
96a Nyrén, P., Hajnal, K. and Baltscheffsky, M. (1984) Biochim. Biophys. Acta. In press.
97 Shakhov, Y.A., Nyrén, P. and Baltscheffsky, M. (1982) FEBS Lett. 146, 177–180.
98 Nyrén, P. and Baltscheffsky, M. (1983) FEBS Lett. 155, 125–130.
99 Kondrashin, A.A., Remennikov, V.G., Samuilov, V.D. and Skullachev, V.P. (1980) Eur. J. Biochem. 113, 219–222.
100 de Meis, L. (1984) J. Biol. Chem. 259, 6090–6097.

Ernster (ed.) Bioenergetics
© 1984 Elsevier Science Publishers B.V.

CHAPTER 7

Mitochondrial nicotinamide nucleotide transhydrogenase

JAN RYDSTRÖM, BENGT PERSSON and HAI-LUN TANG

Department of Biochemistry, Arrhenius Laboratory, University of Stockholm, S-106 91 Stockholm, Sweden

1. Introduction

Nicotinamide nucleotide transhydrogenase (EC 1.6.1.1) catalyzes the reversible reduction of $NADP^+$ by NADH according to the reaction

$$NADP^+ + NADH \rightleftarrows NAD^+ + NADPH \tag{1}$$

The enzyme was originally discovered in extracts of *Pseudomonas fluorescens* by Colowick et al. [1], but was subsequently demonstrated also in mammalian tissues [2] where it was shown to be associated with the mitochondrial inner membrane [3–7]. Although indications for the existence of a control of transhydrogenase by oxidative energy were presented at an early stage [8], which led to the proposal of an energy-linked transhydrogenase by Klingenberg and Slenczka [9] and others [10–27], it was only in 1963 that Danielson and Ernster [28–30] provided direct evidence for the existence of this reaction in submitochondrial particles from beef heart. Energy, generated either by respiration or ATP, was shown to stimulate the forward reaction to a rate comparable to that of the reverse reaction, with a concomitant increase in apparent equilibrium constant from unity to about 500 [31]. It was also shown that both in the absence and in the presence of energy, the mitochondrial enzyme is stereospecific with respect to the 4B-hydrogen of NADPH and the 4A-hydrogen of NADH [32].

Several respiring and photosynthetic bacteria were later also found to contain energy-linked transhydrogenase [33–38]. Hoek et al. [39] showed that the soluble form of bacterial transhydrogenase has a stereospecificity which is different from that of the mitochondrial enzyme and involves the 4B-hydrogen of both NADPH and NADH. The accumulating evidence thus indicate that the transhydrogenases, so far demonstrated, can be divided into two different types, namely AB- and the BB-transhydrogenases, as suggested by Hoek et al. [39]. The BB-transhydrogenase is an exclusively bacterial enzyme. It is easy to solubilize, is a flavoprotein, and has a single catalytic site for NAD(H) and NADP(H). It reacts according to a binary complex ('ping-pong') mechanism. The AB-transhydrogenase occurs in mitochondria and in certain bacteria. It is firmly membrane bound, is not a flavoprotein, has separate catalytic sites for NAD(H) and NADP(H), and reacts according to a short-lived ternary complex mechanism. In spite of these differences, the two enzymes have one feature in common: in the absence of specific effectors, the maximal velocities of the reactions catalyzed by both enzymes are much slower in the forward (left to right, according to reaction 1) than in the reverse reaction. Specific effectors can enhance the formed reaction probably by altering the conformational state of the enzyme. The [NADPH]/[$NADP^+$] ratio, 2′-AMP and Ca^{2+} are such effectors of the BB-transhydrogenases (cf., Ref. 39). In the case of the AB-transhydrogenases, the situation is more complex. The forward maximal velocity and the apparent equilibrium of the reaction appears to be regulated by the prevailing proton motive force (cf., Sections 2.3 and 3.2).

Several recent reviews have dealt with various aspects of AB- and BB-transhydrogenases [40–44]. The present review will, therefore, selectively emphasize aspects which are relevant for the discussion of recent findings and which are important for understanding the function of the mitochondrial transhydrogenase.

2. *Energy-linked transhydrogenase*

2.1. *Relationship to the energy-coupling system*

Energy-linked transhydrogenase is functionally coupled to the energy-transfer system of the membrane in which it is located. This coupling is manifested by an energy-dependent 5–10-fold increase in the rate [28–30] and a 500-fold increased extent [31,45] of the reduction of $NADP^+$ by NADH. In mitochondria the transhydrogenase reaction can be driven by energy generated either by electron transport through any of the coupling sites of the respiratory chain, or by ATP hydrolysis [28–30,46]. It can also be driven by a potassium ion gradient across the mitochondrial inner membrane in the presence of valinomycin [47].

Inhibitors of electron transport do not affect the ATP-driven transhydrogenase reaction [28–30], whereas energy-transfer inhibitors, e.g., oligomycin, do not inhibit the respiration-driven reactions, but inhibit the ATP-driven reaction [28,29,34,48,49]. In so-called non-phosphorylating submitochondrial particles, oligomycin stimulates and may even be obligatory for the energy-linked transhydrogenase reaction driven by either respiration or ATP [41,50–53]. Uncouplers of oxidative phosphorylation abolish the energy-linked transhydrogenase reaction [20,30,33,34,46,54]. Similarly, the reaction is abolished by the combined effects of valinomycin and nigericin in the presence of potassium ions [55]. In submitochondrial particles it has been found that both uncouplers and oligomycin inhibit the ATP-driven transhydrogenase less efficiently than the ATP-driven reduction of NAD^+ by succinate [46,56]. On the other hand, the two reactions are equally sensitive to the mitochondrial ATPase inhibitor [56,57], and it has been suggested that the transhydrogenase and ATPase interact in a direct fashion [56,59]. In view of the apparently close association between the oxidative phosphorylation system and energy-linked transhydrogenation one would expect a competition between the two processes. A competition has indeed been shown to take place under conditions of limiting energy generation [51–53].

The energy expenditure of the mitochondrial ATP-driven transhydrogenase reaction has been estimated to one ATP per NADPH formed [29,30,46,53,60–62]. Since the equilibrium constant of the nonenergy-linked transhydrogenase reaction is 0.79 [2] and that of ATP hydrolysis is about 10^5 M [63], the equilibrium constant of the overall reaction is also of the order of 10^5 M. In spite of the very unfavourable equilibrium a reversibility of the energy-linked transhydrogenase reaction has been demonstrated using ATP production [64] or uptake of lipophilic anions [65–67] as assay.

A possible functional relationship between transhydrogenase and the NADH dehydrogenase complex of the respiratory chain, suggested at an early stage [68], but soon challenged [69], was restressed in recent years, based on an observed NADPH oxidase activity of submitochondrial particles. However, these observations found their explanation in the earlier known [70] ability of NADH dehydrogenase to exhibit some reactivity with NADPH [71], a feature that appears to be common to NAD(H)-specific dehydrogenases especially at slightly acidic pH [72]. The recently reported slow reactivity of the NAD(H)-binding site of transhydrogenase with NADP(H) [73,74] is probably another reflection of this feature.

2.2. Reaction mechanism and regulation

Studies of the steady-state kinetics of the nonenergy-linked and energy-linked transhydrogenase reactions, catalyzed by submitochondrial particles from beef heart, indicate that both reactions proceed by way of a ternary complex of very short lifetime, i.e., a Theorell Chance mechanism [75–77]. This conclusion was based on linear and convergent double reciprocal plots of initial velocities versus substrate concentrations, as well as product inhibition patterns that revealed competitive relationships between NAD^+ and NADH and between $NADP^+$ and NADPH, and noncompetitive relationships between NAD^+ and $NADP^+$ and between NADH and NADPH. This product inhibition pattern indicates that the transhydrogenase has separate binding sites for NADP(H) and NAD(H). Studies with site-specific inhibitors [78,79] suggested that NAD(H) is the first substrate bound to the enzyme. However, a recent reinvestigation of the steady-state kinetics of the purified and reconstituted transhydrogenase reveals that the enzyme most likely follows a random ternary-complex mechanism (see section 3.2).

Michelis constants of the nonenergy-linked beef heart transhydrogenase reaction are 9 μM for NADH, 40 μM for $NADP^+$, 28 μM for NAD^+ and 20 μM for NADPH [75]. Energy-linked changes in the Michaelis constants are particularly apparent with the oxidized nicotinamide nucleotides, giving values of 6.5 and 43.5

μM for $NADP^+$ and NAD^+, respectively [76]; occasionally, the latter value may be considerably higher. Thus, energy favors binding of NADH and $NADP^+$ and dissociation of NAD^+ and NADPH with a simultaneous increase in the maximal activity of the forward reaction. Since the presence of energy only decreases the affinities of the enzyme for NAD^+ and NADPH without effect on the maximal activity of the reverse reaction, it appears that the increase of the forward reaction is mainly due to the increased dissociation of NAD^+ and NADPH, primarily that of NAD^+. The large energy-linked decrease in the K_m for $NADP^+$ in the forward reaction may thus be a secondary effect caused by the increased dissociation of NAD^+.

Energy-linked affinity changes also seem to be of importance for the effect of certain inhibitors of transhydrogenase, e.g., metal ions [45,80]. The energy-linked transhydrogenase reaction catalyzed by submitochondrial particles is known to be inhibited by Mg^{2+} to a lesser extent than the nonenergy-linked reaction [45]. In addition, the effect of Mg^{2+} is pH-dependent with an increasing effect of the inhibitor with increasing pH [45]. Recently, Mg^{2+} was shown to be a competitive inhibitor with respect to $NADP^+$ [80], and is therefore presumably specific for the catalytic NADP(H) site. A close to competitive relationship between solute protons and the extent of inhibition by Mg^{2+} was demonstrated [80], and it is possible that the protons may exert their effect through a proton-dependent increase in the affinity of the transhydrogenase for NADP(H); such affinity changes have indeed been observed [81].

2.3. Energy-coupling mechanism

Early proposals concerning the mechanism of the energy-linked transhydrogenase reaction were based on the "chemical hypothesis of oxidative phosphorylation" [82] and visualized the involvement of high-energy intermediates of the type $1 \sim X$, $NADH \sim I$, $NADP^+ \sim I$, etc. [29,46]. These proposals, however, just as the chemical hypothesis as a whole, had to be abandoned because of lack of experimental evidence.

After the demonstration that mitochondrial transhydrogenase indeed is a proton pump [83,84], the next major question concerned the molecular mechanism by which this is achieved. That the transhydrogenase does not function as a redox loop in the classical chemiosmotic sense [85] can be safely concluded on the basis that transfer of hydrogen between NAD(H) and NADP(H) is direct, without interaction with solute protons [32,86]; that NAD(H) and NADP(H) react with the enzyme on the same (matrix) side of the membrane [87]; that the two substitutes do not cross the membrane because of their high water solubility; and that the enzyme most probably does not contain any endogeneous redox groups involved in catalysis (with the possible exception of firmly-bound nicotinamide nucleotides). It is therefore now generally agreed that the coupling between the catalytic event and the proton translocation in transhydrogenation is indirect. Current proposals visualize that the

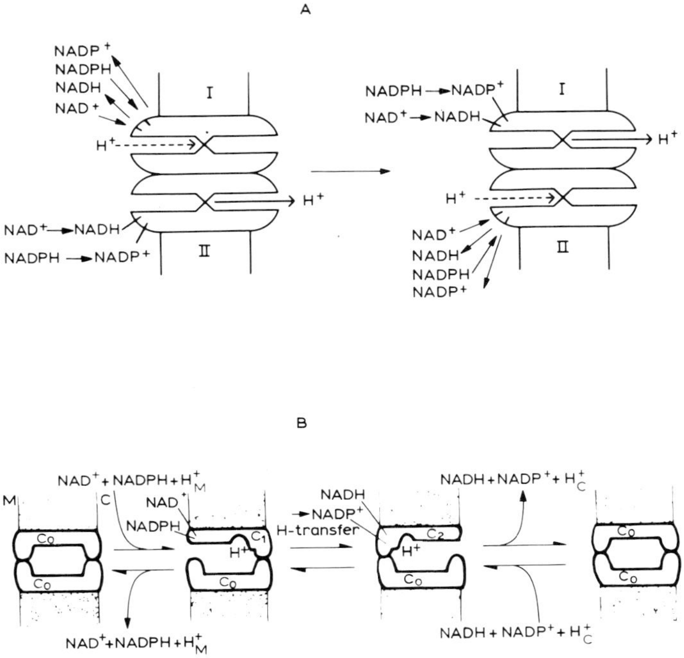

Fig. 7.1. Proposed models of energy-linked transhydrogenation (from Ref. 90 (A) and Ref. 89 (B)).

protons are carried across the membrane either entirely by the transhydrogenase protein itself or, at least partly by charged groups of the substrates. Two recently suggested mechanisms are shown in Figs. 7.1A, B which both involve a dimeric transhydrogenase [88], where the two subunits contain one proton channel each (Fig. 7.1A) or together form a proton channel (Fig. 7.1B).

Certain similarities between mitochondrial transhydrogenase and ATPase have been pointed out in the past [91] and are further underlined by recent knowledge of the modes of action of the two enzymes. From the available data (cf., Refs. 41–44) it appears clear that the thermodynamic effect of energy, i.e., $\Delta\tilde{\mu}_{H^+}$, involves a

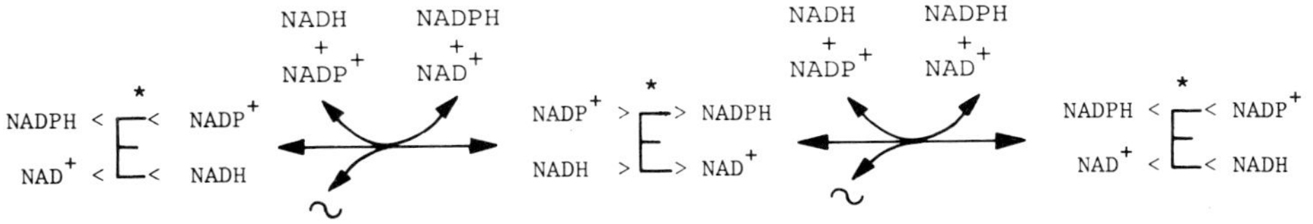

Fig. 7.2. Alternative-site model of energy-linked transhydrogenase. E denotes energized form of transhydrogenase. >E and E<, and <E and E> denote tight and loose binding, respectively, of substrate/products to energized transhydrogenase.

facilitated binding of NADH and $NADP^+$ and release of NAD^+ and NADPH [76]; the actual conversion of the bound substrates to products proceeds without appreciable requirements of energy. Similarly, it has been proposed [92,93] that, in the case of the ATPase system, the available $\Delta\tilde{\mu}_{H^+}$ is used primarily for the binding of ADP + P_i and the release of ATP, while the conversion of bound ADP + P_i to ATP proceeds without further energy requirement. It has also been proposed [94] that the ATPase operates by catalytic cooperativity between identical subunits, where binding of ADP + P_i to one subunit promotes the release from another subunit. This 'alternative-site' mechanism has received considerable experimental support in recent years [95]. A similar mechanism for transhydrogenase, outlined in Fig. 7.1A and shown in more detail in Fig. 7.2, appears to be an attractive possibility in view of the dimeric nature of the enzyme [88].

Another striking similarity between the transhydrogenase and ATPase is the kinetic effect of energy. Thus $\Delta\tilde{\mu}_{H^+}$ acts as an effector of the mitochondrial transhydrogenase, enhancing the maximal velocity of the reaction. Similarly, $\Delta\tilde{\mu}_{H^+}$ can enhance the mitochondrial ATPase activity [56,96] when the latter is suppressed by the ATPase inhibitor [97]. It may thus be concluded that, even though the transhydrogenase and ATPase are widely different in structure, both seem to be regulated by the prevailing $\Delta\tilde{\mu}_{H^+}$.

3. Properties of purified and reconstituted transhydrogenase from beef heart

3.1. Purification and reconstitution

Being a membrane-bound enzyme the purification of transhydrogenase necessitates the use of detergents. Earlier attempts involved the use of digitonin and repeated sucrose-density gradient centrifugation of the digitonin extract [98,99]. However, prolonged exposure to this detergent inactivated transhydrogenase. Several preparations of transhydrogenase were subsequently reported (cf., Ref. 41), neither of which proved to give a preparation which allowed a reliable identification of transhydrogenase and determination of the molecular weight. The first reproducible method for the isolation of transhydrogenase employed solubilization by cholate followed by a classical precipitation by ammonium sulphate in the presence of cholate, ion exchange chromatography and hydroxylapatite chromatography [83,100]. This preparation was the first which demonstrated that the transhydrogenase is composed of a single subunit, the molecular weight of which is about 100 000 (later determined more exactly to 115 000 [42]) It was also demonstrated that when reconsituted in liposomes by the cholate dialysis technique the activity of transhydrogenase shows 'respiratory control', i.e., is enhanced several-fold by uncouplers [83]. Reconstituted transhydrogenase is localized with the active site directed towards the surrounding water phase and generates a membrane potential [83,101], indicating that the enzyme indeed is a transmembrane proton pump. Coreconstitution of purified mitochondrial ATPase from beef heart together with transhydrogenase gave vesicles which contained both ATPase and transhydrogenase, as indicated by the capacity of ATP to drive the reduction of $NADP^+$ by NADH in an oligomycin-sensitive manner [101].

Transhydrogenase was also purified somewhat later and independently by a more complicated procedure which, however, also produced the 115 000 molecular weight polypeptide [102]. Subsequently, additional improved methods have been published, which involve the use of immobilized antibodies [103] and affinity chromatography on immobilized NAD^+ [104] or $NADP^+$ [105]. With all preparations reconstituted transhydrogenase was shown to be a proton pump by both an indirect assay using 9-aminoacridine as pH probe for the interior space of the vesicles [106], and a direct assay using a pH electrode [107].

Most of the procedures published so far have been either unsuitable for larger scale preparations (maximally 1 mg) or irreproducible. In an attempt to overcome these problems a purification procedure has been developed recently [108] which is a modification of a previous method [83]. The modified procedure involves FPLC (fast protein liquid chromatography), is rapid, reproducible and allows the preparation of large amounts of active enzyme. Purified transhydrogenase obtained by this procedure is reconstitutively active and pumps protons as judged by quenching of

TABLE 7.1

Comparison of methods for the purification of mitochondrial nicotinamide nucleotide transhydrogenase from beef heart

Unless otherwise indicated transhydrogenase activity was assayed by reduction of AcPyAD by NADPH and protein was determined by the Lowry method

Preparation	Spec. act. ($\mu mol \cdot min^{-1} \cdot mg^{-1}$)	Purification factor	Yield (%)	Transhydrogenase (mg/prep)	Proton pumping
1. Höjeberg and Rydström [83,100]	4.4 [a]	40	8.7	0.190	+
2. Anderson and Fisher [102]	7	23.3	3.1	0.600	+
3. Anderson et al. [103]	14.2	56.8	15.8	0.275	+
4. Wu et al. [104]	62.3 [b]	115.4	47.4	0.770	+
5. Carlenor et al. [105]	16.8	38.6	8.0	0.219	+
6. Persson et al. [108]	16.2	37.1	10.0	0.293	+
7. Persson et al. [108]	2.9	6.7	3.0	1.910	+

[a] Measured at pH 7.5 and with NAD^+ as hydrogen acceptor.

[b] Protein determined by Coomassie-blue absorption.

9-aminoacridine. A comparison between the various methods of purification is made in Table 7.1 with respect to specific activity, purification factor, yield, amount of transhydrogenase protein per preparation, and capacity for proton pumping.

Most preparations give pure transhydrogenase although, of course, the extent of purity is a matter of detection of impurities. The most recently developed method [108] indeed provides a transhydrogenase that is at least 95% pure, as judged by the high-sensitive silver staining of polyacrylamide slab gels. Silver staining has led to the discovery of several contaminating peptides in earlier procedures (methods 1–5). Previous experiences indicate that larger amounts of transhydrogenase cannot be obtained simply by scaling up the available procedures. Method 7 is designed to give large amounts of less active but pure transhydrogenase (about 2–5 mg) suitable for sequencing studies and protein modification, etc. Specific activities range between 2.9 and 62.3 $\mu mol \cdot min^{-1} \cdot mg^{-1}$ protein. However, these values are difficult to compare since different substrates and assay conditions have been used. For instance, the value 62.3 was obtained using a protein determination method which, uncorrected and with bovine serum albumin as standard, gives a 2.5-fold underestimate of the amount of protein (cf., Ref. 109). All purification procedures give reconstitutively active transhydrogenase, i.e., transhydrogenase vesicles generate an uncoupler-sensitive membrane potential and pH gradient. Triton X-100 was recently shown to be blocking proton translocation unless carefully removed from the enzyme during reconstitution [108], which provides an explanation for the poor proton translocation capacity of previous preparations.

3.2. Catalytic and regulatory properties

As pointed out previously in this review the steady-state kinetics of mitochondrial transhydrogenase, earlier interpreted to indicate a ternary Theorell-Chance mechanism on the basis of competitive relationships between NAD^+ and NADH and between $NADP^+$ and NADPH, and noncompetitive relationships between NAD^+ and $NADP^+$ and between NADH and NADPH, has been reinterpreted in the light of more recent developments in the interpretation of steady-state kinetic data. Thus, although the product inhibition patterns obtained in the earlier reports [75–77] using submitochondrial particles were close to identical to those obtained in a more recent report [90] using purified and reconstituted transhydrogenase, the reinterpretation favors a random mechanism with the two dead-end complexes $NAD^+ \cdot E \cdot NADP^+$ and $NADH \cdot E \cdot NADPH$. A random mechanism is also supported by the observation that the transhydrogenase binds to immobilized NAD^+ as well as $NADP^+$ [105] in the absence of the second substrate.

Reconstituted transhydrogenase has been suggested to catalyze an NADH-NAD^+ exchange via a reduced enzyme intermediate [110]. The proposal is based on the finding that coupled reconstituted vesicles, in the presence of AcPyAD plus NADPH, catalyze a rapid and transient reduction of AcPyAD upon the addition of NADH. It was demonstrated that formation of AcPyADH is stoicheiometrically related to the disappearance of NADH rather than NADPH and that the 4A-hydrogen atom is involved in the reduction of AcPyAD (NAD^+) as well as in the oxidation of NADH [110]. An alternative explanation was provided for the reduction of AcPyAD by NADH in the presence of NADPH which involves energization of the vesicles by reduction of AcPyAD by NADPH, followed by reduction of bound $NADP^+$ by NADH, exchange of NAD^+ for AcPyAD, and subsequently reduction of AcPyAD by the NADPH formed [43,90]. Together, this sequence of reactions results in the apparent reduction of AcPyAD by NADH without a stoicheiometric consumption of NADPH, and without the proposed involvement of a reduced enzyme intermediate. In addition, a role of a reduced enzyme intermediate is unlikely in view of the kinetics of the enzyme [75–77] and the lack of exchange of the 4A-hydrogen of NADH or the 4B-hydrogen of NADPH with water [86].

Since the reduction of AcPyAD by NADH requires energized vesicles it was proposed [90] that an energy-dependent increase in the affinity for NADH and $NADP^+$, as well as an energy-dependent decrease in the affinity for NAD^+ (and $AcPyAD^+$), was necessary for the reaction to occur. The reconstituted transhydrogenase vesicles would therefore be regulated by a proton motive force in a manner similar to that in submitochondrial particles, the transhydrogenase activity of which has been shown to be regulated by energy with respect to the affinities for the oxidized substrates [76,77].

Energy-linked changes in the affinities for the substrates have also recently been demonstrated directly in reconstituted vesicles [90]. By adding increasing amounts of uncouplers it was shown that the K_m for AcPyAd was very high or about 500 μM in partially uncoupled vesicles, and about 150 μM in uncoupled vesicles [90]. A similar change in the K_m for NADPH could not be demonstrated. Thus, it appears that, depending on the direction of the reaction catalyzed, energy regulates transhydrogenase through affinity changes primarily for the oxidized substrates.

3.3. Proton translocation

Transhydrogenase generates a membrane potential [64–67,83,101] and has been shown to be a proton pump in intact mitochondria [111], submitochondrial particles [112], and reconstituted vesicles [106–108], and contains a DCCD-sensitive moiety, presumably a proton channel [89,113]. In the case of the reduction of NAD^+ by NADPH catalyzed by mitochondria, an H^+/H^- ratio of 2 was determined [111], whereas the corresponding value for submitochondrial particles was 0.2 [112]. In contrast, the value obtained with reconstituted vesicles was determined to be one [107]. Recently, a value of 0.5 was reported [114].

Treatment with DCCD inactivates the catalytic activity of transhydrogenase as well as H^+-translocation activity [89]. However, the number of moles of DCCD incorporated per mole of transhydrogenase monomer varies depending on whether inactivation is related to the change in catalytic activity or proton translocation activity. Thus, a ratio of 2 was reported in the former case and one in the latter case [89]. Even though the DCCD-binding domain was proposed to be separate from the active site, DCCD inactivation was influenced by NADP(H) [89]. The conclusion that 1 mol DCCD/mol enzyme monomer is required for inhibition of proton translocation is indeed consistent with a model where both monomers in the active dimer each possess a proton channel [90], rather than a model where a common channel is formed and shared by the two monomers [89] (cf., Fig. 7.1).

Inactivation or modification of two classes of sulfhydryl groups in transhydrogenase, one located in or close to the NADP(H)-binding site and one peripheral to the active sites [100], has allowed the conclusion that these are not involved in proton pumping [44,155]. Also, methane thiolation of both classes of sulfhydryl groups yields an enzyme that still carries out the generation of a membrane potential [115]. Denaturation leads to the exposure of additional sulfhydryl groups [44], which thus may be operative in a recently suggested mechanism involving a thiol-disulfide interchange in transhydrogenase-dependent proton translocation [116]. However, there is presently no experimental support for a mechanism involving sulfhydryl groups.

References

1 Colowick, S.P., Kaplan, N.O., Neufeld, E.F. and Ciotti, M.M. (1952) J. Biol. Chem. 195, 95–105.
2 Kaplan, N.O., Colowick, S.P. and Neufeld, E.F. (1953) J. Biol. Chem. 205, 1–15.
3 Kaplan, N.O., Swartz, M.N., Frech, M.E. and Ciotti, M.M. (1956) Proc. Natl. Acad. Sci. U.S.A. 42, 481–487.
4 Humphrey, G.F. (1957) Biochem. J. 65, 546–550.
5 Devlin, T.M. (1958) J. Biol. Chem. 234, 962–966.
6 Kielley, W.W. and Bronk, J.R. (1958) J. Biol. Chem. 230, 521–533.
7 McMurray, W.C., Maley, G.F. and Lardy, M.A. (1958) J. Biol. Chem. 230, 219–229.
8 Krebs, H.A. (1954) Bull. Johns Hopkins Hosp. 95, 34–44.
9 Klingenberg, M. and Slenczka, W. (1959) Biochem. Z. 331, 486–517.
10 Estabrook, R.W. and Nissley, S.P. (1963) In Funktionelle und Morphologische Organisation der Zelle (Karlsson, P., ed.) pp. 119–131, Springer-Verlag, Berlin.
11 Klingenberg, M. and Schollmeyer (1963) Proc. Int. Congr. Biochem. 6th, 1961, Vol. 5, 46–52.
12 Slater, E.C. and Tager, J.M. (1963) In Energy-Linked Functions of Mitochondria (Chance, B., ed.) pp. 97–113, Academic Press, New York.
13 Tager, J.M. (1963) Biochim. Biophys. Acta 77, 258–265.
14 Estabrook, R.W., Hommes, R.W. and Gonze, J. (1963) In Energy-Linked Functions of Mitochondria (Chance, B., ed.) pp. 143–152, Academic Press, New York.
15 Tager, J.M. and Slater, E.C. (1963) Biochim. Biophys. Acta 77, 227–245.
16 Tager, J.M., Howland, J.L., Slater, E.C. and Snoswell, A.M. (1963) Biochim. Biophys. Acta 77, 266–275.
17 Klingenberg, M. (1963) In Energy-Linked Functions of Mitochondria (Chance, B., ed.) pp. 121–139, Academic Press, New York.
18 Klingenberg, M., von Häfen, H. and Wenske, G. (1965) Biochem. Z. 343, 452–478.
19 Klingenberg, M. (1965) Biochem. Z. 343, 479–503.
20 van Dam, K. and ter Welle (1966) In Regulation of Metabolic Processes in Mitochondria, BBA Library, Vol. 7, pp. 235–245.
21 de Haan, E.J., Tager, J.M. and Slater, E.C. (1967) Biochim. Biophys. Acta 131, 1–13.
22 Papa, S. and Francavilla, A. (1967) In Mitochondrial Structure and Compartmentation (Quagliariello, E. et al., eds.) pp. 363–372, Adriatica Editrice, Bari.

23 Papa, S., Tager, J.M., Francavilla, A., de Haan, E. and Quagliariello, E. (1976) Biochim. Biophys. Acta 131, 14–28.
24 Nicholls, D.G. and Garland, P. (1969) Biochem. J. 114, 215–225.
25 Tager, J.M., Papa, S., de Haan, E., D'Aloya, R. and Quagliariello, E. (1969) Biochim. Biophys. Acta 172, 7–19.
26 Papa, S. and Tager, J.M. (1969) Biochim. Biophys. Acta 172, 20–29.
27 Papa, S. (1969) The Energy Level and Metabolic Control in Mitochondria (Papa, S. et al., eds.) pp. 401–409, Adriatica Editrice, Bari.
28 Danielson, L. and Ernster, L. (1963) Biochem. Biophys. Res. Commun. 10, 91–96.
29 Danielson, L. and Ernster, L. (1963) Biochem. Z. 338, 188–205.
30 Danielson, L. and Ernster, L. (1963) In Energy-Linked Functions of Mitochondria (Chance, B., ed.) pp. 157–175, Academic Press, New York.
31 Lee, C.P. and Ernster, L. (1964) Biochim. Biophys. Acta 81, 187–190.
32 Lee, C.P., Simard-Duquesne, N., Ernster, L. and Hoberman, H.D. (1965) Biochim. Biophys. Acta 105, 397–409.
33 Murthy, P.S. and Brodie, A.F. (1964) J. Biol. Chem. 239, 4292–4297.
34 Fisher, R.J. and Sanadi, D.R. (1971) Biochim. Biophys. Acta 245, 34–41.
35 Bragg, P.D. and Hou, C. (1968) Can. J. Biochem. 46, 631–641.
36 Sweetman, A.J. and Griffiths, D.E. (1971) Biochem. J. 121, 125–130.
37 Keister, D.L. and Yike, N.J. (1966) Biochem. Biophys. Res. Commun. 24, 519–525.
38 Keister, D.L. and Yike, N.J. (1967) Biochemistry 6, 3847–3857.
39 Hoek, J.B., Rydström, J. and Höjeberg, B. (1974) Biochim. Biophys. Acta 333, 237–245.
40 Rydström, J., Hoek, J.B. and Ernster, L. (1976) The Enzymes 13, 51–88.
41 Rydström, J. (1977) Biochim. Biophys. Acta 463, 155–184.
42 Rydström, J. (1981) In Mitochondria and Microsomes (Lee, C.P., Schatz, G. and Dallner, G., eds.) pp. 317–335, Addison-Wesley, Massachusetts.
43 Rydström, J., Lee, C.P. and Ernster, L. (1981) In Chemiosmotic Proton Circuits in Biological Membranes (Skulachev, V.P. and Hinkle, P.C., eds.) pp. 483–508, Addison-Wesley, Massachusetts.
44 Fisher, R.R. and Earle, S.R. (1982) In The Pyridine Nucleotide Coenzymes (Everse, J., Anderson, B. and You, K., eds.) pp. 280–324, Academic Press, New York.
45 Rydström, J., Teixeira da Cruz, A. and Ernster, L. (1970) Eur. J. Biochem. 17, 56–62.
46 Lee, C.P. and Ernster, L. (1966) In Regulation of Metabolic Processes in Mitochondria, BBA Library, Vol. 7, pp. 218–234.
47 Conover, E. (1971) In Energy Transduction in Respiration and Photosynthesis (Quagliariello, E., Papa, S. and Rossi, C.S., eds.) pp. 999–1005, Adriatica Editrice, Bari.
48 Kawasaki, T., Satoh, K. and Kaplan, N.O. (1964) Biochem. Biophys. Res. Commun. 17, 648–654.
49 Lee, C.P., Azzone, G.F. and Ernster, L. (1964) Nature 201, 152–155.
50 Lee, C.P. and Ernster, L. (1967) Methods Enzymol. 10, 543–548.
51 Lee, C.P. and Ernster, L. (1966) Biochem. Biophys. Res. Commun. 23, 176–181.
52 Lee, C.P. and Ernster, L. (1967) In Round Table Discussion on Mitochondrial Structure and Compartmentation (Quagliariello, E. et al., eds.) pp. 353–362, Adriatic Editrice, Bari.
53 Lee, C.P. and Ernster, L. (1968) Eur. J. Biochem. 3, 385–390.
54 Asano, A., Imai, K. and Sato, R. (1967) Biochim. Biophys. Acta 143, 477–486.
55 Montal, M., Chance, B., Lee, C.P. and Azzi, A. (1969) Biochem. Biophys. Res. Commun. 34, 104–110.
56 Ernster, L., Juntti, K. and Asami, K. (1972) J. Bioenerg. 4, 149–159.
57 Asami, K., Juntti, K. and Ernster, L. (1970) Biochim. Biophys. Acta 205, 307–311.
58 Ernster, L., Asami, K., Juntti, K., Coleman, J. and Nordenbrand, K. (1977) In Structure of Biological Membranes (Abrahamsson, S. and Pascher, I., eds.) pp. 135–156, Plenum Press, New York.
59 Nordenbrand, K., Hundal, T., Carlsson, C., Sandri, G. and Ernster, L. (1977) In Bioenergetics of Membranes (Packer, L. et al., eds.) pp. 435–446, Elsevier/North Holland, Amsterdam.
60 Papa, S., Alifano, A., Tager, J.M. and Quagliariello, E. (1968) Biochim. Biophys. Acta 153, 303–305.

61 Haas, D.W. (1964) Biochim. Biophys. Acta 82, 200–202.
62 Tager, J.M., Groot, G.S.P., Roos, D., Papa, S. and Quagliariello, E. (1969) In The Energy Level and Metabolic Control in Mitochondria (Papa, S. et al., eds.) pp. 453–462, Adriatic Editrice, Bari.
63 Rosing, J. and Slater, E.C. (1972) Biochim. Biophys. Acta 267, 275–290.
64 Van de Stadt, R.J., Nieuwenhuis, F.J.R.M. and van Dam, K. (1971) Biochim. Biophys. Acta 234, 173–176.
65 Skulachev, V.P. (1970) FEBS Lett. 11, 301–308.
66 Skulachev, V.P. (1971) Curr. Top. Bioenerg. 4, 127–190.
67 Dontsov, A.E., Grinius, L.L., Jasaitis, A.A., Severina, I.I. and Skulachev, V.P. (1972) J. Bioenerg. 3, 277–303.
68 Hommes, F.A. and Estabrook, R.W. (1963) Biochem, Biophys. Res. Commun. 11, 1–6.
69 Lee, C.P. and Ernster, L. (1964) Annu. Rev. Biochem. 33, 729–788.
70 Rossi, C., Cremona, T., Machinist, M. and Slinger, T.P. (1965) J. Biol. Chem. 240, 2634–2643.
71 Rydström, J., Montelius, J., Bäckström, D. and Ernster, L. (1978) Biochim. Biophys. Acta 501, 370–380.
72 Navazio, F., Ernster, B.B. and Ernster, L. (1957) Biochim. Biophys. Acta 26, 416–421.
73 Hatefi, Y., Phelps, D.C. and Galante, Y.M. (1980) J. Biol. Chem. 255, 9526–9529.
74 Phelps, D.C., Galante, Y.M. and Hatefi, Y. (1980) J. Biol. Chem. 255, 9647–9652.
75 Teixeira da Cruz, A., Rydström, J. and Ernster, L. (1971) Eur. J. Biochem. 23, 203–211.
76 Rydström, J., Teixeira da Cruz, A. and Ernster, L. (1971) Eur. J. Biochem. 23, 212–219.
77 Rydström, J., Teixeira da Cruz, A. and Ernster, L. (1972) In Biochemistry and Biophysics of Mitochondrial Membranes (Azzone, G.F. et al., eds.) pp. 177–200, Academic Press, New York.
78 Rydström, J. (1972) Eur. J. Biochem. 31, 496–504.
79 Rydström, J., Hoek, J.B., Alm, R. and Ernster, L. (1973) In Mechanisms in Bioenergetics (Azzone, G.F. et al., eds.) pp. 579–589, Academic Press, New York.
80 O'Neal, S.G., Earle, S.R. and Fisher, R.R. (1980) Biochim. Biophys. Acta 589, 217–230.
81 Rydström, J. (1974) Eur. J. Biochem. 45, 67–76.
82 Slater, E.C. (1953) Nature 172, 975–978.
83 Höjeberg, B. and Rydström, J. (1977) Biochem. Biophys. Res. Commun. 78, 1183–1190.
84 Earle, S.R., Anderson, W.M. and Fisher, R.R. (1978) FEBS Lett. 91, 21–24.
85 Mitchell, P. (1966) Chemiosmotic Coupling in Oxidative and Photosynthetic Phosphorylation, Glynn Research, Bodmin (U.K.).
86 Ernster, L., Hoberman, H.D., Howard, R.L., King, T.E., Lee, C.P., Mackler, B. and Sottocasa G.L. (1965) Nature 207, 940–941.
87 DePierre, J.W. and Ernster, L. (1977) Annu. Rev. Biochem. 46, 201–262.
88 Anderson, W.M. and Fisher, R.R. (1981) Biochim. Biophys. Acta 635, 194–199.
89 Pennington, R.M. and Fisher, R.R. (1981) J. Biol. Chem. 256, 8963–8969.
90 Rydström, J. and Enander, K. (1983) J. Biol. Chem. 257, 14760–14766.
91 Ernster, L., Lee, C.P. and Torndal, U.B. (1969) In The Energy Level and Metabolic Control in Mitochondria (Papa, S. et al., eds.) pp. 439–451, Adriatic Editrice, Bari.
92 Boyer, P.D. (1974) BBA Library 13, 289–301.
93 Slater, E.C. (1974) BBA Library 13, 1–20.
94 Kayalar, C., Rosing, J. and Boyer, P.D. (1977) J. Biol. Chem. 252, 2486–2491.
95 Boyer, P.D. (1979) In Membrane Bioenergetics (Lee, C.P. et al., eds.) pp. 461–479, Addison-Wesley, Massachusetts.
96 Van de Stadt, R.J., De Boer, B.L. and van Dam, K. (1973) Biochim. Biophys. Acta 292, 338–349.
97 Pullman, M.E. and Monroy, G.C. (1963) J. Biol. Chem. 238, 3762–3769.
98 Kaufmann, B. and Kaplan, N.O. (1961) J. Biol. Chem. 236, 2133–2139.
99 Kaplan, N.O. (1967) Methods Enzymol. 10, 317–322.
100 Höjeberg, B. and Rydström, J. (1979) Methods Enzymol. 15, 275–283.
101 Rydström, J. (1979) J. Biol. Chem. 254, 8611–8619.
102 Anderson, W.M. and Fisher, R.R. (1978) Arch. Biochem. Biophys, 187, 180–190.

103 Anderson, W.M., Fowler, W.T., Pennington, R.M. and Fisher, R.R. (1981) J. Biol. Chem. 256, 1888–1895.
104 Wu, L.N.Y., Pennington, R.M., Everett, T.D. and Fisher, R.R. (1982) J. Biol. Chem. 257, 4052–4055.
105 Carlenor, E., Tang, H.-L. and Rydström, J. (1984) in preparation.
106 Earle, S.R., Anderson, W.M. and Fisher, R.R. (1978) FEBS Lett. 91, 21–24.
107 Earle, S.R. and Fisher, R.R. (1980) J. Biol. Chem. 255, 827–830.
108 Persson, B., Tang, H.-L. and Rydström, J. (1984) J. Biol. Chem., in press.
109 Pollard, H., Menard, R., Brandt, H., Pazoles, C., Cretz, C. and Ramu, A. (1978) Anal. Biochem. 86, 761–763.
110 Wu, L.N.Y., Earle, S.R. and Fisher, R.R. (1981) J. Biol. Chem. 256, 7401–7408.
111 Moyle, J. and Mitchell, P. (1973) Biochem. J. 132, 571–585.
112 Mitchell, P. and Moyle, J. (1965) Nature 208, 1205–1206.
113 Phelps, D.C. and Hatefi, Y. (1981) J. Biol. Chem. 256, 8217–8221.
114 Wu, L.N.Y. and Fisher, R.R. (1982) Biochim. Biophys. Acta 681, 388–396.
115 Earle, S.R., O'Neal, S.G. and Fisher, R.R. (1978) Biochemistry 17, 4683–4690.
116 Robillard, G.T. and Konings, W.N. (1982) Eur. J. Biochem. 127, 597–604.

Ernster (ed.) Bioenergetics
© 1984 Elsevier Science Publishers B.V.

CHAPTER 8

Metabolite transport in mammalian mitochondria

KATHRYN F. LANOUE and ANTON C. SCHOOLWERTH

Departments of Physiology and Medicine, The Milton S. Hershey Medical Center, The Pennsylvania State University, Philadelphia, PA. 17033, U.S.A.

1. Introduction

A steady flow of metabolites both in and out of the mitochondrial matrix space is necessary for mitochondria to perform functions which involve the participation of enzymes inside the membrane permeability barrier. These functions include oxidative phosphorylation and therefore O_2, ADP, phosphate and electron-rich substrates such as pyruvate, fatty acids and ketone bodies must enter the mitochondria, and the products, H_2O, CO_2 and ATP must leave. Although O_2, H_2O and CO_2 are permeable to the inner mitochondrial membrane [1,2], most metabolites are not, because of their highly hydrophilic nature. The outer mitochondrial membrane does not present a significant barrier to hydrophilic metabolites because of the presence of large unregulated channels composed of the membrane protein, porin [3]. The inner mitochondrial membrane has a much larger surface area [4] than the outer membrane and a much higher ratio of protein to lipid [5]. It is composed not only of proteins involved in electron transport and oxidative phosphorylation but also specialized proteins which facilitate, and in many cases provide, directionality to the transport of metabolites [6].

In some mammalian cells, enzymes comprising partial spans of biosynthetic pathways are inside and some outside the mitochondrial matrix space. Therefore, in the liver, six mitochondrial membrane transport proteins are required for urea synthesis, three for gluconeogenesis [7,8], and three others participate in ammoniagenesis [9] in the kidney. The synthesis of neurotransmitter substances such as acetylcholine, glutamate and γ-amino butyric acid requires the participation of metabolite transporters in mitochondrial membranes of nervous tissue [9,10].

To date, twelve transporters with different substrate specificities have been demonstrated in mammalian mitochondrial membranes (cf., Table 8.1). Plant mitochondria have a slightly different set [11]. Early studies of metabolite transport in mitochondria dating from the middle 1960s were concerned with identifying these transporters. More recently, research has been directed toward elucidation of molec-

TABLE 8.1

Mitochondrial metabolite transporters (Data is taken from Refs. 6, 12 and 19)

Category	Name	Physiological substrates	Inhibitors
Electroneutral proton compensated	Glutamate hydroxyl	Glutamate	*N*-Ethylmaleimide, avenaciolide
	Pyruvate	Monocarboxylic acids, ketone bodies, branched-chain ketoacids	α-Cyano-3-hydroxy-cinnamate, organic mercurials, *N*-ethylmaleimide
	Phosphate	Phosphate, arsenate	Organic mercurials, *N*-ethylmaleimide
	Ornithine	Ornithine, citrulline, lysine	
Electroneutral anion exchange	Dicarboxylate	Phosphate, malate, succinate, oxalacetate	Butylmalonate, bathophenanthroline, iodobenzylmalonate, phenylsuccinate, phthalonate, organic mercurials
	α-Ketoglutarate	Malate, α-ketoglutarate, succinate, oxalacetate	Phthalonate, bathophenanthroline, phenylsuccinate, butylmalonate
	Tricarboxylate	Citrate, isocitrate, phosphoenolpyruvate, malate, succinate	1,2,3-Benzenetricar-boxylate, α-acetylcitrate, bathophenanthroline
Neutral	Carnitine	Carnitine, acylcarnitine	Organic mercurials, *N*-ethylmaleimide, sulfobetaines
	Neutral amino acids	Neutral amino acids	Organic mercurials
	Glutamine	Glutamine	Organic mercurials
Electrogenic	Adenine nucleotide	ADP, ATP	Atractyloside, carboxy atractyloside, bongkrekate, long-chain acyl CoA, α-acetylcitrate
	Glutamate/aspartate	Glutamate/aspartate	Glisoxepide (non-specific)

ular mechanisms by which the proteins transport metabolites across the lipid bilayer, and estimations of the extent to which the mitochondrial transporters may limit flux through certain metabolic pathways. This chapter will emphasize recent studies of mammalian mitochondrial transporters while attempting to put these studies in perspective. Reviews of the literature on metabolite transport in plant [11] and mammalian mitochondria [1,6,12–14] have appeared in recent years. Likewise separate reviews of mitochondrial adenine nucleotide transport [15,16], phosphate transport [17] and pyruvate transport [18] have been published. The reader is referred to these for more thorough treatment of the older literature, and of the specific transporters.

2. *Identification of the transporters*

Most metabolites which are substrates of mitochondrial transporters are anions at physiological pH. Therefore, they would be excluded from the matrix space by the large negative electrical potential gradient generated by the operation of the electron transport chain if they were transported electrogenically as anions. The membrane proteins which catalyze the transport of metabolite anions into mitochondria avoid the thermodynamic difficulty by one of several strategies. We have chosen to classify the carriers according to the particular strategy employed to avoid transport against a large electrical gradient. The known mammalian mitochondrial transporters are listed in Table 8.1 according to our system of classification. Many of these transporters were first identified by swelling techniques [20]. Preliminary identification of some carriers was substantiated, and other carriers were first identified by direct measurement of uptake and efflux into the mitochondrial matrix using millipore filtration, direct centrifugation and centrifugal filtration through silicone oils in order to separate mitochondria rapidly from their media [21]. As interest in the field grew, inhibitors specific to some of the transporters were identified and these could then be used to establish the participation of the transporters in metabolic pathways.

The first category of Table 8.1 includes carriers which catalyze net uptake of ions in symport (or antiport) with protons which neutralize the anion's charge. These carriers include the phosphate carrier [22], the monocarboxylate carrier [23] which catalyzes the transport of pyruvate, acetoacetate and possibly branched chain ketoacids [24], and the glutamate carrier [25]. The ornithine carrier is also included in this category although ornithine is a cation at neutral pH. Experiments show that this carrier catalyzes net uptake of ornithine in exchange for a proton [26] making the overall process electroneutral. Recent data [27,28] suggest that uptake and efflux of ornithine is significantly stimulated by the presence of citrulline on the opposite side of the membrane. The ornithine transporter, possibly the same one that catalyzes electroneutral net uptake, appears to facilitate the 1 : 1 exchange of ornithine for citrulline and a proton.

A second category of carriers catalyzes 1 : 1 electroneutral exchanges of anions across the membrane. There have been three such transporters identified [6]: (1) the dicarboxylate carrier, (2) the malate/α-ketoglutarate carrier and (3) the tricarboxylate carrier. The substrates and inhibitors of these transporters are also shown in Table 8.1. The tricarboxylate carrier catalyzes the exchange of each of its substrates for any other. Since some substrates are tricarboxylic acids and some dicarboxylic acids, electrogenic exchange might be anticipated. However, direct measurements [29] of proton movements in conjunction with carrier activity show that net proton transport occurs equivalent to, and in the same direction as, the trivalent ion when tricarboxylates exchange with dicarboxylates. Therefore, the net exchange is electroneutral. Steady-state measurements of the relationship between pH and tricarboxylate gradients also provide convincing evidence of the electroneutral proton-compensated character of the tricarboxylate carrier [30].

A third classification includes those transporters which catalyze transport of metabolites which are neutral at physiological pH. These include the transporters for glutamine [31], and for the other neutral amino acids [32]. It also includes the one required for the oxidation of fatty acids which catalyze the exchange of carnitine for acyl carnitine [33]. This carrier also catalyzes net uptake of carnitine or acyl carnitines into the mitochondrial membrane, but at a rate only 1–2% of the exchange rate [34–36].

The last category includes the electrogenic or, perhaps more correctly, electrophoretic transporters. These transporters catalyze exchanges of metabolites which result in a net movement of electrical charge across the mitochondrial membrane but the charge movement is the one favored by the prevailing electrical potential. The adenine nucleotide carrier catalyzes the exchange of adenine nucleotides across the mitochondrial membrane. It is specific for ADP or ATP and does not transport AMP. Since the carrier catalyzes the exchange of ATP^{4-} for ADP^{3-} without a proton to neutralize the extra negative charge on ATP [37,38], the carrier involves the net efflux of one electron for each molecule of ADP entering the mitochondria in exchange for ATP. At equilibrium in the absence of net synthesis of ATP the following relationship has been demonstrated over a wide range of $\Delta\psi$ (see Refs. 39 and 15 for review).

$$RT \log \frac{(\mathrm{ATP/ADP})_{\mathrm{in}}}{(\mathrm{ATP/ADP})_{\mathrm{out}}} = F\,\Delta\psi \tag{1}$$

where R is the gas constant, T absolute temperature and F the Faraday. The relationship implies that the carrier is fully electrophoretic. The concept that the carrier is electrophoretic has been challenged recently [40,41] and this will be discussed more fully in Section 6.2.

A much slower transport process which catalyzes the net uptake of ATP has been demonstrated in liver [42,43] and heart mitochondria [44,45], and shown to be physiologically important in neonatal liver mitochondria [46,47]. The transporter's activity may be affected by hormones since it has been shown that glucagon treatment increases the total nucleotide content of subsequently isolated liver mitochondria [48]. Although initially thought to be due to rather nonspecific 'leakiness' of mitochondrial membranes or unidirectional transport on the adenine nucleotide carrier, its lack of sensitivity to membrane potential, and its almost absolute dependence on the presence of Mg^{2+} and phosphate suggest that the observed activity is due to a separate transporter [42]. Net nucleotide uptake in plant mitochondria is much more marked but has a similar dependence on Mg^{2+} and phosphate in the media [49].

The activity of this transporter which catalyzes net movements of adenine nucleotides is mentioned here in connection with the electrogenic adenine nucleotide exchange carrier, although it is likely to be a proton-compensated electroneutral exchanger. No counter ion has been identified, nor have compensating proton movements been reported in conjunction with the net nucleotide transport. Also, no influence of ΔpH on steady state total nucleotide levels has been reported.

The glutamate/aspartate carrier, like the adenine nucleotide exchange transporter, is electrophoretic [50]. Although it catalyzes the exchange of two monovalent anions with very similar pK's the exchange is electrophoretic because glutamate is transported with a proton, but aspartate is not. Since there is a large negative potential across the mitochondrial membrane under physiological conditions [51], entry of glutamate and efflux of aspartate from the mitochondria is virtually a unidirectional process [52]. In general, the process appears to be kinetically controlled but, using isolated mitochondria, the system has been allowed to come to equilibrium in the presence of various ratios of (ATP)/(ADP)/(P_i) to vary the energy level and the mitochondrial membrane potential [53]. Under these rather artificial conditions it was demonstrated that:

$$RT \log \frac{(\mathrm{Glu/Asp})_{\mathrm{in}}}{(\mathrm{Glu/Asp})_{\mathrm{out}}} = F\,\Delta\psi \qquad (2)$$

where R, T and F have their usual meanings. As one would expect, in intact hepatocytes under conditions so far studied this relationship does not hold [54].

3. *Distribution*

Not all the transporters discussed above are present in all types of mitochondria; the set of activities present in mitochondria depends on the functional needs of the cells from which the mitochondria are isolated. The adenine nucleotide and phosphate transporters are present in all mitochondria thus far studied. This reflects the fact that the major function of mitochondria is the synthesis of ATP. Even in the rare instances (e.g., brown fat mitochondria [55] and mitochondria in anaerobically growing yeast [56]) where the major function is not ATP synthesis, mitochondria normally have active adenine nucleotide transport. The pyruvate transporter also appears to be ubiquitous. The carnitine transporter has been studied in liver [57], heart [35] and sperm [58] and is probably present in all mitochondria which use long-chain fatty acids.

The α-ketoglutarate/malate exchange carrier and the glutamate/aspartate carriers also have a wide distribution. These two carriers are on the pathway of the malate/aspartate shuttle, which transports reducing equivalents from the cytosol into the mitochondria [6]. Reducing equivalents (NADH) are generated in the cytosol by glycolysis but NADH is impermeable to the mitochondrial membrane in animal cells [59]. Therefore, these two carriers are present in cells with an active glycolytic pathway and which use the cytosolically generated NADH for mitochondrial ATP synthesis. Fat cells do not metabolize reducing equivalents generated in the cytosol for ATP synthesis. Reducing equivalents are retained in the cytosol for synthesis of fatty acids from carbohydrate precursors. Thus, the transaminase pathway for glutamate is absent in white fat mitochondria [60]. In most plant tissues, NADH is permeable to the mitochondrial membrane [11] and the glutamate/aspartate carrier is absent. In these mitochondria, glutamate enters the mitochondrial membrane in exchange for dicarboxylic acids [61].

Some animal mitochondria also use the α-glycerophosphate shuttle to transport cytosolic-reducing equivalents to the mitochondria [62]. The substrate binding site of membrane-bound α-glycerophosphate dehydrogenase faces the cytosol. Therefore, operation of the α-glycerophosphate shuttle requires no substrate transport. In liver, both shuttles are present [63]. However, the malate/aspartate shuttle appears to be most active in the majority of tissues, including tumors [64]. It is likely that the electrophoretic exchange of glutamate for aspartate provides the energy source for the transport of reducing equivalents. An energy source is necessary because the electrical potential of the NADH/NAD couple in the cytosol is about 60 mV more positive than the potential in the mitochondrial matrix [65].

On the other hand, many transporters have very limited distribution. The proton-linked glutamate carrier, sometimes called the glutamate hydroxyl carrier, is active only in liver mitochondria, where its activity is linked to the synthesis of urea and glucose from glutamate [7]. When glutamate is the source of NH_3 for the synthesis of urea, glutamate is metabolized via glutamate dehydrogenase and must be transported into the mitochondria on the glutamate hydroxyl carrier.

In kidney, glutamate dehydrogenase activity is high but the activity of the glutamate hydroxyl carrier is low compared to liver [66,67]. In kidney, glutamate dehydrogenase flux is regulated according to the needs of the organism for ammoniagenesis during metabolic acidosis. The source of ammonia is glutamine and the two intramitochondrial enzymes, glutaminase and glutamate dehydrogenase, are important control sites for ammoniagenesis [9]. Physiologically, glutamate for deamination is generated inside the mitochondria from glutamine, and therefore there is no requirement for high activity of the glutamate hydroxyl carrier in kidney [68]. Indeed, if extramitochondrial glutamate had access to glutamate dehydrogenase it would alter the fine control of ammoniagenesis necessary for acid base balance (c.f., Section 6.4).

The dicarboxylate carrier also has a rather restricted distribution and is present only in gluconeogenic tissues where intramitochondrial malate must be exported into the cytosol for eventual conversion of its carbon atoms to glucose [69]. Malate can be transported out in exchange for phosphate going in. Only the dicarboxylate carrier catalyzes exchange of phosphate for carboxylic acids. Export of malate on another anion exchange carrier such as the α-ketoglutarate or citrate carrier would result in no increase in cytosolic carbon precursors for glucose.

The tricarboxylate carrier also has a rather limited distribution. It is most active in those tissues which synthesize fatty acids [70]. The pathway of fatty acid synthesis from pyruvate involves synthesis of citrate in the mitochondria followed by citrate efflux from the matrix space [71]. Cytosolic citrate is metabolized to cytosolic acetyl CoA by citrate lyase. However, the activity of the tricarboxylate carrier is low in heart, muscle and brain mitochondria.

In summary, the carriers present in different mitochondria determine not only what metabolites are able to move through the mitochondrial membrane but also, due to the specificity of proton cotransport, the direction of metabolite movement. Whether they are also able to modulate fluxes is a matter of current debate.

4. Biosynthesis and insertion into the membrane

Biosynthesis has been studied only in the case of the adenine nucleotide exchange carrier, which has been isolated in pure form from several sources [72–74]. Antibodies have been raised to some of these pure proteins [75], and the availability of antibodies as well as the availability of inhibitors [76] which bind specifically to the nucleotide carrier have permitted studies of its biosynthesis not possible for the other carriers. The immunological work is important since the antibodies can detect the carrier in precursor forms, not yet assembled in the membrane.

An interesting side aspect of the immunological studies is the observation that the antibodies to the exchange carrier isolated from beef heart do not crossreact with the transporter from beef liver [77]. Nevertheless, beef heart adenine nucleotide carrier antibodies crossreact with heart adenine nucleotide carrier isolated from both rat and rabbit. This suggests the presence of different 'isozymes' of the carrier in different organs and raises the possibility that there may be several nuclear genes for the carrier in each species acting as templates for the 'isozymes' of different organs.

The biosynthesis of the adenine nucleotide carrier has been studied most extensively in *Neurospora crassa*. The carrier isolated from *Neurospora* mitochondria is very similar to the carrier isolated from mammalian mitochondria [78]. Its molecular weight, subunit structure, amino acid composition, hydrophobicity and inhibitor specificity are remarkably similar to the mammalian heart and liver carriers. Specific antibodies to the *Neurospora* carrier have been raised in rabbits [79].

Although a few subunits of mitochondrial membrane proteins are coded by mitochondrial DNA and synthesized in the mitochondrial matrix, most membrane proteins including the adenine nucleotide carrier are coded by nuclear genes and synthesized on cytoplasmic ribosomes [80,81]. Chloramphenicol, an inhibitor of mitochondrial protein synthesis, does not inhibit incorporation of radioactive leucine into the carrier in growing *Neurospora crassa*, but cycloheximide, an inhibitor of cytoplasmic protein synthesis, does inhibit leucine incorporation [78]. Also, a yeast nuclear respiratory mutant has been shown to cause a defect in adenine nucleotide transport [81], and the nuclear gene responsible for coding the carrier in yeast is currently being cloned for further studies [82].

A pool of adenine nucleotide carrier precursor protein has been identified in the cytoplasm of *Neurospora crassa* by means of immunoprecipitation techniques [83]. The precursor protein can be synthesized in cell-free lysates of *Neurospora*, or in cell-free lysates of reticulocytes if *Neurospora crassa* RNA is added, and the precursor protein can be identified by immunoprecipitation from these cell-free systems. It has been demonstrated that the precursor is released from the ribosomes into free solution.

The mode of insertion of the free precursor protein into membranes is not clearly understood. Some mitochondrial membrane proteins are synthesized as higher molecular weight precursors and are inserted into or through membranes where endogenous proteases clip the 'leader sequence' to form the mature membrane protein [84]. In contrast, the adenine nucleotide carrier precursor has the same molecular weight and N-terminal amino acid sequence as the mature protein [83]. The precursor evidently has a conformation which differs somewhat from the final form. The precursor does not bind carboxyatractyloside, whereas the carrier in the membrane, and in Triton X-100 extracts of the membrane, does. Insertion into the membrane in the functional conformation requires the presence of an electrical potential gradient across the membrane [85]. In the absence of an electrical potential, the precursor binds to specific sites on the *Neurospora crassa* mitochondrial membrane and remains bound without conversion to the functional carboxyatractylate binding form. Addition of respiratory substrate to the mitochondria with bound precursor protein produces an electrical potential across the membrane and conversion of the precursor to its functional carrier form [86].

5. Molecular mechanism

Although transport of metabolites and ions through membranes has been actively studied for several decades, the molecular mechanism by which carrier proteins catalyze the movement of hydrophilic metabolites through natural lipid bilayers has never been elucidated. A thermodynamically reasonable model mechanism has been proposed by Singer on theoretical grounds [87]. According to this model, the carrier may be composed of two subunits. Both subunits span the membrane and interact with each other. The substrate binding site is in a crevice between the subunits. The subunits can 'rock' against each other, changing sites of direct protein-protein interaction. In this way, the crevice might be altered and in one configuration exposed to the outer side of the membrane, and in a second configuration exposed to the inner side. Since the substrate is bound in the crevice, the 'rocking' phase corresponds to the movement of the metabolite through the lipid bilayer. A change in configuration of the crevice could in principle result in a change in the binding constant of the substrates within the crevice. However, in that case, a conformational change of the substrate carrier complex from the tighter binding form to the looser binding form would not proceed spontaneously in the absence of a source of energy [88]. The conformational change may occur only when substrate is bound to the carrier as in the case of the obligatory exchange transporters, or alternatively, the rate of the change of conformation may be a function of substrate binding so that net transport is possible but exchange more rapid.

Studies of the molecular mechanism of transport of the carriers of the mitochondrial membrane can be divided into two categories. One includes kinetic studies and the other includes structural studies, which rely either on the availability of pure isolated carrier protein or involve careful and detailed studies of substrate specificity and the effects of minor modifications of substrate structure in an attempt to specify the structural requirements of the 'catalytic' site.

Although some kinetic studies have provided evidence consistent with the Singer model [87], others have been more equivocal. Insofar as the availability of pure carriers have permitted data collection, the structural studies have been consistent with the Singer model.

"The site that was out
Becomes in without doubt.
Conformations are changed
And sites rearranged.
Done exceedingly quick
Like a magical trick.
The kinetics are right
For defining the site,
And the odds are all long
That the mode is ping-pong.
It's a model supreme
That wise heads did dream;
It stands to the test
And so far it's best."

Rothstein (1980)

5.1. Kinetic studies

The above poem [89] reflects the notion that transporters operating according to the Singer model should exhibit ping-pong kinetics. Ping-pong kinetics are observed in

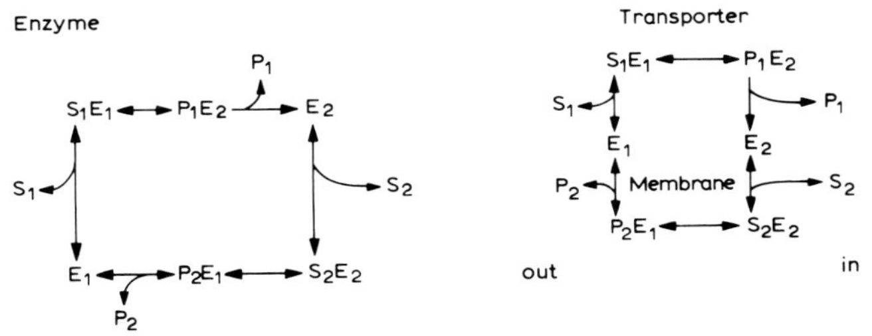

sequential enzymatic reactions of the transaminase type [90]. In the case of transaminase, E_1 and E_2 are the pyridoxal and pyridoxamine forms of the enzyme, whereas in the case of a transporter, E_1 and E_2 correspond to inward or outward facing configurations of the carrier. In the scheme depicted, the carrier catalyzes the exchange of substrates S_1 and S_2 across a membrane, and S_1 and P_1 are the same metabolite but on opposite sides of the membrane. The key to obtaining ping-pong kinetics is the irreversibility of the reaction generating the free E_1 or E_2 form of the enzyme. Ping-pong kinetics can only be generated by measuring true initial rates because significant accumulation of P_1 causes reversal of E_2 to E_1 without reaction with S_2 or production of P_2.

Ping-pong kinetics would be expected, not only for carriers catalyzing exchange of two substrates but also for transporters which catalyze net transport of a single substrate (S) such as the phosphate, pyruvate and glutamate hydroxyl carriers, if the return of the free carrier to the original side of the membrane is rate limiting (i.e., $E_2 \rightarrow E_1$) and one examines the kinetic consequences of an effect which changes the rate of $E_2 \rightarrow E_1$.

In this case, the reaction sequence is:

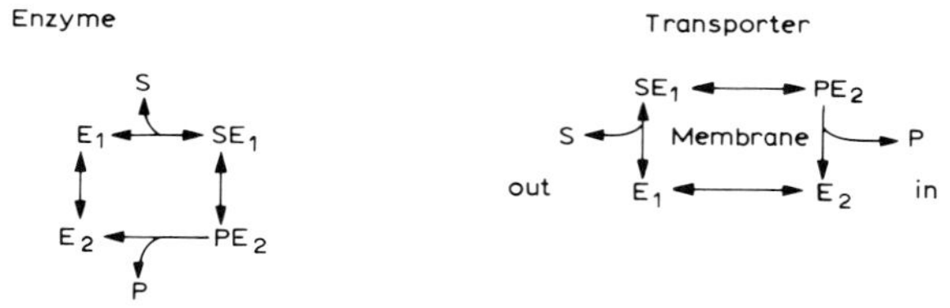

If the ratio of K_m/V_{max} for substrate (S_1) remains constant when K_m and V_{max} change due to a change of concentration of substrate (S_2), the mechanism is judged to be ping pong [90]. For a single substrate carrier transported via the Singer model, a change in the 'rate-limiting' rate of return of the free carrier to its original conformation should produce proportional changes of K_m and V_{max}. On the other hand, a manipulation which alters the rate of release of product (P) from the carrier, thus speeding transport, should produce a change of V_{max}, but not of the K_m, of S.

Studies of the mitochondrial transporters have been hampered from the outset by technical difficulties inherent in measuring initial rates of uptake of metabolites from the large volume of external media into the very small volume of the mitochondrial matrix space. A number of methods are currently in use. Changes in light scattering due to mitochondrial swelling in high concentrations of permeant compounds were used in early studies to identify different carrier systems [20]. This method is a qualitative one and cannot be used to obtain precise kinetic constants needed for tests of molecular models. Centrifugal filtration is used to measure kinetics of those transporters with relatively slow kinetics, since each sample can be taken no faster than 15–20 s apart, and must be pipetted over silicone oil and then centrifuged to quench the reaction. The method most frequently used in kinetic studies is the inhibitor stop technique [21]. Using this technique, a specific rapid inhibitor of the transporter is added to the mitochondria at various time intervals after substrate. In principle, with an adequate inhibitor, time resolution is limited only by the techniques used for additions and mixing. Using automated stop-flow equipment, time resolution in the millisecond range is possible. The method, however, is limited to those transport processes which can be rapidly and completely inhibited in the necessary time frame.

It is important, but technically difficult in many cases, to obtain unequivocal initial rates. However, product inhibition and reaction reversal are difficult to avoid in transport studies when measuring uptake into a very small space. The classical studies of K_m and V_{max} of mitochondrial transporters, quoted extensively in review articles, have been obtained not from initial rates, but from measurements of the first order approach to an equilibrium value for the internal concentration of permeable metabolites [91–94].

According to this method, "the first order rate constant of uptake" is measured by plotting ln $C_{eq}/C_{eq} - C_i$ vs time, where C_{eq} is the concentration of metabolite at steady state and C_i is the concentration at time, t. The slope of the line obtained is the rate constant and this, multiplied by C_{eq}, is the initial rate of entry of the metabolite C.

When the mathematics are derived for a metabolite such as glutamate or pyruvate which experience net uptake, it is apparent that k is the first order rate constant of efflux of C from the mitochondria. At equilibrium, the rate of uptake is equal to the rate of efflux and efflux is equal to k (C_{eq}). It is assumed that the rate of uptake is a constant with time and efflux of C is first order [67,95]. In contrast to the situation when measuring net uptake or exchange of two different metabolites, the mathematics are straightforward and few assumptions need to be made when analyzing isotopic exchange of a single metabolite at equilibrium. However, under other conditions the method should not be applied in a general way, as it has been, without testing the assumptions. An assumption made in using the method to obtain the exchange rate of different metabolites across the membrane is that uptake of the external metabolite is either a first order function of the internal metabolite, or that the internal face of the carrier cannot distinguish between the two metabolites.

5.1.1. Proton co-transporters

The pyruvate, glutamate and phosphate transporters catalyze net uptake and release of their substrates with stoicheiometric amounts of protons [6]. Early evidence for the electroneutrality of the process was the good inverse correlation between the H^+ gradient across the mitochondrial membrane and the gradients of these permeant anions, especially at equilibrium and at low metabolite concentrations [96,97]. At equilibrium the rate of inward transport should equal the rate of efflux and the distribution of permeant anion should be proportional to the ΔpH since:

$$k(C_e^{n-})(H_e^+)^n = k(C_i^{n-})(H_i^+)^n \quad (3)$$

$$\log(C_i^{n-})/(C_e^{n-}) = n\,\Delta\text{pH} \quad (4)$$

where k = rate constant of the carrier; n = anionic charge; C_e^{n-} = external concentration of the anion; C_i^{n-} = matrix concentration of the anion; H_e^+ = external hydrogen ion concentration; and H_i^+ = matrix hydrogen ion concentration.

It has been shown experimentally that relationship (4) is maintained whether the internal or external pH is varied in order to change ΔpH. Relationship (4) could also be derived by assuming that the free acid is completely permeable and the anion impermeable to the membrane. At equilibrium the concentration of free acid should be equal inside and out, and the amount of internal and external anions determined by pK_a and ΔpH.

However, since the proportion of most metabolites present as free acid at pH 7 is low, it is likely that the carriers have separate binding sites for protons and anions. The relationships between proton concentrations, substrate concentrations and rates of transport have been examined in order to gain insight into the molecular mechanisms of transport for the glutamate, pyruvate and phosphate carriers.

5.1.1.1. Glutamate transporter. Uptake kinetics have been measured in rat liver mitochondria using the inhibitor stop technique at temperatures varying from 20 to 28 °C [98–100]. At 1 mM glutamate, well below the measured K_m of 4 mM, the transport exhibits an acid pH optimum between 6 and 6.5. Since V_{max} measured by swelling is not stimulated by protons [101], it can be assumed that K_m is lowered by H^+. However, no systematic study has been made of the effect of external pH on the V_{max} and K_m of glutamate uptake using more quantitative methods of measuring transport. There is good agreement about K_m, at pH 7.0, and fair agreement about the value for V_{max} (between 10 and 20 nmol · min^{-1} · mg^{-1} at 25 °C) among most workers [95,98,99]. One group [100] has obtained much higher values for V_{max} using very early (4 s) initial rates. However, a later study by the same group shows that the inhibitors used to stop transport in the first 4 s do not react rapidly and completely with the transporter in that time span. Recent detailed studies [95] of the carrier, which agree with the two earlier studies with regard to the V_{max} of uptake, have been done without inhibitors using centrifugal filtration. This is possible at 28 °C because the transporter is slow. In these studies it was found that variations of pH between 8.5 and 7.5, at constant external pH, resulted in stimulation of inward transport by H^+ in the matrix, while the ratio of K_m/V_{max} remained constant. This suggests that the return of free carrier to a configuration facing the outside of the membrane is optimal at neutral rather than alkaline pH, and that the return to the outward facing configuration in the free form is rate limiting for transport. This was compatible with the observation that glutamate/glutamate exchange is 3–4-times faster than net uptake. Other studies [95,102,103] of the effect of matrix pH on glutamate efflux have been carried out at both 15 and 28 °C. Two laboratories [95,103] have shown that matrix H^+ decreases the K_m of glutamate in the matrix, in symmetry with its proposed effect on the external face of the carrier. The pK of this effect on glutamate binding in rat liver mitochondria was estimated to be 6.2.

The effect of external pH on glutamate efflux has also been studied in both liver and kidney mitochondria [103,104]. External protons decrease the V_{max} of efflux without altering the K_m of matrix glutamate. If protons are viewed as preventing the return of the free carrier to the inward facing configuration, the data are at variance with the Singer model. However, if protons prevent the release of product (glutamate) on the external face, the data can be reconciled with a sequential model. If glutamate remains bound to the carrier until the transported proton is released, the carrier could reverse its configuration without achieving net transport, and the effect of an increase in external H^+ would be a decrease in the V_{max} of efflux without effecting the K_m of matrix glutamate.

The transporter has much lower activity in kidney [66], but the kinetic patterns are similar to those observed with liver mitochondria. Prevention of glutamate efflux from mitochondria by H^+ in kidney tubular cells is probably a very important aspect of control of ammoniagenesis in kidney (c.f., Section 6.4).

5.1.1.2. Pyruvate transporter. Kinetic studies of the pyruvate transporter have been hampered by technical difficulties [5]. Numerous and conflicting kinetic parameters have been reported [105–107]. This is because (1) the transport is faster than the glutamate carrier [108], (2) pyruvate binds nonspecifically to mitochondrial proteins [109], and (3) the quench rate due to compounds used for inhibitor stop depends on specific reaction conditions and are not fast enough to allow the necessary kinetic resolution in seconds [106]. Also, (4) pyruvate is notoriously unstable [110] and (5) at mM concentrations pyruvate can be transported across the mitochondrial membrane by diffusion without carrier mediation [105]. These problems have been meticulously documented by Vaartjes et al. in a recent study [110]. It was concluded that the inhibitor stop technique using cyanocinnamate is unreliable, and that α-cyano-3-hydroxycinnamate is a faster, more potent reagent but still not suitable for time resolution in the 5–10 s range. Centrifugal filtration appeared to be the method of choice but since time resolution is poor, experiments can only be performed at very low temperatures.

Halestrap has suggested a way of measuring carrier-mediated transport at higher temperatures [106,111]. Uncoupled pyruvate respiration is measured as a function of the concentration of α-cyano-3-hydroxycinnamate. Since the transporter is not rate limiting for respiration, especially above pH 7.0 [106,112], low concentrations of inhibitors do not inhibit respiration but inhibition is linear at higher concentrations. If the reciprocal of respiration is plotted as a function of inhibitor concentration, the resulting Dixon plot extrapolated to zero inhibition provides a measure of transport in the absence of inhibition. Using this method Halestrap demonstrated that the V_{max} of transport is stimulated by an alkaline matrix pH, and he therefore proposed that release of the proton from the carrier inside the matrix is rate limiting for pyruvate influx. This is in contrast to the results with the glutamate transporter. The more acidic pK of the proton binding site of the glutamate carrier may favor release of the proton following transport. The rate of pyruvate uptake (42 nmol · min^{-1} · mg^{-1}) measured at 37 °C and pH 7, using 1 mM pyruvate, is compatible with those measured by Vaartjes et al. [110] using centrifugation at 27 °C (23 nmol · min^{-1} · mg^{-1} at pH 6.8). Additionally the studies of Papa and Paradies [113] and of Brouwer et al. [101] show that lowering the pH of the external media stimulates transport by lowering the K_m for pyruvate as in the case of glutamate. Efflux is also inhibited by H^+ in the external media but the apparent pK (8.3) for this effect is in a much more alkaline region than for the glutamate carrier.

The K_m and V_{max} of exchange transport have been measured and the major laboratories obtain fairly consistent values [5,18]. Pyruvate exchanges for monocarboxylic acids including acetoacetate and possibly the branched chain keto acids [24]. It has been demonstrated in perfused rat livers that exchange of internal acetoacetate for external pyruvate is important in the control of ketogenesis [114]. The measured V_{max} of exchange appears to be independent of the nature of the metabolites exchanged. At 4°C and pH 7.0, the value is 16 nmol · min^{-1} · mg^{-1} and the K_m for external pyruvate is 0.11 mM, whereas the V_{max} of net uptake is 0.6 nmol · min^{-1} · mg^{-1} [18].

Careful analysis of the effect of the concentrations of the counter ions on transport parameters K_m and V_{max} have not been performed as a test for ping pong kinetics. However, one can compare K_m and V_{max} of net uptake with the exchange transport parameters at 4°C. Since V_{max} is about 20-fold lower for net uptake, the measured uptake K_m should be proportionately lower if the kinetics were ping pong. This is clearly not the case since reliable estimates of K_m for net transport vary between 0.1 and 0.6 mM [106,107,115–117].

5.1.1.3. Phosphate transporter. Mitochondrial phosphate transport is the fastest of the mitochondrial transport processes. The speed of the process has precluded detailed kinetic studies although phosphate/phosphate exchange kinetics have been measured by the inhibitor stop technique using an automated timing device in rat liver mitochondria with *N*-ethylmaleimide as the inhibitor [118]. The V_{max} at 0°C was 205 nmol · min^{-1} · mg^{-1}, and the K_m, 1.6 mM at pH 7.4. Although effects of pH and substrate concentration on net transport have not been studied by direct methods, conclusions can be drawn from studies of equilibrium distribution. Studies of the distribution of phosphate at equilibrium at different medium pH values, as a function of external phosphate, show that a maximum internal phosphate concentration is achieved at lower external phosphate when the H^+ ion concentration of the media is increased [119]. This suggests that the K_m of phosphate is lowered by H^+. The conclusion is strengthened by recent studies of the reconstituted isolated phosphate transporter [120–125]. In reconstituted liposomes phosphate efflux was stimulated over 10-fold by intraliposomal H^+ (pH 6.8 vs 8.0) when the phosphate concentration was below the K_m (2.3 mM). A similar increase in the H^+ of the external media caused only a slight (25%) inhibition of efflux [125].

An interesting study of the substrate specificity of the carrier utilizing swelling techniques has been carried out by Frietag and Kadenbach [126]. The study shows that monofluoro, but not difluorophosphate, can be transported on the phosphate transporter in an *N*-ethylmaleimide sensitive way. This suggests that the substrate of the carrier is divalent phosphate, in which case two proton binding sites on the carrier would be required to maintain the electroneutrality of transport. Data from a number of laboratories suggest, however, that under special conditions phosphate may be transported electrophoretically down an electrochemical potential gradient without the accompanying protons [127–130].

A pattern emerges from kinetic studies of the glutamate, pyruvate and phosphate transporters, all of which catalyze proton symport. Proton binding to the carrier lowers the K_m of the metabolite on the same side of the membrane. The glutamate data suggest that this is true on both the cytosolic and matrix sides. However, proton binding to the carrier on one side of the membrane inhibits initial rates of transport from the opposite face. Since this is due to a decrease in V_{max} without a change of K_m on the opposite side, it is tentatively concluded that the proton prevents release of 'product' from the carrier. This would obviously decrease the K_m for transport on the same side of the membrane.

5.1.2. Neutral exchange carriers

5.1.2.1. α-Ketoglutarate / malate carrier. Some years ago the V_{max} and external K_m of this carrier were measured using rat liver mitochondria [94]. The activity in liver is high, and in order to obtain good kinetic resolution the whole time course of the approach to equilibrium was used, rather than an initial rate. The mitochondria were preloaded with an unreported amount of malate, and phenylsuccinate was used to quench transport rapidly. The half time ($T_{1/2}$) to equilibrium (about 15 nmol · mg^{-1} protein of intramitochondrial α-ketoglutarate) was 32 s at 9 °C. A double reciprocal plot of the rate vs external α-ketoglutarate concentration was linear in the range 0.01–0.33 mM α-ketoglutarate. The V_{max} at 9 °C was 43 nmol · min^{-1} · mg^{-1} and the K_m of α-ketoglutarate was 46 μM.

More recently, extensive kinetic studies of the same carrier have been carried out by Sluse and co-workers [131–133] using rat heart mitochondria. The authors have measured the K_m's of the internal substrates as well as the effect of the internal substrate concentration on the V_{max} and the external K_m's. Early experiments which employed a narrow range of substrate concentrations demonstrated a kinetic pattern incompatible with the ping-pong kinetics expected from a simple gated-pore model [129,130]. The calculated K_m values of the internal substrates were 1.48 mM (α-ketoglutarate) and 3.60 mM (malate). These values were found to be independent of external substrate. The K_m's of the external substrates (α-ketoglutarate and malate) were also independent parameters and V_{max} was 1–2 nmol · min^{-1} · mg^{-1} at 4 °C. Later studies [133] using phenylsuccinate as the quenching agent for the inhibitor stop method have included a much wider range of internal and external substrate concentrations. Only the exchange of α-ketoglutarate$_{(out)}$ for malate$_{(in)}$ was studied. The previous conclusions concerning internal malate were verified, since changing the concentration of external α-ketoglutarate did not change the observed K_m of internal malate, although a lower K_m for malate (1.3 mM) was obtained than in the previous studies. Significant deviations from the expected rates (assuming K_m of 1.3 mM for internal malate) were obtained when external α-ketoglutarate levels over 60 μM were employed. It is, moreover, disturbing that when the external α-ketoglutarate levels were varied over a wide range at constant internal malate, large deviations from simple Lineweaver-Burk saturation curves were obtained. The kinetic behavior was so complex that it defied interpretation. The authors suggest that there may be several substrate and regulatory sites for α-ketoglutarate on the external face of the carrier.

However, data obtained at higher temperatures also using rat heart mitochondria [134] suggest that the complexity of the kinetic pattern may be an artifact of the temperature or the techniques employed. At 28°C mitochondria metabolizing glutamate, generate a constant internal concentration of α-ketoglutarate (0.4 mM), and α-ketoglutarate efflux is a function of the added external malate. Initial rates of α-ketoglutarate efflux from the mitochondria were measured over a wide range of malate concentrations. A K_m of 0.2 mM for external malate and a V_{max} of 50 nmol $\cdot$ min^{-1} $\cdot$ mg^{-1} at 28°C were obtained, and double reciprocal plots of efflux of α-ketoglutarate vs external malate concentrations did not deviate significantly from linearity. The discrepancies in the two sets of data may be methodological or may relate to the differences in temperature.

5.1.2.2. Other electroneutral exchange transporters. Although it may be tempting to discount the Sluse studies due to the kinetic complexity of the data obtained and obvious technical problems, other workers studying exchange transporters have concluded that substrate binding sites on either side of the membrane are independent of each other and that the ratio K_m/V_{max} is not a constant. Variation of the second substrate concentration, on the opposite side of the membrane usually produces a V_{max} change but no change of K_m for the first substrate. Data of this pattern have been observed for the dicarboxylate carrier [135] and for the carnitine-acyl carnitine exchange carrier. At least in ox heart mitochondria the carnitine-acyl carnitine exchange carrier appears to have symmetrical, but independent, binding sites with a K_m of ~5 mM on both sides of the membrane [136]. However, one worker [36] studying the kinetics of carnitine transport in rat heart mitochondria, found ping-pong behavior when the internal concentration of carnitine is varied from 0.6 to 1.5 mM. Although the authors did not point out the significance of the data, calculation of the ratio of the K_m's to V_{max}'s measured over this rather narrow range of internal carnitine levels showed that the ratio was a constant.

5.1.3. The electrogenic carriers

Deviation from expected kinetic behavior for the gated pore model has also been observed in the case of the two electrogenic transporters, the glutamate/aspartate and the adenine nucleotide carriers.

5.1.3.1. Glutamate/aspartate carrier. Aspartate from the external media does not penetrate the mitochondrial membrane readily and, therefore, mitochondria cannot be directly loaded with aspartate [137]. This hampered the study of the kinetics of the glutamate/aspartate exchange transporter until an indirect procedure was found for loading liver mitochondria with high levels of aspartate [138]. This is accomplished by incubating glutamate-loaded mitochondria with oxalacetate. Intramitochondrial oxalacetate transaminates the glutamate to aspartate. The aspartate remains inside the mitochondrial matrix in the absence of external glutamate and, on addition of external glutamate, the aspartate leaves the mitochondria in 1:1 exchange with glutamate. The rate of the process can be studied using centrifugal filtration through silicone oil, and the measured disappearance of aspartate from the matrix as a function of time after glutamate addition has been used to estimate intramitochondrial aspartate K_m's [102,139]. This is possible in a single time course since the exchange is unidirectional, and product inhibition by intramitochondrial glutamate is negligible [52]. Sensitive measurements of pH changes during exchange of glutamate for aspartate have shown that equivalent amounts of protons enter the mitochondria with glutamate rendering the overall exchange electrogenic [140]. This provides an explanation for the well-documented observation that formation of aspartate during glutamate metabolism is inhibited by uncoupling agents which decrease $\Delta\psi$ in liver and heart mitochondria [134,141]. In the presence of a physiologically high electrical membrane potential ($\Delta\psi \sim 180$ V) the forward exchange is rapid and the rate of the reverse exchange (aspartate$_{(in)}$, glutamate$_{(out)}$) is negligible [52]. When the membrane potential is collapsed with uncoupling agents or valinomycin in the presence of high K^+, the forward exchange rate is decreased 10-times or more, but the reverse exchange is only slightly stimulated [52].

Detailed kinetic studies have been performed in an effort to gain information about the molecular mechanism of transport. Williamson and co-workers [102,139] have found that increases of external glutamate and $\Delta\psi$ increase V_{max} but do not influence the K_m of internal aspartate. Williamson suggests that despite the notable absence of ping-pong kinetics, the data are compatible with a sequential, gated-pore model if there are two substrate sites on the inner surface of the carrier, and if aspartate binding to one internal site facilitates the release of glutamate from the other. The suggestion is compatible with the existence of two possibly identical subunits exhibiting negative cooperativity.

Williamson's experiments were carried out in the presence of transaminase inhibitors, and his kinetic parameters are inconsistent with rates of aspartate transport measured under more physiological conditions, using isolated mitochondria [134,140,142]. Although consistent with fluxes and metabolite levels measured in hepatocytes under a limited range of conditions [102,139], examination of a broader range of conditions reveals order of magnitude discrepancies between predicted and observed rates [143].

Studies performed by the authors of this review [142,144], suggest that the discrepancies are due to functional microcompartmentation between the aspartate aminotransferase and the aspartate transporter. The apparent K_m for aspartate efflux can be dramatically decreased by generation of intramitochondrial aspartate by the aminotransferase reaction. Detailed isotopic studies using labelled matrix aspartate in liver mitochondrial [142] and labelled intramitochondrial glutamate in kidney mitochondria [144] confirmed the initial suggestion.

In contrast to the result of Williamson [139], the present authors find that efflux of aspartate from the mitochondria is a first order process. Since no saturation of the carrier can be detected, determination of effects of $\Delta\psi$ and external glutamate on the K_m of matrix aspartate cannot be measured, even at the highest levels of aspartate loading possible.

In order to circumvent the problem, kinetics of transport were measured in glutamate-loaded submitochondrial particles from liver mitochondria, where the matrix surface of the membrane is exposed to the external media. These data were consistent with a ping-pong mechanism but not sufficiently precise to allow a firm judgement [145–147].

It is apparent that methodological problems limit the usefulness of the kinetic approach to studying mechanisms. However, data of an entirely different nature argue against a sequential model for the aspartate transporter. Studies of the metabolism of cysteine sulfinic acid show that it is transported by the glutamate/aspartate transporter and that it transaminates with α-ketoglutarate or oxalacetate to yield glutamate or aspartate and β-sulfinyl pyruvate, which spontaneously hydrolyzes into sulfite and pyruvate [148,149].

When mitochondria are loaded with glutamate, cysteine sulfinic acid exchanges with glutamate, and efflux of protons can be observed. No proton movements are observed in conjunction with the cysteine sulfinic acid/aspartate exchange. Therefore, cysteine sulfinic acid must be transported electrogenically as the anion.

When mitochondria are incubated with cysteine sulfinic acid and malate, malate is rapidly oxidized, and aspartate production occurs as rapidly as in the presence of glutamate and malate. Collapse of $\Delta\psi$ with uncoupling agents inhibits aspartate production in the presence of glutamate, as noted above, but aspartate formation is not inhibited by uncouplers in the presence of cysteine sulfinic acid. Viewed superficially this is not surprising, since the exchange of one monovalent cation for another is electroneutral. However, if the process is sequential, the exchange is composed of two electrogenic steps, not one electroneutral step. Therefore, the lack of influence of $\Delta\psi$ is puzzling, though the data could be explained by a concerted mechanism.

5.1.3.2. The adenine nucleotide carrier. The kinetics of the adenine nucleotide carrier have great physiological relevance since it is likely that the kinetic parameters of the carrier are major determinants of the rate of in situ cytosolic ATP generation. Accordingly the rate of transport using intact mitochondria has been studied extensively. Most kinetic studies employ isotopically labelled substrates of the carrier, and measure uptake or release of isotopes from the mitochondrial matrix. Some workers attempt to obtain initial rates directly from very early time points [150,151]; others use a longer time course and assume that there is a smooth first order approach to an equilibrium value [152]. Problems may be encountered in attempting to interpret data from both types of studies. Direct accurate measurements of initial rates are almost impossible to achieve because the transport is rapid, the exchangeable pools are small, and because the inhibitor stop technique inhibits at variable rates dependent on conditions of energization and substrate concentration [153].

The other kinetic method [153], in which the rate constant of the first order approach to equilibrium is determined, provides ambiguous results because of uncertainty in the size of the exchangeable pool of nucleotides. The natural log plot of the percentage equilibrated is not linear regardless of the value chosen as 100% equilibrated. This is probably due to heterogeneity of the exchangeable intramitochondrial pool. The intramitochondrial adenine nucleotide pool is composed of ATP, ADP and AMP. AMP is not a substrate of the carrier, yet radioactive ATP and ADP on entering the mitochondria can be converted to AMP, thus providing a slow, artifactual increase in the apparent size of the exchangeable pool. The two intramitochondrial substrates, ATP and ADP, are transported at different rates [152,154] and, therefore, should not be treated as a single pool. Additionally, Ca^{2+} and Mg^{2+} form complexes with ATP and ADP which are not substrates of the carrier [37,155]. Therefore, the free Mg^{2+} concentration and free Ca^{2+} concentrations (both of which are difficult to measure) may influence both transport and the size of the rapidly exchanging pool [45,154,156].

Despite these methodological problems certain generalities about transport can be made because they have been repeatedly observed by different laboratories using a variety of techniques [15,16,152]. The only physiological substrates of the carrier are ATP and ADP. The rate of the $ADP_{(out)}$, $ATP_{(in)}$ exchange is faster than the reverse process in the presence of a high $\Delta\psi$, but not when $\Delta\psi = 0$. The K_m of external ADP is very low, in the range 1–10 μM and is relatively independent of $\Delta\psi$. In the absence of a significant $\Delta\psi$, the K_m's for ATP and ADP are similar but the K_m of ATP increases with respect to ADP when $\Delta\psi$ is increased. The magnitude of this effect is rather variable as measured by different laboratories ranging from 10- to 100-fold [150,157,158].

Recently, the effect of internal substrate level has been systematically studied at different concentrations of external nucleotide using intact rat liver mitochondria [150] and rat heart mitochondria [151]. Ping-pong kinetics were not observed in either study. In both cases, initial rates of uptake of radiolabelled substrates were measured at 2°C. Non-linear biphasic Lineweaver-Burk plots, obtained by Barbour and Chan [150] using rat liver mitochondria, provided K_m values for two classes of binding sites on the external face. Decreasing the matrix exchangeable nucleotides decreased V_{max} but did not change K_m. On the other hand, studies of rat mitochondrial adenine nucleotide transport carried out by Sluse and co-workers [151] demonstrated monophasic kinetics. In contrast to the results of Barbour and Chan [150], lowering the internal substrate increased the K_m of the external substrate without altering V_{max}. The result implies that a depletion of endogenous nucleotides would not effect transport at high concentrations of external nucleotides. This result is incompatible with those of Barbour and Chan [150], and also with results obtained by other workers [42–48] who have studied the effects of varying the endogenous nucleotide levels on steady-state rates of ATP synthesis and nucleotide transport in the presence of saturating levels of external ADP. It is clear from these studies, carried out at 20–28°C, that steady-state rates of ATP synthesis in the presence of excess ADP are a function of the endogenous levels of nucleotides in the range of 2–6 $nmol \cdot mg^{-1}$ mitochondrial protein. The fact that ping-pong kinetics were not observed in either of the kinetic studies discussed above [150,151] has been quoted extensively as evidence against a sequential model of transport. Since the two studies agree with each other only on this one point, one should perhaps view the evidence with some skepticism.

In order to circumvent the problems inherent in measuring kinetics in intact mitochondria, Krämer and Klingenberg [159,160] measured adenine nucleotide transport kinetics in a reconstituted system composed of artificial liposomes and the isolated carrier protein. Exchange transport is slower in the reconstituted system since there is less active carrier per nmol of exchangeable intravesicular substrate. Carboxyatractyloside was used to quench transport at 0, 10, 20 and 30 s. The system offers advantages over intact mitochondria since substrate levels on both sides of the membrane can be independently varied in the absence of metabolic dismutation, substrate binding to divalent cations and to other cellular proteins, etc. Unfortunately, these authors have not examined the kinetic effects of altering the concentration of intravesicular substrate, which they maintain at high saturating levels in all experiments. The $\Delta\psi$ was altered by applying different K^+ concentration gradients across the liposomal membrane in the presence of valinomycin. The values of $\Delta\psi$ reported are probably overestimated since the gradient diminishes with time and only the initial K^+ gradient present at the time of formation of the liposomes was used to assess $\Delta\psi$. The effect of the applied $\Delta\psi$ on the external K_m and V_{max} of the four different exchanges catalyzed by the carrier were examined. The V_{max} of the exchange of ADP for ADP was not altered by $\Delta\psi$, but a change of $\Delta\psi$ from 0 to 120 mV caused a 50% decrease in K_m. The V_{max} of the exchange of $ATP_{(out)}$ for $ATP_{(in)}$ was decreased 2–3-fold by the increase in $\Delta\psi$ from 0 to 180 mV while the K_m of external ATP was increased from 56 to 95 μM qualitatively, but not quantitatively, similar to intact mitochondria.

The heteroexchanges $ATP_{(out)}$ for $ADP_{(in)}$, and $ADP_{(out)}$ for $ATP_{(in)}$, were also studied, but under conditions where both substrates were present in equimolar concentration on both sides of the membrane. The rate of entry of radiolabelled ADP under these conditions was studied at saturating substrate levels at variable $\Delta\psi$ from +120 to −180 mV with respect to the external side of the membrane. The ratio of the rate of ADP entry to ATP entry was plotted as a function of $\Delta\psi$ and the ratio varied from 5 to 0.2 as a linear function of $\Delta\psi$ from +120 to −90 mV.

The linearity of the influence of $\Delta\psi$ on velocity, suggests that the effect of $\Delta\psi$ is not mediated by a discrete change in the conformation of the carrier and implies that the carrier catalyzes an electrophoretic movement of ions through the membrane. When the data were analyzed by several models, the one which best fit the data assumed that the distribution of the substrate carrier complex was determined by the $\Delta\psi$. However, since the rate-limiting step in a sequential mechanism should be the transport of ATP into the mitochondria against the potential gradient, it is not obvious why the ATP/ATP exchange should be less sensitive to $\Delta\psi$ than the heteroexchanges. The difference between the effect of $\Delta\psi$ on the ATP/ATP exchange compared to the ADP/ATP exchange was particularly striking when comparing the effect of $\Delta\psi$ on the four possible exchanges in an experiment in which both substrates were present on both sides.

The data can perhaps be reconciled with a sequential model if one assumes that, under all conditions, both steps in the sequence are partially rate controlling. Thus, in the presence of an electrical potential across the membrane favoring ATP efflux, ATP/ATP exchange could be faster than the exchange of external ATP for internal ADP, if the return of the carrier to the outward facing direction is partially rate limiting and facilitated when ATP, rather than ADP, is bound at the inner surface.

The data reported could not be explained by the concerted mechanism suggested by Sluse and co-workers [151] since a concerted mechanism involving exchange of ATP for ATP would be electroneutral and insensitive to $\Delta\psi$.

5.2. Structural studies

5.2.1. The adenine nucleotide carrier

Studies of the adenine nucleotide carrier have been greatly facilitated by the early identification of two inhibitors of the transport process, which bind specifically and tightly to the carrier protein, atractyloside and bongkrekic acid [76,161]. Atractyloside [162], which is a plant toxin obtained from a thistle binds to the carrier from the cytosolic side of membrane and is competitive with the nucleotide substrates [163]. Bongkrekic acid is a mold toxin [66] and, like atractyloside, it inhibits the transport in a highly specific manner [165,166]. However, studies first conducted by Klingenberg and co-workers, and later confirmed by other laboratories [16], showed that this compound is an effective inhibitor only when bound to the matrix surface of the membrane. The demonstration that these two inhibitors react with the carrier from opposite sides of the membrane was the first clear demonstration that the carrier is asymmetrically oriented in the membrane.

Largely on the basis of these data, the original gated-pore model was suggested [152]. Two conformations of the carrier were hypothesized, each having a single substrate binding site. In the so-called 'm'-configuration the substrate binding site faces the matrix and interacts with bongkrekic acid, whereas in the 'c' configuration the carrier substrate site faces the cytosol and binds atractylate. Significant controversy has arisen about the details of this model, but it appears to be well supported by structural evidence, although the kinetic studies are ambiguous [150,151].

Structural studies have been made possible because the carrier has been isolated in pure form. Isolation of the protein was facilitated by the availability of radiolabelled carboxyatractyloside, a derivative of atractyloside [76,162], which binds to the carrier even more tightly than atractyloside. After treatment with [^{35}S]carboxyatractyloside in the native membrane the protein can be followed through various isolation steps by means of the assay of radioactivity [167,168]. The tritiated bongkrekic-carrier complex has also been isolated [73] and antibodies prepared to both inhibitor protein complexes [77]. The free carrier has also been isolated [169] and can be reconstituted into artificial liposomes in an active form [159,160]. Some of the studies of the reconstituted carrier were described above. Sodium dodecyl sulfate gel electrophoretic studies indicate that the carrier has a molecular weight of about 30000, whereas in non-denaturing detergents, both small angle neutron scattering measurements of the carrier [170], and measurements of the sedimentation coefficient of the isolated carrier in detergent phospholipid micelles [171], suggest that the molecular weight of the protein is between 61000 and 63000. There is only one half a binding site for carboxyatractylate or for bongkrekate per

30 000 monomer [152,172]. The data indicate, therefore, that the carrier is a dimer composed of two identical 30 000 subunits.

Antibodies to the carboxyatractylate protein complex do not crossreact with the bongkrekate protein complex [75], suggesting that the two inhibitors react with different conformations of the protein. A particularly convincing demonstration that the isolated protein-inhibitor complexes represent different conformations of the same protein was carried out by showing that it is possible to convert the isolated carboxyatractylate form to the bongkrekate form in solution. Conversion required the presence of both ADP and bongkrekate, and could be demonstrated by precipitation of the antigen antibody complex of the bongkrekate form of the carrier [73].

The concept that the protein exists in two conformations is also supported by the observation that the protein's reactivity to various reagents changes under conditions thought to change the carrier from one configuration to the other. According to the theory [15,172], the carrier in the natural mitochondrial membrane, suspended in a media lacking adenine nucleotides is largely in the 'c' state. Addition of external nucleotides converts some of the carriers to the 'm' state. Carboxyatractyloside binding converts all the carrier molecules to the 'c' state. Conversely, bongkrekic acid causes complete conversion to the 'm' state. The sulfhydryl group reagent, *N*-ethylmaleimide, does not react with the carrier in the presence of carboxyatractyloside, but bongkrekic acid addition to mitochondria facilitates the reaction of a sensitive sulfhydryl group with the reagent. Additionally, the carrier does not react with *N*-ethylmaleimide when mitochondria are suspended in buffer containing no adenine nucleotides, but does when the nucleotides are added [173,174].

Similar studies suggest that the carrier is more subject to digestion by trypsin and to ultraviolet irradiation damage in the 'm' than the 'c' state [73,175,176]. Conversely, tyrosine groups on the carrier are more accessible to iodine surface labelling in the 'c' than in the 'm' state [175]. Phenylglyoxal has also been used as a site-specific amino acid reagent in structural studies of the transporter. The reagent is relatively specific for arginine, and it is likely that arginine cationic groups are involved in the binding of the substrate. Studies of the effect of phenylglyoxal on carboxyatractylate and bongkrekate binding to mitochondria have been carried out [177–179], and these studies show that the reagent inhibits carboxyatractylate binding, but not bongkrekate binding. The results can be interpreted in two ways. The phenylglyoxal reacts with the substrate binding site only when the site is exposed to the cytosolic face (i.e., in the 'c' state). One interpretation suggests that the site exposed to the cytosolic side is quite different from the inner site, i.e., that there may be very little overlap of the amino acid chains involved [177,178]. Another interpretation [179] of the data, however, is that the phenylglyoxal does not interact at the substrate binding site, but elsewhere on the protein; but it does react only in the 'm' state. Since it may lock the protein into that state, it would prevent any reaction with carboxyatractylate without interacting directly with the carboxyatractylate binding site.

Controversy over the nature and number of the adenine nucleotide binding sites is ongoing. A significant unanswered question is whether there is a single reorienting binding site per dimeric protein formed at an interface of the monomers, or whether each monomer has an adenine nucleotide binding site, which exhibits negative cooperativity to such an extent that binding at one site virtually excludes binding at the second site (for review see Ref. 16). If release of substrate from the carrier requires binding at the site on the other monomer, the carrier would not be expected to exhibit ping-pong kinetics, thus explaining the controversial kinetic data.

Attempts made to identify more than one binding site per dimer have been largely unsuccessful. One approach has been to study the binding of fluorescent analogues of ADP. One such study has been carried out with formycin diphosphate, a transportable analogue which differs from ADP by an exchange of C for N between positions 7 and 8 on the purine moiety. Fluorescence emission is used as a measure of binding to the carrier in the membrane, and it was shown that carboxyatractyloside, bongkrekic acid and ADP each completely displaced the analogue from the carrier implying that there is only one substrate site per dimeric protein [180]. More detailed and quantitative studies [16,181,182] have been performed with 3′-*O*-naphthoyl ADP (N-ADP) a fluorescent analogue which binds, but is not transported. The analogue binds to the carrier from both surfaces. Binding is associated with a fluorescence decrease and, therefore, the extent of binding and release can be quantitated using intact mitochondria or submitochondrial particles. The studies reveal two types of binding sites corresponding to the distribution of carrier molecules in the 'c' or 'm' state, but the binding sites add up to only one per dimeric carrier.

The carrier binding site has also been probed in detailed surveys of the structure of analogues which bind to the carrier and of the structure of those which can both bind and be transported [183,184]. The findings provide a detailed description of the steric, contact and structural elements which are necessary for binding and additional structural prerequisites that are necessary for subsequent transport. The data suggest that a non-fixed orientation (either syn or anti) of the heterocyclic ring structure and the ribose base is important for transport. ADP and ATP are non-fixed, but exist mainly with the adenine base positioned in the anti orientation. Synthesis of analogues which were conformationally restricted to either the syn or anti orientation were not transported. Only the analogues fixed in the syn-conformation could bind. It is tempting to speculate that a change from syn to anti might occur during transport. The studies described were carried out using intact mitochondria. Therefore, only the 'c' state binding site was examined and it would be interesting to know if the 'm' state binding site had the same or opposite specificity for the heterocyclic ring/ribose orientation.

The complete 297 amino acid sequence of the adenine nucleotide transporter has been determined [185] and deductions have been made about the two-dimensional structure in the plane perpendicular to the membrane. Some of these deductions have been made using computer analyses of the amino acid sequence [186]. Using a computer program to locate sequence homologies, three domains have been identified within the sequence with significant sequence overlap. The three repeating areas are characterized by a conserved region centered on a cysteine residue found in three hydrophilic segments. Some of these segments, and perhaps all, are linked by hydrophobic sequences more than 20 amino acids long. Non-polar stretches of amino acids (25–28 long) are prominent features of many internal membrane proteins. It has been clearly demonstrated in several such proteins [187] that these stretches are α-helical, are buried within the hydrophobic region of the lipid bilayer, and traverse the membrane. The translocase is strikingly polar for an integral membrane protein. Only three clearly hydrophobic segments of sufficient length to span the membrane can be identified, but three others are likely candidates. In Fig. 8.1, all six have been incorporated into a hypothetical model. Segments III, IV and V are most hydrophobic, whereas I, II and IV, although of sufficient length, have several acidic and basic residues. The hydrophobic segments of bacteriorhodopsin whose three-dimensional structure is known from X-ray diffraction studies have charged groups and these groups form the interface with an H^+ channel down the center of the protein [188].

The three conserved cysteine residues in the translocase are marked by black dots on the figure. We have also incorporated into the figure information gleaned from

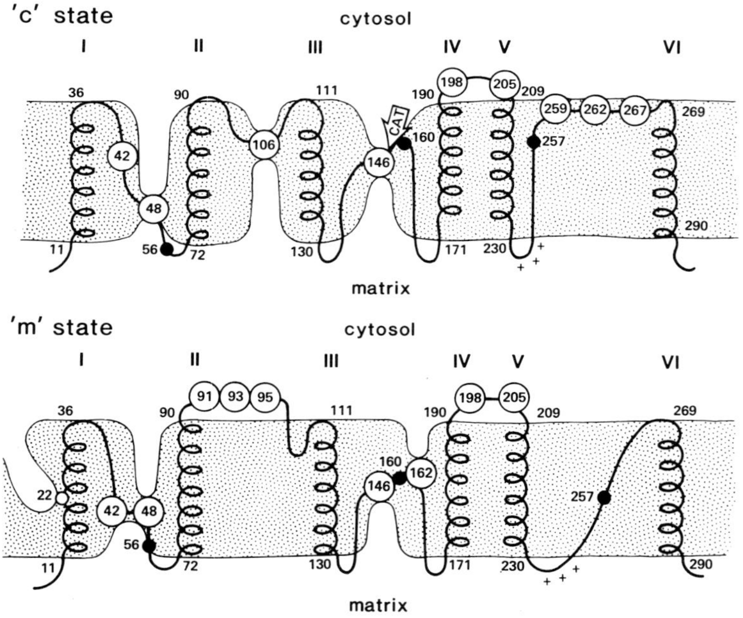

Fig. 8.1. Two-dimensional schematic representation of the structure of the adenine nucleotide carrier. The line represents the amino acid chain of the protein and all numbers on or within the line represent the number of the amino acids in the linear sequence. The black dots are cysteine residues, about which there is significant sequence homology [186]. The helical regions are segments of hydrophobic amino acids thought to span the membrane [186]. The CAT arrow represents the site of photoaffinity labelling of an azido derivative of atractyloside [189]. The open circles are lysine residues which react with pyridoxal phosphate in intact mitochondria or submitochondrial particles [190,191].

reactions observed with a photoaffinity label of atractyloside [189] and with pyridoxal phosphate, a non-penetrant reagent which forms covalent linkages with exposed lysine residues [190,191]. In order to identify the site on the carrier which binds atractyloside, radiolabelled azido derivatives of atractyloside were prepared and used to photolabel the carrier [189]. The carrier was subsequently isolated and chemically cleaved. Even though both short- and long-chain azido derivatives were used in these experiments, all derivatives labelled one specific section of the protein. The labelling site is between cysteine 159 and methionine 200, as shown on the figure.

Pyridoxal phosphate forms pyridoxal lysine linkages with the carrier in the membrane. Since the reagent does not penetrate the membrane, the sidedness of the lysine residues can be determined by treating mitochondria with the reagent [190]. The matrix surface of the mitochondria is exposed to the external media in submitochondrial particles. Additional information is obtained by studying the carrier either in the 'm' or the 'c' state by exposing the mitochondrial or submitochondrial particle to carboxyatractyloside or bongkrekic acid prior to pyridoxal phosphate treatment [191]. The 'c' state data are shown in the A panel of the figure and the 'm' state data are in the B panel. The lysine groups are shown in the figure as open circles, and their sidedness is indicated by their placement in relationship to the cytosol or matrix spaces. A curious feature of the data is that in the 'c' state, many lysines are exposed to the cytosolic surface only (lysines 42, 198, 205, 259, 262 and 267). A few are exposed to both surfaces (lysines 48, 106 and 146). Initially, it was thought that 146 was exposed only to the matrix surface, but later data using atractyloside rather than the more tightly bound carboxyatractyloside to fix the carrier in the 'c' state showed that the carboxyatractyloside was protecting the lysine 146 from the cytosolic surface. This is in good agreement, as shown, with the photoaffinity labelling study.

On the other hand, in the 'm' state, the exposure of some lysines change as would be expected for a 'gated pore' carrier. Thus, lysine 42 switches from the cytosolic side to the matrix side. Lysines 91, 93 and 95 appear on the cytosolic face, and 106 disappears. Lysine 146 is exposed only on the matrix side and 162 appears from the cytosolic side. A rather unsatisfactory aspect of the model is the necessity for the hydrophilic segment 230–259 to cross the membrane. This appears necessary if hydrophobic segment V spans the membrane, 198 and 205 are cytosolic, and 259, 262 and 267 are also cytosolic. Since these last lysines are very weakly reactive it is possible they are deep in a pocket, not shown in the model.

Nevertheless, most aspects of the data are in excellent accord with a very asymmetric gated-pore model. It is tempting to imagine a third dimension by curving this two-dimensional model out of its plane in such a way that lysine residues 42, 48, 106 and 146 are all contained within the same channel. It is apparent that the data are most satisfactorily explained by channels in the structure. An important unanswered question is whether each monomer forms an independent channel with an independent substrate binding area, or whether the dimeric structure is required for formation of a single channel.

5.2.2. The phosphate transporter

A particularly reactive sulfhydryl group (or groups) on the phosphate transporter react with relatively low concentrations of maleimide derivatives and organic mercurials to inhibit the function of the transporter [17]. This sulfhydryl group is probably near, or a part of, the substrate binding site because phosphate protects the carrier activity from inhibition by *N*-ethylmaleimide. Fonyo has written a comprehensive review on the subject of sulfhydryl sensitivity of the phosphate carrier and its implications concerning the mechanism of transport [192].

Sulfhydryl reagents have been used in attempts to distinguish the cytoplasmic from the matrix surfaces of the transporter, and to determine whether the phosphate transporter has a single reorienting substrate binding site. Pedersen and co-workers have carefully examined the ability of various sulfhydryl reagents to penetrate the mitochondrial membrane and find that *N*-ethylmaleimide penetrates the membrane, whereas the organic mercurial, mersalyl, does so very poorly [129]. Prior treatment of intact mitochondria with mersalyl could protect the phosphate transporter from subsequent irreversible inactivation by *N*-ethylmaleimide. The initial inactivation by mersalyl could be completely reversed by dithiothreitol. Since mersalyl is a non-penetrant the data suggest that a critical sulfhydryl group is on the cytosolic surface or can be completely 'pulled' to that surface. The fact that mersalyl also inhibits transport in submitochondrial particles implies that the critical sulfhydryl group has a dual exposure.

Recently Wohlrab has used the isolated reconstituted phosphate transporter to study this problem [121–125]. The isolated phosphate transporter has a molecular weight of 34 000 as judged by sodium dodecyl sulfate gel electrophoretic studies. It appears [126] that there is less than one reactive SH group which binds *N*-ethylmaleimide per 34 000 Da monomer [125]. Therefore, the data imply that the active carrier may be a dimer (with half the sites' reactivity) having a molecular weight of 68 000. Other data obtained from the same laboratory show that if the isolated carrier is not stored in the presence of dithiothreitol it is autooxidized to a form with a much lower V_{max} and higher K_m. The autooxidized carrier can be reactivated by dithiothreitol, and in the autooxidized form the protein is protected from (irreversible) inhibition by *N*-ethylmaleimide. Taken together these data suggest that there are two SH groups on the active transporter which are sufficiently close to each other to form a disulfide bond. They also suggest that these SH groups are part of a reorienting substrate binding site.

A number of workers have been able to isolate a protein, covalently linked to radioactive *N*-ethylmaleimide, identified as the mitochondrial phosphate transporter [122,193–196]. The isolation and identification of the transporter was based for the most part on the maleimide and mersalyl reactivity of the protein. The molecular weight of the protein isolated from different sources varies from 27 000 to 34 000. Because of the covalent linkage to the irreversible inhibitor, reconstitution of transport was not feasible.

Three laboratories have been able to isolate a protein free of maleimide labelling which will catalyze phosphate transport in reconstituted proteoliposomes

[122,197–199]. In one case, isolation was based on the ability to reconstitute transport [198,199]. In two other cases, the isolation has been based on methods developed for isolation of the *N*-ethylmaleimide-labelled protein [124,197].

The similarities between the phosphate transporter and the adenine nucleotide transporter are striking. Both have monomeric molecular weights in the region 30 000–34 000 and it is likely that the phosphate transporter is a dimer, like the adenine nucleotide transporter. Physical characteristics are so similar that it has been difficult to separate the two. The initial step in isolation is based on elution from hydroxyapatite [122,197,200]. The phosphate transporter and the adenine nucleotide transporter both have extraordinarily low affinity for hydroxyapatite. Kadenbach and co-workers have been able to separate the two carriers, obtaining a pure active transport protein, using affinity chromatography with an organomercurial agarose gel and a mercaptoethanol step gradient to elute the carrier [197]. Wohlrab on the other hand has been able to isolate a reconstitutively active protein preparation only when the adenine nucleotide carrier is still present in the preparation, except in the case of insect flight muscle mitochondria where pure, active carrier has been isolated, free of the adenine nucleotide carrier [124]. The preparation from beef heart recently described by Wohlrab, appears to be the most highly active, and has a turnover number for net transport at 22°C of $1.4 \times 10^4 \cdot \mathrm{min}^{-1}$ for a 68 000 dimer [125].

5.2.3. Other transporters

The pyruvate transporter [201] and the carnitine translocase [202] have both been isolated but not characterized in any detail. The pyruvate transporter and the carnitine translocase, like the phosphate transporter, are inhibited by maleimide derivatives and mercurials, although at higher concentrations of the sulfhydryl reagents. The pyruvate transporter has been isolated in inactive form covalently linked to phenyl maleimide. Identification was based on the correlation of labelling of the protein with inhibition of transport, and by the fact that mercurials prevented the labelling. The molecular weight of the isolated monomeric protein is surprisingly low, approximately 15 000.

Several additional mitochondrial carrier systems have been reconstituted into active form in proteoliposomes, using as starting material a crude neutral detergent mixture of membrane proteins from submitochondrial particles. These include the citrate transporter [203], the dicarboxylate carrier [204], and the carnitine transporters [202]. These reconstitution activities could be used as a basis for further purification and structural studies, but such studies have not yet been reported.

6. The influence of mitochondrial transporters on metabolic fluxes

6.1. Overview and definitions

Although the complexities involved in the interrelationships of various metabolic pathways within the intact cell have made it difficult to evaluate the role of metabolite transporters in modulating overall flux, considerable progress has been made over the last several years. Recent advances have been made possible by the development of several techniques for measuring metabolite concentrations in the cytoplasmic and mitochondrial compartments separately, as opposed to total tissue measurements utilized in the past. These methods [205–207] involve fractionation of intact cells utilizing organic solvents, detergents or mechanical forces. To evaluate the influence of the mitochondrial metabolite transporters, the kinetic characteristics of the translocators must be taken into account, with respect to substrate and product concentrations on either side of the mitochondrial membrane and to the characteristics of the interrelated enzymatic steps. Although each of these methods has been subject to criticism, they have offered considerable advantages over previously used indirect methods for calculating metabolite distribution from total tissue levels.

It has been suggested that a way to determine whether or not a reaction may be regulatory (and displaced from thermodynamic equilibrium) is to compare the mass action ratio of the reactants and the products with the known equilibrium constant. Although most authors subscribe to the notion that the rate-controlling step (or steps) in a pathway operate far from thermodynamic equilibrium [208–210], the reverse is not always the case. The observation that an enzyme or transporter is removed from equilibrium does not prove its regulatory role. For the purposes of this discussion we will utilize those definitions reviewed by Newsholme and Start [210] and Newsholme [211] with regard to equilibrium and nonequilibrium conditions. A nonequilibrium condition obtains if the rate of the reverse component is very much less than the rate of the forward component. Moreover, for a process that is removed from equilibrium, a negative free energy change will occur as it proceeds toward equilibrium. A near-equilibrium state obtains if the rate of the reverse component is similar to that of the forward component, and the rates of each are very much greater than the overall flux through the reaction step.

Since considerable confusion has arisen in the literature because of non-uniformity in utilization of terms, certain terms which will be used in this discussion will be defined initially. As pointed out by Duszynski et al. [212], a rate-controlling step should be distinguished from a rate-limiting step. The term rate limitation is restricted to situations in which only one step controls flux through a metabolic pathway. In contrast to the rate-limiting steps, there may be several rate-controlling steps within a given pathway. A rate-controlling step is defined as a reaction in which a change in activity leads to a change in net flux through the pathway. For example, in muscle tissue in the absence of insulin, glucose transport is likely to be limiting for glycolysis, whereas in the presence of insulin, glucose transport, the phosphofructokinase step and pyruvate kinase may all be rate controlling. A rate-controlling step is likely to be equivalent to what Newsholme and Crabtree have termed a regulatory step, which catalyzes a non-equilibrium reaction and is regulated by factors other than the pathway substrate [211,213]. The definitions as originally stated are slightly different, but rate-controlling and regulatory steps may be used synonymously. The term rate-controlling step will be used in this review. A different term 'flux-generating step' defined [211] as the first reaction in a pathway that is saturated with the pathway substrate, does not signify a reaction which is *necessarily* rate-controlling or rate-limiting. For this reason, the term flux-generating will not be used in the current review.

Since the terms rate-controlling and rate-limiting are qualitative in nature, several investigators have attempted to devise more quantitative approaches to the problem of metabolic control. Higgins [214] developed the concept of control strength. This was later restructured and expanded by Kacser and Burns [215,216] and Heinrich and Rapoport [217]. More recently, as discussed below, the term 'control strength' has been utilized by several workers [212,218] in an attempt to explore and quantitate the role of mitochondrial metabolite transporters on metabolic flux. The control strength is defined as the fractional change in the steady-state flux through the pathway produced by a fractional change in the activity of the enzyme or reaction step. Thus, control strength is the rate of fractional change in steady-state flux and fractional change in activity of a particular enzyme. According to this concept, the sum of the control strengths of all steps in a pathway is equal to 1.0, which implies that each step is independent and makes no allowance for steps which may be obligatorily linked by physical interaction between enzymes or enzymes and transporters. The terminology does not appear to allow assessment of the interrelationships of branching or related pathways that are not linear. An enzyme with a high capacity (V_{max}) and a relatively high K_m could be rate controlling if it were at a metabolic branch point. The observed effect of a change of V_{max}, brought about by inhibitor addition, could underestimate the actual controlling influence which might be exerted by a change of K_m. Nevertheless, utilization of this approach has yielded interesting information which will be reviewed below. To evaluate the control strength, the effect of specific, irreversible and rapid acting inhibitors on the particular reaction step is studied. There must be a specific quantitative relationship between the amount of added inhibitor and total enzyme activity (V_{max}). Flux through the pathway is titrated using different levels of inhibitor and the control strength of the step is calculated from the initial slope of the curve relating flux to amount of added inhibitor. It has been generally assumed that the shape of the inhibitor titration curve suggests whether or not the reaction studied is rate controlling. That is, it has been assumed that a step is unlikely to be rate controlling if the inhibitor titration curve is sigmoidal rather than hyperbolic in shape. However, more detailed analysis [219] has suggested that sigmoidicity of the curve does not eliminate the possibility that the titrated reaction is rate controlling. By using the initial slope of the inhibitor titration curve an estimate of the extent of sigmoidicity is obtained and this is used to calculate control strength.

The mitochondrial translocators which have been most carefully assessed with respect to their role in control of metabolism are: (1) the adenine nucleotide translocator with respect to its role in the control of respiration; (2) the liver pyruvate transporter and the control of gluconeogenesis; and (3) kidney glutamate and glutamine transport and their control of ammoniagenesis.

6.2. Control of respiration by the adenine nucleotide carrier

The mechanism of control of mitochondrial respiration remains an important question of mitochondrial bioenergetics. It was initially proposed by Chance and Williams [220] that respiration is kinetically controlled by ADP availability, an hypothesis which has received renewed support by the studies of Jacobus et al. [221]. However, the major controversy has centered on whether or not the adenine nucleotide translocator is rate limiting, or even rate controlling, for respiration. As a consequence of this controversy, two hypotheses have emerged over the past several years, one of which has been recently modified. Wilson and Ericinska [222] have proposed a 'near-equilibrium hypothesis', whereas others have advocated variations of a translocase hypothesis in which the adenine nucleotide translocator is either rate limiting or rate controlling for respiration.

In a recent review, Vignais and Lauquin [223] argued against a rate-limiting role for the adenine nucleotide carrier. This conclusion was based on the following observations: (1) the K_m for ADP is an order of magnitude less than the estimated concentration of free ADP in the cytosol; (2) the capacity of ADP/ATP transport measured in isolated liver mitochondria is greater than the estimated rate at which ATP is delivered to the cytosol in situ; (3) a state of near-equilibrium exists between the redox span NADH to cytochrome *c* and the cytosolic phosphorylation state. Therefore, all intermediate reactions, including adenine nucleotide translocation, should also be at near-equilibrium. For these reasons, these reviewers felt that rate-limitation of respiration by the adenine nucleotide translocator was extremely unlikely. Interpretation of the findings and the methodological problems involved in obtaining the data reviewed by Vignais and Lauquin constitute major issues in the controversy on the role for the translocator in controlling respiration.

The near-equilibrium hypothesis has been championed by the studies of Wilson and co-workers. Erecinska et al. [224] have shown that respiration coupled to ATP synthesis is proportional to the inorganic phosphate (P_i) concentration as well as the ADP/ATP ratio. According to the near-equilibrium theory of control, respiration should respond to changes in the extramitochondrial phosphorylation potential as a whole and not to alterations in the concentrations of individual reactants, such as ATP/ADP ratios. Furthermore, at a constant NAD^+/NADH ratio, respiration should respond to the phosphorylation potential and, if this latter term remains constant, respiration also should remain constant.

Interpretation of the validity of the near-equilibrium concept is dependent on the accuracy of intramitochondrial free NAD/NADH measurements and the difference between extra- and intramitochondrial phosphorylation potentials. In a series of studies, Wilson and associates [40,41,225,226] have presented evidence in support of their hypothesis. Utilizing rat liver mitochondria, Forman and Wilson [225] compared the mass action ratios to calculated equilibrium constants under conditions promoting either forward (net ATP synthesis) or reversed (net ATP hydrolysis) electron transport. Since the mass action ratios calculated under various conditions were similar to the calculated K_{eq}, these findings were said to support a near-equilibrium between the reactions of the electron transport chain between NADH and cytochrome *c* and the external phosphorylation potential. Wilson and co-workers [40] have suggested that the findings of others demonstrating that the ratio of extramitochondrial ATP/ADP exceeds, by more than an order of magnitude, the ratio in the intramitochondrial space is explained by significant binding of intramitochondrial ADP which can lead to erroneously low calculated intramitochondrial ATP/ADP ratios. On this basis they reject the concept that the carrier is electrogenic and far from equilibrium. In order to determine the ratio of free adenine nucleotides within the mitochondria, and to evaluate the binding of adenine nucleotides, measured total ATP/ADP ratios were compared to those calculated from equilibrium with nucleotide substrates for which binding should be minimal. In one such study, detergent solubilized mitochondria were used, supplemented with nucleoside diphosphokinase and creatine kinase. Addition of creatine phosphate, UDP and GDP generated a high ratio of endogenous ATP/ADP which then achieved near equilibrium with the UTP/UDP and GTP/GDP pools, via the added nucleoside diphosphokinase. At high ratios of ATP/ADP (and, therefore, low levels of ADP), evidence for ADP binding was observed and thus the ratio of free ATP/ADP attained values as high as 4-times the total ATP/ADP.

The experiment is not absolutely convincing, since many more protein binding sites are available to ADP in detergent-solubilized mitochondria than there are inside the matrix. Also, in studies of this sort it is uncertain what the effect of the detergent itself might be.

In another study, Wilson and co-workers [41] have attempted to estimate the free ATP/ADP ratio in the matrix from the matrix free GTP/GDP ratio, which is in turn estimated from the mass action ratio of the phosphoenolpyruvate carboxykinase reaction. Intramitochondrial GTP generation via substrate level phosphorylation is inhibited with arsenite and GTP is supplied to the kinase from intramitochondrial ATP generated via oxidative phosphorylation. It is assumed that, when the system reaches steady state, ATP/ADP and GTP/GDP are in equilibrium. Mitochondria were used from sources with unusually high activities of phosphoenolpyruvate carboxykinase and nucleoside diphosphokinase. The data suggest that the ratio of free ATP/ADP is higher than the total ATP/ADP in the matrix. However, convincing proof that the system is near equilibrium is lacking. The arsenite concentrations used (0.25–0.5 mM) were unusually low so that significant generation of intramitochondrial GTP might have occurred through the block, allowing disequilibrium between the ATP and GTP. The ATP and GTP systems are usually far from equilibrium in the mitochondrial matrix because of the relatively low activity of nucleoside diphosphokinase [227]. Also, comparison of initial and final rates of phosphoenolpyruvate synthesis or degradation would have provided more convincing proof that the phosphoenolpyruvate system was in near equilibrium. Free adenine nucleotide levels within the mitochondrial matrix are undoubtedly significantly different from the total measured metabolite levels, and it may be possible to explain a higher ATP/ADP ratio outside metabolizing mitochondria than inside, by ADP binding to matrix protein.

However, other types of data support the concept that the carrier is electrogenic and these data are not readily explained by intramitochondrial ADP binding. The data include the observations that the large difference between the intramitochondrial and extramitochondrial ATP/ADP ratios disappears when uncoupling agents are added to mitochondria [15,157], and the observation that there is a linear relationship with a slope of 0.85 relating the mitochondrial membrane-electrical membrane potential as measured by $^{86}Rb^{+}$ distribution in the presence of valinomycin to the log $(ATP/ADP)_{external}/(ATP/ADP)_{internal}$ [39]. No obligatorily linked proton co-transport or any other ion co-transport can be observed in conjunction with the exchange of ATP^{4-} for ADP^{3-} [38]. In oligomycin inhibited mitochondria, entry of ATP^{4-} and efflux of ADP^{3-} causes a statistically significant increase in measured $\Delta\mu H^{+}$ [228]. Using a dye which is a membrane potential probe, it is possible to observe an increase in $\Delta\psi$ when ATP is added to depolarized, non-respiring, oligomycin-inhibited mitochondria [229]. Moreover, Klingenberg and Krämer [159,160] have demonstrated the electrogenic character of the transport process in reconstituted liposomes (c.f., Section 5.1), where the only protein present is the translocase and ADP binding is certainly minimal. It is difficult to reconcile all these observations with the claim for a near-equilibrium character of an electroneutral translocator as suggested by Wilson and his associates.

At the present time, it is the opinion of the present reviewers that Wilson and collaborators have not provided sufficient evidence to conclude that the adenine nucleotide carrier is electroneutral and near equilibrium, nor have they conclusively shown that the first two sites of oxidative phosphorylation are in near equilibrium, especially in view of the likelihood that the P/O ratio of the first two sites is 1.5 rather than 2 [230,231]. Therefore, we conclude that their studies do not convincingly exclude a role for the adenine nucleotide translocator in the control of mitochondrial respiration.

As reviewed previously [6], Davis and co-workers [232,233] have provided evidence for a rate-limiting role for the adenine nucleotide translocator based on the observation that the extramitochondrial ATP/ADP ratio correlates with respiratory rate as long as phosphate is not limiting for phosphorylation. Additional support for this thesis was provided by others [231,234–239]. However, more recent studies from two laboratories, utilizing control theory principles, have altered this hypothesis to suggest that the adenine nucleotide translocator is one of several rate controlling rather than a single rate limiting step. Much of the data with respect to a rate-controlling role of the adenine nucleotide translocator derive from studies by Tager and co-workers. For example, the relationship between oxygen utilization and the intra- and extramitochondrial phosphate potentials was assessed by incubating rat liver mitochondria with an extramitochondrial (glucose plus hexokinase) or an intramitochondrial (citrulline synthesis) ATP-utilizing system. These studies [240,241] indicate that the respiration rate was directly related to the intra- rather than extramitochondrial ATP/ADP ratio. Moreover, they revealed no direct, unequivocal control of respiration by the extramitochondrial ATP/ADP ratio and demonstrated that the adenine nucleotide translocator was displaced from equilibrium, in contrast to the contention of Wilson et al.

Additional studies [212,218,219,242,243] to quantitate the role of the adenine nucleotide translocator in the control of mitochondrial respiration have been performed utilizing inhibitor titrations with carboxyatractyloside. The results indicated that in State 4 (no ADP), no control was exerted by the translocator. However, as the rate of respiration was increased up to State 3 (excess ADP), the control strength of the carrier increased to a maximum value of 30%, at 80% of State 3 respiration. These studies indicate that the adenine nucleotide translocator cannot be considered to be the only rate-controlling step in oxidative phosphorylation. However, they do provide experimental support for a controlling role for the carrier at intermediate to maximal levels of respiration. An important corollary of these studies is that the reaction rate may be altered by a change in substrate concentration (elasticity). It is also clear that to confirm these studies quantitatively, they must be extended to intact cells. Although such studies have been more difficult, the results are compatible with the conclusion reached by Tager et al. [212].

In contrast to the results of Tager and associates, Forman and Wilson [226] obtained different results using CAT inhibitor titrations in both rat heart and liver mitochondria. The initial slopes of their titration curves are horizontal, suggesting that the control strength is zero. It is not clear why their results are so different, although ATP, phosphate and $MgCl_2$ concentrations were not the same as in the studies of Tager and co-workers.

Neither of these two groups determined whether the lowest concentrations of carboxyatractyloside caused alterations of the ratios of ATP/ADP inside and outside the mitochondria. If the carrier has a large excess capacity, it may be unidirectional, but controlled by the concentrations of substrate on one or both sides of the membrane. If it is far from equilibrium, and net flux is controlled by substrate, substrate concentrations would have to change to maintain flux even when a small fractional change in the activity of the carrier occurred. On the other hand, if the carrier maintains the substrates in near-equilibrium, a small fractional change of the activity should have no effect on the concentrations of the substrates. If a small change in the activity of the carrier causes imperceptible changes in flux, but does significantly alter the internal and external ATP/ADP ratios, according to the formal definition of rate control, the carrier could not be considered rate controlling. Nevertheless, the metabolic consequences of changes in both the intra- and extramitochondrial adenine nucleotide ratios are so significant that one would have to consider the influence of the carrier in metabolism an important one. Recent studies [244] of ATP turnover in perfused working hearts, by means of saturation transfer nuclear magnetic resonance, suggest that the turnover of ATP is only 3-times the rate of O_2 consumption. Since the P/O ratio of oxidative phosphorylation is approximately three, the data implies unidirectionality of ATP synthesis and transport under in situ conditions which approach mitochondrial State 3. Also, it is difficult to understand in the context of the equilibrium theory why matrix levels of total adenine nucleotides in the presence of saturating external ADP levels should influence respiration [42–47].

Studies performed by Kunz and collaborators [245–249] have provided further insight into the complexities in the control of respiration. These workers altered the rate of extra- and intramitochondrial ATP utilization in rat liver mitochondria [245,246] and observed that the adenine nucleotides in the immediate neighborhood of the ATPase were not in equilibrium with the extramitochondrial adenine nucleotide pool as suggested by Erecinska et al. [222]. Inhibitor titrations with carboxyatractyloside with varying rates of extra- and intramitochondrial ATP utilization also indicated that the degree of control exerted by the adenine nucleotide translocator on respiration was variable and dependent upon the complexity of the metabolic system studied [249]. For example, utilizing a hexokinase-glucose system, little control of respiration was observed by the translocator until rates of respiration exceeded 60 nmol $O_2 \cdot ml^{-1} \cdot min^{-1}$. As the respiration rate rose, the control strength increased up to a maximum value of 45% in the fully active state. At lower rates of respiration, the proton leak and the mitochondrial ATPase activity presumably controlled respiration. However, utilizing a more complex system in which phosphoenolpyruvate and pyruvate kinase were added in order to stabilize the external ATP/ADP ratio, a higher degree of control strength was observed for the translocator except at high respiration rates when the same control strength was observed as with the hexokinase system alone. In this system, pyruvate kinase competes for ADP with the mitochondrial oxidative phosphorylation system without affecting the rate of glucose-6-phosphate formation from hexokinase.

From these studies and those of Tager, an intermediate view is emerging with respect to the role of the adenine nucleotide translocator in the control of mitochondrial respiration. This view suggests that a variable degree of control is exerted depending upon the metabolic state, substrate availability, etc., such that control may be considerable or minimal. Such a view also implies that the control strength may vary from tissue to tissue and from cell to cell within the tissue.

At the present time, it is not possible to resolve the controversies that exist with regard to factors controlling the rate of respiration. However, it is our view that the available data do not provide any support for the extreme views. It seems likely that the adenine nucleotide translocator is not the sole controlling or rate-limiting step for respiration. Nor does it appear likely that it has no role. The degree of control exerted is a variable, but important factor.

6.3. Gluconeogenesis and the pyruvate transporter

Some years ago, the interesting observation was made that mitochondria isolated from glucagon-treated animals carboxylate pyruvate at twice the rate of control mitochondria [250]. The effect can be detected within a few minutes after intravenous injection of glucagon [251]. This implies that the hormone induces a rapid, but stable modification of the mitochondria which persists through the isolation procedure. Since pyruvate carboxylation may be rate controlling for gluconeogenesis, the observation was pursued with the thought that the mitochondria might be an important target of hormone action. Further studies have shown that pyruvate oxidation, as well as carboxylation, is stimulated by pretreatment of either intact animals or isolated hepatocytes with glucagon [252,253]. The phenomenon was attributed to an increase in the activity of the pyruvate transporter [254], but direct measurements of pyruvate transport indicated that the rate of pyruvate entry was faster in treated mitochondria only when transport was supported by a ΔpH generated by substrate oxidation [18,255]. Further studies demonstrated that State 3 (ADP stimulated) oxidation of most of the substrates of mitochondrial dehydrogenases was faster in the hormone-treated mitochondria [18,252,256]. Also, rates of uncoupler-stimulated ATPase [251,257], rates of citrulline synthesis from ornithine [258], and reverse electron transport in submitochondrial particles were all stimulated [256]. Measurements of ion and metabolite gradients showed that adenine nucleotide, K^+ and Mg^{2+} levels were elevated in the treated mitochondria [259,260]. Significantly, it was found that the mitochondrial proton motive force was higher [255].

Halestrap initially concluded that the increases in mitochondrial pyruvate transport and carboxylation were due to an increase in ΔpH secondary to stimulation of the electron transport chain in the cytochrome bc_1 region [255]. The conclusion was based largely on spectral measurements of the redox state of these cytochromes in the control and stimulated states. The spectral measurements were later found to be artifactual due to low amplitude Ca^{2+} swelling of the mitochondria. Halestrap then suggested that the stable changes in the mitochondria might reside in the lipid components of the membrane due to phospholipase A_2 activity [261,262], but he has been unable to confirm this with lysophospholipid measurements [263]. On the other hand, using an EPR spin label probe of the lipid environment of the isolated mitochondria, Hoek has found differences between control and treated mitochondria [264].

Serious questions have arisen about whether or not the observations are an isolation artifact of the mitochondria involved. When control mitochondria are isolated in a mannitol based media, rather than a sucrose-based media, their properties are very similar to those observed in mitochondria isolated from glucagon-treated animals [265,266]. Also, when the mitochondria are isolated in mannitol, the differences between control and treated mitochondria virtually disappear. Additionally, although uncoupler-stimulated respiration from durohydroquinone is faster in mitochondria isolated from hepatocytes exposed to glucagon, the respiration of intact hepatocytes in the presence of uncouplers and durohydroquinone was not affected by hormone pretreatment [267]. Also, measurements of ΔpH in situ using a combined indicator dye and radioactive permeant anion method revealed no increase in ΔpH after addition of glucagon or phenylephrine [268]. Measurements of in situ $\Delta\psi$ also showed no changes [269]. Halestrap has recently suggested that the observed changes may be due to in situ volume changes, but the size of the observed changes seems too small to be responsible for sizeable increases in pyruvate oxidation [263].

Despite these findings, other data suggest that mitochondrial membranes may be a target for glucagon and phenylephrine action. Measurements of metabolite gradients in situ have shown that endogenous adenine nucleotide levels are increased by glucagon and the internal ATP/ADP ratio is higher in cells treated with the hormone [143]. Cytosolic glutamate decreases, and cytosolic aspartate increases, while measurements of gluconeogenesis show that flux through the transaminase/glutamate/aspartate carrier pathway increases. Also, hepatocytes permeabilized after glucagon treatment continue to show differences in pyruvate carboxylation compared to controls [112], and a very recent paper from Haynes' group shows that mitochondria rapidly isolated from hepatocytes in a mannitol media continue to exhibit differences from controls [270].

Inhibitor titration curves using α-cyanohydroxycinnamate, which is a specific and non-competitive inhibitor of pyruvate transport, have been performed in an effort to resolve the problem [271,272]. When rates of gluconeogenesis in rat hepatocytes were measured as a function of added α-cyanohydroxycinnamate, a hyperbolic curve was obtained in the presence of 10 mM lactate and 1 mM oleate, suggesting that pyruvate transport is completely rate-limiting for gluconeogenesis [271]. When glucagon was added, the rates were all faster, but the curve remained hyperbolic suggesting that pyruvate transport is the important rate-limiting step stimulated by glucagon. These results are startling in view of the known important regulatory features of unidirectional enzymes such as pyruvate carboxylase and pyruvate kinase known to be involved in the control of gluconeogenesis.

The titration curves have been repeated and confirmed recently by Rognstad [272] who examined only the control situation but in somewhat more detail than Thomas and Halestrap [271]. Rognstad evaluated flux through the pyruvate transporter as ^{14}C incorporation into CO_2 and [^{14}C]glucose. Omission of oleate, which inactivates pyruvate carboxylase by lowering matrix acetyl CoA, causes the titration curve to become sigmoidal. Rognstad [272], in agreement with Thomas and Halestrap, finds that there is an α-cyanohydroxycinnamate-insensitive component of the titration curve which he attributes to passive diffusion of pyruvate. At 2 mM lactate, the inhibitor curves are hyperbolic in the absence of oleate and no inhibitor insensitive flux can be detected.

Strongly sigmoidal titration curves were obtained when the gluconeogenic substrate was pyruvate and it is important to note, in this context, that inhibitor titration curves of pyruvate oxidation and carboxylation in isolated mitochondria using α-cyanohydroxycinnamate are routinely sigmoidal, when the media pH is more alkaline than 7.0 [106,112]. Measurements of the effects of hormone on the inhibitor titration curves over a wider range of conditions, deliberately varying the activities of pyruvate carboxylase, pyruvate kinase and PEP carboxykinase, might have shed some light on this intriguing problem.

6.4. Ammonia formation by the kidney

The kidney plays a major role in the maintenance of acid-base homeostasis, particularly with respect to metabolic acidosis. In response to metabolic acidosis, the kidney is able to increase its production of ammonia resulting in enhanced urinary ammonium excretion, a process linked to proton excretion and the generation of bicarbonate by the kidney. Recent experimental work has focused on the mitochondria since the primary precursor of renal ammoniagenesis is glutamine, and the enzymes responsible for the degradation of glutamine, glutaminase and glutamate dehydrogenase, are located within the mitochondrial matrix space.

The increases of ammonia formation due to increased acid loads can be divided into two categories, acute and chronic. There is an acute response which occurs in the proximal tubule of the kidney when the cell cytosol H^+ concentration increases. The acute response to a blood pH drop of 0.3 units causes a 2–3-fold increase in ammonia formation [273]. A much more pronounced increase occurs when the kidney has been chronically exposed to an acid load for a number of days [274]. Good progress has been made recently in elucidating factors which are important in acute regulation. Chronic regulation is as yet rather poorly understood. Of particular interest for both types of regulation are the interrelationships between the mitochondrial metabolite transport and the enzymatic reactions leading to enhanced ammonia formation [6,9,274–277].

6.4.1. Acute regulation

Until recently, researchers interested in the regulation of acute acidosis were puzzled by the paradoxical difference between results obtained in whole tissue, and those obtained using isolated mitochondria. Due to the alkaline pH optimum of glutaminase, isolated mitochondria utilize glutamine more rapidly at alkaline pH, whereas glutamine is utilized more rapidly in intact tissue at acid pH.

The paradox has been explained recently by elucidation of the important role of α-ketoglutarate as a feedback inhibitor of ammonia formation by glutamate dehydrogenase and indirectly, glutaminase. At acid pH, three laboratories have demonstrated an accelerated rate of α-ketoglutarate oxidation [278–280]. The increase in α-ketoglutarate metabolism can be explained by an H^+-induced stimulation of α-ketoglutarate dehydrogenase activity [278,281], resulting from a decrease in the K_m for α-ketoglutarate at more acid pH values. Moreover, in intact mitochondria, Tannen et al. [279] and Schoolwerth and LaNoue [280] have demonstrated increased flux through α-ketoglutarate dehydrogenase leading to reduced levels of α-ketoglutarate in the mitochondrial matrix space. The pH-related decreases in α-ketoglutarate levels which occur in the tissue, and which can be demonstrated in isolated mitochondria, are important because α-ketoglutarate is a strikingly potent competitive inhibitor of glutamate dehydrogenase. The conclusion was initially drawn that lowered matrix levels of α-ketoglutarate were solely responsible for the acute increase in kidney ammoniagenesis. Very recent studies from the laboratory of the present reviewers suggest that one other factor is involved.

Due to the kinetic properties of the glutamate hydroxyl carrier, acid medium pH decreases the efflux of glutamate from the matrix space in kidney [104] as in liver [103]. Although medium pH has no effect on glutamate uptake in kidney [66], as opposed to liver mitochondria [95], the decrease in glutamate efflux leads to an increased concentration of matrix glutamate. Thus, at acid pH, the matrix glutamate/α-ketoglutarate ratio rises substantially and provides the driving force for augmented glutamate dehydrogenase flux [280]. The increase in glutamate deamination results in the increased ammonia formation noted initially at acid pH with no change in glutaminase flux. The augmented α-ketoglutarate metabolism, and decrease in matrix glutamate efflux, can also explain the observed decreases in total tissue concentrations of these two metabolites in acute acidosis in vivo.

Taking a somewhat different view, Goldstein [282] and Goldstein and Boylan [31] have noted that α-ketoglutarate at physiological concentrations inhibits uptake and deamidation of glutamine in isolated mitochondria. They have postulated that, since this occurs within seconds, the effects are due to α-ketoglutarate outside rather than inside the mitochondria, implying α-ketoglutarate inhibition of the glutamine transporter. However, Strzelecki and Schoolwerth [283] have demonstrated that the results of Goldstein can be explained by the formation of glutamate from α-ketoglutarate within the matrix space. These studies suggest that the inhibition by α-ketoglutarate of glutamine metabolism in rotenone-inhibited rat kidney mitochondria can be explained by glutamate inhibiting glutaminase, and do not require a separate effect of α-ketoglutarate on the glutamine carrier.

Baverel and Lund [284] demonstrated recently that bicarbonate, independent of pH, led to a decrease in ammonia formation by rat kidney cortical tubules. The mechanism by which this effect occurs remains to be clarified, although Baverel and Lund suggested that the effect might occur at the level of glutamate dehydrogenase. However, in light of the fact that the specific activities of the α-ketoglutarate pools in those studies was not constant, this explanation does not appear to be warranted. Moreover, recent studies by Tager and associates [285] suggested that bicarbonate inhibits glutamate metabolism in rat liver mitochondria at the level of succinate dehydrogenase. Whether the situation is different in kidney remains to be determined.

Simpson and co-workers have proposed that an increase in the pH gradient (ΔpH) across the inner mitochondrial membrane may contribute to the metabolic alterations which occur in acute metabolic acidosis. In studies performed with rotenone-inhibited mitochondria from rat and rabbit [275,287,290], they noted that, particularly with a bicarbonate buffer system, acidification of the medium led to an increase in the ΔpH. Since many metabolites distribute across this membrane in response to the ΔpH, they proposed that in acute acidosis a generalized increase in the transport of metabolites from cytosol to mitochondria occurs, with a resultant decrease in the cytosolic, and an increase in the mitochondrial, metabolite pools. Subsequent enhanced mitochondrial metabolism results in a return of the mitochondrial pool to its previous steady-state levels. They further suggest that the net effect is a decrease in the total tissue content of many metabolites which has been noted in renal cortical tissue in vivo. Although this hypothesis may, in part, explain some of the alterations, it is clearly too generalized to apply to all metabolites, even those transported in symport with protons. For example, other factors such as intramitochondrial metabolism (lowering the K_m for α-ketoglutarate) and the balance between uptake and efflux will determine the net effect of acidification on metabolite gradients across the mitochondrial membrane. Nevertheless, the theory proposed by Simpson and associates may provide an explanation for the decrease in cytosolic content of some metabolites in acute acidosis. In general, most studies have supported a close interrelationship between mitochondrial glutamate transport and the enhanced rate of glutamate deamination which appears to contribute substantially to the increase in ammonia formation observed acutely at acid pH. Many of these studies have been recently reviewed by Tannen [276] and Tannen and Sastrasinh [277]. They emphasize a role for a stimulation in α-ketoglutarate metabolism as well as alterations in glutamate transport across the inner mitochondrial membrane.

6.4.2. Chronic acidosis

The mechanism by which ammoniagenesis is stimulated in chronic acidosis is demonstrably different from the one which prevails in acute acidosis. The metabolic alteration which occurs in the chronic situation involves a stable change of the mitochondria themselves. Mitochondria isolated from kidney cortex of chronically acidotic animals exhibit rates of ammonia formation from glutamine more than 2-fold higher than control mitochondria. It has been demonstrated in vivo, and using isolated mitochondria, that the increase in ammonia formation from glutamine is secondary to enhanced flux through both glutamate dehydrogenase and glutaminase [68,288]. Although, in the rat, this may be explained in part by enhanced activity of glutaminase and glutamate dehydrogenase, this cannot explain the increase observed in dogs in which induction of these enzymes does not occur. Although the mechanisms may well be different in the two species, if a unifying explanation for the marked increase in ammonia synthesis exists, additional explanations must be brought to bear.

It has been suggested that transport of glutamine into mitochondria might be a factor limiting its metabolism [289,290], and that the activity of the transporter might be higher in chronic acidosis, thus accounting for the differences in metabolism of control and experimental mitochondria. This hypothesis has been developed and advanced by Simpson and co-workers [275,289–291]. Adam and Simpson [289,290] first noted that the accumulation of [U-^{14}C]glutamine by rat and dog kidney mitochondria was increased in both acute and chronic metabolic acidosis. These studies were performed with rotenone to prevent glutamate metabolism and to evaluate the uptake and deamidation steps of glutamine metabolism. In both acute and chronic metabolic acidosis increased counts from labelled glutamine were found in the matrix space. However, the counts were all in the form of glutamate and, since no glutamine was detected in the matrix, it was concluded that the transport rather than the deamidation of glutamine was rate limiting. Subsequently, Curthoys and Shapiro [292] came to the opposite conclusion and suggested that glutaminase, not glutamine transport, was the rate-limiting step. In contrast to the studies of Adam and Simpson, these workers demonstrated that if the mitochondria were purified to remove contaminating phosphate-independent glutaminase, and if phosphate, a potent activator of glutaminase, was removed from the incubation medium, glutamine was detectable in the matrix space. Moreover, in chronic metabolic acidosis the matrix levels of glutamine did not change, an observation which they interpreted to support a rate-controlling role for glutaminase rather than for the glutamine carrier. Subsequently, in additional experiments Simpson and Hecker [293,294] have evaluated the distribution of glutamine in mitochondria from normal and acidotic rat kidneys. By carefully removing the 'fluffy' layer from the mitochondrial preparations they demonstrated that no significant phosphate-independent glutaminase activity persisted. Moreover, by performing maneuvers designed to reduce the activity of glutaminase, such as low temperature and addition of inhibitors of glutaminase including glutamate and α-ketoglutarate, they concluded that the glutamine distribution space did not consistently exceed the mannitol distribution space.

Discrepancies among different laboratories may relate to the rapidity with which the mitochondria are separated from the incubation medium by silicone oil centrifugation. The levels of glutamine within the matrix space may be normally low. These low amounts of glutamine may be metabolized during separation of mitochondria from the media if separation is comparatively slow.

Thus, part of the difficulty in determining a potential role for the glutamine transporter in glutamine metabolism has been the lack of an available specific inhibitor of glutaminase. Recent studies by Kovacevic and co-workers have demonstrated that it is possible to achieve satisfactory inhibition of glutaminase by performing studies at 0 °C in the absence of phosphate. Utilizing this methodology, Kovacevic and Bajin [295,296] were able to load mitochondria with [^{14}C]glutamine and to measure the kinetics of efflux from the matrix space. Matrix levels in excess of 10 nmol · mg^{-1} protein were obtained by this method. Kovacevic and Bajin also compared the V_{max} of efflux to the activity of glutaminase at 0 °C and found that the rate of transport was an order of magnitude greater than the activity of glutaminase, suggesting that the capacity of the transporter may not limit glutamine metabolism in normal kidney mitochondria.

Since Kovacevic and Bajin were able to show high levels of glutamine in the matrix space, when mitochondria were loaded with glutamine under conditions in which the enzyme was inhibited, it is unlikely that the glutamine carrier is the same molecule as glutaminase, as suggested by Simpson and co-workers [294]. Additional studies are required to fully characterize the glutamine carrier. Sufficient studies exist to indicate that glutamine uptake does occur by a carrier-mediated process, and that it is inhibited by sulfhydryl reagents [295,296]. Further characterization of the kinetics of uptake is required.

In contrast to the studies of glutamine transport, several studies have suggested a potential role for the malate/α-ketoglutarate transporter in chronic acidosis [135,297]. Although Cheema-Dhadli and Halperin [135] were unable to demonstrate activation of the dicarboxylate (malate/phosphate) transporter in kidney cortex mitochondria from rats with chronic metabolic acidosis, Brosnan et al. [297] demonstrated that the malate/α-ketoglutarate carrier was activated in chronic acidosis. Additional studies are required to characterize this effect fully and to determine its role in the overall process of augmented ammonia formation in metabolic acidosis. However, it is interesting to speculate that an increased activity of the malate/α-ketoglutarate carrier in chronic acidosis could lead to lower levels of matrix α-ketoglutarate, since α-ketoglutarate is generated inside the mitochondria. Thus, control of ammoniagenesis in both chronic and acute acidosis could be a result of regulation of matrix α-ketoglutarate levels. The direct effect of α-ketoglutarate is to increase the K_m of matrix glutamate for glutamate dehydrogenase [144,298]. In this context, it may be significant that detailed studies of fluxes and substrate concentrations during metabolism of glutamine in isolated mitochondria showed that there was a dramatic decrease in the apparent K_m of matrix glutamate for glutamate dehydrogenase in mitochondria from chronically acidotic rats compared to controls [144,288].

7. *Conclusion*

In general, mechanistic studies of the mitochondrial metabolite transporters carried out in the last five years have provided evidence supporting a gated-pore, sequential model for transport. Structural studies of the adenine nucleotide carrier are more advanced than studies of the other transporters, and these provide the most clear cut evidence.

Nevertheless, viewed superficially, kinetic studies of the exchange transporters do not appear to lend strong support to the gated-pore model. Although hampered by grave technical difficulties, most of these studies are in agreement, and show that the carriers do not exhibit the ping-pong kinetics expected from a simple sequential system.

This could be due to the technical problems inherent in measuring initial rates. Nevertheless, the sheer weight of the evidence and the fact that the K_m's of the pyruvate transporter for exchange and for net uptake are similar, in the face of a 20-fold difference in V_{max}, argues against the results being purely artifactual. It may be important metabolically to maintain relatively constant cytosolic K_m's when intramitochondrial metabolite levels change. This could be achieved by negative cooperativity between identical subunits of the exchange transporters. As discussed above, each subunit might have a substrate binding site but negative cooperativity between subunits might prevent binding of more than one substrate at a time. Thus, binding at the empty substrate site would be necessary before the transported substrate could be released from the carrier. This phenomenon would preclude ping-pong kinetics. Negative cooperativity between potentially alternating sites has been proposed as a mechanism for several ATPase transport proteins [299–301].

Studies of the role of metabolic transporters in the control and modulation of metabolic fluxes have provided somewhat ambiguous results in the case of respiration and gluconeogenesis. The role of the glutamate transporter in the acute control of ammonia formation seems clear at this time, and a proposed role of the malate/α-ketoglutarate carrier in chronic acidosis is an intriguing possibility.

References

1 Klingenberg, M. (1970) Essays Biochem. 6, 119–159.
2 Massari, S., Frigeri, L. and Azzone, G.F. (1972) J. Membr. Biol. 9, 57–70.
3 Frietag, H., Neupert, W. and Benz, R. (1982) Eur. J. Biochem. 123, 629–636.
4 David, H. (1975) Biol. Zentralbl. 94, 129–153.
5 Parsons, D.F. and Yano, Y. (1967) Biochim. Biophys. Acta 135, 362–364.
6 LaNoue, K.F. and Schoolwerth, A.C. (1979) Ann. Rev. Biochem. 48, 871–922.
7 Meijer, A.J., Gimpel, J.A., Deleeuw, G.A., Tager, J.M. and Williamson, J.R. (1975) J. Biol. Chem. 250, 7728–7738.
8 Meijer, A.J., Gimpel, J.A., Deleeuw, G., Tischler, M.E., Tager, J.M. and Williamson, J.R. (1978) J. Biol. Chem. 253, 2308–2320.
9 Kovacevic, Z. and McGivan, J.D. (1983) Physiol. Rev. 63, 547–605.
10 Marchbanks, R., Fonnum, F., Grahame-Smith, D.G., Balazs, R., Machiyama, Y. and Patel, A.J. (1973) In Metabolic Compartmentation on the Brain (Balazs, R. and Cremer, J.E., eds.) pp. 21–70, MacMillan Press, London.
11 Wiskich, J.T. (1977) Annu. Rev. Plant Physiol. 28, 45–69.
12 Meijer, A.J. and Van Dam, K. (1981) In Membrane Transport (Bonting, S.L. and De Pont, J.J.H.H.M., eds.) pp. 235–254, Elsevier, Biomedical Press, Amsterdam.
13 Meijer, A.J. and Van Dam, K. (1974) Biochim. Biophys. Acta 346, 213–244.
14 Fonyó, A., Palmieri, F. and Quagliariello, E. (1976) In Horizons in Biochemistry and Biophysics (Quagliariello, E., Palmieri, F. and Singer, T.P., eds.) Vol. 2, pp. 60–105, Addison-Wesley, Massachusetts.
15 Klingenberg, M. (1980) J. Membr. Biol. 56, 97–105.
16 Vignais, P.V., Block, M.R., Boulay, F., Brandolin, G. and Lauquin, G.J.M. (1983) In Physical Chemistry of Transmembrane Ion Motions (Spach, G., ed.) Elsevier Science Publishers, Amsterdam.
17 Pedersen, P.L. and Wehrle, J.P. (1982) In Membranes and Transport (Martonosi, A.N., ed.) pp. 645–663, Plenum Publishing Co., New York.
18 Halestrap, A.P., Scott, R.D. and Thomas, A.P. (1980) Int. J. Biochem. 11, 97–105.
19 Parvin, R., Goswani, T. and Pande, S.V. (1980) Can. J. Biochem. 58, 822–830.
20 Chappell, J.B. (1968) Br. Med. Bull. 24, 150–157.
21 Palmieri, F. and Klingenberg, M. (1979) Methods Enzymol. 56, 279–301.
22 McGivan, J.D. and Klingenberg, M. (1971) Eur. J. Biochem. 20, 392–399.
23 Papa, S., Francavilla, A., Paradies, G. and Meduri, B. (1971) FEBS Lett. 12, 285–288.
24 Patel, T.B., Waymack, P.P. and Olson, M.S. (1980) Arch. Biochem. Biophys. 201, 629–635.
25 Debise, R., Brand, Y., Durand, R., Gachon, P. and Jeminet, G. (1977) Biochemie 59, 497–508.
26 McGivan, J.D., Bradford, N.M. and Beavis, A.D. (1977) Biochem. J. 162, 147–156.
27 Bradford, N.M. and McGivan, J.D. (1980) FEBS Lett. 113, 294–298.
28 Bryla, J. and Harris, E.J. (1976) FEBS Lett. 72, 331–336.
29 Papa, S., Lofrumento, N.E., Kanduc, D., Paradies, G. and Quagliariello, E. (1971) Eur. J. Biochem. 22, 134–143.
30 Stucki, J.W. (1976) FEBS Lett. 61, 171–175.
31 Goldstein, L. and Boylan, J.M. (1978) Am. J. Physiol. 234, F514–F521.
32 Cybulski, R.L. and Fisher, R.R. (1977) Biochemistry 16, 5116–5120.
33 Pande, S.V. (1975) Proc. Natl. Acad. Sci. U.S.A. 72, 883–887.
34 Ramsay, R.R. and Tubbs, P.K. (1976) Eur. J. Biochem. 69, 299–303.
35 Pande, S.V. and Parvin, R. (1980) J. Biol. Chem. 255, 2994–3001.
36 Idell-Wenger, J.A. (1981) J. Biol. Chem. 256, 5597–5603.
37 Pfaff, E. and Klingenberg, M. (1968) Eur. J. Biochem. 6, 66–79.
38 LaNoue, K.F., Mizani, S.M. and Klingenberg, M. (1978) J. Biol. Chem. 253, 191–198.
39 Klingenberg, M. and Rottenberg, H. (1977) Eur. J. Biochem. 73, 125–130.
40 Wilson, D.F., Nelson, D. and Erecinska, M. (1982) FEBS Lett. 143, 228–232.
41 Wilson, D.F., Erecinska, M. and Shramm, V.L. (1983) J. Biol. Chem. 258, 10464–10473.
42 Aprille, J.R. and Austin, J. (1981) Arch. Biochem. Biophys. 212, 689–699.

43 Austin, J. and Aprille, J.R. (1983) Arch. Biochem. Biophys. 222, 321–325.
44 LaNoue, K.F., Watts, J.A. and Koch, C.D. (1981) Am. J. Physiol. 241, H663–H671.
45 Asimakis, G.K. and Sordahl, L.A. (1981) Am. J. Physiol. 241, H672–H678.
46 Aprille, J.R. and Asimakis, G.K. (1980) Arch. Biochem. Biophys. 201, 564–575.
47 Pollak, J.K. and Sutton, R. (1980) Biochem. J. 192, 75–83.
48 Hamman, H.C. and Haynes, R.C. (1983) Arch. Biochem. Biophys. 223, 85–94.
49 Abou-Khalil, S. and Hanson, J.B. (1977) Arch. Biochem. Biophys. 183, 581–587.
50 LaNoue, K.F. and Tischler, M.E. (1974) J. Biol. Chem. 249, 7522–7528.
51 Nicholls, D.G. (1974) Eur. J. Biochem. 50, 305–315.
52 Tischler, M.E., Pachence, J., Williamson, J.R. and LaNoue, K.F. (1976) Arch. Biochem. Biophys. 173, 448–462.
53 Davis, E.J., Bremer, J. and Akerman, K.E. (1980) J. Biol. Chem. 255, 2277–2283.
54 Williamson, J.R. and Viale, R.O. (1979) In Methods in Enzymology (Fleischer, S. and Packer, L., eds.) Vol. 56, pp. 252–278, Academic Press, New York.
55 LaNoue, K.F., Koch, C.D. and Meditz, R.B. (1982) J. Biol. Chem. 257, 13740–13748.
56 Gbelska, Y., Subik, J., Suoboda, A., Goffeau, A. and Kovac, L. (1983) Eur. J. Biochem. 130, 281–286.
57 Parvin, R. and Pande, S.V. (1979) J. Biol. Chem. 254, 5423–5429.
58 Calvin, J. and Tubbs, P.K. (1976) J. Reprod. Fert. 48, 417–420.
59 Purvis, J.L. and Lowenstein, J.M. (1961) J. Biol. Chem. 236, 2794–2803.
60 Robinson, B.H. and Halperin, M.L. (1970) Biochem. J. 116, 229–233.
61 Day, D.A. and Wiskich, J.T. (1981) Arch. Biochem. Biophys. 211, 100–107.
62 Jakob, A., Williamson, J.R. and Asakuro, T. (1971) J. Biol. Chem. 246, 7623–7631.
63 Williamson, J.R., Jakob, A. and Refino, C. (1971) J. Biol. Chem. 246, 7632–7641.
64 Greenhouse, W.V.V. and Lehninger, A.L. (1976) Cancer Res. 36, 1392–1396.
65 Williamson, D.H., Lung, P. and Krebs, H.A. (1967) Biochem. J. 103, 514–527.
66 Schoolwerth, A.C., LaNoue, K.F. and Hoover, W.J. (1983) J. Biol. Chem. 258, 1735–1739.
67 Kovacevic, Z., McGivan, J.D. and Chappell, J.B. (1970) Biochem. J. 118, 265–274.
68 Kunin, A.S. and Tannen, R.L. (1979) Am. J. Physiol. 237, F55–F62.
69 Sluse, F.E., Meijer, A.J. and Tager, J.M. (1971) FEBS Lett. 18, 149–153.
70 Robinson, B.H. and Gei, J. (1975) Can. J. Biochem. 53, 643–647.
71 Lowenstein, J.M. (1968) In Metabolic Roles of Citrate (Goodwin, T.W., ed.) p. 61, Academic Press, New York.
72 Klingenberg, M., Riccio, P. and Aquila, H. (1978) Biochim. Biophys. Acta 503, 193–210.
73 Aquila, H., Eiermann, W., Babel, W. and Klingenberg, M. (1978) Eur. J. Biochem. 85, 545–560.
74 Boulay, F., Brandolin, G., Lauquin, G.J.M., Jolles, J., Jolles, P. and Vignais, P. V. (1979) FEBS Lett. 98, 161–164.
75 Buchanan, B.B., Eiermann, N., Riccio, P., Aquila, H. and Klingenberg, M. (1976) Proc. Natl. Acad. Sci. U.S.A. 73, 2280–2284.
76 Vignais, P.V. (1976) Biochim. Biophys. Acta 456, 1–38.
77 Eiermann, W., Aquila, H. and Klingenberg, M. (1977) FEBS Lett. 74, 209–214.
78 Hackenberg, H., Riccio, P. and Klingenberg, M. (1978) Eur. J. Biochem. 88, 373–378.
79 Hallermayer, G., Zimmermann, R. and Neupert, W. (1977) Eur. J. Biochem. 81, 523–532.
80 Harney, M.A., Hallermayer, G., Korb, H. and Neupert, W. (1977) Eur. J. Biochem. 81, 533–544.
81 Kovac, L., Kolarov, J. and Subik, J. (1977) Mol. Cell. Biochem. 14, 11–14.
82 Lauqin, G.J.M., Block, M.R., Boulay, F., Brandolin, G. and Vignais, P.V. (1982) In EBEC Lyon Reports, Vol. 2, pp. 449–450.
83 Zimmermann, R., Paluch, U., Sprinzl, M. and Neupert, W. (1979) Eur. J. Biochem. 99, 247–252.
84 Maccechini, M.-L., Rudin, Y., Blobel, G. and Schatz, G. (1979) Proc. Natl. Acad. Sci. U.S.A. 76, 343–347.
85 Zimmermann, R. and Neupert, W. (1980) Eur. J. Biochem. 109, 217–229.
86 Zwizinski, C., Schleyer, M. and Neupert, W. (1983) J. Biol. Chem. 258, 4071–4074.
87 Singer, S.J. (1974) Ann. Rev, Biochem. 43, 805–833.
88 Jencks, W.P. (1982) In Membranes and Transport (Martonosi, A.N., ed.) pp. 515–520, Plenum Publishing Co., New York.

89 Rothstein, A. (1980) Ann. N.Y. Acad. Sci. 358, 96–102.
90 Cleland, W.W. (1963) Biochim. Biophys. Acta 67, 104–137.
91 Quagliariello, E., Palmieri, F., Prezioso, G. and Klingenberg, M. (1969) FEBS Lett. 4, 251–254.
92 Palmieri, F., Stipani, I., Quagliariello, E. and Klingenberg, M. (1972) Eur. J. Biochem. 26, 587–594.
93 Palmieri, F., Prezioso, G., Quagliariello, E. and Klingenberg, M. (1971) Eur. J. Biochem. 22, 66–74.
94 Palmieri, F., Quagliariello, E. and Klingenberg, M. (1972) Eur. J. Biochem. 29, 408–416.
95 LaNoue, K.F., Schoolwerth, A.C. and Pease, A.J. (1983) J. Biol. Chem. 258, 1735–1739.
96 Palmieri, F. and Quagliariello, E. (1969) Eur. J. Biochem. 8, 473–481.
97 Mitchell, P. and Moyle, J. (1969) Eur. J. Biochem. 9, 149–155.
98 Meyer, J. and Vignais, P.M. (1973) Biochim. Biophys. Acta 325, 375–384.
99 Bradford, N.M. and McGivan, J.D. (1973) Biochem. J. 134, 1023–1029.
100 Hoek, J.B. and Njogu, R.M. (1976) FEBS Lett. 71, 341–346.
101 Brouwer, A., Smits, G.G., Tas, J., Meijer, A.J. and Tager, J.M. (1973) Biochemie 55, 717–725.
102 Williamson, J.R., Hoek, J.B., Murphy, E., Coll, K.E. and Njogu, R.M. (1980) Ann. N.Y. Acad. Sci. 341, 593–608.
103 Hoek, J.B., Coll, K.E. and Williamson, J.R. (1983) J. Biol. Chem. 258, 54–58.
104 Schoolwerth, A.C., LaNoue, K.F. and Hoover, W.J. (1984) Am. J. Physiol. 246, F266–F271.
105 Pande, S.V. and Parvin, R. (1978) J. Biol. Chem. 253, 1563–1573.
106 Halestrap, A.P. (1978) Biochem. J. 172, 377–387.
107 Titherage, M.A. and Coore, H.G. (1975) Biochem. J. 150, 553–556.
108 Paradies, G. and Papa, S. (1978) In Bioenergetics at Mitochondrial and Cellular Levels (Wojtczak, L., Lenartowicz, F. and Zborowski, J., eds.) pp. 39–77, Nenchi Institute of Experimental Biology, Warsaw.
109 Zahlten, R.N., Hochberg, A.A., Stratman, F.W. and Lardy, H.A. (1972) FEBS Lett. 21, 11–13.
110 Vaartjes, W.J., Geelen, M.J.H. and Van den Berg, S.G. (1979) Biochim. Biophys. Acta 548, 38–47.
111 Halestrap, A.P. (1977) Biochem. Soc. Trans. 5, 216–219.
112 Martin, A.D. and Titherage, M.A. (1983) Biochem. Soc. Trans. 11, 78–81.
113 Papa, S. and Paradies, G. (1974) Eur. J. Biochem. 49, 265–274.
114 Zwiebel, F.M., Schwabe, U., Olson, M.S. and Scholz, R. (1982) Biochemistry 21, 346–358.
115 Halestrap, A.P. (1975) Biochem. J. 148, 85–96.
116 Paradies, G. and Papa, S. (1975) FEBS Lett. 52, 149–152.
117 Paradies, G. and Papa, S. (1976) FEBS Lett. 62, 318–321.
118 Coty, W.A. and Pedersen, P.L. (1974) J. Biol. Chem. 249, 2593–2598.
119 Quagliariello, E. and Palmieri, F. (1967) Eur. J. Biochem. 4, 20–27.
120 Wohlrab, H. (1980) In First European Bioenergetics Conference, pp. 301–302, Patron, Editore, Bologna.
121 Wohlrab, H. (1980) J. Biol. Chem. 255, 8170–8173.
122 Wohlrab, H. (1980) Ann. N.Y. Acad. Sci. 358, 364–367.
123 Wohlrab, H., Brigida, M., Flowers, N. and Costello, D. (1981) In Calcium and Phosphate Transport Across Biomembranes (Bronner, F. and Peterlik, M., eds.) pp. 99–107, Academic Press, New York.
124 Wohlrab, H. and Flowers, N. (1982) J. Biol. Chem. 257, 28–31.
125 Wohlrab, H., Collins, A. and Costello, D. (1984) Biochemistry, in press.
126 Freitag, H. and Kadenbach, B. (1978) Eur. J. Biochem. 83, 53–57.
127 Azzone, G.F., Massari, S. and Pozzan, T. (1976) Biochim. Biophys. Acta 423, 15–26.
128 Wehrle, J., Citron, N. and Pedersen, P.L. (1978) J. Biol. Chem. 253, 8598–8603.
129 Wehrle, J. and Pedersen, P.C. (1979) J. Biol. Chem. 254, 7269–7275.
130 Lotscher, H.R., Schwerzmann, K. and Carafoli, E. (1979) FEBS Lett. 99, 194–197.
131 Sluse, F.E., Ranson, M. and Liebecq, C. (1972) Eur. J. Biochem. 25, 207–217.
132 Sluse, F.E., Goffart, G. and Liebecq, C. (1973) Eur. J. Biochem. 32, 283–291.
133 Sluse, F.E., Duychaerts, C., Liebecq, C. and Sluse-Goffart, C.M. (1979) Eur. J. Biochem. 100, 3–17.
134 LaNoue, K.F., Walajtys, E.I. and Williamson, J.R. (1973) J. Biol. Chem. 248, 7171–7183.
135 Cheema-Dhadli, S. and Halperin, M.L. (1978) Can. J. Biochem. 56, 23–28.
136 Tubbs, P. and Ramsay, R. (1979) In Function and Molecular Aspects of Biomembrane Transport (Quagliariello, E., Palmieri, F., Papa, S. and Klingenberg, M., eds.) pp. 279–286, Elsevier Biomedical Press, Amsterdam.

137 LaNoue, K.F. and Williamson, J.R. (1971) Metabolism 20, 119–140.
138 LaNoue, K.F., Meyer, A.J. and Brouwer, A. (1974) Arch. Biochem. Biophys. 161, 544–550.
139 Murphy, E., Coll, K.E., Viale, R.O., Tischler, M.E. and Williamson, J.R. (1979) J. Biol. Chem. 254, 8369–8376.
140 LaNoue, K.F., Bryla, J. and Bassett, D.J.P. (1974) J. Biol. Chem. 249, 7514–7521.
141 DeHaan, E.J. and Tager, J.M. (1968) Biochim. Biophys. Acta 153, 98–112.
142 Duszynski, J., Mueller, G. and LaNoue, K.F. (1978) J. Biol. Chem. 253, 6149–6157.
143 Siess, E.A., Brocks, D.G., Lattke, H.K. and Wieland, O.H. (1977) Biochem. J. 166, 225–235.
144 Schoolwerth, A.C. and LaNoue, K.F. (1980) J. Biol. Chem. 255, 3403–3411.
145 LaNoue, K.F., Duszynski, J., Watts, J.A. and McKee, E. (1979) Arch. Biochem. Biophys. 195, 578–590.
146 LaNoue, K.F. and Duszynski, J. (1977) In Mechanism of Proton and Calcium Pumps (Azzone, G.F., Avron, M., Metcalfe, J.C., Quagliariello, E. and Siliprandi, N., eds.) pp. 297–307, Elsevier-North Holland, Amsterdam.
147 LaNoue, K.F. and Watts, J.A. (1979) In Function and Molecular Aspects of Biomembrane Transport (Quagliariello, E., Palmieri, F., Papa, S. and Klingenberg, M., eds.) pp. 345–353, Elsevier-North Holland, Amsterdam.
148 Palmieri, F., Stipani, I. and Iacobazzi, V. (1979) Biochim. Biophys. Acta 555, 531–546.
149 Palmieri, F., Stipani, I., Iacobazzi, V. and Quagliariello, E. (1979) In Function and Molecular Aspects of Biomembrane Transport (Quagliariello, E., Palmieri, F., Papa, S. and Klingenberg, M., eds.) pp. 335–344, Elsevier-North Holland, Amsterdam.
150 Barbour, R.I. and Chan, S.H.P. (1981) J. Biol. Chem. 256, 1940–1948.
151 Duyckaerts, C., Sluse-Goffart, C.M., Fux, J.-P., Sluse, F.E. and Liebecq, C. (1980) Eur. J. Biochem. 106, 1–6.
152 Klingenberg, M. (1976) In The Enzymes of Biological Membranes: Membrane Transport (Martonosi, A.N., ed.) Vol. 3, pp. 383–438, Plenum, New York.
153 Nohl, H. and Klingenberg, M. (1978) Biochim. Biophys. Acta 503, 155–169.
154 Duszynski, J., Savina, H.Z. and Wojtczak, L. (1978) FEBS Lett. 86, 9–13.
155 Verdouw, H. and Bertina, R.M. (1973) Biochim. Biophys. Acta 325, 385–396.
156 Zoccarato, F., Ruglo, M., Siliprandi, D. and Siliprandi, N. (1981) Eur. J. Biochem. 114, 195–199.
157 Souverijn, J.H.M., Huisman, L.A., Rosing, J. and Kemp, A., Jr. (1973) Biochim. Biophys. Acta 305, 185–198.
158 Klingenberg, M. and Pfaff, E. (1968) In Metabolic Roles of Citrate (Goodwin, T.W., ed.) pp. 105–127, Academic Press, New York.
159 Krämer, R. and Klingenberg, M. (1980) Biochemistry 19(3), 556–560.
160 Krämer, R. and Klingenberg, M. (1982) Biochemistry 21, 1082–1089.
161 Vignais, P.V., Vignais, P.M., Lauquin, G. and Morel, F. (1973) Biochemie 55, 763–778.
162 Vignais, P.V., Vignais, P.M. and DeFaye, G. (1973) Biochemistry 12, 1508–1519.
163 Weidemann, M.J., Erdelt, H. and Klingenberg, M. (1970) Eur. J. Biochem. 16, 313–335.
164 Henderson, P.J.F. and Lardy, H.A. (1970) J. Biol. Chem. 245, 1319–1326.
165 Klingenberg, M. and Buckholz, M. (1973) Eur. J. Biochem. 38, 346–358.
166 Erdelt, H., Weidemann, M.J., Buckholz, M. and Klingenberg, M. (1972) Eur. J. Biochem. 30, 107–122.
167 Klingenberg, M., Riccio, P., Aquila, H., Schmidt, B., Brebe, K. and Toptsch, P. (1974) In Membrane Proteins in Transport and Phosphorylation (Azzone, G.F., Avron, M., Metcalfe, J.C., Quagliariello, E. and Siliprandi, N., eds.) pp. 229–243, Elsevier-North Holland, Amsterdam.
168 Brandolin, G., Meyer, C., DeFaye, G., Vignais, P.M. and Vignais, P.V. (1974) FEBS Lett. 46, 149–153.
169 Krämer, R. and Klingenberg, M. (1977) FEBS Lett. 82, 363–367.
170 Block, M.R., Zaccai, G., Lauquin, G.J.M. and Vignais, P.V. (1982) Biochem. Biophys. Res. Commun. 109, 471–477.
171 Hackenberg, H. and Klingenberg, M. (1980) Biochemistry 19, 548–555.
172 Klingenberg, M. (1981) In Mitochondria and Microsomes (Lee, C.P., Schatz, G. and Dallner, G., eds.) pp. 293–316, Addison-Wesley Inc., New York.

173 Aquila, H. and Klingenberg, M. (1982) Eur. J. Biochem. 122, 141–145.
174 Aquila, H., Eiermann, W. and Klingenberg, M. (1982) Eur. J. Biochem. 122, 133–139.
175 Klingenberg, M., Hackenberg, H., Krämer, R., Lin, C.S. and Aquila, H. (1980) Ann. N.Y. Acad. Sci. 358, 83–95.
176 Lauquin, G.J.M., Villier, C., Michejda, J., Brandolin, G., Boulay, F., Cesarini, R. and Vignais, P.V. (1978) In The Proton and Calcium Pumps (Azzone, G.F., Avron, M., Metcalfe, J.M., Quagliariello, E. and Siliprandi, N., eds.) pp. 251–262, Elsevier/North Holland, Amsterdam, New York.
177 Block, M.R., Lauquin, G.J.M. and Vignais, P.V. (1981) Biochemistry 20, 2692–2699.
178 Block, M.R., Lauquin, G.J.M. and Vignais, P.V. (1981) FEBS Lett. 131, 213–218.
179 Klingenberg, M. and Appel, M. (1980) FEBS Lett. 119, 195–199.
180 Grabe, C. and Klingenberg, M. (1979) Biochim. Biophys. Acta 546, 539–550.
181 Block, M.R., Lauquin, G.J.M. and Vignais, P.V. (1983) Biochemistry 22, 2202–2208.
182 Block, M.R., Lauquin, G.J.M. and Vignais, P.V. (1982) Biochemistry 21, 5451–5457.
183 Boos, K.S., Schlimme, E. and Ikehara, M. (1978) Z. Naturforsch. 33C, 552–556.
184 Boos, K.S. and Schlemme, E. (1979) Biochemistry 18, 5304–5309.
185 Aquila, H., Misra, D., Eulitz, M. and Klingenberg, M. (1982) Hoppe-Seyler's Z. Physiol. Chem. 363, 345–349.
186 Saraste, M. and Walker, J.E. (1982) FEBS Lett. 144, 250–254.
187 Engelman, D.M., Henderson, R., McLachlan, A.D. and Wallace, B.A. (1980) Proc. Natl. Acad. Sci. U.S.A. 77, 2023–2027.
188 Stoeckenius, W. and Bogomolni, R.A. (1982) Ann. Rev. Biochem. 52, 587–616.
189 Boulay, F., Lauquin, G.J.M., Tsugita, A. and Vignais, P.V. (1983) Biochemistry 22, 477–484.
190 Bogner, W., Aquila, H. and Klingenberg, M. (1982) FEBS Lett. 146, 259–261.
191 Bogner, W., Aquila, H. and Klingenberg, M. (1983) In Structure and Function of Membrane Proteins (Quagliariello, E. and Palmieri, F., eds.) pp. 145–156, Elsevier Science Publishers, Amsterdam.
192 Fonyó, A. (1979) Pharmacol. Ther. 7, 627–645.
193 Coty, W.A. and Pedersen, P.L. (1975) J. Biol. Chem. 250, 3515–3521.
194 Hadvary, P. and Kadenbach, B. (1976) Eur. J. Biochem. 67, 573–581.
195 Breand, Y., Touraille, S., Debise, R. and Durand, R. (1976) FEBS Lett. 65, 1–7.
196 Palmieri, F., Genchi, G., Stipani, I., Francia, F. and Quagliariello, E. (1974) In Membrane Proteins in Transport and Phosphorylation (Azzone, G.F., Klingenberg, M., Quagliariello, E. and Siliprandi, N., eds.) pp. 245–256, Elsevier/North Holland, Amsterdam.
197 dePinto, V., Tommasino, M., Palmieri, F. and Kadenbach, B. (1982) FEBS Lett. 148, 103–106.
198 Banerjee, R.K., Shertzer, H.G., Kanner, B.I. and Racker, E. (1977) Biochem. Biophys. Res. Commun. 75, 772–778.
199 Banerjee, R.K. and Racker, E. (1979) Membr. Biochem. 2, 203–225.
200 Mende, P., Hüther, F.-J. and Kadenbach, B. (1983) FEBS Lett. 158, 331–334.
201 Thomas, A.P. and Halestrap, A.P. (1981) Biochem. J. 196, 471–479.
202 Schulz, H. and Racker, E. (1979) Biochem. Biophys. Res. Commun. 89, 134–140.
203 Stipani, I., Krämer, R., Palmieri, F. and Klingenberg, M. (1980) Biochem. Biophys. Res. Commun. 97, 1206–1214.
204 Saint-Macary, M. and Foucher, B. (1983) Biochem. Biophys. Res. Commun. 113, 205–211.
205 Elbers, R., Heldt, H.W., Schmucker, P., Soboll, S. and Wiese, H. (1974) Hoppe-Seyler's Z. Physiol. Chem. 355, 378–393.
206 Zuurendonk, P.F. and Tager, J.M. (1974) Biochim. Biophys. Acta 333, 393–399.
207 Akerboom, T.P.M., van der Meer, R. and Tager, J.N. (1979) Techniques in Metabolic Research B205, pp. 1–33, Elsevier/North-Holland, Amsterdam.
208 Krebs, H.A. (1967) Adv. Enzymol. Regul. 5, 409–434.
209 Rolleston, F.S. (1972) Curr. Top. Cell. Regul. 5, 47–75.
210 Newsholme, E.A. and Start, C. (1973) In Regulation in Metabolism, pp. 1–33, Wiley, London.
211 Newsholme, E.A. (1980) FEBS Lett. 117, K121–K134.
212 Duszynski, J., Groen, A.K., Wanders, R.J.A., Vervoorn, R.C. and Tager, J.M. (1982) FEBS Lett. 146, 262–266.

213 Newsholme, E.A. and Crabtree, B. (1979) J. Mol. Cell. Cardiol. 11, 839–856.
214 Higgins, J. (1965) In Control of Energy Metabolism (Chance, B., Estabrook, R.K. and Williamson, J.R., eds.) pp. 13–46, Academic Press, New York.
215 Kacser, H. and Burns, J.A. (1973) Symp. Soc. Exp. Biol. 32, 65–104.
216 Kacser, H. and Burns, J.A. (1979) Biochem. Soc. Trans. 7, 1149–1161.
217 Heinrich, R. and Rapoport, T.A. (1974) Eur. J. Biochem. 42, 97–105.
218 Tager, J.M., Wanders, R.J.A., Groen, A.K., Kunz, W., Bohnensack, R., Küster, U., Letko, G., Bohme, G., Duszynski, J. and Wojtczak, L. (1983) FEBS Lett. 151, 1–9.
219 Groen, A.K., Wanders, R.J.A., Westerhoff, H.V., van der Meer, R. and Tager, J.M. (1982) J. Biol. Chem. 257, 2754–2757.
220 Chance, B. and Williams, G.R. (1956) Adv. Enzymol. 17, 65–134.
221 Jacobus, W.E., Moreadith, R.W. and Vandegaer, K.M. (1982) J. Biol. Chem. 257, 2397–2402.
222 Ericinska, M. and Wilson, D.F. (1982) J. Mem. Biol. 70, 1–14.
223 Vignais, P.V. and Lauquin, G.J.M. (1979) Trends Biochem. Sci. 4, 90–92.
224 Erecinska, M., Stubbs, M., Miyata, Y., Ditre, C.M. and Wilson, D.F. (1977) Biochim. Biophys. Acta 462, 20–35.
225 Forman, N.G. and Wilson, D.F. (1982) J. Biol. Chem. 257, 12908–12915.
226 Forman, N. and Wilson, D.F. (1983) J. Biol. Chem. 258, 8649–8655.
227 Klingenberg, M. and Heldt, H.W. (1982) In Metabolic Compartmentation (Sies, H., ed.) pp. 101–122, Academic Press, London.
228 Zoratti, M., Pietrobon, D. and Azzone, G.F. (1983) Biochim. Biophys. Acta 723, 59–70.
229 Laris, P.D. (1977) Biochim. Biophys. Acta 459, 110–118.
230 Wikstrom, M., Krab, K. and Saraste, M. (1981) Ann. Rev. Biochem. 50, 623–655.
231 Lemasters, J.J., Grunwald, R. and Emaus, R.K. (1984) J. Biol. Chem., in press.
232 Davis, E.J. and Lumeng, L. (1975) J. Biol. Chem. 250, 2275–2282.
233 Davis, E.J. and Davis-van Thienen, W.I.A. (1978) Biochem. Biophys. Res. Commun. 83, 1260–1266.
234 Küster, U., Bohnensack, R. and Kunz, W. (1976) Biochim. Biophys. Acta 440, 391–402.
235 Soboll, S., Scholz, R. and Heldt, H.W. (1978) Eur. J. Biochem. 87, 377–390.
236 Akerboom, T.P.M., Bookelman, H. and Tager, J.M. (1977) FEBS Lett. 74, 50–54.
237 Akerboom, T.P.M., Bookelman, H., Zuurendonk, P.F., van der Meer, R. and Tager, J.M. (1978) Eur. J. Biochem. 84, 413–420.
238 Van der Meer, R., Akerboom, T.P.M., Groen, A.K. and Tager, J.M. (1978) Eur. J. Biochem. 84, 421–428.
239 Lemasters, J.J. and Sowers, A.E. (1979) J. Biol. Chem. 254, 1248–1251.
240 Wanders, R.J.A., Groen, A.K., Meijer, A.J. and Tager, J.M. (1981) FEBS Lett. 132, 201–206.
241 Tager, J.M., Wanders, R.J.A., Groen, A.K., van der Meer, R., Akerboom, T.P.M. and Meijer, A.K. (1981) Acta Biol. Med. Ger. 40, 895–906.
242 Groen, A.K., van der Meer, R., Westerhoff, H.V., Wanders, R.J.A., Akerboom, T.P.M. and Tager, J.M. (1982) In Metabolic Compartmentation (Sies, H., ed.) pp. 9–37, Academic Press, New York.
243 Tager, J.M., Groen, A.K., Wanders, R.J.A., Duszynski, J., Westerhoff, H.V. and Vervoorn, R.C. (1983) Biochem. Soc. Trans. 11, 40–43.
244 Matthews, P.M., Bland, J.L., Gadian, D.G. and Radda, G.K. (1981) Biochem. Biophys. Res. Commun. 103, 152–159.
245 Letko, G. and Küster, U. (1979) Acta Biol. Med. Ger. 38, 1379–1385.
246 Kunz, W., Bohnensack, R., Böhme, G., Küster, U., Letko, G. and Schönfeld, P. (1981) Arch. Biochem. Biophys. 209, 219–229.
247 Gellerich, F. and Saks, V.A. (1982) Biochem. Biophys. Res. Commun. 105, 1473–1481.
248 Bohnensack, K., Küster, U. and Letko, G. (1982) Biochim. Biophys. Acta 680, 271–280.
249 Gellerich, F.N., Bohnensack, R. and Kunz, W. (1983) Biochim. Biophys. Acta 722, 381–391.
250 Adam, P.A.J. and Haynes, R.C. Jr. (1969) J. Biol. Chem. 244, 6444–6450.
251 Yamazaki, R.K., Sax, R.D. and Hauser, M.A. (1977) FEBS Lett. 75, 295–299.
252 Yamazaki, R.K. (1975) J. Biol. Chem. 250, 7924–7930.
253 Titherage, M.A. and Coore, H.G. (1976) FEBS Lett. 63, 45–50.
254 Titherage, M.A. and Coore, H.G. (1976) FEBS Lett. 71, 73–78.
255 Halestrap, A.P. (1978) Biochem. J. 172, 389–398.

256 Titherage, M.A., Binder, S.B., Yamazaki, R.K. and Haynes, R.C. (1978) J. Biol. Chem. 253, 3356–3360.
257 Titherage, M.A. and Haynes, R.C. (1980) J. Biol. Chem. 255, 1471–1477.
258 Bryla, J., Harris, E.J. and Plumb, J.A. (1977) FEBS Lett. 80, 433–448.
259 Haynes, R.C. (1976) Metabolism 25, 1361–1363.
260 Hughes, B.P. and Barrett, G.J. (1978) Biochem. J. 176, 295–304.
261 Halestrap, A.P. (1982) Biochem. J. 204, 37–47.
262 Armston, A.E., Halestrap, A.P. and Scott, R.D. (1982) Biochim. Biophys. Acta 681, 429–439.
263 Quinlan, P.T., Thomas, A.P., Armston, A.E. and Halestrap, A.P. (1983) Biochem. J. 214, 395–404.
264 Hoek, J.B., Moehren, G. and Waring, A.J. (1983) In Isolation, Characterization and Use of Hepatocytes (Harris, R.A. and Cornell, N.W., eds.) pp. 245–250, Elsevier Biomedical Press, Amsterdam.
265 Siess, E.A., Fahimi, F.M. and Wieland, O.H. (1981) Hoppe-Seyler's Z. Physiol. Chem. 362, 1643–1651.
266 Siess, E.A. (1983) Hoppe-Seyler's Z. Physiol. Chem. 364, 279–290.
267 LaNoue, K.F., Strzelecki, T. and Finch, F. (1984) J. Biol. Chem. 259, 4116–4121.
268 Strzelecki, T., Thomas, J.A., Koch, C.D. and LaNoue, K.F. (1984) J. Biol. Chem. 259, 4122–4129.
269 LaNoue, K.F. and Strzelecki, T. (1983) Fed. Proc. 42, Abstr. 2265.
270 Jensen, C.B., Sistare, F.D., Hamman, H.C. and Haynes, R.C. (1983) Biochem. J. 210, 819–827.
271 Thomas, A.P. and Halestrap, A.P. (1981) Biochem. J. 198, 551–564.
272 Rognstad, R. (1983) Int. J. Biochem. 15, 1417–1421.
273 Narins, R.G., Emmett, M., Rascoff, J., Jones, E.R. and Relman, A.S. (1982) Contrib. Nephrol. 31, 47–52.
274 Tannen, R.L. (1978) Am. J. Physiol. 235, F265–F277.
275 Simpson, D.P. (1983) Kidney Int. 23, 785–793.
276 Tannen, R.L. (1983) Med. Clin. N. Am. 67, 781–798.
277 Tannen, R.L. and Sastrasinh, S. (1983) Kidney Int., in press.
278 Lowry, M. and Ross, B.D. (1980) Biochem. J. 190, 771–780.
279 Tannen, R.L. and Kunin, A.S. (1981) Am. J. Physiol. 240, F120–F126.
280 Schoolwerth, A.C. and LaNoue, K.F. (1983) Am. J. Physiol. 244, F399–F408.
281 Schoolwerth, A.C., Strzelecki, T., LaNoue, K.F. and Hoover, W.J. (1982) Contrib. Nephrol. 31, 127–133.
282 Goldstein, L. (1976) Biochem. Biophys. Res. Commun. 70, 1136–1141.
283 Strzelecki, T. and Schoolwerth, A.C. (1981) Biochem. Biophys. Res. Commun. 102, 588–593.
284 Baverel, G. and Lund, P. (1979) Biochem. J. 184, 599–606.
285 Wanders, R.J.A., Meijer, A.J., Groen, A.K. and Tager, J.M. (1983) Eur. J. Biochem. 133, 245–254.
286 Simpson, D.P. and Hager, S.R. (1979) J. Clin. Invest. 63, 704–712.
287 Hager, S.R. and Simpson, D.P. (1982) Mol. Physiol. 2, 203–211.
288 Schoolwerth, A.C., Nazar, B.L. and LaNoue, K.F. (1978) J. Biol. Chem. 253, 6177–6183.
289 Adam, W. and Simpson, D.P. (1974) J. Clin. Invest. 54, 165–174.
290 Simpson, D.P. and Adam, D.P. (1975) Med. Clin. North Am. 59, 555–567.
291 Simpson, D.P. (1980) J. Biol. Chem. 255, 7123–7128.
292 Curthoys, N.D. and Shapiro, R.A. (1978) J. Biol. Chem. 253, 63–68.
293 Simpson, D.P. and Hecker, J. (1982) Kidney Int. 21, 774–779.
294 Simpson, D.P. and Hecker, J. (1982) Contrib. Nephrol. 31, 105–110.
295 Kovacevic, Z. and Bajin, K. (1982) Contrib. Nephrol. 31, 111–114.
296 Kovacevic, Z. and Bajin, K. (1982) Biochim. Biophys. Acta 687, 291–295.
297 Brosnan, J.T., Redmond, W., Morgan, D. and Whalen, P. (1980) Int. J. Biochem. 12, 131–133.
298 Schoolwerth, A.C., Hoover, W.J., Daniel, C.H. and LaNoue, K.F. (1980) Int. J. Biochem. 12, 145–149.
299 Kurobe, Y., Nelson, R.W. and Ikemoto, N. (1983) J. Biol. Chem. 258, 4381–4389.
300 Gresser, M.J., Myers, J.A. and Boyer, P.D. (1982) J. Biol. Chem. 257, 12030–12038.
301 Grubmeyer, C., Cross, R.L. and Penefsky, H.S. (1982) J. Biol. Chem. 257, 12092–12100.

Ernster (ed.) Bioenergetics

© *1984 Elsevier Science Publishers B.V.*

CHAPTER 9

The uptake and the release of calcium by mitochondria

ERNESTO CARAFOLI [a] and GIANLUIGI SOTTOCASA [b]

[a] *Laboratory of Biochemistry, Swiss Federal Institute of Technology (ETH), Universitätsstr. 16, 8092 Zurich, Switzerland and* [b] *Istituto di Chimica Biologica, Università di Trieste, via A. Valerio 32, 34127 Trieste, Italy*

1. Early history

The first indication that mitochondria can absorb large amounts of Ca^{2+} from the surrounding medium came from a study published by Slater and Cleland in 1953 [1]. Since the absorption took place at 0 °C, it was concluded that the process did not involve mitochondrial energy transformations. Three years later, Chance [2] published the very interesting observation that Ca^{2+} uncoupled respiration in isolated liver mitochondria in a transient way, and observed that the duration of the uncoupled period was proportional to the amount of Ca^{2+} added to the medium. This suggested to him that Ca^{2+}, unique among uncouplers, was somehow 'consumed' in the process. Although it was not concluded that Ca^{2+} was accumulated into mitochondria in a respiration-dependent way, the experiment was a clear indication that this was indeed the case. That respiring mitochondria are able to accumulate Ca^{2+} from the surrounding medium was stated directly by Saris in 1959 [3]. In his experiments the acidification of the external medium upon addition of Ca^{2+} to mitochondria was related to the formation of an H^+ gradient in exchange for the accumulated Ca^{2+}. The evidence for the uptake of Ca^{2+} was once again indirect, but the experiments of Saris and especially the concept of an H^+/Ca^{2+} exchange are remarkable nevertheless, and pioneer future themes and problems. It is perhaps appropriate to mention here that the chemiosmotic hypothesis, which propelled these concepts into the center of the scientific stage had not yet been proposed, and was thus still a very long way from being accepted by the majority of the specialists.

Direct evidence for the ability of mitochondria to take up Ca^{2+} from the medium in a respiration-dependent process was provided in 1961–1962 by Vasington and Murphy [4]. De Luca and Engström [5] extended the finding by demonstrating that respiration was not an absolute requirement for the uptake reaction, provided that

ATP was added to the medium. Thus, the concept emerged that the uptake of Ca^{2+} could be driven alternatively by respiration or ATP hydrolysis. In a comprehensive study published in 1962, Vasington and Murphy [6] established, however, that the uptake of Ca^{2+} required *both* respiration and ATP (or ADP), a finding that was not understood at that moment, and which could be rationalized only by later work with the protective effect at ATP against the structural damage to mitochondria induced by the accumulation of excess Ca^{2+}. Vasington and Murphy could make mitochondria accumulate very large amounts of Ca^{2+}, up to 2.6 nmol per mg of protein, and made a number of other important observations. The process was inhibited by uncoupling agents, indicating its active nature, but was insensitive to the classical inhibitor of ATP synthesis and hydrolysis, oligomycin. The observation, together with the finding that during the uptake of Ca^{2+} no phosphorylation of ADP took place, established the process as an alternative to oxidative phosphorylation, and indicated that the reaction harvested the energy made available by the respiratory chain at a 'level' prior to the point where oligomycin acts in the ATP synthesis process. In addition to coupled respiration and ATP (ADP), the Ca^{2+}-uptake process also required inorganic phosphate, a finding which was soon rationalized by Lehninger et al. [7], who found that inorganic phosphate was accumulated by mitochondria together with Ca^{2+}. In the very limited aqueous space within mitochondria the solubility product of Ca^{2+} and phosphate was soon exceeded, and precipitates of insoluble Ca-phosphate salts formed. This removed Ca^{2+} from solution, effectively decreasing the transmembrane gradient of ionized Ca^{2+}, and thus permitting the accumulation of additional Ca^{2+}. Electronmicroscopic observations of mitochondria after the accumulation of large amounts of Ca^{2+} (and phosphate) indeed revealed numerous electron-opaque masses within the profiles of the organelles [8]. These masses were isolated [9], and shown to contain, as their main components, Ca^{2+} and phosphate in the molar ratio of hydroxyapatite (1.67 : 1). That the main precipitate formed inside mitochondria was hydroxyapatite had been suggested by Rossi and Lehninger in 1963 [10], based on the measurement of the stoicheiometry of the accumulated Ca^{2+} and phosphate. One finding that relates to the requirement for ATP (ADP) in the respiration-supported Ca^{2+} uptake process, and explains why oligomycin, the classical inhibitor of ATP hydrolysis by mitochondria, had no effect on the uptake reaction, was made by Carafoli et al. in 1965 [11]. ATP (ADP) penetrated into mitochondria during the accumulation of Ca^{2+} and phosphate, most likely to stabilize the insoluble Ca-phosphate granules, or even to prime their precipitation. It is well known that adenine nucleotides are adsorbed by growing hydroxyapatite crystals. However, the deposits inside mitochondria do not show the typical X-ray diffraction pattern of crystalline hydroxyapatite, and remain amorphous indefinitely. This is a remarkable phenomenon, which suggests the existence, in mitochondria loaded with Ca^{2+} and phosphate, of substances that prevent the crystalline transformation of hydroxyapatite. Phosphocitrate has recently been identified [12] as a possible candidate for such an inhibitory role.

2. *The 'limited loading' of mitochondria with Ca^{2+}*

In the early studies mentioned in the preceding section, mitochondria were normally exposed to concentrations of Ca^{2+} in excess of 1 mM, and the uptake process was allowed to continue until massive amounts of Ca^{2+} (and phosphate) were accumulated and precipitated inside. Under these conditions, in spite of the frequent presence of ATP/ADP and Mg^{2+} as protecting agents, mitochondria became structurally deranged, and no longer suited for fine studies on the mechanism of the uptake reaction. Therefore, once the process of energy-linked uptake of Ca^{2+} was clearly established as a reality, it became obvious that it was preferable to study it under conditions of 'limited loading', i.e., conditions under which mitochondria were exposed to concentrations of Ca^{2+} not exceeding 50–100 μM, in the absence of phosphate. These conditions are presumably non-injuring, lead to the accumulation of 100–150 nmol of Ca^{2+} per mg protein maximally, and are essentially still used in the majority of studies today. Based on 'limited loading' studies a number of properties of the uptake process have been established, mechanistic and otherwise. In the discussion to follow, emphasis will be placed on those among the developments that appear, in the light of what has since become known, particularly significant. A number of comprehensive reviews have appeared and they can be consulted for complete information [13–16]. It is appropriate to point out at this point that the ability to accumulate Ca^{2+} is common to mitochondria from nearly all sources examined. However, two sources, yeasts and blowfly flight muscles at certain stages of the development of the insect, produce mitochondria with very limited ability to take up Ca^{2+} [17–19].

In a study published in 1964, Rossi and Lehninger [20] extended the observations made by Chance in 1956 on the stimulation of oxygen consumption by mitochondria, and established that approximately $2Ca^{2+}$ were transported into mitochondria as a couple of electrons traversed each one of the respiratory chain segments which were called, in the terminology of the 1960's, 'coupling sites'. They thus concluded that the Ca^{2+}/oxygen ratio for the full respiratory chain span was 6. In other words, mitochondria use the same amount of respiratory energy to phosphorylate one ADP molecule or to transport across the membrane $2Ca^{2+}$.

The matter of the ejection of H^+ during Ca^{2+} uptake was reinvestigated by a number of authors [21–24], who found ratios between H^+ ejection and Ca^{2+} uptake approaching 2. Concomitant with the ejection of H^+, increase in the titratable alkalinity of the mitochondria was measured [25–28].

Depending on whether the source of energy is the respiratory chain or added ATP, the uptake of Ca^{2+} is inhibited by respiratory chain inhibitors or oligomycin, respectively. As expected, protonophoric uncouplers inhibit in both cases. The competitive inhibition by Mg^{2+} [29], is of particular interest, since it may have physiological significance due to the presence of Mg^{2+} in the cytosol of cells. Mg^{2+}, however, is not transported by mitochondria through the Ca^{2+}-uptake system which, on the other hand, translocates Mn^{2+}, Ba^{2+} and Sr^{2+} (see Refs. 13–15 for reviews).

At mM concentrations Mg^{2+} decreases the affinity of the uptake system for Ca^{2+} very significantly, and transforms the rate versus concentration curve for Ca^{2+} uptake from hyperbolic to sigmoidal [30] (see below). The two most useful inhibitors of Ca^{2+} uptake, however, are La^{3+} [31] and ruthenium red [32,33], which are both active maximally at μM concentrations or less. Ruthenium red is a hexavalent cation, formerly thought to be a specific reagent for mucopolysaccharides and glycoproteins, but now known to interact with a number of other compounds as well [34]. Ruthenium red has now become an indispensable tool in studies of mitochondrial Ca^{2+} uptake. Since the amount that produces maximal inhibition corresponds to about one mol per mol of respiratory chain carriers, it is normally assumed that ruthenium red interacts with the putative Ca^{2+} carrier of the inner membrane (see below). Recent work [35] has indeed shown that the inhibition is not due to non-specific surface charge effects, but to the interaction of ruthenium red with a specific component of the inner membrane.

One important problem, particularly if one wants to involve mitochondria in the regulation of intracellular Ca^{2+}, is the affinity and the maximal rate of the uptake system. The latter is now accepted to be as high as 10 nmol · mg^{-1} protein · s^{-1} at 25 °C in the presence of phosphate [36], (see Ref. 14 for a comprehensive discussion), but considerably lower in its absence. At the physiological temperature of about 37 °C, this rate is approximately doubled. As mentioned before, phosphate increases the total amount of Ca^{2+} that can be accumulated, but its stimulating effect on the initial rate of uptake must evidently be explained differently. It may be mentioned here that the penetration of phosphate into mitochondria lowers the intramitochondrial pH, which increases after the penetration of Ca^{2+}. This may remove a hypothetical limiting step in the Ca^{2+} uptake process, but it must be emphasized here that the precipitation of Ca^{2+} and phosphate within the matrix as amorphous hydroxyapatite involves *net* acidification, and thus complicates the matter considerably.

As for the affinity of the mitochondrial uptake system for Ca^{2+}, this has been a matter for debate for a long time (see Ref. 14 for a review). It is now generally agreed that the K_m of the Ca^{2+} uptake system is in the range of 1–15 μM depending on the tissue and, most likely, also on the method of measurement chosen. Values much in excess of these, as can be found in the earlier literature, appear to have been due to the inclusion of the inhibitor Mg^{2+} in the reaction medium. Mg^{2+} competes very efficiently with Ca^{2+} for the transport system, particularly at the very low Ca^{2+}/Mg^{2+} ratios normally employed in experiments of this type. As mentioned before, in the presence of Mg^{2+} the curve relating the velocity of transport to the Ca^{2+} concentration has a pronounced sigmoidal character. Although Crompton et al. [29] have shown that the curve becomes hyperbolic if Mg^{2+} is eliminated from the medium, other authors [37] have found sigmoidal kinetics even in the absence of Mg^{2+}, and have concluded from their data that the putative Ca^{2+} carrier has two binding sites for Ca^{2+}. More recent data, however, have shown that the kinetics of Ca^{2+} influx into mitochondria is invariably hyperbolic in experiments carried out under non-limiting conditions [38,39].

3. *Mechanism of the Ca^{2+} uptake process*

The first attempts to rationalize the mechanism of the Ca^{2+} uptake process produced reaction schemes which were naturally based on the conceptions and terminology of the so-called 'chemical hypothesis' of energy coupling [40]. In 1970, however, it was found that Ca^{2+} uptake could be driven, in the absence of respiratory activity or ATP hydrolysis, by diffusion potentials created by a gradient of H^+ in the presence of protonophoric uncouplers, by the permeant anion thiocyanate, or by K^+ in the presence of valinomycin [41,42]. Clearly, these results were in line with the general principles of the chemiosmotic theory of energy coupling [43], indicating that the process of Ca^{2+} uptake had to be rationalized according to these principles. These observations, in fact, indicated that the process was electrophoretic in nature, a suggestion which was supported and extended by Rottenberg and Scarpa in 1974 [44]. These authors carried out quantitative measurements in which the distribution of Ca^{2+} across the membrane of energized mitochondria was compared to that of Rb^+. In the presence of valinomycin, the latter moves across the membrane electrophoretically, and distributes according to the membrane potential. In the case of Ca^{2+}, Rottenberg and Scarpa had to use the permeant anion acetate to reduce the binding of Ca^{2+} to membrane sites, and to minimize the difference between total Ca^{2+} taken up and Ca^{2+} free in the matrix. They found that the distribution of Ca^{2+} is also a function of the transmembrane potential, and occurs with a net charge transfer of 2.

$$E = \frac{RT}{nF} \ln \frac{[Rb_i^+]}{[Rb_o^+]}$$

$$\frac{^{86}Rb_i}{^{86}Rb_o} = 13.3 \qquad \text{log ratio} = 1.24$$

$$\frac{^{45}Ca_i}{^{45}Ca_o} = 230 \qquad \text{log ratio} = 2.36$$

$$\text{thus, } \log \frac{[Ca_i^{2+}]}{[Ca_o^{2+}]} = 2 \log \frac{[Rb_i^+]}{[Rb_o^+]}$$

Thus, it can be concluded that the driving force for Ca^{2+} uptake is the electrical component (the $\Delta\psi$) of the proton motive force created by respiration or by ATP hydrolysis. Since the accepted value for the $\Delta\psi$ in energized mitochondria is of the order of -180 mV, it follows that the purely electrophoretic influx of Ca^{2+} would lead to a gradient of Ca_i/Ca_o equal to 10^6. This follows from the development of the Nernst equation against a potential of -180 mV

$$-180 = \frac{RT}{2F} \cdot \ln \frac{[Ca_i^{2+}]}{[Ca_o^{2+}]}$$

$$-180 = 0.03 \cdot \log 10^6$$

As will be discussed later, however, the gradient of ionized Ca^{2+} across the membrane of energized mitochondria seems to be only a fraction of the value of 10^6 predicted on the basis of the equilibrium of the Nernst equation. This unexpected observation seemed paradoxical at the beginning, but it has later been instrumental in steering thoughts and experiments towards the possibility of separate routes for Ca^{2+} release from mitochondria.

4. *Molecular components of the calcium uptake system*

That Ca^{2+} transport could be the result of a carrier-mediated process was already suggested by Lehninger and Carafoli in 1969 [45] and a similar suggestion was repeatedly made later [37,46]. One obvious development was the attempt to solubilize and isolate the putative carrier molecule. Criteria used were the ability to bind Ca^{2+} with high affinity and the sensitivity of the binding to inhibitors such as lanthanides and ruthenium red. The first report on the solubilization of high-affinity lanthanum-sensitive Ca^{2+} binding activity from mitochondria appeared in 1971 [47] and Table 9.1 lists a number of preparations from mitochondria reported to contain high affinity binding components for Ca^{2+}. Many of these preparations contained, in addition to protein, a certain amount of sugars, and have thus been regarded for a long time as glycoproteins. The fractions have been characterized to different extents

TABLE 9.1

Calcium binding preparations from mitochondria

Origin	Nature	Mol. weight	Affinity for Ca^{2+}	Reference
Liver	Glycoprotein	33 000	High	Sottocasa et al. [48]
Liver	Protein	?	High	Evtodienko et al. [49]
Liver	Glycolipoprotein	150–200 000	High	Gómez-Puyou et al. [50]
Adrenal C.	Glycoprotein	16 000	Low	Kimura et al. [51]
Liver	Glycopeptide	5000	High	Tashmukhamedov et al. [52]
Liver	Glycolipoprotein	70 000	High	Utsumi and Oda [53]
Liver	Protein	35–40 000	High	Carafoli et al. [54]
Heart	Peptide	3000	High	Jeng et al. [55]
Liver	Glycoprotein	32 000	High	Happel et al. [56]
Liver	Lipopeptide	3000 (?)	High	Sokolove and Brenza [57]

but it is reasonable to assume that preparations obtained by similar methods and having similar properties contain the same active component. Thus, the preparations by Evtodienko et al. [49], and Tashmukhamedov et al. [52] are likely to contain the same active component as that first solubilized by Lehninger in 1971 [47] and subsequently purified to electrophoretic homogeneity by Sottocasa et al. [48]. The same probably holds for the preparation from adrenal cortex mitochondria [51].

Among the preparations capable of high affinity lanthanum- and ruthenium red-sensitive Ca^{2+} binding, the best defined is that originally isolated by Sottocasa et al. [48,58]. The preparation was originally obtained from a variety of animal mitochondria (see Ref. 59 for a review) by osmotic shock followed by two electrophoretic steps and desalting by gel filtration. The preparation showed a single protein band in polyacrylamide gel electrophoresis, had a molecular weight of about 30 kDa, and stained in the gels with specific reagents for both protein and sugars. Chemical analysis revealed the presence of typical glycoprotein sugar components such as neutral- and aminosugars and sialic acid. Early amino acid analysis [60,61] indicated the presence of a high percentage of dicarboxylic amino acids. The preparation contained variable amounts of bound Ca^{2+} and Mg^{2+}. If the desalting step was omitted, preparations with high lipid contents were obtained. The preparation consistently bound Ca^{2+} with high affinity (K_d of the order of 0.1 μM). In addition, a variable number of low affinity binding sites for Ca^{2+} were also present. A recent development [60] of the technique for the preparation of this Ca^{2+}-binding component has led to the isolation of a protein whose SDS-polyacrylamide gel molecular weight was about 14000. The minimum molecular weight calculated from amino acid composition was 15577. The amino acid analysis confirmed the high content of glutamic and aspartic acid, but no sugars were found. The conclusion by Panfili et al. [62] is that this protein component, which has been named calvectin, is identical to the protein moiety of previous 'Ca^{2+} binding glycoprotein' preparations. The evidence in favour of the conclusion may be summarized as follows: (1) the old and the new preparations are electrophoretically indistinguishable; (2) the two preparations behave in the same way in electrofocusing runs, indicating the presence of two very closely migrating protein components in each preparation with only a small difference in isoelectric point (around pH 3.4); (3) when immobilized on biogel P-2 columns, calvectin specifically absorbs antibodies prepared against the old 'Ca^{2+} binding glycoprotein' (these antibodies may be subsequently eluted and shown to inhibit mitochondrial Ca^{2+} uptake); (4) high affinity Ca^{2+} binding to calvectin occurs with the same K_d (0.8×10^{-7} M) as that published for the old 'Ca^{2+}-binding glycoprotein' [58]. Calvectin is thus to be considered as the active component of the previous preparation and it is therefore legitimate to propose that some carbohydrates and, under some conditions, lipids co-purified with the old 'Ca^{2+}-binding glycoprotein'. Native preparations of calvectin showing an electrophoretic mobility and gel permeation properties compatible with a molecular weight around 33000 could be regarded as dimers, with a binding capacity of $2Ca^{2+}$ per dimer. Interestingly, extensively purified calvectin has virtually no low affinity binding sites for Ca^{2+}. Concerning the intramitochondrial location of calvectin, early data suggested that this component [48] could be easily extracted from the intermembrane space. Subsequently [63], the conclusion was reached that calvectin could be present also in outer and inner membrane preparations. Later [64,65] it was shown that calvectin could be *reversibly* and *specifically* associated to the inner mitochondrial membrane. Its presence in the intermembrane space, therefore, could be the result of loose association with the inner membrane. Its presence in outer membrane preparations could conceivably be accounted for by calvectin entrapped in outer membrane vesicles or associated to contaminating inner membrane vesicles. The finding that the association to the inner mitochondrial membrane can be modulated by the presence of Ca^{2+} and/or Mg^{2+}, as well as by respiration-driven Ca^{2+} uptake [66], may be physiologically relevant. Experiments supporting the concept that calvectin may be a necessary component of the mitochondrial Ca^{2+} permeabilization system were presented by Sandri et al. [66] (see also Ref. 59) who could show that Ca^{2+}-preloaded mitochondria became impermeable to Ca^{2+} when treated with the uncoupler pentachlorophenol under appropriate conditions. The treatment released calvectin from mitochondria and a negative linear correlation was found between the amount of protein released and the residual rate of Ca^{2+} transport. Thus, this component was visualized as necessary, although perhaps not sufficient, for Ca^{2+} transport across the inner mitochondrial membrane. Additional evidence came from immunological studies by Panfili et al. [67], in which antibodies raised by injecting rabbits with a preparation of the 'Ca^{2+} binding glycoprotein' were shown to inhibit respiratory chain-driven Ca^{2+} uptake in mitochondria and mitoplasts. 50% inhibition was obtained with either 10 μg of antibody per mg mitochondrial protein or 2 μg antibody per mg mitoplast protein. At these concentrations no effects were observed on electron transport or respiratory control ratios. These results have recently been confirmed with antibodies against purified calvectin (Panfili and Bernardi, personal communication). The study of the effect of anticalvectin antibodies has now been extended to a number of experimental situations [68]. The inhibited systems were: (a) the uptake of Ca^{2+} coupled to ATP hydrolysis; (b) the uptake coupled to K^+ extrusion; (c) the efflux of Ca^{2+} induced by uncouplers [69]; (d) the efflux induced by EDTA or EGTA; (e) the efflux induced by the addition of acetoacetate or oxaloacetate [70] (see later). The specificity of the effect was confirmed by the lack of inhibitory activity towards Ca^{2+} movements which do not involve the electrophoretic-uptake system such as that coupled to Na^+ influx or Ca^{2+}/H^+ exchange [71] (see later).

Further evidence in favour of a direct involvement of calvectin in mitochondrial Ca^{2+} uptake came from experiments in which calvectin-depleted mitochondria were found to have only about 10% of the original initial rate of Ca^{2+} uptake. These depleted mitochondria could be restored to the original Ca^{2+} transporting capacity (higher than 1 $\mu mol\ Ca^{2+} \cdot min^{-1} \cdot mg^{-1}$ protein) either by readdition of a mitochondrial extract or by treatment with very low amounts (0.1 $\mu g \cdot mg^{-1}$ mitochondrial protein) of purified calvectin in the presence of 0.1 mM Mg^{2+} [72]. The effect obtained by the readdition of the crude extract could be abolished by pretreatment with anticalvectin antibodies. Since calvectin is apparently not an integral protein, the possibility that a Ca^{2+} ionophore could be present in the inner mitochondrial membrane, operating in series with a peripheral recognition site such as calvectin, has been considered in the past [61,73]. Evidence in literature for the presence of compounds acting as Ca^{2+} ionophores has been provided by Jeng et al. [55,74], by isolating from heart mitochondria a peptide of about 3 kDa, which was named calciphorin, and which promoted Ca^{2+} transport through an organic phase in a ruthenium red-sensitive way. The specificity for the transported species and the affinity and inhibitor sensitivity were all compatible with the putative function as a Ca^{2+} ionophore [57,74]. The peptide has been recently isolated also from liver mitochondria [57]. The problem of the possible interplay of the 3 kDa peptide with calvectin, and the related matter of the possible function of the two components, requires further study. A puzzling aspect is the apparent existence of two ruthenium red-sensitive steps, which would be present in a system consisting of calvectin plus calciphorin operating in series.

5. The reversibility of the Ca^{2+} influx system and the problem of a separate route for Ca^{2+} efflux from mitochondria

It is easy to see that the electrophoretic process of Ca^{2+} uptake, if permitted to attain equilibrium against a transmembrane potential of -180 mV, would lead to either an intolerably high concentration of ionized Ca^{2+} in the mitochondrial matrix, or to the extreme impoverishment of the cytosolic Ca^{2+}. The latter condition is ruled out by the existence, in the cytoplasm of all cells, of numerous reactions that are obligatorily modulated by Ca^{2+} in the μM or sub-μM range. In fact, whenever direct measurements have been possible, it has been found that the concentration of free Ca^{2+} in the cytosol oscillates around 0.1 μM [75–78].

As a result, then, if the electrophoretic influx route would be allowed to reach equilibrium against a transmembrane potential of -180 mV, and in the face of a free Ca^{2+} concentration in the cytoplasm of 0.1 μM, the matrix concentration of free Ca^{2+} would reach the value of 100 mM. This is clearly impossible in view of the existence, in the mitochondrial matrix, of a number of key enzymes that are modulated by Ca^{2+} around the μM level [79]. One way out of the dilemma would be by assuming equilibration of the Ca^{2+}-uptake system against a transmembrane potential lower than -180 mV, i.e., by postulating that the transmembrane potential is allowed to fluctuate widely towards substantially lower values. This, however, is extremely unlikely, since the membrane potential is a fundamental element in mitochondrial energy coupling, and its wide fluctuation would have serious consequences on all aspects of mitochondrial function. Another, and more logical, way out of the dilemma would be by postulating that the electrophoretic uptake route does not reach equilibrium, i.e., it operates irreversibly, never mediating the efflux of Ca^{2+}. Efflux of Ca^{2+} would be catalyzed by an independent route, which would have to be electrically silent, or even driven in the direction of Ca^{2+} efflux by the negative potential inside the inner membrane. Depending on the relative efficiencies of the uptake and release routes, the gradient of free Ca^{2+} across the membrane of energized mitochondria could thus, in principle, attain any value between 1 and 10^6, except that, for the reasons mentioned before, high values will never be permitted.

The first direct indication for the existence of a Ca^{2+} release route separated from the uptake pathway came from the observation by Rossi et al. [80] that ruthenium red, which completely blocks the electrophoretic uptake of Ca^{2+}, fails to inhibit the release of accumulated Ca^{2+} induced by uncouplers. Evidently, since the uptake route was blocked in this experiment by ruthenium red (and later experiments have shown that ruthenium red remains bound in the presence of uncouplers [81]), the exit of Ca^{2+} had to occur in this case through an independent pathway. Uncouplers, however, are unphysiological means of inducing Ca^{2+} release, and have no relevance to the release mechanism operating in vivo, which is known to proceed at a much slower rate [82].

Much emphasis has been placed in recent years on possible mechanisms for Ca^{2+} release. Work in several laboratories has now led to the identification of one release route that is now accepted as physiologically significant and of a number of other mechanisms that, although not yet conclusively accepted, may also play a role in vivo. One of them, induced by the transition of the redox state of mitochondrial pyridine nucleotides towards oxidation, has led to very interesting developments and will thus be discussed in some detail even if its in vivo role is disputed. Other Ca^{2+}-release processes, e.g., those induced by inorganic phosphate and by fatty acids [83–85] are at the moment highly controversial from the standpoint of their physiological significance, and will thus not be discussed in this chapter.

6. *The Na^+-activated Ca^{2+} release route*

In 1974, Carafoli et al. [86] discovered that Na^+ discharged Ca^{2+} from heart mitochondria under fully energized conditions, i.e., under conditions in which the electrophoretic Ca^{2+} influx route was fully operative. The experiment suggested very strongly the 'cycling' of Ca^{2+} across the membrane, since the addition of ruthenium red before the Na^+ pulse greatly accelerated the rate of Ca^{2+} release. Evidently, Ca^{2+} left mitochondria on the route promoted by Na^+, and re-entered on the electrophoretic uniporter. Blockade of the latter with ruthenium red unmasked the full activity of the release route (i.e., it accelerated the efflux of Ca^{2+}), since the process was now uncomplicated by the re-uptake of the lost Ca^{2+}. The phenomenon was Na^+ specific: with the partial exception of Li^+, no other cation induced Ca^{2+} release. The last finding was of great interest indeed, since Na^+ had been specifically tested with regard to release-inducing agents of possible physiological significance. It was felt that Na^+, which is present in the cytosol, and probably undergoes fluctuations during the functional cycle of cells (at least excitable cells like those of heart), could provide a link between events at the plasma membrane and the mitochondrial Ca^{2+} store. In the original experiments by Carafoli et al. [86] very high concentrations of Na^+ had been used. In a series of studies carried out subsequently by Crompton, and Carafoli et al. [87–90], the process was studied in great detail using much lower concentrations of Na^+, well within the physiological range. Crompton et al. [87] found that the rate of Ca^{2+} release from heart mitochondria was linked to the concentration of added Na^+ by a markedly sigmoidal dependence, with Hill coefficients between 2 and 3, suggesting the involvement of two or more Na^+ in the process. The effect by Na^+ was interpreted to indicate that the ruthenium red insensitive system catalyzed an exchange between Na^+ outside and Ca^{2+} inside, providing a mechanism for the exit of Ca^{2+} against its electrochemical gradient. The Na^+/Ca^{2+} exchanger thus provides the second leg of a cycle in which Ca^{2+} enters mitochondria electrophoretically on the ruthenium red-sensitive component, and exits in exchange for Na^+. The Na^+ that has entered mitochondria in exchange for Ca^{2+} would then return to the extramitochondrial space in exchange for H^+ on the well known Na^+/H^+ antiporter described by Mitchell and Moyle [91] (Fig. 9.1).

Nicholls [92] was able to rule out the alternative possibility of a primary Ca^{2+}/H^+ exchange, coupled to Na^+ influx to collapse the pH gradient created by the efflux of Ca^{2+}, by showing that the dissipation of the pH gradient with nigericin in the presence of K^+ did not promote Ca^{2+} release from Ca^{2+}-loaded mitochondria. Interestingly, the Na^+-induced efflux process attained half-maximum velocity at about 8 mM Na^+; this is not far from the value for intracellular Na^+ activity in heart [93], which was the source of mitochondria for the original experiments on the Na^+-induced release route. Given the pronounced sigmoidal character of the kinetics of the process, it is clear that in the range of ionic concentrations which may be considered physiological (4–10 mM) even a comparatively small variation of the intracellular Na^+ may produce significant changes in the rate of Ca^{2+} efflux. In turn, this would correspond to significant variations in the cytosolic Ca^{2+} concentration, at least in tissues of high mitochondrial content (e.g., heart, which contains about 10% net weight in mitochondria [94]). The maximal velocity of the Na^+-induced release process, calculated by Crompton et al. [87] for heart mitochondria at

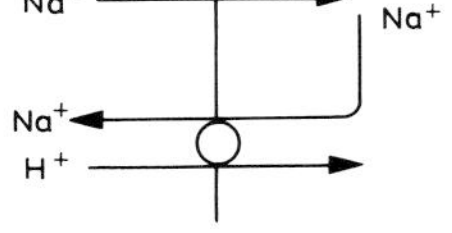

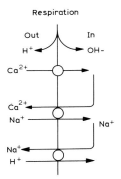

Fig. 9.1. The Na^+/Ca^{2+} cycle of mitochondria.

25 °C, is of the order of 0.25 nmol $Ca^{2+} \cdot mg^{-1}$ protein $\cdot s^{-1}$ (probably 0.5 at 37 °C). The affinity of the system for Ca^{2+}, evaluated by measuring the activity of the exchanger in the Ca^{2+}-Ca^{2+} exchange mode for external Ca^{2+} [88] is expressed by a K_m of 13 μM. Clearly, the rate of Na^+-induced Ca^{2+} efflux is inadequate to compete with that of the electrophoretic influx uniporter at concentrations of Ca^{2+} optimal for the latter. Under physiological conditions, however, i.e., sub-μM external Ca^{2+} and mM Mg^{2+}, the two systems may efficiently compete. The general problem of the integration of the Na^+-induced release route and of the electrophoretic uniporter into the physiological framework will be considered in more detail at the end of this chapter. At this point, suffice it to say that additional complicating factors must also be considered, among them the stimulation of the Na^+/Ca^{2+} exchanger by K^+ [95] and its inhibition by Mg^{2+} [94].

As mentioned repeatedly above, the Na^+/Ca^{2+} exchanger is insensitive to ruthenium red. It is, however, sensitive to La^{3+}, but in a range of concentration that is much higher than for the electrophoretic uniporter. In addition, whereas the uniporter is maximally inhibited by Tm^{3+}, and minimally by La^{3+}, the exchanger is inhibited by them in the opposite order [90]. One point of importance is the stoicheiometry of the Na^+/Ca^{2+} exchange. It follows from what has been discussed in the preceding paragraphs that the exchanger, if operational in the presence of a normal membrane potential negative inside, must be either electrically silent, or exchange > 2 external Na^+ per one internal Ca^{2+}. When tested in the Ca^{2+}-Ca^{2+} exchange mode, the system is indeed electrically neutral (i.e., the $Ca^{2+}:Ca^{2+}$ stoicheiometry is one). In the conventional Na^+/Ca^{2+} exchange mode, it also exchanges at least $2Na^+$ per $1Ca^{2+}$ [96], as was already indicated by the fact that Hill coefficients approaching 2 could be calculated from the sigmoidal curve linking the rate of Ca^{2+} release to the external Na^+ concentration [86].

The Na^+-induced route, originally described in heart mitochondria, is very active in mitochondria isolated from a number of excitable tissues [88]. Crompton et al. [89] had originally concluded that the exchanger was inactive in some mitochondrial types (e.g., liver), but more recent work has shown that it operates, although at a rate that is very low, also in mitochondria that were previously thought to be insensitive (see, e.g., Ref. 97). It is clear, however, that tissue differences exist, and that the exchanger operates as a very active route only in the category of mitochondria which Crompton et al. [89] had generalized under the heading 'excitable tissues'.

One logical consequence of the Na^+/Ca^{2+} cycle, which is a futile cycle, is energy dissipation. Crompton et al. [88] have indeed demonstrated that Na^+, added to heart mitochondria after the accumulation of a pulse of Ca^{2+}, uncouples respiration in a ruthenium red-sensitive way. The level of energy-dissipation due to the Na^+/Ca^{2+} cycle in vivo, however, is most likely marginal, if one considers that the Ca^{2+} uptake leg of the cycle, which determines the actual uncoupling of mitochondrial respiration, probably functions at a fraction of its maximal capability (see above). It has been mentioned already that the calculated rate of Ca^{2+} recycling, admittedly in liver mitochondria where the Na^+-promoted release route is very poorly represented, corresponds to the re-uptake of only about 0.05 nmol $Ca^{2+} \cdot mg^{-1}$

protein · s^{-1} [82]. This value can be compared to the figure of 3–10 nmol Ca^{2+} for the maximal rate of uptake on the electrophoretic uniporter.

7. Calcium movements evoked by changes in the redox state of pyridine nucleotides

An observation made by Lehninger et al. in 1978 [98] showed that Ca^{2+}-preloaded mitochondria from various tissues released Ca^{2+} in response to the addition of acetoacetate or oxaloacetate. The release was reversed by a subsequent addition of β-hydroxybutyrate. The interpretation of this phenomenon was that the redox state of pyridine nucleotides controlled the rate of entry and exit of Ca^{2+}, thus regulating its concentration in the external medium. In the experiment, mitochondria were maintained in an energized state throughout the experiment by the addition of succinate plus rotenone or other pyridine nucleotide-independent substrates. The phenomenon observed was not ruthenium red-sensitive, suggesting that the efflux pathway for Ca^{2+} activated by the shift in pyridine nucleotide redox state was different from the uptake pathway. It was not decided whether the NADH/NAD couple or the NADPH/NADP couple was involved in the regulation. The dependence of calcium movements on the pyridine nucleotide redox level was further investigated by Panfili et al. [70], who observed that the acetoacetate- (or oxaloacetate)-induced calcium efflux was inhibited by anti-calvectin antibodies. This suggested that calvectin was involved in the efflux pathway and indicated that pyridine nucleotides could somehow interact with this protein. Direct measurements showed that calvectin indeed binds NAD and NADH (four sites per protein molecule). The binding shows a positive cooperativity for NAD (Hill coefficient = 1.9). Dissociation constants for the complex were 1.2×10^{-4} M for NAD and 6.6×10^{-4} M for NADH. Under the same conditions NADPH and NADP were bound only marginally. Interestingly, calvectin bound to NAD showed a lower affinity for Ca^{2+}. The interaction between NAD and calvectin, as well as the effects of this interaction on Ca^{2+} binding, have been studied also by other methods. In agreement with the data mentioned above, the circular dichroism spectrum of NAD-calvectin complex was influenced by the addition of Ca^{2+} [99]. The conclusion from these experiments was that calvectin was involved not only in the Ca^{2+}-uptake pathway, but also in the efflux promoted by the oxidation of pyridine nucleotides. The results implied that calvectin had access to the internal mitochondrial pool of pyridine nucleotides. In principle, it could be suggested that the Ca^{2+}-efflux from mitochondria evoked by addition of acetoacetate occurs *via* the reversal of the electrophoretic uniporter, thus explaining the involvement of calvectin in both the uptake and the release. This was proposed by Nicholls and Brand [100], who considered the efflux as the result of the collapse of the membrane potential. Under their experimental conditions, the addition of ATP plus oligomycin prevented Ca^{2+} efflux. Lötscher et al. [101,102] have used hydroperoxides to induce oxidation of pyridine nucleotides. The addition of these compounds to Ca^{2+}-preloaded mitochondria induced a Ca^{2+} efflux which was electroneutral (i.e., it did not occur on the electrophoretic uniporter). The effect, which under appropriate conditions was reversible, was accompanied by the cleavage of nucleotides with release of nicotinamide from mitochondria. The problem of whether this Ca^{2+}-releasing process is physiologically meaningful, or only linked to non-specific mitochondrial damage is still debated. Indeed, other groups [103,104] have concluded that there is no direct correlation between the ability of the mitochondria to retain Ca^{2+} and the redox state of their pyridine nucleotides. One interesting finding [105] is that the release of Ca^{2+} requires not only the oxidation of pyridine nucleotides but also their hydrolysis. A related finding which may in the long run become highly interesting is the ADP-ribosylation of a protein of the inner mitochondrial membrane during the release of Ca^{2+} induced by transitions in the redox state of pyridine nucleotides [106].

8. Regulation of the mitochondrial Ca^{2+} transport process

It follows from what has been discussed in the preceding sections that the Ca^{2+} balance between mitochondria and the extramitochondrial space results from the algebraic sum of the separate influx and efflux pathway(s) of the Ca^{2+} cycle. The next logical question is whether the cycle is regulated and, if so, whether its uptake or release portion, or both, are the target of regulatory systems. Experiments in several laboratories have now shown that respiring mitochondria maintain a steady-state external free Ca^{2+} concentration of about 1 μM, irrespective of the amount of Ca^{2+} accumulated in the matrix and of the magnitude of the membrane potential. The last statements, however, are valid within limits. Nicholls [107] has shown that the ability of mitochondria to maintain a limit external Ca^{2+} concentration of about 1 μM becomes impaired if either loads of Ca^{2+} in excess of about 60 $nmol \cdot mg^{-1}$ protein are accumulated in the matrix, or if the membrane potential decreases to values low enough to permit the reversal of the electrophoretic uniporter. If the latter situation prevails, the distribution of Ca^{2+} becomes determined by its thermodynamic equilibration across the uniporter. At normal values of membrane potential, and for amounts of Ca^{2+} accumulated that are within limits that can be expected in mitochondria in situ, Nicholls [107] has made the reasonable proposal that the regulation of the Ca^{2+} balance between mitochondria and the external medium is a kinetic phenomenon, in which an essentially constant efflux rate is superimposed on a rate of influx which varies with the concentration of Ca^{2+} in the ambient surrounding mitochondria. In situ, of course, this simple kinetic picture would be complicated by the presence, outside mitochondria, of concentrations of Mg^{2+} able to inhibit substantially the influx route. Abstracting from Mg^{2+}, however, as the extramitochondrial free Ca^{2+} concentration approaches the 'set point' of 1 μM the rate of influx decreases gradually until, at the steady state, it balances that of the efflux route(s). At this point, which corresponds to the maintenance of an external free Ca^{2+} concentration of about 1 μM, Nicholls [107] has calculated that Ca^{2+} will cycle across the inner mitochondrial membrane at a rate of about 5 nmol · mg^{-1} protein · min^{-1}. Should, somehow, the extramitochondrial Ca^{2+} concentrations be permitted to decrease below 1 μM, the efflux pathway(s) will become predominant and restore the extramitochondrial Ca^{2+} to the steady-state concentration of about 1 μM. Increased uniporter activity will on the other hand correct situations where the extramitochondrial free Ca^{2+} is allowed to fluctuate above 1 μM. Perhaps it may be mentioned at this point that the kinetic regulation discussed here does not place any demands on the type of the Ca^{2+} efflux route in operation. In tissues where the Na^+-promoted system is optimally active (e.g., heart, brain) the level of energy-dissipating Ca^{2+} cycling alluded to above (about 5 nmol · mg^{-1} protein · min^{-1}) will most likely be attained. For tissues that rely on mitochondrial Ca^{2+} release systems different from the Na^+ route, and conceivably slower than it, it is likely that the rate of steady-state Ca^{2+} cycling will be lower. In all cases, however, the set-point of about 1 μM external Ca^{2+} will be reached.

The type of regulation envisaged above, based on the kinetics of the influx and efflux pathways, is by definition relatively rapid (short term), and continuously operative during the normal cell cycle. Additional possibilities for regulation are offered by the activity of certain hormones, which may add additional complexities to the relatively simple and predictable picture outlined above. Their effects on the Ca^{2+} cycle, however, are likely to be less rapid, i.e., of the long term type. Several hormones have been proposed to influence mitochondrial Ca^{2+} transport, the first studies on steroids having appeared almost immediately after the discovery of the process [108,109]. Stimulatory effects of steroids were also found in later studies [110], and several authors have reported on stimulatory effects of insulin and glucagon [111–113]. The matter, particularly that of the latter hormones, is controversial, since the effect of glucagon is obviously related to receptor interaction at the plasma membrane, and to the liberation of cAMP to influence mitochondria, a problem that is at the moment considered in the negative (see below). As for insulin, its direct effects on mitochondria bear directly on the controversial matter of the physiological meaning of the process of insulin internalization, and/or on the matter of the putative messenger of insulin action. cAMP has been reported to release Ca^{2+} from several mitochondrial types [114] but the finding has later been discounted [115]. Later work, however, has shown that cAMP may yet have, under particular experimental conditions, some Ca^{2+}-releasing effect [116].

One matter that has been very much debated in recent years is that of the response of the mitochondrial Ca^{2+} pool in liver to α-adrenergic stimulation. Whereas some authors [117,118] maintain that the documented increase in cytosolic Ca^{2+}, consequent upon α-adrenergic stimulation of various liver preparations, is due to mobilization of Ca^{2+} from mitochondria, others have concluded that the main source of the mobilized Ca^{2+} is elsewhere, possibly in the plasma membrane [119,120].

Recent interesting information on this problem has come from work by Crompton and his associates [121,122] who found that α-adrenergic stimulation of rats increased the velocity of Ca^{2+} influx via the electrophoretic uniporter in isolated heart mitochondria. β-adrenergic stimulation, by contrast, had no effect. They have also found that the Na^+-promoted efflux pathway, in mitochondria isolated from the livers of rats treated with β-adrenergic agonists, was increased with respect to control rats. The effect was specific for the Na^+-promoted pathway, since the Na^+-independent portion of the Ca^{2+} efflux was unaffected. It may be appropriate to point out again that the Na^+-promoted efflux in liver mitochondria is much slower than in heart (about one fifth in the experiments by Crompton and his associates). α-Adrenergic stimulation had no effect on either the Na^+-induced or on the Na^+-independent efflux. These results, particularly the stimulation of the Ca^{2+} influx process, and the lack of effect on the release of Ca^{2+} by α-adrenergic stimulation, are clearly relevant to the proposal (see above) that α-adrenergic stimulation mobilizes the mitochondrial Ca^{2+} pool, and militates against it.

Very recently, Hayat and Crompton [123] have presented additional results that bear on the matter of the regulation of the Na^+/Ca^{2+} exchanger. They have found that extramitochondrial Ca^{2+} competes with Na^+ for binding sites, presumably on the exchanger, whereas at concentrations below 1 μM it inhibits as a partial non-competitive antagonist. The inhibition of the Na^+/Ca^{2+} exchange by Ca^{2+} may reach 70%, and is characterized by the binding of Ca^{2+} to sites that are half-maximally saturated at 0.7–0.8 μM free Ca^{2+}. The data indicate the existence, on the exchanger, of sites which confer to it sensitivity to extramitochondrial Ca^{2+}, and thus provide the system with additional possibilities for regulation.

9. *Mitochondria in the intracellular homeostasis of* Ca^{2+}

The transfer of the information described in the preceding sections of this chapter to the in vivo situation is a matter where opinions are sharply divided, even if more than 20 years have elapsed since the discovery by Vasington and Murphy [4]. One key problem, naturally, is the impossibility of reproducing the composition and the conditions of the cytosol in in vitro experiments. The above mentioned effect of Mg^{2+} on the rate of Ca^{2+} influx into mitochondria is but one striking example of the difficulties inherent to the extrapolation to the in situ conditions. Of interest in this respect are recent experiments [124,125] in which methods have been devised to estimate simultaneously the membrane potential across the plasma membrane and the mitochondria of nerve endings in situ. The conclusion of this work has been that the concentration of free Ca^{2+} in the cytosol correlates directly to the membrane potential across the mitochondrial membrane, and is maintained at a steady-state level below 1 μM. Simulation of the in situ conditions has also been the aim of studies [126] in which isolated liver endoplasmic reticulum has been added to media in which isolated liver mitochondria were made to take up Ca^{2+}, or in which liver cells have been treated with digitonin to abolish the permeability barrier of the plasma membrane. It was found that respiring mitochondria lower the external Ca^{2+} concentration to about 0.5 μM. The addition of endoplasmic reticulum vesicles produces a further decrease of the external Ca^{2+} to about 0.2 μM. Thus, mitochondria and endoplasmic reticulum apparently cooperate in the control of intracellular Ca^{2+}, the latter being active at lower Ca^{2+} concentrations in the cytosol than the former. Results which confirm and extend this concept have been reported by Orrenius and his coworkers [127–129]. They have used intact hepatocytes as experimental tools and have influenced the intracellular homeostasis of Ca^{2+} by manipulating either the redox state of mitochondrial pyridine nucleotides and/or the efficiency of the ATP-driven Ca^{2+}-uptake system of the endoplasmic reticulum by changes in the reduced/oxidized glutathione levels as induced by *t*-butyl hydroperoxide or menadione.

The matter of the velocity of Ca^{2+} transport by mitochondria under physiological conditions is also of importance from the standpoint of a possible role in the rapid regulation of cell Ca^{2+}. It follows from a consideration of the respective velocities of Ca^{2+} movements in mitochondria, sarco(endo)plasmic reticulum, and across plasma membranes (see Ref. 130) that mitochondria play but a minor role in the intracellular Ca^{2+} homeostasis under physiological conditions. If, however, intracellular Ca^{2+} deviates above the normal 0.1–1.0 μM level, the activation of the quantitatively powerful mitochondrial uptake system will buffer the situation and return it to normal through gradual release once the emergency has passed. The buffering reaction will eventually have to be completed by the increased activity of the outwardly directed Ca^{2+} transporting systems. But it is not surprising that mitochondria under conditions of intracellular Ca^{2+} overload appear replenished with electron-opaque masses identical to those observed in the isolated organelles after accumulation of massive amounts of Ca^{2+} and phosphate (see Ref. 131 for review). The concept that mitochondria may act as long-term Ca^{2+} sinks able to store very impressive amounts of Ca^{2+} demands that their machinery be able to function at a relatively normal level even after accumulation of large loads of Ca^{2+} and phosphate. This seems indeed to be so, since a number of cells exist (e.g., chondrocytes, osteoblasts, osteocytes, etc.) where the unusually high Ca^{2+} traffic across the cytoplasm produces Ca^{2+}-loading of mitochondria in vivo. The organelles, however, appear structurally well preserved and, in spite of some biochemical anomalies, still able to function adequately [131].

The conclusion that mitochondria may be essentially inactive in the intracellular Ca^{2+} homeostasis during normal conditions, and only become activated when emergency situations must be controlled (or, possibly, in response to pharmacological influences), may sound excessively negative. Conceptually, it would seem difficult to justify the development of such a sophisticated machinery as that for the transport of Ca^{2+} in and out of mitochondria if its use would only be limited to the improbable cases where cytosolic Ca^{2+} is allowed to fluctuate widely out of the normal limits. It may well be, however, that the primary reason for the development of the systems that constitute the Ca^{2+} cycle of mitochondria has not been the control of cytosolic Ca^{2+} under normal physiological conditions. The kinetic limitations of the mitochondrial system, and the existence in cells of other membrane structures capable of transporting Ca^{2+} back and forth more efficiently than mitochondria under normal in situ conditions, are undisputable facts. Even the role of mitochondria as long-term Ca^{2+} buffers, able to store large amounts of precipitated Ca^{2+} and phosphate, may have been a convenient byproduct of the essential reason for the development of the mitochondrial Ca^{2+} transport system: this may well have been the necessity of controlling Ca^{2+} not in the cytoplasm, but *within* mitochondria themselves. The suggestion [132,133] is based on the work by Denton and his co-workers [134–137] which has shown that several key matrix enzymes (pyruvate dehydrogenase kinase, pyruvate dehydrogenase phosphate phosphatase, isocitric dehydrogenase, α-ketoglutarate dehydrogenase) are modulated by Ca^{2+} in the μM range. Although the concentration of free Ca^{2+} in the matrix is not yet known with certainty, it is certainly impressive that mitochondria accumulate both Ca^{2+} and phosphate, thus greatly buffering the otherwise very large variations of free Ca^{2+} in the matrix that could be expected.

References

1 Slater, E.C. and Cleland, K.W. (1953) Biochem. J. 55, 566–580.
2 Chance, B. (1956) In Proc. 3rd Intern. Congr. Biochem. (Liebecq, C., ed.) pp. 300–304, Academic Press, New York.
3 Saris, N.E. (1959) Fin. Kemistsamfundets Medd. 68, 98–107.
4 Vasington, F.D. and Murphy, J.V. (1961) Fed. Proc. 20, 146.
5 De Luca, H.F. and Engstrom, G. (1961) Proc. Natl. Acad. Sci. U.S.A. 47, 1744–1750.
6 Vasington, F.D. and Murphy, J.V. (1962) J. Biol. Chem. 237, 2670–2677.
7 Lehninger, A.L., Rossi, C.S. and Greenawalt, J.W. (1963) Biochem. Biophys. Res. Commun. 10, 444–448.
8 Greenawalt, J.W., Rossi, C.S. and Lehninger, A.L. (1964) J. Cell Biol. 23, 21–38.
9 Weinbach, E.C. and Von Brand, T. (1965) Biochem. Biophys. Res. Commun. 19, 133–137.
10 Rossi, C.S. and Lehninger, A.L. (1963) Biochem. Z. 338, 698–713.
11 Carafoli, E., Rossi, C.S. and Lehninger, A.L. (1965) J. Biol. Chem. 240, 2254–2261.
12 Tew, W.P., Mahle, C., Benavides, J., Howard, J.E. and Lehninger, A.L. (1980) Biochemistry 19, 1983–1988.
13 Lehninger, A.L., Carafoli, E. and Rossi, C.S. (1967) Adv. Enzymol. 29, 259–320.
14 Carafoli, E. and Crompton, M. (1976) In Calcium in Biological Systems (Duncan, C.J., ed.) pp. 89–115, Cambridge Univ. Press, Cambridge.
15 Bygrave, F.L. (1977) Curr. Top. Bioenerg. 6, 259–318.
16 Carafoli, E. (1982) In Membrane Transport of Calcium (Carafoli, E., ed.) pp. 109–139, Academic Press, London.
17 Carafoli, E., Balcavage, W.X., Lehninger, A.L. and Mattoon, J.R. (1970) Biochim. Biophys. Acta 205, 18–26.
18 Carafoli, E., Hansford, R.G., Sacktor, B. and Lehninger, A.L. (1971) J. Biol. Chem. 246, 964–972.
19 Bygrave, F.L., Daday, A.A. and Doy, F.A. (1975) Biochem. J. 146, 601–608.
20 Rossi, C.S. and Lehninger, A.L. (1964) J. Biol. Chem. 239, 3971–3980.
21 Chappell, J.B., Cohn, M. and Greville, G.D. (1963) In Energy Linked Functions of Mitochondria (Chance, B., ed.) pp. 219–231, Academic Press, New York.
22 Chance, B. (1965) J. Biol. Chem. 240, 2729–2748.
23 Rossi, C.S., Bielawski, J. and Lehninger, A.L. (1966) J. Biol. Chem. 241, 1919–1921.
24 Schäfer, G. and Bojanowski, D. (1972) Eur. J. Biochem. 27, 364–375.
25 Chance, B. and Mela, L. (1966) Proc. Natl. Acad. Sci. U.S.A. 55, 1243–1251.
26 Gear, A.L., Rossi, C.S., Reynafarje, B. and Lehninger, A.L. (1967) J. Biol. Chem. 242, 3403–3413.

27 Addanki, S., Cahill, F.D. and Sotos, J.F. (1968) J. Biol. Chem. 243, 2337–2348.
28 Gunter, T.E. and Puskin, J.S. (1972) Biophys. J. 12, 625–635.
29 Crompton, M., Sigel, E., Salzmann, M. and Carafoli, E. (1976a) Eur. J. Biochem. 69, 453–462.
30 Affolter, H. and Carafoli, E. (1980) Biochem. Biophys. Res. Commun. 95, 193–196.
31 Mela, L. (1968) Arch. Biochem. Biophys. 123, 286–293.
32 Moore, C.L. (1970) Biochem. Biophys. Res. Commun. 42, 298–305.
33 Vasington, F.D., Gazzotti, P., Tiozzo, R. and Carafoli, E. (1972) Biochim. Biophys. Acta 256, 43–54.
34 Luft, J.H. (1971) Anat. Rec. 171, 347–365.
35 Niggli, V., Gazzotti, P. and Carafoli, E. (1978) Experientia 34, 1136–1137.
36 Vercesi, A., Reynafarje, B. and Lehninger, A.L. (1978) J. Biol. Chem. 253, 6379–6385.
37 Reed, K.C. and Bygrave, F.L. (1975) Eur. J. Biochem. 55, 497–504.
38 Affolter, H. and Carafoli, E. (1981) Eur. J. Biochem. 119, 199–201.
39 Bragadin, M., Pozzan, T. and Azzone, G.F. (1979) Biochemistry 18, 5972–5978.
40 Rasmussen, H., Chance, B. and Ogata, E. (1965) Proc. Natl. Acad. Sci. U.S.A. 53, 1069–1074.
41 Selwyn, M.J., Dawson, A.P. and Dunnet, S.J. (1970) FEBS Lett. 10, 1–5.
42 Scarpa, A. and Azzone, G.F. (1970) Eur. J. Biochem. 12, 328–335.
43 Mitchell, P. (1966) Chemiosmotic Coupling in Oxidative and Photosynthetic Phosphorylation, Glynn Research Ltd., Bodmin, U.K.
44 Rottenberg, H. and Scarpa, A. (1974) Biochemistry 13, 4811–4819.
45 Lehninger, A.L. and Carafoli, E. (1969) In Biochemistry of the Phagocytic Process (Schultz, J., ed.) pp. 922–931, North Holland, Amsterdam.
46 Vinogradov, A. and Scarpa, A. (1973) J. Biol. Chem. 248, 5527–5531.
47 Lehninger, A.L. (1971) Biochem. Biophys. Res. Commun. 42, 312–318.
48 Sottocasa, G.L., Sandri, G., Panfili, E. and de Bernard, B. (1971) FEBS Lett. 17, 100–105.
49 Evtodienko, J.V., Peskova, L.V. and Shchipakin, W.N. (1971) Ukr. J. Biochem. 43, 98–104.
50 Gómez-Puyou, A., Gomez-Puyou, M.T., Becker, G. and Lehninger, A.L. (1972) Biochem. Biophys. Res. Commun. 47, 814–819.
51 Kimura, T., Chu, J.-W., Mukai, K., Ishizuka, I. and Iamakawa, T. (1972) Biochem. Biophys. Res. Commun. 49, 1678–1683.
52 Tashmukhamedov, B.A., Gagelgans, A.I., Mamatkulov, K. and Makhmudova, E.M. (1972) FEBS Lett. 28, 239–245.
53 Utsumi, K. and Oda, T. (1974) In Organization of Energy-Transducing Membranes (Nakao, M. and Packer, L., eds.) pp. 265–267, University Park Press, Baltimore.
54 Carafoli, E., Schwerzmann, K., Roos, I. and Crompton, M. (1978) In Transport by Proteins (Blauer, G. and Sund, H., eds.) pp. 171–186, Walter de Gruyter, Berlin.
55 Jeng, A.Y., Rayan, T.E. and Shamoo, A.E. (1978) Proc. Natl. Acad. Sci. U.S.A. 75, 2125–2129.
56 Happel, R.D. and Krall, A.R. (1979) Biochem. Soc. Trans. 7, 1311.
57 Sokolove, P.M. and Brenza, J.M. (1983) Arch. Biochem. Biophys. 221, 404–416.
58 Sottocasa, G.L., Sandri, G., Panfili, E., de Bernard, B., Gazzotti, P., Vasington, F.D. and Carafoli, E. (1972) Biochem. Biophys. Res. Commun. 47, 808–813.
59 Sottocasa, G.L., Panfili, E. and Sandri, G. (1977) Bull. Mol. Biol. Med. 2, 1–28.
60 Carafoli, E., Gazzotti, P., Saltini, C., Rossi, C.S., Sottocasa, G.L., Sandri, G., Panfili, E. and de Bernard, B. (1973) In Mechanisms in Bioenergetics (Azzone et al., eds.) pp. 293–307, Academic Press, New York.
61 Carafoli, E. and Sottocasa, G.L. (1974) In Dynamics of Energy-Transducing Membranes (Ernster et al., eds.) pp. 293–307, Academic Press, New York.
62 Panfili, E., Sandri, G., Liut, G., Staucher, B. and Sottocasa, G.L. (1983) In Calcium Binding Proteins (de Bernard et al., eds.) pp. 347–354, Elsevier Science Publishers, Amsterdam.
63 Sottocasa, G.L., Sandri, G., Panfili, E., Gazzotti, P. and Carafoli, E. (1974) In Calcium Binding Proteins (Drabikowski et al., eds.) pp. 855–874, Elsevier Science Publishers, Amsterdam.
64 Sottocasa, G.L., Panfili, E. and Sandri, G. (1977) Bull. Mol. Biol. Med. 2, 1–28.
65 Sandri, G., Panfili, E. and Sottocasa, G.L. (1978) Bull. Mol. Biol. Med. 3, 179–188.

66 Sandri, G., Panfili, E. and Sottocasa, G.L. (1976) Biochem. Biophys. Res. Commun. 68, 1272–1279.
67 Panfili, E., Sandri, G., Sottocasa, G.L., Lunazzi, G., Liut, G. and Graziosi, G. (1976) Nature 264, 185–186.
68 Sottocasa, G.L., Sandri, G., Panfili, E., Liut, G. and Saris, N.E. (1981) In Vectorial Reactions in Electron- and Ion-Transport in Mitochondria and Bacteria (Palmieri et al., eds.) pp. 319–322, Elsevier, Amsterdam.
69 Sottocasa, G.L., Panfili, E., Sandri, G., Liut, G., Tiribelli, C., Luciani, M. and Lunazzi, G.C. (1979) In Macromolecules in the Functioning Cell (Salvatore et al., eds.) pp. 205–218, Plenum Press, New York.
70 Panfili, E., Sottocasa, G.L., Sandri, G. and Liut, G. (1980) Eur. J. Biochem. 105, 205–210.
71 Panfili, E., Crompton, M. and Sottocasa, G.L. (1981) FEBS Lett. 123, 30–32.
72 Sandri, G., Sottocasa, G.L., Panfili, E. and Liut, G. (1979) Biochim. Biophys. Acta 558, 214–220.
73 Carafoli, E. (1975) Mol. Cell Biochem. 8, 133–140.
74 Jeng, A.Y. and Shamoo, A.E. (1980) J. Biol. Chem. 255, 6897–6903.
75 Lew, V.L., Tsien, R.Y., Miner, C. and Boockchin, R.M. (1982) Nature 298, 478–481.
76 Pozzan, T., Arslan, P., Tsien, R.Y. and Rink, T.J. (1982) J. Cell Biol. 94, 335–340.
77 Tsien, R.Y., Pozzan, T. and Rink, T.J. (1982) Nature 295, 68–71.
78 Pozzan, T., Lew, D.P., Wollheim, C.B. and Tsien, R.Y. (1982) Science 221, 1413–1415.
79 Denton, R.M. and McCormack, J.G. (1980) Trans. Biochem. Soc. 8, 266–268.
80 Rossi, C.S., Vasington, F.D. and Carafoli, E. (1973) Biochem. Biophys. Res. Commun. 50, 846–852.
81 Caroni, P., Schwerzmann, R. and Carafoli, E. (1978) FEBS Lett. 961, 339–342.
82 Stucki, J.W. and Ineichen, E.A. (1974) Eur. J. Biochem. 48, 365–375.
83 Roos, I., Crompton, M. and Carafoli, E. (1980) Eur. J. Biochem. 110, 319–325.
84 Siliprandi, N., Rugolo, M., Siliprandi, D., Toninello, F. and Zoccarato, F. (1979) In Function and Molecular Aspects of Biomembrane Transport (Quagliariello et al., eds.) pp. 146–156, Elsevier/North Holland, Amsterdam.
85 Roman, I., Gmaj, P., Nowicka, C. and Angielski, S. (1979) Eur. J. Biochem. 104, 615–623.
86 Carafoli, E., Tiozzo, R., Lugli, G., Crovetti, F. and Kratzing, C. (1974) J. Mol. Cell. Cardiol. 6, 361–371.
87 Crompton, M., Capano, M. and Carafoli, E. (1976b) Eur. J. Biochem. 69, 453–462.
88 Crompton, M., Kunzi, M. and Carafoli, E. (1977) Eur. J. Biochem. 79, 549–558.
89 Crompton, M., Moser, R., Lüdi, H. and Carafoli, E. (1978a) Eur. J. Biochem. 82, 25–31.
90 Crompton, M. and Heid, I. (1978b) Eur. J. Biochem. 91, 599–608.
91 Mitchell, P. and Moyle, J. (1967) Biochem. J. 109, 1147–1162.
92 Nicholls, D.G. (1978) Biochem. J. 176, 463–474.
93 Lee, C.O. and Fozzard, H.G. (1975) J. Gen. Physiol. 65, 695–708.
94 Scarpa, A. and Graziotti, P. (1973) J. Gen. Physiol. 62, 765–772.
95 Crompton, M., Heid, I. and Carafoli, E. (1980) FEBS Lett. 115, 257–259.
96 Affolter, H. and Carafoli, E. (1980) Biochem. Biophys. Res. Commun. 95, 193–196.
97 Haworth, R.A., Hunter, D.R. and Berkoff, H.A. (1980) FEBS Lett. 110, 216–218.
98 Lehninger, A.L., Vercesi, A. and Bababunmi, E.A. (1978) Proc. Natl. Acad. Sci. U.S.A. 75, 1690–1694.
99 Quadrifoglio, F., Panfili, E. and Sottocasa, G.L. (1981) unpublished results.
100 Nicholls, D.G. and Brand, M.D. (1980) Biochem. J. 188, 113–118.
101 Lötscher, H.R., Winterhalter, K.H., Carafoli, E. and Richter, C. (1979) Proc. Natl. Acad. Sci. U.S.A. 76, 4340–4344.
102 Lötscher, H.R., Winterhalter, K.H., Carafoli, E. and Richter, C. (1980) J. Biol. Chem. 255, 9325–9330.
103 Prpic', V. and Bygrave, F.L. (1980) J. Biol. Chem. 255, 6193–6199.
104 Wolkowicz, P.E. and McMillin-Wood, J. (1980) J. Biol. Chem. 255, 10348–10353.
105 Richter, Ch., Winterhalter, K., Baumhütter, S., Lötscher, H.-R. and Moser, B. (1983) Proc. Natl. Acad. Sci. U.S.A. 80, 3188–3192.

106 Hofstetter, W., Mühlebach, T., Lötscher, H.-R., Winterhalter, K. and Richter, Ch. (1981) Eur. J. Biochem. 117, 361–367.
107 Nicholls, D.G. (1978) Biochem. J. 176, 463–474.
108 Kimberg, D.V. and Goldstein, S.A. (1966) J. Biol. Chem. 241, 95–103.
109 Kimberg, D.V. and Goldstein, S.A. (1967) Endocrinology 80, 89–98.
110 Kimura, S. and Rasmussen, H. (1977) J. Biol. Chem. 252, 1217–1225.
111 Dorman, D.M., Barrit, G.J. and Bygrave, F.L. (1975) Biochem. J. 50, 389–395.
112 Yamazaki, R.K. (1975) J. Biol. Chem. 250, 7924–7930.
113 Hughes, B.P. and Barrit, G.J. (1978) Biochem. J. 176, 295–304.
114 Borle, A. (1975) J. Membr. Biol. 16, 221–236.
115 Scarpa, A., Malmström, K., Chiesi, M. and Carafoli, E. (1976) J. Membr. Biol. 29, 205–208.
116 Arshad, J.H. and Holdsworth, E.S. (1980) J. Membr. Biol. 57, 207–212.
117 Chen, J.L.J., Babcock, D.F. and Lardy, N.A. (1978) Proc. Natl. Acad. Sci. U.S.A. 75, 2234–2238.
118 Blackmore, P.F., Brumley, F.T., Marks, J.L. and Exton, J.H. (1978) J. Biol. Chem. 253, 4851–4858.
119 Claret-Berthon, B., Claret, M. and Mazet, J.L. (1977) J. Physiol. (London) 272, 529–552.
120 Althaus-Salzmann, M., Carafoli, E. and Jakob, A. (1980) Eur. J. Biochem. 106, 241–248.
121 Kessar, P. and Crompton, M. (1981) Biochem. J. 200, 379–388.
122 Goldstone, T.P. and Crompton, M. (1982) Biochem. J. 204, 369–371.
123 Hayat, L.H. and Crompton, M. (1982) Biochem. J. 202, 509–518.
124 Scott, I.D. and Nicholls, D.G. (1980) Biochem. J. 186, 21–33.
125 Nicholls, D.G. and Åkerman, K. (1981) In Calcium and Phosphate Transport across Membranes (Bronner, F. and Peterlik, M., eds.) pp. 83–86, Academic Press, New York.
126 Becker, G.L., Fiskum, G. and Lehninger, A.L. (1980) J. Biol. Chem. 255, 9009–9012.
127 Jewell, S.A., Bellomo, G., Thor, H., Orrenius, S. and Smith, M.T. (1982) Science 217, 1257–1259.
128 Bellomo, G., Jewell, S.A., Thor, H. and Orrenius, S. (1982) Proc. Natl. Acad. Sci. U.S.A. 79, 6842–6846.
129 Bellomo, G., Jewell, S.A. and Orrenius, S. (1982) J. Biol. Chem. 257, 11558–11562.
130 Carafoli, E., Caroni, P., Chiesi, M. and Famulski, K. (1982) In Metabolic Compartmentation (Sies, H., ed.) pp. 521–547, Academic Press, London.
131 Carafoli, E. and Roman, I. (1980) Mol. Aspects Med. 3, 297–429.
132 Denton, R.M. and McCormack, J.G. (1980) Trans. Biochem. Soc. 8, 266–268.
133 Carafoli, E. (1980) In Exercise Bioenergetics and Gas Exchange (Cerretelli, P. and Whipp, B.J., eds.) pp. 3–12, Elsevier/North Holland, Amsterdam.
134 Denton, R.M., Randle, P.J. and Martin, B.R. (1972) Biochem. J. 128, 161–163.
135 Denton, R.M., Richards, D.A. and Chin, J.G. (1978) Biochem. J. 176, 899–906.
136 Cooper, R.H., Randle, P.J. and Denton, R.M. (1974) Biochem. J. 143, 625–641.
137 McCormack, J.G. and Denton, R.M. (1979) Biochem. J. 180, 533–544.

Ernster (ed.) Bioenergetics
© 1984 Elsevier Science Publishers B.V.

CHAPTER 10

Thermogenic mitochondria

JAN NEDERGAARD and BARBARA CANNON

The Wenner-Gren Institute, University of Stockholm, Norrtullsgatan 16, S-113 45 Stockholm, Sweden

1. Introduction

Only in three tissues — in two plant tissues (the Arum lily flower and the skunk cabbage spadix) and in mammalian brown adipose tissue — has a direct thermogenic role of mitochondria been substantiated.

In the thermogenic plant mitochondria the heat evolution has been claimed to be accomplished by a non-energy conserving system (alternate oxidase system) which is not coupled to proton extrusion and ADP phosphorylation (for review see Ref. 1).

Although peroxisomes do exist in brown adipose tissue, with the ability to β-oxidise fatty acids without energy conservation (at least not in the FAD/H_2O_2 step) [2], there is no doubt that the dominating part of the enormously high heat production in brown adipose tissue results from *mitochondrial* oxidation of substrates. Indeed this tissue can produce as much as about 400 W of heat per kg [3,4], several orders of magnitude more than the thermogenic capacity of mammalian tissues in general (e.g., about 1 W per kg body weight in adult resting man). Further, this mitochondrial heat production is carefully adjusted to both long-term and short-term physiological demands of the individual, with an increased capacity being recruited for newborn mammals and for mammals living in the cold; for arousal from hibernation; and for metabolic regulation during changes in food quantity or quality.

Within the last decade we have obtained a tentative concept of the molecular basis for this mammalian mitochondrial thermogenesis, and we know that in contrast to the thermogenic plant mitochondria, substrate oxidation in brown adipose tissue mitochondria is basically energy conserving, with proton extrusion occurring [5], with respiratory control, and with an ability, in principle, to capture the chemical energy in the form of ATP.

In this review we have not attempted to provide a historical account of the development of this concept (for such an account see Ref. 6), but rather attempted to show how this concept has enabled a series of apparently disparate observations on brown adipose tissue mitochondria to be unified.

This review is updated to autumn 1983. For other reviews of brown adipose tissue mitochondria see Refs. 7–9, for a review on the integration of mitochondrial function in the brown fat cell see Ref. 3, and for the function of brown adipose tissue as such, see e.g., Refs. 4, 10–12.

2. *The thermogenin concept*

With our present understanding, the thermogenic qualities of brown adipose tissue mitochondria are a consequence of the existence in the mitochondrial inner membrane of a polypeptide, *thermogenin,* uniquely [13–15] found in brown adipose tissue. (For technical and historical reasons, thermogenin is also known under several other names, such as the GDP-binding protein, the M_r 32000 protein, the purine-nucleotide-binding protein (NbP), the uncoupling protein (UCP), the proton conductance pathway, etc.)

As illustrated in Fig. 10.1, thermogenin is probably a dimer, consisting of two identical subunits of 32000 Da. Thermogenin functions as an anion translocator over the mitochondrial membrane. The anions which can be transported include Cl^-, Br^- and probably OH^-, and it is the permeation of this latter anion which is the basis for the thermogenic — uncoupling — property of thermogenin (see also Section 5.2.1). Thermogenin has a regulatory binding site for purine nucleotides, to which e.g., ATP, ADP, GTP and GDP can bind with relatively high affinity. The binding of these nucleotides to thermogenin decreases the activity of thermogenin, i.e., the anion permeability.

Thus, when thermogenin is in its active state (Fig. 10.1) the proton gradient created by the respiratory chain is annihilated by the hydroxide ions leaking out

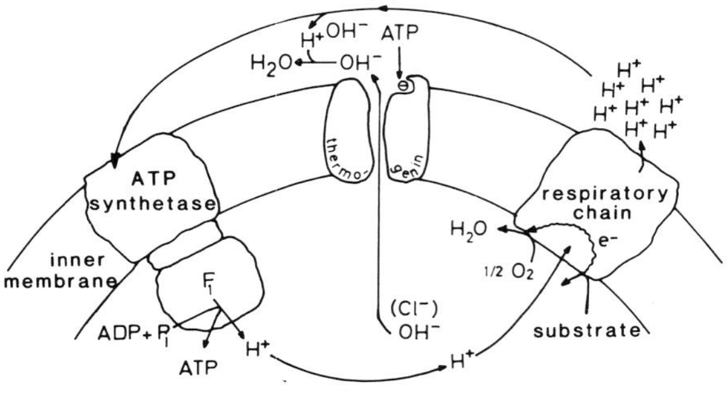

Fig. 10.1. The thermogenin concept. In the model for mitochondrial thermogenesis presented here, the thermogenesis is assumed to originate from the action of the brown fat-specific protein, thermogenin. Thermogenin acts as an OH^- conductor, regulated by cytosolic nucleotides (here shown as ATP) and by the so-called mediator (see Section 5). The OH^- neutralizes the H^+ electrochemical gradient created by respiration, and substrate oxidation occurs unhampered by this gradient and without energy conservation. (Adapted from Ref. 6.)

through thermogenin, and an opposing proton-motive force never develops suffi-ciently to diminish substrate utilization. Thus, substrate combustion proceeds un-hampered by ATP demands, and no energy is conserved: thermogenesis occurs. This thermogenin concept can be used to explain a series of properties (Sections 2.1–2.6) of brown adipose tissue mitochondria.

2.1. The uncoupled state of traditionally isolated and tested brown adipose tissue mitochondria

When mitochondria are isolated and tested in media which only contain osmotic support (sucrose or KCl) and a buffer (and Mg^{2+}, P_i and EDTA, if necessary), they respire rapidly on substrates such as succinate or glycerol-3-phosphate (citrate, 2-oxoglutarate and malate are poor substrates in brown fat mitochondria from most species, as the substrate permeases are poorly developed [16]). This rapid respiration is seen in Fig. 10.2. This observation was initially made even before the thermogenic function of brown adipose tissue was known [17]. When R. Em. Smith had established that heat production was the function of the tissue [18], an intrinsic uncoupled state of the mitochondria [19,20] could be understood as the means of heat production, the intensity of which would be limited only by substrate supply [21]. However, Horwitz et al. [21a] found that addition of the artificial uncoupler DNP could potentiate the respiration of the tissue, and the conclusion had to be that the mitochondria — although uncoupled when isolated — were coupled in situ.

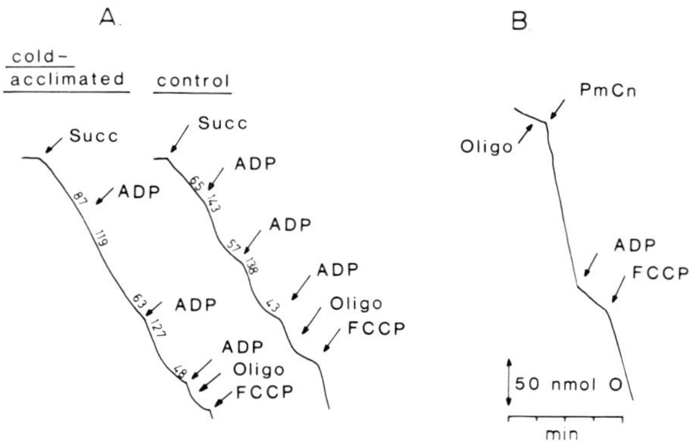

Fig. 10.2. The respiratory pattern of isolated brown fat mitochondria. A. When substrate (succinate) is added to brown fat mitochondria (here isolated from cold-acclimated or control guinea-pigs), they respire rapidly. Upon ADP addition the rate is initially increased (normal State 2–3 transition), but the ensuing State 4 rate is lower than State 2. Successive ADP additions result in a successively decreased State 4 rate. Numbers indicate respiratory rates in nmol oxygen·min^{-1}·mg^{-1} protein. (Adapted from Pedersen and Flatmark [93]; for details see this paper.) B. The specific coupling effect of purine nucleotides (here ADP) can be demonstrated after addition of oligomycin so that the respiratory stimulation due to ATP synthesis is eliminated. Addition of the uncoupler FCCP results in a respiratory rate identical to that prior to ADP, indicating that the ADP effect is on coupling, and not due to inhibition of substrate oxidation. (Adapted from Cannon et al. [23]; for details see this paper.)

2.2. The coupling effects of purine nucleotides

That a specific 'coupling' effect of purine nucleotides — different from anything seen in other tissues — exists in brown adipose tissue was realized in the early 1970's. The interpretation of this phenomenon was hampered by several complications — one being the multiple roles of the nucleotides. If, e.g., ADP is added to brown fat mitochondria, it has both its conventional stimulatory effect on respiration *and* the unique inhibitory effect. This is well illustrated by the effect of successive additions of ADP (Fig. 10.2). Although Hohorst and Rafael [22] already in 1968 had observed very conspicuous coupling effects of GTP on brown fat mitochondria, the significance of these observations was not immediately realized by the international scientific community — perhaps mainly because this work was not published in English, but also because the substrate used was 2-oxoglutarate, and the observations were therefore interpreted as being related to substrate level phosphorylation. Not until Cannon et al. [23] further analysed the coupling effect of adenine nucleotides was it realized that a unique mechanism was involved. Thus, it was found that the 'coupling' effect of ADP was on the outer side of the mitochondrial membrane, and it was not related to substrate level phosphorylation or to the ATP synthesizing system, nor to the activation of free fatty acids [23].

Although it is in many circumstances sufficient to distinguish between a 'coupled' and an 'uncoupled' state of mitochondria — the latter being generally characterized by an absence of effect of added artificial uncoupler to well-respiring mitochondria

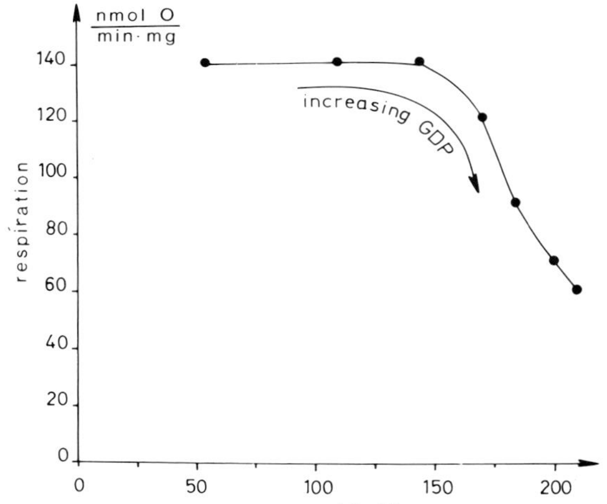

Fig. 10.3. The relationship between the proton motive force (pmf) and the respiration (thermogenesis) of brown fat mitochondria; the effect of GDP. The membrane potential and proton gradient were determined by ion distribution methods. It is seen that, with increasing concentrations of GDP, the pmf is increased. However, not until the pmf exceeds about 150 mV is the respiration inhibited. (Adapted from Nicholls [5]; for details see this paper.)

— the development of methods for determination of the components of the proton motive force (pmf, $\Delta\tilde{\mu}_{H^+}$), i.e., the mitochondrial membrane potential and the ΔpH, enabled Nicholls to characterize the effect of purine nucleotides (GDP) in a more quantitative way [5] (Fig. 10.3). It is seen that only a small decrease in proton motive force is necessary to transfer brown fat mitochondria from a fully coupled to a maximally respiring (fully thermogenic) state. This means that brown fat mitochondria can maintain other membrane potential-dependent functions while performing thermogenesis. These functions include the control of cytosolic Ca^{2+}; thus, as seen in Fig. 10.4, Ca^{2+} control is lost at lower GDP concentrations (i.e., lower pmf) than those which induce maximal thermogenesis [24]. This means that the mitochondria can be independently controlled by β_1-adrenergic stimuli (for thermogenesis) and by α_1-adrenergic stimuli (for Ca^{2+} control [25]).

2.3. The high (but regulated) halide permeability

When brown fat mitochondria are incubated with isoosmotic solutions of KCl or KBr they swell, provided that valinomycin is added to make K^+ permeable. This halide permeability is inhibitable by purine nucleotides, just as is respiration [26].

In the chemiosmotic theory, uncoupled respiration is due to a high H^+ permeability over the membrane. This permeability can be followed by swelling in KAc in the presence of valinomycin. Acetate enters as HAc by diffusion (it is too large an anion to enter via the thermogenin pathway), with K^+ entering via valinomycin. In order for swelling to occur the extra charge (H^+) must be counterbalanced, and as this can happen by the exit of H^+, the endogenous H^+ permeability becomes the rate-limit-

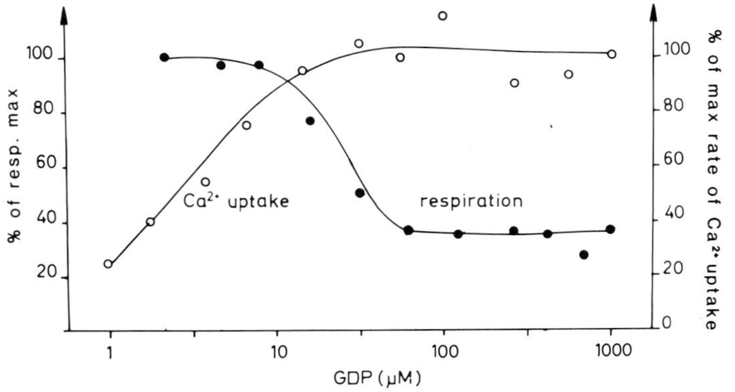

Fig. 10.4. The relationship between respiration (thermogenesis) and Ca^{2+} control in brown fat mitochondria. Ca^{2+} uptake was measured with the arsenazo technique. It is seen that when the GDP concentration is reduced below about 50 μM, respiration is evoked, and maximal thermogenesis occurs at, or below, 10 μM Ca^{2+}. The rate of Ca^{2+} uptake (as well as other parameters of mitochondrial Ca^{2+} metabolism) is unaffected by a lowered GDP concentration until the GDP concentration is reduced to well below 10 μM. Thus, even when the mitochondrial energetic state is sufficiently reduced to allow maximal thermogenesis to occur, Ca^{2+} control is unaffected. (Adapted from Nedergaard [24]; for further details see this paper.)

ing step under these conditions, i.e., swelling can be considered a measure of H^+ permeability. Also this permeability shows sensitivity to purine nucleotides (as would be expected by the chemiosmotic theory from the coupling effect of the nucleotides). Moreover, the inhibitory potency of different nucleotides on Cl^- permeability, H^+ permeability and respiration is so closely similar that one gets the impression that it is the same process which is under study [27] (Fig. 10.5).

In the chemiosmotic theory, H^+ and OH^- gradients and movements are indistinguishable [28], and it was suggested by Nicholls [29] that in reality it is OH^- (not H^+) which passes through the thermogenin pathway, annihilating the proton gradient (Fig. 10.1) and in this way allowing for thermogenesis. According to this model, there is no proton conductance associated with thermogenin, and the H^+ permeability measured is due to a counterflow of OH^-. (Apparently, in an unknown way, the GDP inhibition of permeability may, under certain conditions, be counteracted by respiration [27] (Fig. 10.6).)

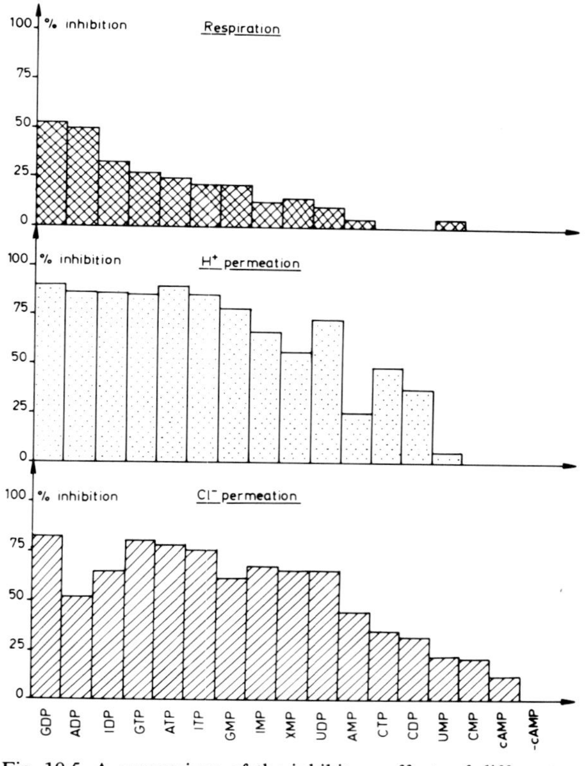

Fig. 10.5. A comparison of the inhibitory effects of different nucleotides on respiration, H^+ permeation and Cl^- permeation. Respiration was measured with glycerol-3-phosphate as substrate. Permeabilities were measured as swelling in KAc (for H^+) or KCl (for Cl^-) in the presence of valinomycin. All nucleotides were added at 2 mM. The figure is based on the values reported by Nicholls et al. [27].

2.4. The matrix condensation during mitochondrial isolation

When brown fat mitochondria are isolated in a nonionic medium (sucrose) without any additions, thermogenin activity will be uninhibited during the isolation, and small anions will be lost from the matrix. K^+ may be lost through the 'endogenous' nigericin-like antiporter found in these membranes (Fig. 10.6), acting in concert with thermogenin as an OH^- conductor, as discussed above. Indeed, the K^+ content of freshly isolated brown fat mitochondria is very low compared to, e.g., liver mitochondria [24,30]. Thus, due to the lack of osmotic matrix support, routinely isolated brown fat mitochondria are condensed and expand slowly when exposed to ionic media, this giving rise to steadily increasing respiratory rates [31] (Fig. 10.7). This problem is normally circumvented by allowing the mitochondria to swell in KCl before performing respiratory experiments. A more expensive way to obtain approximately the same situation is to have a nucleotide present during isolation (Fig. 10.7); such a procedure has also been observed to increase oxidation rates of acyl-carnitine substrates [32], but see Fig. 10.7.

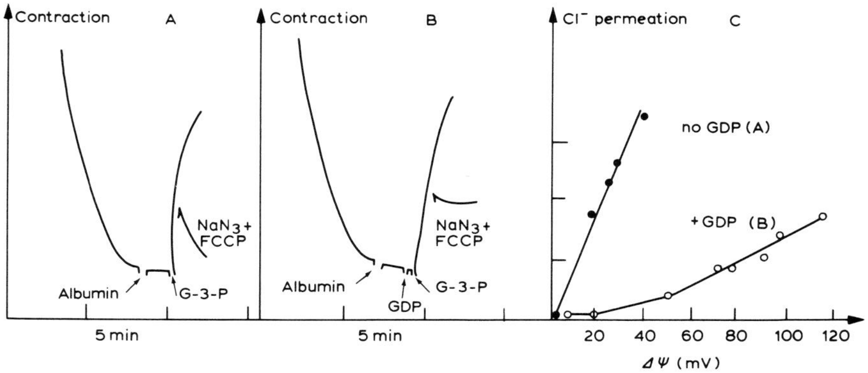

Fig. 10.6. The effect of respiration and membrane potential ($\Delta\psi$)) on Cl^- permeation in brown adipose tissue mitochondria. When brown fat mitochondria were incubated in KCl in the presence of the K^+/H^+ ionophore, nigericin, they swelled (A, B). If a respiratory substrate (here G-3-P: glycerol-3-phosphate) was added to the expanded mitochondria, they contracted, and this contraction ceased immediately and swelling was reintroduced if azide (NaN_3) and an uncoupler (FCCP) were added (Fig. A). The passive halide ion permeability can be inhibited by GDP (cf., Fig. 10.5), but respiration-driven contraction in KCl-expanded mitochondria was only partially inhibited by the presence of GDP (Fig. B); if again azide and uncoupler were added during the contraction, the mitochondria did not swell, indicating that the thermogenin channel was closed by GDP. This behaviour can partly be explained by the fact that the Cl^- permeation is driven by the membrane potential. Indeed, when, under similar conditions, the rate of contraction was plotted as a function of the membrane potential, it was seen that the rate was membrane potential dependent. It should, however, be noted that at low membrane potentials GDP nearly totally abolished the Cl^- permeation but when the membrane potential was increased above 30 mV, the inhibitory effect of GDP was apparently partially lost. The basis for this phenomenon is not understood; it is not even known if there is a lower affinity of thermogenin for GDP in the energized membrane, as measurements of GDP affinities always refer to the non-energized situation. (Adapted from Nicholls et al. [27] (A, B) and Nicholls [94] (C).)

2.5. The existence of a purine nucleotide binding site on brown fat mitochondria

Since it was shown by Cannon et al. [23] that ADP acted from the outside of the mitochondria to induce respiratory control, a specific site of interaction could be envisaged. Such a site was characterized by Nicholls [33] by binding of [^{3}H]GDP. Further, by labelling brown fat mitochondria with [^{32}P]azido-ATP, Heaton et al. [34] demonstrated that — besides the ATP/ADP-translocase at 30 kDa — a specific band with a molecular weight of 32 000 was labelled, and this was identical with the GDP-binding site. This protein (i.e., thermogenin) had already been observed by Ricquier and Kadér as the only protein the concentration of which was markedly altered in brown fat mitochondria isolated from cold-acclimated animals [35] (Fig. 10.8).

2.6. The ability of brown fat mitochondria to alter their capacity for heat production

Alterations in thermogenic capacity can be accomplished by altering the amount of thermogenin in the mitochondrial membrane (generally without altering, e.g., the

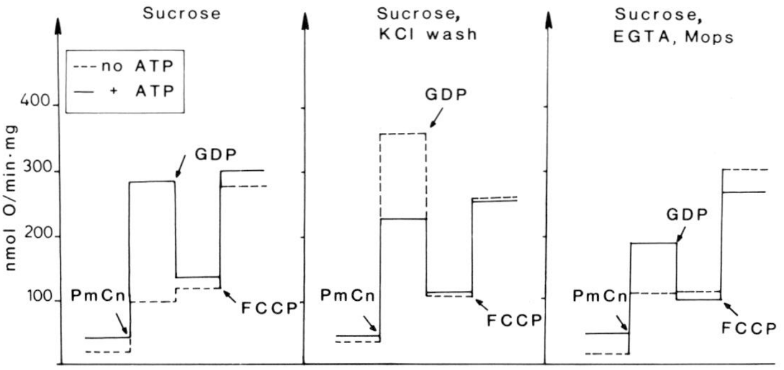

Fig. 10.7. The effect of incubation conditions on the respiratory capacity of brown fat mitochondria. Pooled brown fat from cold -acclimated hamsters was divided into parts. Some fractions were isolated in 250 mM sucrose, others in sucrose, EGTA and Mops, as described by LaNoue et al. [32]. The mitochondria were isolated either in the presence or the absence of 10 mM ATP. Some of the pure sucrose-isolated mitochondria were washed with 100 mM KCl, 20 mM K-TES (pH 7.2) before being tested in the oxygen electrode. All preparations were tested in the oxygen electrode in a standard medium containing 100 mM KCl, 20 mM K-TES, 4 mM KH_2PO_4, 2 mM $MgCl_2$, 1 mM EDTA, 2% bovine serum albumin (pH 7.2). Respiration, after the addition of 40 μM palmitoyl carnitine (PmCn), and the further addition of 1 mM GDP and an uncoupler (4 μM FCCP), was measured and plotted here as formalized trace. It is seen (*left*) how in the sucrose buffer the respiratory rate is slow and slowly increasing (due to matrix expansion (Nicholls et al. [31]) if the isolation is performed in the absence of ATP; however, the presence of ATP during isolation and storage does lead to the mitochondria being optimally active from the start of the incubation. Further, (*middle*) a KCl wash had the same effect as the presence of ATP, and — in our hands — a more complicated isolation buffer system (*right*) did not improve respiratory qualities (contrast LaNoue et al. [32]). It was concluded that isolation in 250 mM sucrose with the last wash in KCl is an easy and adequate way to obtain respiratorily competent brown fat mitochondria (our unpublished results).

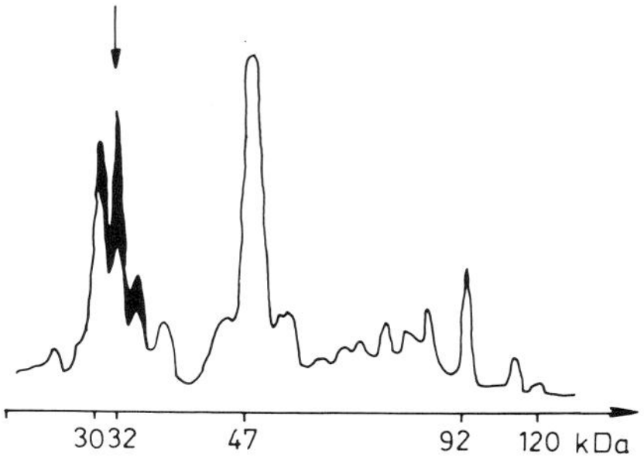

Fig. 10.8. The effect of cold acclimation on the polypeptide composition of rat brown fat mitochondria. Densitometric tracings of SDS polyacrylamide gels with mitochondrial membranes from control and cold-acclimated rats were superimposed, and the areas where the peak from the cold-acclimated animal exceeded that of the control is indicated in black (for no peak was the inverse true). Only a band at 32 kDa (arrow) was increased by cold acclimation (due to unresolved peaks, the adjacent peaks seem also to be increased, but this is an effect of the base-line broadening of the 32 kDa peak). The 32 kDa peak was later identified with thermogenin. (Adapted from Ricquier and Kadér [35].)

concentration of cytochromes or dehydrogenases). Indeed, by the utilization of the methods discussed in the next section, an increased concentration of thermogenin in the mitochondrial membrane has been found in each circumstance where an increased activity of brown adipose tissue is implied: perinatally in guinea-pig [36], rat [37] and hamster [38], during cold acclimation of adult rats [37,39] and hamsters [40], and when brown fat is stimulated by foods (by overeating a palatable diet [41], by lack of protein [42], or by a surplus intake of essential fatty acids [43] or glucose [44]). Similarly, where decreased activity is implied, thermogenin concentration is reduced, e.g., in the *ob/ob* mouse [45], (even before obesity is apparent [46]) and in the *fa/fa* rat (the obesity of these animals may in fact be due to the inactivation of brown adipose tissue) [47], and in the thyroxine-treated rat [48] which probably has a diminished requirement for brown adipose tissue heat production.

The possibility to induce an animal to rapidly alter the concentration of a mitochondrial inner membrane protein by such a simple physiological stimulus as cold exposure makes the control of the synthesis and expression of thermogenin in brown adipose tissue an interesting object for biochemical studies.

3. The manifestations and measurements of thermogenin

Due to the interest in relating the degree-of-activation of brown fat mitochondria to different physiological states, including obesity, the evaluation of thermogenin concentration in brown fat mitochondria has both biochemical and physiological interest. As thermogenin manifests itself as a (a) GDP-binding, (b) 32 kDa, (c) polypeptide (antigen) which (d) conducts Cl^- and OH^- through the membrane, methods have been developed which are based on each of these properties.

3.1. GDP-binding

The method most commonly used for determination of the amount of thermogenin is the binding of [^{3}H]GDP to mitochondrial membranes, as described by Nicholls [33] or Sundin and Cannon [37]. Such binding studies are preferably performed in a buffer with [^{14}C]sucrose as a marker of the extramitochondrial space. Under these conditions Scatchard analysis tends to indicate only one site for [^{3}H]GDP binding, with traditional Michaelis-Menten kinetics and with an affinity for GDP of about 0.5 μM, both in rats [37] and in hamsters [38,49]. Physiologically induced changes are characterized by an increased number of binding sites, and not by a change in affinity (Fig. 10.9). From these analyses, it would seem to be both theoretically and practically adequate to estimate the amount of thermogenin by measuring the amount of [^{3}H]GDP-binding at 10 μM GDP (20 $\times$ K_d) (and multiplying by two, if it is assumed that there is only one binding site per thermogenin dimer (see Section 4 and Ref. 61).

There are, however, two possible complications: the existence of multiple affinity sites (or unspecific sites), and the question of unmasking.

A '*very-high-affinity*' site ($K_m \leqslant 0.1$ μM) has been claimed to exist by Rothwell

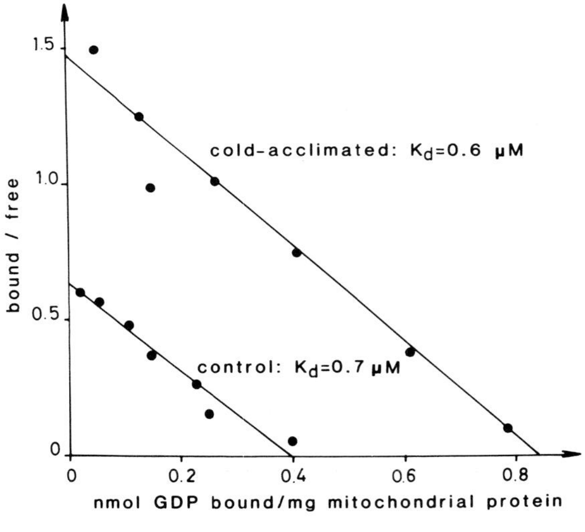

Fig. 10.9. Effect of cold acclimation on GDP binding to brown fat mitochondria. Brown adipose tissue mitochondria from control and cold-acclimated hamsters were incubated with increasing concentrations of [^{3}H]GDP (0.1–10 μM) and the binding determined (after compensation for trapped buffer by dual labelling with [^{14}C]sucrose). The result was plotted as a Scatchard plot. It is seen that no unspecific GDP binding is observable and that only one site with one affinity is involved (straight lines on the Scatchard plot). Further, it can be seen that a single incubation concentration of 10 μM (highest point) gives, in both cases, a very good approximation of the total specific binding (B_{max}) determined by Scatchard analysis. Finally, it is evident that cold acclimation results in more binding sites for GDP on the mitochondria, but the affinity of these sites for GDP is unchanged. (Adapted from Sundin et al. [40] see also Ref. 40a.)

and Stock [50], but the suggestion of this site is based on curvilinear Scatchard plots, and it is perhaps not certain that full equilibration has been obtained.

A '*low-affinity site*' is proposed by Rial and Nicholls [51], but as this 'site' has an 'affinity' for GDP of about 500 μM (several orders of magnitude higher than that of the thermogenin GDP-binding site normally referred to), it is indistinguishable from what in receptor studies would have been called 'unspecific binding'. This 'site' causes absolutely no problem in the determination of thermogenin if this determination is performed at 10 μM GDP. This is also seen from the fact that if a 100-fold excess of unlabelled GDP is added to a mitochondrial incubation with [^{3}H]GDP, [^{3}H]GDP-binding is totally competed out, in brown fat mitochondria from both rats and hamsters (Ref. 40 and our unpublished observations). (If such a competition is performed with less than a 100-fold excess, or if a less efficient nucleotide is used (e.g., ADP), residual binding will be observed and the true content of thermogenin will therefore be underestimated. The residual binding does not however represent unspecific binding, but is the residual amount expected from Michaelis-Menten kinetics).

Unmasking [52] is the term used for an apparent increase in GDP-binding without a parallel increase in amount of thermogenin. It is still not certain whether this

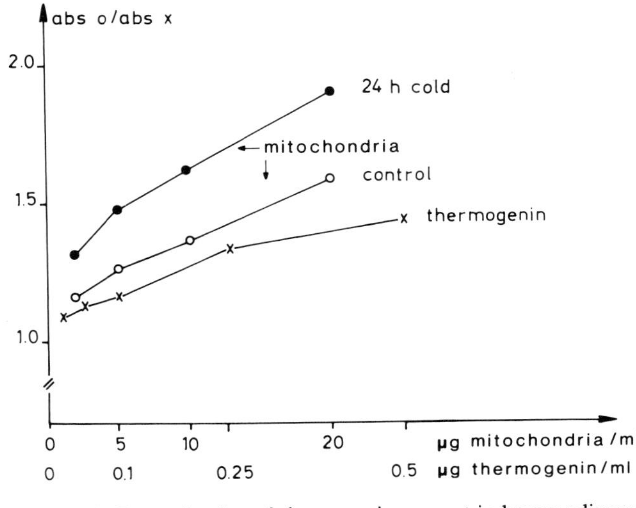

Fig. 10.10. Determination of thermogenin amount in brown adipose tissue mitochondria by the enzyme-linked immunosorbent assay (ELISA) system. The amount of thermogenin was determined as elsewhere described (Cannon et al. [13]; Sundin et al. [40]; Hansen et al. [56]) in an assay system based on the competition between absorbed and added thermogenin for rabbit *anti*-rat-thermogenin antibodies. The interaction was followed with a sheep *anti*-rabbit-IgG antibody conjugated to alkaline phosphatase. The reaction was linearized as indicated (abs 0 is the absorbance developed in the absence of competing thermogenin). It is seen that this assay can detect less than 0.25 μg thermogenin, i.e., the content in less than 10 μg of mitochondria. It is also seen that the thermogenin content of rat brown fat mitochondria is approximately doubled after a 24 h cold stress. (Our unpublished observations.)

phenomenon exists, and what the significance — if any — is of the phenomenon.

As the amount of thermogenin protein has been traditionally estimated from measurements on SDS-PAGE (Section 3.2), and as this method is inherently less sensitive than GDP-binding, it is possible that changes in thermogenin occasionally may be too small to be detected on the gels, but sufficient to be seen by GDP-binding. Indeed, when the ELISA technique is used (see Section 3.3), increases in amount of thermogenin protein per mg mitochondrial protein are seen already after one day of cold exposure (Fig. 10.10), or after cafeteria feeding [52a].

Further, manipulations of the mitochondria, such as osmotic swelling or shrinkage (Fig. 10.11), or a pH cycle [53], may change the number of GDP binding sites found. This indicates that if non-GDP-binding thermogenin may be found, it is not necessarily a chemically modified form of thermogenin, but rather it may be physically inaccessible to the assay.

As the degree-of-activation of mitochondria is said to influence their physical appearance, it may be suggested that some of the observed 'unmasking' phenomena may be due to the actual state of the brown fat mitochondria at the time of isolation, and/or to the incubation procedure, rather than reflecting true changes in the in situ qualities of thermogenin itself. This, however, remains to be investigated.

In conclusion it would seem that GDP-binding is an adequate way of determining thermogenin concentration in brown fat mitochondria. It should however be added that for physiological studies, this is not the only relevant parameter. As brown adipose tissue hypertrophies when stimulated, increases in thermogenin *content* per animal are often markedly greater than increases in thermogenin *concentration* in

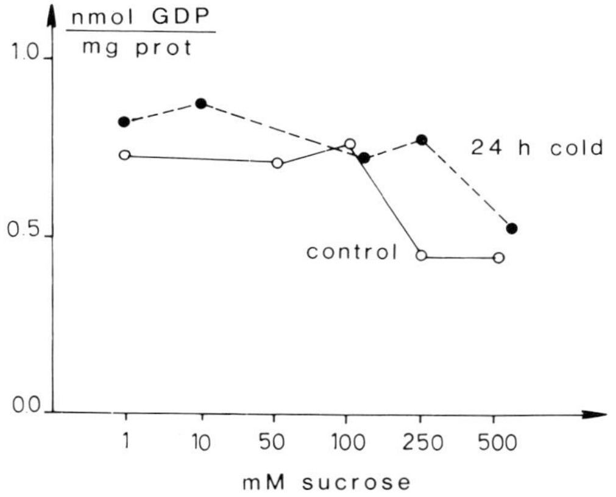

Fig. 10.11. The effect of osmolarity on the apparent number of GDP binding sites in brown adipose tissue mitochondria. The number of GDP-binding sites was measured in mitochondria from control and cold-exposed (24 h at 4°C) rats as earlier described (Sundin and Cannon [37]), but in media with the indicated concentrations of sucrose. Note that, if iso-osmotic sucrose is used (250 mM), a low GDP binding can be observed, especially in control rats. This may be related to the condensation phenomenon discussed in section 2.4. However, if 100 mM sucrose is used, the mitochondria swell, and the full number of binding sites is determined. (Our unpublished observations.)

mitochondria, and it is undoubtedly the total thermogenin content which is the most physiologically relevant parameter [37,54,55].

3.2. *Gel electrophoresis*

As thermogenin is visible as a band at 32 kDa on SDS-PAGE gels (as first observed by Ricquier and Kadér [35]), it is possible to observe changes in thermogenin amount by measuring directly on gels (Fig. 10.8). This is only a relative measure, and it is difficult to establish the base line as, even in brown fat mitochondria, not all protein with a molecular weight of 32 000 is thermogenin. This problem could be resolved by using two-dimensional gel electrophoresis, but no such studies have so far been published, and the p*I* of thermogenin is not known. The gel-electrophoretic method of measurement is undoubtedly going to be replaced by immunological techniques.

3.3. *Immunoassays*

As it is now possible to purify thermogenin (see Section 4) it has become possible to produce antibodies against thermogenin. The antibodies have been used in qualitative [15] and quantitative studies. The quantitative assays (radioimmunoassay [14] and enzyme-linked immunosorbent assays (ELISA) [13,52a,56]) have a very high sensitivity, requiring very little material, and can detect small amounts of, and small changes in, thermogenin (Fig. 10.10). By the use of these techniques it has been possible to establish that thermogenin is a unique feature of brown adipose tissue [13–15]. They can also be applied to, e.g., cell preparations [13,57] and probably to more crude tissue extracts. Unfortunately, the antibodies seem rather species specific [56], and a new antibody may have to be developed against thermogenin from each species. This species specificity must be due to a rather large variability of thermogenin, which is a mitochondrial enzyme that has appeared relatively late in evolution (apparently first with the occurrence of eutherian mammals). It may perhaps be possible to develop a monoclonal antibody which can interact with well-conserved critical determinants on thermogenin and which may thus have a broad species specificity.

3.4. *GDP-sensitive permeabilities*

Whereas the methods mentioned above all determine thermogenin amounts, the functional activity of thermogenin can be measured — at least in relative terms — by measuring GDP-sensitive Cl^- permeability [38,48].

Similarly, GDP-sensitive respiration (which is a measure of thermogenin-associated H^+ (OH^-) permeability) is an estimate of functionally active thermogenin — provided that a sufficiently efficient substrate is used (palmitoyl-carnitine) such that respiration is thermogenin-limited (i.e., that an artificial uncoupler can stimulate respiration above the initial level).

When changes in thermogenin concentration due to cold acclimation of hamsters were measured by GDP-binding, ELISA, GDP-sensitive swelling, and GDP-sensitive respiration, it was found that these parameters increased in parallel, indicating that all thermogenin was functionally active [38,57].

4. The thermogenin molecule

It is possible to purify thermogenin to (near) homogeneity by a rather simple procedure. The task is facilitated by the high concentration of thermogenin in the mitochondrial membrane — as much as about 15% of the protein may be thermogenin. Although it initially seemed natural to utilize the GDP affinity to purify thermogenin, attempts to do this yielded very low amounts [58]. Rather, it was noted by Lin and Klingenberg that thermogenin had some similarities to the ATP/ADP translocase, and by utilizing an isolation procedure similar to that developed for the translocase, these authors obtained large yields of pure thermogenin [59]. In this procedure, brown fat mitochondria are first lysed in a hypotonic, Lubrol-containing buffer, and the washed membranes are then suspended in 5% Triton X-100. This solution is applied to a hydroxyapatite column, and the break-through fraction is already nearly pure thermogenin (the ATP/ADP translocase is apparently easily degraded if this elution is performed at room temperature). The thermogenin can be further purified on a sucrose gradient [60].

Some of the molecular qualities of thermogenin, determined mainly from studies on the isolated molecule, are summarized on Fig. 10.12 and below [59–63]. It is likely that active thermogenin is a dimer, because in its isolated form it behaves

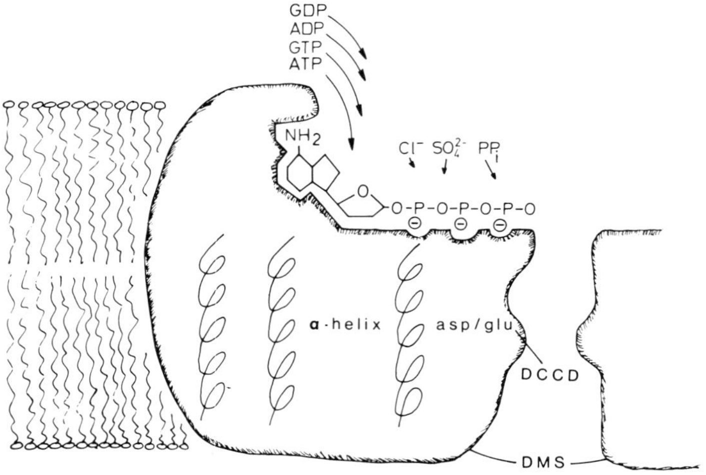

Fig. 10.12. The molecular properties of thermogenin. The figure summarizes the molecular properties of thermogenin, as deduced from studies on the isolated molecule. See Section 4 for further discussion.

hydrodynamically with an apparent molecular weight of 64 000 [61]; on standing it also tends to associate to dimers (E. Steen Hansen, personal communication), and because the reagent dimethyl suberimidate (DMS) crosslinks some of the thermogenin [60].

As not more than 16 nmol of GDP can be bound per mg isolated thermogenin, it would seem that only one binding site is exposed per dimer. This is similar to what is known about the ATP/ADP translocase, but it is perhaps difficult to visualize (Fig. 10.12).

It seems that a lysine and a tryptophane residue are closely spatially related to the binding site. In the isolated thermogenin system, the binding site shows the highest affinity for ATP and then for GTP, ADP, GDP and ITP (at neutral pH). However, at pH above 7.3, the affinity for ATP drops very steeply (2 pK_D per pH). This phenomenon is only observed for ATP and may indicate that ATP is the nucleotide which binds physiologically. The binding and dissociation of purine nucleotides is a rather slow process, which may indicate the occurrence of a conformational change. The very high concentration of thermogenin units (about one per 'coupling site', three per cytochrome aa_3) indicates that thermogenin should not be considered simply as a 'pore' or 'hole' for OH^-, but rather that a quite elaborate mechanism is involved in the transport of OH^- over the membrane. Again the similarity with the ATP/ADP-translocase is striking. The translocase is present in, e.g., heart mitochondria at approximately the same concentration as thermogenin is present in brown fat mitochondria, and at each turn of the translocase, one electrical charge is carried back over the mitochondrial membrane. It is tempting to suggest that a mechanism somewhat similar to that of the translocase is involved for thermogenin, but presently there is no evidence for any involvement of the nucleotide binding site in the actual transport process of OH^- over the membrane.

If brown fat mitochondria are incubated with high concentrations of [^{14}C]DCCD (dicyclohexylcarbodiimide), a very prominent incorporation of radioactivity is seen at a band at 32 kDa [64]. If thermogenin is subsequently isolated, this label is recovered in thermogenin. These results should not be interpreted to mean that thermogenin and the so-called DCCD-binding protein of the F_0 part of the ATP-synthase complex are identical; very high levels of DCCD are needed to obtain incorporation into thermogenin, and when DCCD is used in the concentrations where the binding is specific for 'the' DCCD-binding protein, the incorporation is very low [65], in agreement with the very low amount of ATP-synthase found in brown fat mitochondria of most species [66]. It is, however, interesting that, in parallel with the inhibitory effect of DCCD-binding to the F_0, DCCD-binding to thermogenin results in an inhibition of the Cl^- permeability of the mitochondria [64]. It can thus reasonably be assumed that the binding occurs to an acidic amino acid (aspartate or glutamate) which is closely adjacent to the 'channelling' part of thermogenin.

The amino acid content of thermogenin is not very nonpolar, but it contains unusually much threonine (7–10%, both in hamsters [60] and in rats [67]). More than 40% of the peptide chain is in the form of an α-helix. The complete amino acid sequence of thermogenin has not yet been published.

More information on the properties of thermogenin could undoubtedly be learned by the use of a reconstituted system. However, despite attempts in several laboratories, no successful incorporation of thermogenin into phospholipid vesicles has yet been reported, and it has thus not to date been unequivocally demonstrated that thermogenin in itself has the uncoupling qualities we have ascribed to it in this review.

5. The regulation of thermogenin activity

When brown fat cells are stimulated by norepinephrine, they respond by producing heat [68]. One step leading towards heat production is undoubtedly the liberation of substrate (free fatty acids) from the triglyceride stores within the cell [69,70]. However, since thermogenin activity is inhibited by cytosolic nucleotides (both ATP and ADP), thermogenin would be expected to be chronically deactivated within the cell [23]. Thus, the procuration of substrate alone cannot be considered a sufficient message to increase heat production, and it is necessary to postulate a physiological intracellular mediator, i.e., a substance or process that will activate thermogenin, even in the presence of cytosolic nucleotides.

The nature of this mediator is not presently known. Amongst the candidates discussed, two main groups can be discerned; one group being free fatty acids or their derivates (implying that activation of thermogenin occurs as a consequence of free fatty acid release) and one group where the implication is that another 'signal' leads from β-adrenergic stimulation to thermogenin activation. We shall first briefly discuss the latter, more heterogeneous group. (Suggested mediation of thermogenesis which is not via thermogenin activation will not be discussed here.)

5.1. Suggested non-free fatty acid mediators

Although thermogenin is often referred to as 'the GDP-binding protein', the affinity of thermogenin for GDP in isolated mitochondria is only slightly higher than for ATP [34], and even this may be an artefact, because on isolated thermogenin, ATP is the nucleotide with the highest affinity [60]. There is thus little reason to implicate changes in GDP levels as physiological in situ regulators of activity because any such changes will be overruled by ATP.

Changes in ATP level are more promising as candidates. In cells stimulated by norepinephrine, ATP levels were initially reported to drop [71], but this occurred after thermogenesis had peaked. Further, under incubation conditions where thermogenesis was both higher and more prolonged, ATP levels remained high [72]. Although it has been suggested that the cytosolic amount of ATP becomes very low during norepinephrine stimulation [32], it is difficult to visualize how this occurs without a parallel increase in ADP levels, and as ATP and ADP are almost equipotent in their inhibition of thermogenin [23,27], the thermogenic effects of such shifts would be marginal.

The observation that thermogenin precipitously loses its affinity for ATP when pH is increased above 7.3 [62] could suggest that an internal alkalinization may play a role in the activation of thermogenin (first suggested in another context by Chinet et al. [73]). However, no such alkalinization has so far been reported, and no mechanism which would lead to an alkanization has been substantiated. Before ATP binding to thermogenin would become negligible, quite alkaline internal conditions would have to be created (pH 8.3 would still only mean a K_d of about 1 mM), and it is equally feasible that the observed decrease in ATP affinity above pH 7.3 is a coincidence which would not physiologically be brought into action.

5.2. Mediators secondary to free fatty acid release

There are both teleological and experimental reasons to consider relevant the possibility that mediation (activation of thermogenin) is a consequence of the release of free fatty acids.

Experimentally, the most simple argument comes from the fact that addition of free fatty acids to isolated cells leads to a thermogenesis which, in all measurable characteristics, is indistinguishable from that observed after norepinephrine stimulation [74,75]. This 'minimal theory' thus states that "a rise in the intracellular free fatty acid concentration is probably the only stimulus needed to increase the rate of respiration" [75].

Although it was initially felt that the free fatty acids acted simply as uncouplers on the mitochondria, this is today considered to be much too simple an explanation. A sketch of the possible interactions of free fatty acids and their derivatives with the mitochondria is shown in Fig. 10.13. Although a series of derivatives are at hand, only extramitochondrial free fatty acids and acyl-CoAs have been more seriously investigated as candidates for the mediator.

5.2.1. Free fatty acids

There is no doubt that the addition of free fatty acids to brown fat mitochondria results in a stimulation of respiration. The reason for this response is however not unequivocally clear. As seen in Fig. 10.13, there are at least four possible sites of interaction of free fatty acids with brown fat mitochondria: **(1)** competitively with purine nucleotides on the binding site on thermogenin; **(2)** on another site on thermogenin; **(3)** on another protein site on the membrane; or **(4)** directly with the membrane.

(1). There is no evidence for a significant interaction with the nucleotide binding site on thermogenin. The amount of nucleotide bound is (almost) unaffected by the addition of free fatty acids [76,77], and free fatty acids can stimulate respiration even in the presence of high ATP levels [78].

(2). It has been suggested that free fatty acids in some way stimulate thermogenin by acting on a (putative) site on thermogenin other than the GDP-binding site. The argument for this has until now been rather indirect and simply related to the fact that brown fat mitochondria are more sensitive to free fatty acids as uncouplers than are, e.g., rat liver mitochondria. This high sensitivity was observed by Rafael et al. in 1969 [79]. The high sensitivity has later been shown with palmitoyl-carnitine as substrate [80], but in this system it is difficult to distinguish from differences due to different substrate utilization efficiencies. However, even with an FAD-linked substrate (succinate), brown fat mitochondria are some 8-fold more sensitive to free fatty acids than are liver mitochondria [81] (Fig. 10.14). As one difference between liver and brown fat mitochondria is the presence of thermogenin only in the latter, it has been implied that the increased fatty acid sensitivity is caused by thermogenin.

Recently, it has been implied on more direct grounds that the effect of free fatty acids is directly on thermogenin. Thus, if $H^+(OH^-)$ permeability is increased by free fatty acid addition, this can be counteracted by addition of GDP [77]. This effect is peculiar in that free fatty acids do not in themselves increase Cl^- permeability [76,77], nor (as stated above) is the free fatty acid-induced increase in respiration sensitive to GDP inhibition. Thus, a very complicated scheme must be proposed [82] in which there (again) are two permeabilities associated with thermogenin: an anion permeability of unknown physiological significance, and an H^+ permeability, only measurable in swelling experiments, not presently as thermogenesis.

Two other possible sites of interactions of free fatty acids with brown fat mitochondria remain: interaction with other proteins (**3**) or directly with the

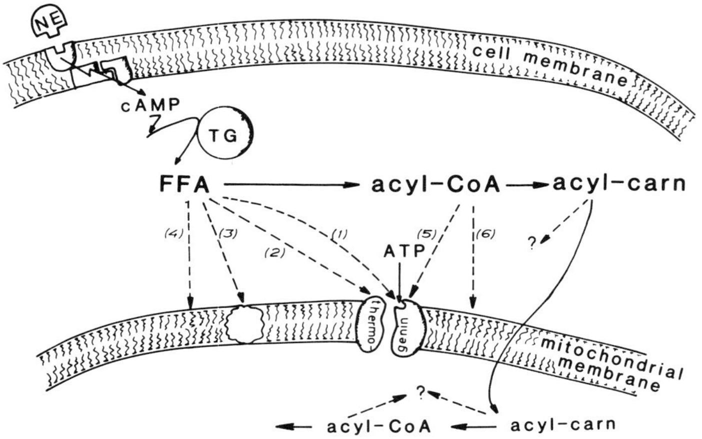

Fig. 10.13. A sketch of the possible interactions of free fatty acids and their derivatives with brown fat mitochondria. The sketch illustrates some of the candidates for the 'mediator' of thermogenesis (i.e., the substance or process that will activate thermogenin (alt. another site of the mitochondrial membrane) even in the presence of the inhibitory cytosolic nucleotides). Common for the candidates shown here is that they are formed subsequent to the activation of lipolysis of the stored triglycerides (TG) by norepinephrine (NE) via cAMP-dependent processes. The candidates illustrated are free fatty acids (FFA), interacting (**1**) with the purine-nucleotide binding site on thermogenin, (**2**) with another site on thermogenin, (**3**) with another protein than thermogenin, or (**4**) directly with the membrane, and the acyl-CoAs, interacting (**5**) specifically with the purine-nucleotide binding site on thermogenin, or (**6**) unspecifically with the membrane. For discussion, see Section 5.

membrane (**4**). As brown fat mitochondria from cold-acclimated animals are more sensitive to free fatty acids than mitochondria from control animals [81] (Fig. 10.14), and as no protein component of the membrane other than thermogenin changes markedly upon cold acclimation (Fig. 10.8), free fatty acid interaction with a protein other than thermogenin is less likely.

What remains is the possibility (**4**) that free fatty acids, also in brown fat mitochondria (under experimental conditions) function as classical uncouplers (i.e., as weak lipophilic acids). The reason for the higher sensitivity to free fatty acids of brown fat than of liver mitochondria — as well as the higher sensitivity of cold-acclimated than of control brown fat mitochondria — may then simply reside in the fact that the more sensitive mitochondria have more mitochondrial inner membrane [83] with which the free fatty acids can interact. This hypothesis has as yet not been experimentally tested.

5.2.2. Acyl-CoA

Due to the presence of an ADP-like moiety in the coenzyme-A part of acyl-CoA, it

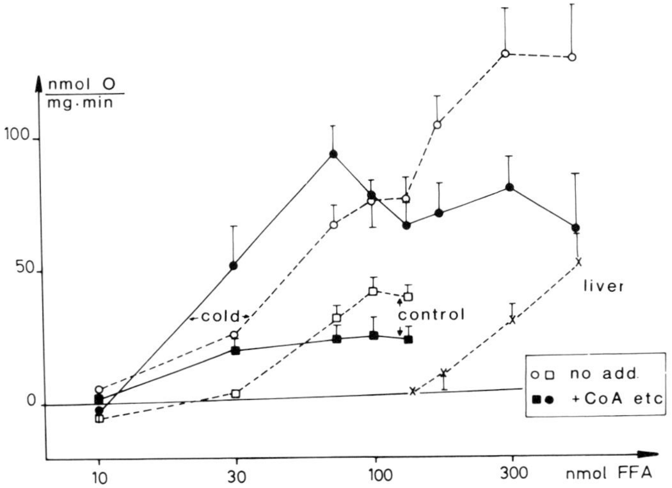

Fig. 10.14. The uncoupling effect of free fatty acids and their CoA derivatives on mitochondria. Brown fat mitochondria from *cold*-acclimated or *control* rats, as well as *liver* mitochondria, all at a protein concentration of 0.5 $mg \cdot ml^{-1}$, were studied in the oxygen electrode in a medium containing 0.05% albumin and with 20 mM succinate as substrate. The stimulation of respiration caused by the addition of the indicated amount of palmitic acid to 1 ml incubation medium was measured (------). It is seen that brown fat mitochondria from control rats were about 3-times more sensitive to fatty acids than were liver mitochondria, and that cold-acclimation further increased the sensitivity by a factor of 3. If, however, CoA and an ATP-regenerating system were added to the mitochondrial incubation (———), the sensitivity was increased for the *low* additions of free fatty acids, indicating that it may not be the fatty acids themselves which, in these concentrations, are the uncoupling agents, but perhaps the acyl-CoAs being generated. At high concentrations, this effect is not seen, indicating that the direct effect of free fatty acids ((**4**) in Fig. 10.13) may dominate — this effect is not seen with liver mitochondria (not shown). (Unpublished observations in our laboratory, by H. Barré et al.)

was suggested by Cannon et al. [76] that acyl-CoAs — which would be produced after norepinephrine stimulation of brown fat cells — could interfere directly with the GDP/ATP-binding site on thermogenin ((**5**) on Fig. 10.13). Acyl-CoA can compete with GDP for the binding site and — functionally — acyl-CoA (palmitoyl-CoA) can reintroduce Cl^- permeability in brown fat mitochondria in which this Cl^- permeability has first been limited by low amounts of GDP. More importantly, this effect of palmitoyl-CoA can be overcome by the presence of a high amount of GDP, again indicating an interaction between palmitoyl-CoA and GDP on the same site [76]. (This latter effect could apparently not be reproduced by Nicholls and Locke [9] under rather similar, although not identical, conditions.) There is no unequivocal demonstration of the effect of acyl-CoA on the respiration (thermogenesis) of isolated mitochondria, although there have been implications that palmitoyl-CoA uncouples its own oxidation [6,84,85].

It has been criticized that the effect of acyl-CoA is unspecific and due to a general detergent action. However, acyl-CoA does not induce K^+ permeability in parallel with Cl^- permeability [85], and there are clearly effects of palmitoyl-CoA on brown fat mitochondria at concentrations below those that yield (definitely 'unspecific') uncoupling in rat liver mitochondria. Further, the 'uncoupling' effects of free fatty acids are clearly — at low concentrations — 'better' if acyl-CoA formation is allowed to proceed than when it is hampered [81] (Fig. 10.14).

Thus, whereas there is no doubt that acyl-CoAs will interact unspecifically with the membrane if added at very high concentrations ((**6**) on Fig. 10.13), the possibility remains that acyl-CoAs may interact in a more specific way at low concentrations with thermogenin, and thus be the physiological mediators of thermogenesis.

6. The regulation of thermogenin amounts

As mentioned above, the concentration of thermogenin (per mg mitochondrial protein) is under physiological control. The increase seen in thermogenin amount per mitochondrion can either be due to insertion of thermogenin into pre-existing mitochondria and/or to the synthesis of a new generation of mitochondria with a higher thermogenin content (an increased mitochondrial turnover has been observed in cold-exposed animals [86]).

The gene for thermogenin is found in the nucleus of brown fat cells. This was initially implicated from studies with protein synthesis inhibitors [52], and recently Ricquier et al. [87] have found that a polyA-RNA fraction (i.e., cytosolic messenger-RNA) isolated from brown adipose tissue contained the messenger for thermogenin (i.e., thermogenin could be synthesized from this RNA in a reticulocyte-lysate system). Similarly, Freeman et al. [88] found that isolated polysomes from brown adipose tissue contained messenger-RNA for thermogenin. These authors were also able to show that if the newly synthesized thermogenin was incubated with mitochondria (mitochondria from Chinese hamster ovary cells were used), thermogenin was taken up into a space where it could not be destroyed by trypsin. Whether thermogenin was actually functionally incorporated into these mitochondria (i.e., if thermogenic ovary mitochondria were constructed) was not tested, but the experiments imply that it is possible to insert thermogenin into pre-existing mitochondria.

Thermogenin synthesized in vitro has an apparent molecular weight which is identical to the 'mature' thermogenin found in mitochondria. Thus, thermogenin is presently, together with the ATP/ADP translocase from *Neurospora crassa* [86] (but perhaps not the mammalian translocase [90]) and subunit VI of the cytochrome bc_1 complex (in *N. crassa*) [91], one of the few integral mitochondrial inner membrane proteins which are synthesized without a 'signal' peptide. It seems that thermogenin nonetheless recognises where it is to be incorporated.

6.1. The expression of thermogenin

The ability to express thermogenin is found only in brown fat cells. Thus, even in mitochondria from white fat, there is absolutely no evidence of any thermogenin being present [13]. Even in undifferentiated brown fat cell precursors, the gene for thermogenin has apparently already been 'opened' (is in its active conformation) — i.e., the cells are 'committed' to become brown fat cells — whereas in white fat cell precursors the gene is 'closed' (is in its inactive conformation) and thermogenin cannot be expressed in these cells [57]. The molecular mechanism behind this 'commitment' is still totally unknown.

Similarly, we have no knowledge about what causes the brown fat cells to increase the synthesis of thermogenin when an increased capacity for heat production is needed. The neurohumoral signals for increased thermogenin synthesis are presently being searched for, both by in-vivo experiments, where animals are treated with different pharmacological agents, e.g., norepinephrine [91a] and by in-vitro studies, where cultured brown fat cells [92] are exposed to different hormones and biological extracts, and where homeostatic interactions are less of a problem than in vivo. These studies have not progressed far, and we have at present no clear idea of what the intracellular messengers are which regulate the synthesis of thermogenin, both in proportion to the amount of mitochondria and to the physiological demand.

It may thus be envisaged, that for the general understanding of the regulation of the biogenesis of mitochondria, research on thermogenin may — like the metabolism of brown adipose tissue itself — be inefficient, but never futile.

Acknowledgements

Our research on brown adipose tissue is supported by the Swedish Natural Science Research Council. Unpublished experiments mentioned here were performed by Agneta Bergström and Barbro Svensson. Illustrations were drawn by Elisabeth Palmér. The authors are grateful to E. Rial and D.G. Nicholls for making available a manuscript prior to publication.

References

1 Moore, A.L. and Rich, P.R. (1980) Trends Biochem. Sci. 5, 284–288.
2 Nedergaard, J., Alexson, S. and Cannon, B. (1980) Am. J. Physiol. (Cell Physiol.) 239, C208–C216.
3 Nedergaard, J. and Lindberg, O. (1982) Int. Rev. Cytol. 74, 187–286.
4 Girardier, L. (1983) In Mammalian Thermogenesis (Girardier, L. and Stock, M., eds.) pp. 50–98, Chapman and Hall, London.
5 Nicholls, D.G. (1974) Eur. J. Biochem. 49, 573–583.
6 Lindberg, O., Cannon, B. and Nedergaard, J. (1981) In Mitochondria and Microsomes (Lee, C.P. et al., eds.) pp. 93–119, Addison-Wesley, Reading, Massachusetts.
7 Cannon, B. and Lindberg, O. (1979) Methods Enzymol. 55F, 65–78.
8 Nicholls, D.G. (1979) Biochim. Biophys. Acta 549, 1–29.
9 Nicholls, D. and Locke, R. (1983) In Mammalian Thermogenesis (Girardier, L. and Stock, M.J., eds.) pp. 8–49, Chapman and Hall, London.
10 Cannon, B. and Johansson, B.W. (1980) In Molecular Aspects of Medicine (Baum, H. and Gergely, J., eds.) Vol. 3, pp. 119–223, Pergamon Press, Oxford.
11 Cannon, B. and Nedergaard, J. (1984) Essays Biochem. In press.
12 Cannon, B. and Nedergaard, J. (1984) In New Perspectives in Adipose Tissue (Van, R. and Cryer, A., eds.) Butterworth. In press.
13 Cannon, B., Hedin, A. and Nedergaard, J. (1982) FEBS Lett. 150, 129–132.
14 Lean, M.E.J., Branch, W.J., James, W.P.T., Jennings, G. and Ashwell, M. (1983) Biosci. Rep. 3, 61–71.
15 Ricquier, D., Barlet, J.-P., Garel, J.-M., Combes-Georges, M. and Dubois, M.P. (1983) Biochem. J. 210, 859–866.
16 Cannon, B., Bernson, V.M.S. and Nedergaard, J. (1984) Biochim. Biophys. Acta. In press.
17 Lepkovsky, S., Wang, W., Kaike, T. and Dimick, M.K. (1959) Fed. Proc. 18, 272, abstr. no. 1075.
18 Smith, R.E. and Hock, R.J. (1963) Science 140, 199–200.
19 Smith, R.Em., Roberts, J.C. and Hittelman, K.J. (1966) Science 154, 653–654.
20 Lindberg, O., DePierre, J., Rylander, E. and Afzelius, B.A. (1967) J. Cell Biol. 34, 293–310.
21 Smith, R.E. and Horwitz, B.A. (1969) Physiol. Rev. 49, 330–425.
21a Horwitz, B.A., Herd, P.A. and Smith, R.Em. (1968) Can. J. Physiol. Pharmacol. 46, 897–902.
22 Hohorst, H.-J. and Rafael, J. (1968) Hoppe-Seyler's Z. Physiol. Chem. 349, 268–270.
23 Cannon, B., Nicholls, D.G. and Lindberg, O. (1973) In Mechanisms in Bioenergetics (Azzone, G.F. et al., eds.) pp. 357–364, Academic Press, New York and London.
24 Nedergaard, J. (1983) Eur. J. Biochem. 133, 185–191.
25 Nedergaard, J., Connolly, E., Nånberg, E. and Mohell, N. (1984) Biochem. Soc. Trans. 12, 393–396.
26 Nicholls, D.G. and Lindberg, O. (1973) Eur. J. Biochem. 37, 523–530.
27 Nicholls, D.G., Cannon, B., Grav, H.J. and Lindberg, O. (1974) In Dynamics of Energy-Transducing Membranes (Ernster, L., Estabrook, R.W. and Slater, E.C., eds.) pp. 529–537, Elsevier, Amsterdam.
28 Mitchell, P. (1966) Chemiosmotic Coupling in Oxidative and Photosynthetic Phosphorylation. Glynn Res. Ltd., Bodmin, Cornwall, England.
29 Nicholls, D.G. (1976) FEBS Lett. 61, 103–110.
30 Drahota, Z. (1979) In Brown Adipose Tissue (Lindberg, O., ed.) pp. 225–244, American Elsevier, New York.
31 Nicholls, D.G., Grav, H.J. and Lindberg, O. (1972) Eur. J. Biochem. 31, 526–533.
32 LaNoue, K.F., Koch, C.D. and Mechitz, R.B. (1982) J. Biol. Chem. 257, 13740–13748.
33 Nicholls, D.G. (1976) Eur. J. Biochem. 62, 223–228.
34 Heaton, G.M., Wagenvoord, R.J., Kemp, A. Jr. and Nicholls, D.G. (1978) Eur. J. Biochem. 82, 515–521.
35 Ricquier, D. and Kader, J.-C. (1976) Biochem. Biophys. Res. Commun. 73, 577–583.
36 Rafael, J. and Heldt, H.W. (1976) FEBS Lett. 63, 304–308.
37 Sundin, U. and Cannon, B. (1980) Comp. Biochem. Physiol. 65B, 463–471.

38 Sundin, U., Herron, D. and Cannon, B. (1981) Biol. Neonate 39, 141–149.
39 Desautels, M., Zaror-Behrens, G. and Himms-Hagen, J. (1978) Can. J. Biochem. 56, 378–383.
40 Sundin, U., Moore, G. and Cannon, B. (1983) Submitted for publication.
40a Desautels, M. and Himms-Hagen, J. (1981) Can. J. Biochem. 59, 619–625.
41 Brocks, S.L., Rothwell, N.J., Stock, M.J., Goodbody, A.E. and Trayhurn, P. (1980) Nature 286, 274–276.
42 Rothwell, N.J., Stock, M.J. and Tyzbir, R.S. (1982) J. Nutr. 112, 1663–1672.
43 Nedergaard, J., Becker, W. and Cannon, B. (1983) J. Nutr. 113, 1717–1724.
44 Sundin, U. and Néchad, M. (1983) Am. J. Physiol. 244, C142–C149.
45 Himms-Hagen, J. and Desautels, M. (1978) Biochem. Biophys. Res. Commun. 83, 628–634.
46 Goodbody, A.E. and Trayhurn, P. (1982) Biochim. Biophys. Acta 680, 119–126.
47 Holt, S. and York, D.A. (1982) Biochem. J. 208, 819–822.
48 Sundin, U. (1981) Am. J. Physiol. 241, C134–C139.
49 Sundin, U. (1981) Brown Adipose Tissue. Control of Heat Production. Development During Ontogeny and Cold Adaptation. University of Stockholm. ISBN 91-7146-152-3.
50 Rothwell, N. and Stock, M.J. (1984) In Thermal Physiology (Hales, J.R.S., ed.) Raven Press, New York, pp. 145–153.
51 Rial, E. and Nicholls, D.G. (1983) FEBS Lett. 161, 284–288.
52 Desautels, M. and Himms-Hagen, J. (1979) Can. J. Biochem. 57, 968–976.
52a Nedergaard, J., Raasmaja, A. and Cannon, B. (1984) Biochem. Biophys. Res. Commun. In press.
53 Stribling, D. (1983) In Mammalian Thermogenesis (Girardier, L. and Stock, M.J., eds.) pp. 321–354, Chapman and Hall, London.
54 Cannon, B., Nedergaard, J. and Sundin, U. (1981) In Survival in Cold (Musacchia, X.J. and Jansky, L., eds.) pp. 99–120, Elsevier/North Holland, Amsterdam.
55 Cannon, B. and Nedergaard, J. (1983) J. Thermal Biol. 8, 85–90.
56 Hansen, E.S., Nedergaard, J., Cannon, B. and Knudsen, J. (1984) Comp. Biochem. Physiol. In Press.
57 Nedergaard, J. and Cannon, B. (1984) In Thermal Physiology (Hales, J.R.S., ed.) Raven Press, New York, pp. 169–173.
58 Ricquier, D., Gervais, C., Kader, J.C. and Hemon, Ph. (1979) FEBS Lett. 101, 35–38.
59 Lin, C.S. and Klingenberg, M. (1980) FEBS Lett. 113, 299–303.
60 Lin, C.-S. and Klingenberg, M. (1982) Biochem. 21, 2950–2956.
61 Lin, C.S., Hackenberg, H. and Klingenberg, E.M. (1980) FEBS Lett. 113, 304–306.
62 Klingenberg, M. and Lin, C.-S. (1982) EBEC Reports 2, 447.
63 Klingenberg, M. (1984) Biochem. Soc. Transac. 12, 390–393.
64 Kolarov, J., Houstek, J., Kopecký, J. and Kuzela, S. (1982) FEBS Lett. 144, 6–10.
65 Svoboda, P., Houstek, J., Kopecký, J. and Drahota, Z. (1981) Biochim. Biophys. Acta 634, 321–330.
66 Cannon, B. and Vogel, G. (1977) FEBS Lett. 76, 284–289.
67 Ricquier, D., Lin, C.-S. and Klingenberg, M. (1982) Biochem. Biophys. Res. Commun. 106, 582–589.
68 Nedergaard, J., Cannon, B. and Lindberg, O. (1977) Nature, 267, 518–520.
69 Bieber, L.L., Pettersson, B. and Lindberg, O. (1975) Eur. J. Biochem. 58, 375–381.
70 Nedergaard, J. and Lindberg, O. (1979) Eur. J. Biochem. 95, 139–145.
71 Pettersson, B. and Vallin, I. (1976) Eur. J. Biochem. 62, 383–390.
72 Pettersson, B. (1977) Eur. J. Biochem. 72, 235–240.
73 Chinet, A., Friedli, C., Seydoux, J. and Girardier, L. (1978) In Effectors of Thermogenesis (Girardier, L. and Seydoux, J., eds.) Experientia (Suppl.) 32, ISBN 3-7643-1002-2, pp. 25–32.
74 Reed, N. and Fain, J.N. (1968) J. Biol. Chem. 243, 6077–6083.
75 Prusiner, S.B., Cannon, B. and Lindberg, O. (1968) Eur. J. Biochem. 6, 15–22.
76 Cannon, B., Sundin, U. and Romert, L. (1977) FEBS Lett. 74, 43–46.
77 Rial, E., Poustie, A. and Nicholls, D.G. (1983) Eur. J. Biochem. 137, 197–203.
78 Locke, R., Rial, E., Scott, I.D. and Nicholls, D.G. (1982) Eur. J. Biochem. 129, 373–380.
79 Rafael, J., Ludolph, H.-J. and Hohorst, H.-J. (1969) Hoppe-Seyler's Z. Physiol. Chem. 350, 1121–1131.

80 Heaton, G.M. and Nicholls, D.G. (1976) Eur. J. Biochem. 67, 511–517.
81 Barré, H., Nedergaard, J. and Cannon, B. (1984) Submitted for publication.
82 Nicholls, D.G., Snelling, R. and Rial, E. (1984) Biochem. Soc. Trans. 12, 388–390.
83 Lindgren, G. and Barnard, T. (1972) Exp. Cell Res. 70, 81–90.
84 Cannon, B. (1971) Eur. J. Biochem. 23, 125–135.
85 Cannon, B., Nedergaard, J. and Sundin, U. (1980) Adv. Physiol. Sci. 32, 479–481.
86 Himms-Hagen, J., Dittmar, E. and Zaror-Behrens, G. (1980) Can. J. Biochem. 58, 336–344.
87 Ricquier, D., Thibault, J., Bouillard, F. and Kuster, Y. (1983) J. Biol. Chem. 258, 6675–6677.
88 Freeman, K.B., Chien, S.-M., Litchfield, D. and Patel, H.V. (1983) FEBS Lett. 158, 325–330.
89 Zimmermann, R., Paluch, U., Sprinzl, M. and Neupert, W. (1979) Eur. J. Biochem. 99, 247–252.
90 Chien, S.-M. and Freeman, K.B. (1983) Fed. Proc. FASEB 42, 2125.
91 Teintze, M., Slaughter, M., Weiss, H. and Neupert, W. (1982) J. Biol. Chem. 257, 10364–10371
91a Mory, G., Bouilland, F., Combes-George, M. and Ricquier, D. (1984) FEBS Lett. 166, 393–396.
92 Néchad, M., Kuusela, P., Carneheim, C., Björntorp, P., Nedergaard, J. and Cannon, B. (1983) Exp. Cell Res. 149, 105–118.
93 Pedersen, J. and Flatmark, T. (1973) Biochim. Biophys. Acta 305, 219–230.
94 Nicholls, D.G. (1974) Eur. J. Biochem. 49, 585–593.

Ernster (ed.) Bioenergetics
© 1984 Elsevier Science Publishers B.V.

CHAPTER 11

Bacteriorhodopsin and related light-energy converters

JANOS K. LANYI

Department of Physiology and Biophysics, University of California, Irvine, CA 92717, U.S.A.

1. Introduction

The halobacteria are among Nature's most unusual creatures. They grow, and indeed survive, only in nearly saturated brines; thus their commitment to this extreme environment is complete and seemingly irreversible [1,2]. Enzymes and membranes prepared from the organisms are stable only in several molar salts [2–4], in accordance with the fact that the intracellular concentration of salt is very high. Indeed, the salt content of the cells is high enough to make salt the principal osmoregulatory substance. The ionic composition, however, is not the same inside and outside the cells. While the typical environment in which the halobacteria find themselves contains 3–4 M NaCl, the internal Na concentration is only 0.5–1 M, the rest being made up of KCl. Although several possible physiological reasons have been proposed for the lowered internal sodium concentration [5,6], it is clear that the ability to actively transport large amounts of ions across the cytoplasmic membrane is one of the adaptive features which enable these organisms to deal with high salinity. Given the low solubility of oxygen in saturated brines, the existence of light-driven transport systems in halobacterial membranes seems to make sense, since the utilization of sunlight is an excellent strategy for slowly growing organisms. It is therefore perhaps not surprising that we find characteristic and unique retinal proteins in these organisms, which energize ion translocations. Although the halobacteria grow very well in the dark on oxidative [7,8] or fermentative [9] metabolism, growth under strictly anaerobic conditions was demonstrated to be supported by illumination as the sole source of energy [9,10]. But beyond these physiological reasons, the widespread interest in the halobacterial pigments lies in the fact that they represent well-defined and simple systems for the study of transport physiology and the description of structure-function relationships in membrane proteins.

The purple membrane of *Halobacterium halobium* was discovered in the late 1960's when Stoeckenius and co-workers described sheet-like structures which could be prepared from cell envelopes of this organism by dialysis against low ionic strength buffer [11,12]. Under these conditions the protein-containing outer envelope was lost and a large part of the cytoplasmic membrane itself disintegrated, leaving behind sedimentable purple sheets. Later, in collaboration with Oesterhelt, they found that the 'purple membrane' contained a single hydrophobic protein of about 25 kDa, termed bacteriorhodopsin [13,14], which was arranged in a crystalline array in the plane of the membrane [15]. The purple color originated from a retinal bound to this protein in such a way as to shift its absorption maximum from 380 nm (free retinal) or 440 nm (retinal in a protonated Schiff's base linkage) to 570 nm. The intimate interaction between retinal and protein implied that the complex was a functional structure, in contrast with the red carotenoids of the halobacteria, whose absorption band was unchanged by any association with proteins and whose main role had been suggested to be merely photoprotective. Indeed, illumination of whole *H. halobium* cells [14] with light of wavelengths absorbed by bacteriorhodopsin caused pH changes in the medium, inhibition of respiration, increased ATP synthesis, and phototaxis. Since the ATP synthesis could be brought about by illumination in the absence of respiration, and was sensitive to proton conductors [16], it was concluded that bacteriorhodopsin functioned as a proton pump, in accordance with Mitchell's, then highly controversial, chemiosmotic hypothesis. Thus, it seemed that illumination of these cells would set up the circulation of protons, through bacteriorhodopsin (outward directed pump) and the ATPase (inward directed chemiosmotic device driven by the proton motive force). The elegance of this system was in the fact that the bacteriorhodopsin, located in distinct patches of about 0.5 μm diameter in the cytoplasmic membrane, was to a large extent physically removed from the ATPase molecules, which were thought to be located in the cytoplasmic membrane regions between the patches, i.e., in the red membrane (named for carotenoids). Communication between these two components would be, therefore, largely via the transmembrane gradients, as predicted by the chemiosmotic hypothesis. When purple membrane patches and mitochondrial ATPase were incorporated together into liposomes, illumination was indeed found to induce ATP synthesis [17,18], although with very low efficiency. Direct evidence for electrogenic proton translocation by bacteriorhodopsin was also provided by these reconstitution experiments.

Other results indicated that complex spectroscopic changes followed the absorption of a photon by the pigment. Spectroscopic studies in the visible region, carried out with sustained illumination at temperatures down to $-196\,^{\circ}$C or with flashes and time-resolved measurements at temperatures near ambient, showed that absorption of light set off the rise and decay of at least five photointermediates, designated by their absorption maxima as in rhodopsin [19–26]. While a single kinetic scheme which rigorously excludes all other possibilities has been very difficult to obtain, it was clear that a linear sequence of the intermediates K-590, L-550, M-412, N-530 and O-640, which recycle back to the original BR-570 pigment within about 10 ms, was consistent with many of the results. The manner in which these changes are coupled to proton translocation within the molecule was not known and still continues to be a mystery, however. The involvement of tyrosine groups in the functioning of bacteriorhodopsin was suggested by measurements of absorption changes in the UV [27,28], intrinsic protein fluorescence [19,27], and determinations of the consequences of chemical modification [29,30]. Theoretical models have tended to assume proton conduction via an ice-type structure in the protein [31–33], and a large pK change of various functional groups near the retinal, leading to unidirectional transfer of the proton into a high proton activity location [34]. Based on resonance Raman measurements [35,36], the reversible deprotonation of the retinal Schiff's base was subsequently thought to be a key feature of the proton transfer. While the Schiff's base certainly plays an important role, early models based solely on this idea are confounded by the observation that at salt concentrations high enough to overcome surface charge the maximum number of protons transported per photocycle is not one but two [37–40].

Because of the ease of isolation and its crystalline structure, purple membrane proved to be valuable material for structural studies. X-ray diffraction data indicated that the protein was rich in helical content [41,42]. Electron diffraction reflections from single sheets yielded a three-dimensional image in which seven apparently helical segments were arranged at roughly right angles to the plane of the membrane [43,44]. The bacteriorhodopsin molecules formed trimeric units, consistent with CD spectra which were interpreted as originating from exciton interaction among three chromophore groups [45–48]. The primary structure of the protein was determined, first by chemical sequencing of the amino acids [49,50], then by cloning the gene for bacteriorhodopsin and sequencing the DNA [51]. Fitting the amino acid sequence onto the structural model proved difficult, and various models are still disputed. Higher resolution of structure [52,53], as well as studies of the access to specific residues on the protein surface [30,54–59] are some of the approaches taken recently to decide among the structural possibilities.

The retinal was first thought [60] to be on a lysine, subsequently identified as lysine 41, but later found [58,61–64] to be in a different part of the molecule, on lysine 216. The angle of the absorption vector of the chromophore was determined to be about 23° inclined from the membrane surface [34,65,66], as expected for a linear molecule attached to a helical peptide segment. The position of the retinal in the protein was located from differential neutron scattering by bacteriorhodopsin in which the retinal was labeled with deuterium [67,68].

Concurrently with these studies, the photophysiology of whole *H. halobium* cells and cell envelope vesicles derived from them was described. Illumination was found to increase proton motive force, largely in the form of higher membrane potential [69–72], and in addition to ATP formation, sodium was found to be actively extruded [72] and amino acids taken up [73] by the cells. Sodium extrusion was shown to be via an electrogenic sodium/proton antiporter [74–77], which in the presence of potassium as counterion created a concentration gradient for sodium (out > in). The latter, together with the membrane potential, activated sodium symport systems for 19 amino acids [78–81]. The various ionic translocations induced by light were quite confusing in the case of whole cells, however. In particular, the reasons underlying the complex pattern of pH changes caused by illumination of whole cells [71,82–85] were not understood. The discovery [86–91] of a second light-driven transport system, halorhodopsin, in the halobacteria, provided a reasonable explanation, according to which active proton efflux during illumination is caused by bacteriorhodopsin, and passive proton influx is caused by increased membrane potential due to active sodium extrusion [88] by halorhodopsin. The sum of these opposite proton movements provided a good approximation of the pH changes observed with the cells. Halorhodopsin proved to be also a retinal protein, with spectroscopic and photocycling properties somewhat different from those of bacteriorhodopsin [92,93]. Labeling of the apoprotein of halorhodopsin with radioactive retinal showed it to be a protein of similar size to bacterio-opsin [94–96]. Recent results indicate that halorhodopsin is not an outward sodium pump, but rather an inward chloride pump [97]. Chloride indeed was found to have profound effects on the chromophore and the photochemical events in this pigment [98,99].

In addition to the above described pigments, bacteriorhodopsin and halorhodopsin, another, unidentified, retinal protein had long been implicated in photosensory effects in the halobacteria. Increased reversed swimming upon step-down in the intensity of yellow light or step-up in the intensity of blue light was observed, which cause the cells to accumulate in regions of yellow light but avoid regions of blue light [100,101]. A third retinal pigment, termed 'slowly cycling rhodopsin', has been described in the halobacterial membranes, and this pigment has some properties which make it a candidate for the light-acceptor for the photosensory effect [102].

The retinal proteins of halobacteria constitute a unique set of light energy transduction devices, based on similar chemistry but designed to perform different functions. The contributions of bacteriorhodopsin to our understanding of the structure and function of membrane proteins have been, and will no doubt continue to be, spectacular. As descriptions of the properties of the other two halobacterial retinal pigments are now becoming available, they promise to provide further insights into how membrane proteins function.

2. *Bacteriorhodopsin*

2.1. *Structure*

The so-called 'purple membrane' consists of patches of approx. 0.5 μm diameter in the cytoplasmic membrane of the halobacteria. It can be purified by the simple expedient of dialyzing the cells against distilled water, a treatment which disintegrates the rest of the cytoplasmic membrane but leaves the purple membrane apparently intact [103–105]. The isolated purple membrane sheets, recovered by density gradient sedimentation, contain virtually a single protein species, bacteriorhodopsin [13], and lipids nearly representative of the rest of the cytoplasmic membrane [12,13,106], although a characteristic glycosulfolipid is found only in the purple membrane [107]. Bacteriorhodopsin has a molecular weight of about 26 000, as determined by amino acid analysis [13,54,60] and sedimentation [108] in Triton X-100. It migrates anomalously in SDS-acrylamide gels [60]. The pigment contains one retinal per apoprotein molecule [13]. Because of its reactions with hydroxylamine, which removes the retinal [109], and with borohydride, which produces a stable covalent bond [13,60,110], it can be concluded that, as in rhodopsin, the retinal is in a Schiff's base linkage with a lysine.

Bacteriorhodopsin in the purple membrane is arranged in a regular hexagonal pattern, which is readily apparent on viewing freeze-fracture patterns on the surface of whole cells or membrane vesicles, or electron micrographs of dried, negatively stained purple membrane sheets [15,111,112]. In the latter case a cracking pattern at characteristic angles of 60° appears on the sides corresponding to the cytoplasmic surfaces, but not on the outer surfaces. The asymmetry of the two surfaces is revealed also in heavy-metal decorated sheets [113], where a hexagonal arrangement of grains is seen on one surface only. From thin sections it is evident that the thickness of the membrane is 50–60 Å, although the usual three-track appearance of bilayer membranes is not seen. Electron density profiles calculated from X-ray diffraction patterns yield two somewhat asymmetric peaks, which should correspond to the headgroup spacing, about 40 Å apart [114]. X-ray patterns of purple membrane suspensions show sharp reflections from the lattice structure in the plane of the membrane [15,105], in addition to the diffuse reflections expected from the bilayer profile and from lipid chain packing. In the hexagonal lattice the unit cell of three-fold symmetry contains three protein molecules and ten lipids/protein [15]. The X-ray patterns reveal high helical content, with α-helical segments of about 40 Å length spanning the bilayer [41,42]. This is in accord with the electron density profile of these membranes, showing higher density near the center of the bilayer than would be predicted from the hydrocarbon core [114]. A high helical content, with the helices oriented roughly normal to the plane of the membrane, is consistent with dichroism measurements of the IR absorption bands due to amide vibrations in oriented purple membrane films [115]. On the other hand, the amide I band is somewhat blue-shifted, a finding attributed to distortion of the helices [115,116]. Hydrogen/tritium exchange experiments show that only 25% of the protein is available for exchange [117,118], a finding also consistent with high helical content.

Electron diffraction from single sheets could be analyzed with special image enhancing techniques, keeping the exposure at very low intensities, so as to prevent damage by the electron beam [43]. Electron density projection maps of seven Å resolution [44], and those obtained later with somewhat better resolution [119], indicate that the protein is in clusters of three units in the membrane, a pattern repeated to form extended arrays. The centers of the clusters define the unit cell of 3-fold symmetry [44]. In the projection maps each protein molecule contains seven electron-dense regions, identified as helical segments of the polypeptide chain. A three-dimensional structure, obtained from diffraction data of tilted samples [44], consists of three inner columns (helices) at right angles to the plane of the membrane, and four outer columns at angles somewhat deviating from 90°. The interpretation of the electron density maps is aided by the availability of a second kind of bacteriorhodopsin array, produced by dialysis of detergent-solubilized membranes [120], with orthorhombic arrangement of the protein. In the latter pattern (which does not occur naturally) the protein molecules face alternate sides, and from the two different kinds of crystalline arrangements of the bacteriorhodopsin the assignment of the helical segments to individual molecules can be made with less ambiguity. Three-dimensional bacteriorhodopsin crystals have also been obtained [121,122]. The principle for growing such crystals is that some of the gaps between the proteins must be filled with amphoteric molecules instead of the water found in crystals of soluble proteins. The nature of the detergent used is therefore of great importance, octylglucoside being one of those suitable [121]. The crystals obtained show strong birefringence in the absorption range of bacteriorhodopsin, and promise to offer greater resolution than the electron scattering of single sheets.

The location of the retinal in this structure is identified from neutron diffraction differences between two samples of bacteriorhodopsin, both bleached but one reconstituted with H-containing retinal and the other with D-containing retinal [67,68]. The differences in the bilayer density profiles place the β-ionone ring of the retinal near the center of the width of the membrane [67]. The angle of the polyene chain of the retinal (parallel with its transition moment) to the plane of the membrane is about 23°, as determined from the observed CD band splitting, which implies exciton interaction among the three retinals in a cluster [65,66], and from linear dichroism of oriented bacteriorhodopsin multilayers [38,68,123–125], and of purple membrane oriented by electric fields [126–128]. The Schiff's base linkage therefore must be nearer to one side of the membrane surface; arguments based on the location of lysine-216 (see below) place it near the cytoplasmic side. The projection of the retinal in the plane of the membrane takes into account that the densest deuterium label is on the ring, with a third of the density being on the hydrocarbon chain. At an angle of 23° the expected image of the retinal is a strong peak connected to a smaller peak, 12 Å removed. Such an image is indeed found in difference density maps [68], and places the retinal between the inner and outer helices of the molecule. The position of the Schiff's base nitrogen in these projections is somewhat ambiguous, with two or three possibilities open, for example near helix 7. Illumination of bacteriorhodopsin reconstituted with a photosensitive analogue of retinal produces [129] crosslinking between the ionone ring and residues in helix 6, suggesting that the retinal is inclined toward this helix and thus toward the extracellular side.

Neutron diffraction of samples with added deuterated water [130], as well as hydrogen exchange studies [118], suggest that no aqueous channels exist between the protein molecules in purple membranes, or between the helices, i.e., the lipids completely fill the spaces. As expected from the crystalline structure and the relatively low lipid/protein ratio, the packing of protein in purple membrane is quite rigid, and the mobility of the protein in the lattice appears low, as determined by flash dichroism [131,132]. On the other hand, bacteriorhodopsin incorporated into liposomes at high lipid/protein ratios exists as a monomeric molecule [133], and shows rapid rotation by this criterion (relaxation time 15 μs) above the phase transition of the lipids [123,134–136]. The effective diameter of the protein calculated [137] from the rotational motion is about 43 Å, a value consistent with the physical dimensions of bacteriorhodopsin. Little rotation is seen below the phase transition, suggesting reversible two-dimensional crystallization [138] of the molecules at the transition temperature. This crystallization might be the consequence of the insolubility of the protein in the ordered lipid regions, which results in its segregation into the remaining fluid regions during the phase transition [136]. Bacterio-opsin does not readily form the lattice until retinal is added [139,140], although a naturally occurring 'white membrane' has been described, which contains bacterio-opsin as the major protein situated, in electron micrographs, in what appears to be the characteristic hexagonal array [141]. Crosslinking of bacteriorhodopsin with azidophenylisothiocyanate produces dimers and trimers, suggesting close proximity of the polypeptide chains in the lattice [142]. As expected, stearic acid type ESR probes, which partition into the lipid regions, show highly restricted motion in the purple membrane [143]. However, it is possible that the spin labels associate preferentially with the protein and thus do not reflect the behavior of the lipid domains [144]. No lipid phase transition is seen [145] in the purple membrane below 60°. Within the protein, however, fast rotation of some residues, notably tyrosine, has been observed [146,147] with ^{2}H-NMR.

Amino acid analysis of bacteriorhodopsin indicates [13] that the content of hydrophobic amino acids is high. Cysteine and histidine are absent [60,148]. The amino acid sequence was determined by chemical sequencing of the protein, using a strategy based mainly on cyanogen bromide cleavage [49,50,149]. The bacteriorhodopsin gene was sequenced [51] using a probe from a unique sequence segment near the N-terminus. The hydrophobic amino acids cluster around seven regions, which have been assigned [49,150] roughly to the seven transmembrane helical segments. The exact mapping of the sequence onto such a structure has not yet been done, but the plausible models now available are based on the idea that sequences containing charged amino acids project out of the membrane surface. Access to these sequences by proteolytic enzymes and other membrane-impermeable reagents [55,59,151–153] have been studied. The most obvious of these exposed sequences are located at the C-terminal end (on the cytoplasmic side) and at the N-terminal end (on the extracellular side), which can be split off with proteolytic digestion of purple membrane sheets or purple membranes in enclosed membrane systems of the appropriate surface orientation [54,154,155]. Most of the polypeptide chain, other than a continuous segment of about 20 residues from the C-terminus [156], must be located inside the bilayer, however [41]. No model can be constructed which avoids placing charged amino acids into the interior of the bilayer. Although such an arrangement would seem to be, at first sight, thermodynamically highly unfavorable, charge pairing inside the protein may stabilize the structure. In fact, it has been proposed [31] that it is residues interacting by polar forces inside the protein molecule which confer functional properties, such as proton conductivity, to bacteriorhodopsin. Upon exposure to the bilayer environment, the polypeptide chain will spontaneously refold to produce a functional molecule even after complete denaturation, e.g., in formic acid [157].

The assignment of the postulated helical segments in the amino acid sequence to the seven columns in the three-dimensional structure (i.e., the connectivity) has been attempted:

partly by elimination of unlikely alternatives on the basis of the required lengths for the loops and the calculated electron density in the projections of the helices in purple membrane labeled with ^{2}H-labeled amino acids [52,158];

partly by selecting the best fit to chromophore location [58] and chemical reactivity to membrane impermeable reagents [159];

partly by obtaining higher resolution structures with image enhancement techniques [53];

and partly by determining difference diffraction patterns from bacteriorhodopsin modified at various specified positions [57,160]. By these methods most of the 5040 (7!) possible helix assignments have been eliminated, leaving five [53] or eleven [158] possibilities, over which there is some disagreement. Further elimination of structures ought to come from chemical modifications which reveal neighbor relationships among residues [129]. Difference projections from neutron diffraction of specifically deuterated bacteriorhodopsin suggest that hydrophobic residues, such as valine, project from this structure to the exterior, i.e., toward the lipid chains, while the hydrophilic residues are located on the inside [161,162].

The existence of additional DNA sequence in the bacteriorhodopsin gene, corresponding to residues preceding the N-terminus, is demonstrated when cDNA is synthesized with reverse transcriptase, using partially purified mRNA from halobacteria as template and a primer sequence for residues 9–12 in bacteriorhodopsin [163]. The additional sequence, 13 residues long, agrees with a partial sequence given earlier [164] for a presumed precursor of bacteriorhodopsin, but does not contain the usual residues for a signal sequence. Its function is therefore unknown at present.

The location of the retinal on the polypeptide chain is ascertained by cleavage of the protein, after reduction of the Schiff's base so as to produce a covalent bond, followed by sequencing of the fragments. The retinal is followed either with ^{3}H-label [63], from reduction with [^{3}H]$NaBH_4$, or by fluorescence of the reduced product or UV absorption [62]. Chymotrypsin cleavage produces a large fragment of a molecular weight about 15 000, which corresponds to residues 72–248 and includes the C-terminus. This fragment, which contains lysines 129, 159, 172 and 216, carries virtually all of the retinal. Cyanogen bromide cleavage localizes [63] the retinal in the C-terminal fragment, which contains only lysine 216. The residues near the retinal are further identified [62] by sequencing of the CNBr fragments or of subtilisin-cleaved fragments, and the results confirm lysine 216 as the site of retinal. The possibility of retinal migration between this site and the earlier suggested site, lysine 41, either during the reduction or as part of the photochemistry of bacteriorhodopsin, is ruled out by resonance Raman results with bacteriorhodopsin specifically labeled with [^{15}N]lysine in the larger chymotrypsin residue [165]. Interaction of the retinal with both lysines is likewise ruled out by investigations of the ^{15}N isotope effect on the Schiff's base vibrations [166]. The position of the retinal is determined also in experiments where bacteriorhodopsin, reconstituted with ^{3}H-labeled retinal, is reduced with borohydride so as to fix the retinal, and then ozonolyzed to remove all but the radioactively labeled segment of the isoprenoid chain [64]. Trypsin digestion and cyanogen bromide cleavage each produce a single distinct radiolabeled peptide, corresponding to residues near lysine 216. Thus, the site of retinal is now firmly established to be on lysine 216.

The purple membrane lattice can be dissociated with mild detergents to yield bacteriorhodopsin monomers [45,167–169]. Dissociation will take place [167,169] without loss of the chromophore in both Triton X-100 and octylglucoside at pH 5, and virtually all of the lipids can be removed after this treatment by gel filtration in deoxycholate [170]. The significance of the lattice structure in the purple membrane is not very clear, but it appears from whole cell studies that crystalline bacteriorhodopsin is more effective in photophosphorylation than the monomeric pigment [171]. Remarkably, sodium dodecyl sulfate-denatured bacteriorhodopsin, with extensive loss of secondary structure, could be renatured to yield a product similar to native bacteriorhodopsin, which will spontaneously recrystallize [172].

2.2. The chromophore

The absorption band of free retinal in ethanol is at 380 nm, and at about 440 nm when it forms a protonated Schiff's base. The principal absorption band of bacteriorhodopsin in the visible range is red-shifted much farther, to 568 nm, indicating that the retinal is in close interaction with neighboring residues in the protein [173,174]. Tyrosine 26 may have a special role in this interaction, as its selective nitration shifts the absorption maximum to 532 nm, and after the nitration the p*K* of a reversible blue-shift (to 510 nm) is considerably lowered [30]. The involvement of one or two carboxylate groups as point charges near the ring of the retinal has also been proposed [175–180]. Arguments have been made for a special role of water in the chromophore, from the effects of high pressure and its sudden release on bacteriorhodopsin, which cause spectral shifts [181].

Protonation of bacteriorhodopsin between pH 2 and 3 leads to [182–184] a shift to 605 nm, and a shift back to 580 nm occurs upon adding acid to lower the pH further, to one. The latter effect is anion specific and may reflect the binding of an anion near the retinal. The formation of the 605 nm species proceeds in several kinetic steps, possibly reflecting internal proton migration to the bacteriorhodopsin trimers [185]. The 605 nm form of bacteriorhodopsin retains the trimer structure [183,186], and contains primarily all-*trans* retinal [183,187]. Illumination of the 605 nm species gives rise to other spectroscopic forms which contain 9-*cis* retinal [182,188]. The properties of the 605 nm form of bacteriorhodopsin have been utilized [125,182,189] to propose a model of the chromophore in which a negatively charged residue causes a 35 nm blue-shift to produce the 570 nm band of bacteriorhodopsin. Acetylated bacteriorhodopsin forms the 605 nm species at higher pH, at 4.8, and the low pH reversal of the red-shift does not occur [190], suggesting the involvement of a positively charged residue as well in these effects. Both this [191] and the anionic residue [192] behave as though they were on the cytoplasmic surface of the bacteriorhodopsin molecule. Modification of anionic residues with water-soluble carbodiimides prevents the formation of the 605 nm species [193].

Denaturation of bacteriorhodopsin with detergents or organic solvents leads [12] to a shift of the absorption band to 370 nm, which represents an unstable retinal-protein complex. Some solvents, such as dimethylsulfoxide or dimethylformamide [194], and detergents, such as sodium dodecyl sulfate [195], produce partial shifts, e.g., to 460 nm. In the completely or partially denatured forms of bacteriorhodopsin the Schiff's base becomes very sensitive, even in the dark, to hydroxylamine, with which it reacts to form the retinal oxime [194,196], and to sodium borohydride, which reduces the bond to form a stable covalent linkage [194,195]. Such reactions are seen in the native pigment only during extensive illumination with those reagents. Illumination of the dimethylsulfoxide-produced species gives rise to a retidinyl-protein similar to the M-412 form in the normal photocycle [110].

Bacteriorhodopsin exists in two states: the light-adapted and the dark-adapted forms. These are readily interconverted into one another during continued illumination (light-adaptation) and upon incubation in the dark (dark-adaptation) [13,194], and the change can be followed by a small absorption shift in the visible absorption band. The light-adaptation is a photoisomerization process which yields [194,197–199] a pigment containing 100% all-*trans* retinal and absorbing at 568 nm. During dark-adaptation an equilibrium state is produced for the chromophore, which absorbs at 553 nm and contains about equal amounts of all-*trans* and 130-*cis* retinal [194,198–200]. At very high light-intensities, however, the steady state contains, in addition to all-*trans* retinal, 20% of the 13-*cis* isomer, which is thought to be produced under these conditions with low quantum efficiency from one of the photointermediates of the all-*trans* photocycle [201]. Dark-adaptation is much accelerated [202] at acid pH. Reconstitution of bacterio-opsin with all-*trans* and 13-*cis* retinals produces pigments with the expected 568 and 548 nm absorption bands, respectively [199,200,202–204].

The apoprotein of bacteriorhodopsin can be prepared from hydroxylamine-bleached purple membrane [109,205], from halobacteria grown in the presence of nicotine [206] which inhibits retinal synthesis or from the cytoplasmic membrane of certain retinal-negative *H. halobium* strains. The latter produce patches containing only the colorless apoprotein, or 'white membrane', which can be purified by the same methods as purple membrane [141,207]. Addition of all-*trans* retinal to the apoprotein causes first the formation of various blue-shifted intermediates [196], then the formation of the 568 nm species, indistinguishable from bacteriorhodopsin. The first intermediate, BR-400, will be produced by 13-*cis*, 11-*cis* retinal and by retinol. Following this reaction (but only when all-*trans* or 13-*cis* retinal is used), a BR-430/460 is produced, before the final reconstituted product. This finding led to the conclusion that only the naturally occurring isomers, i.e., the 130-*cis* and the all-*trans* retinal can form the Schiff's base. Many substituted retinal analogues reconstitute the chromophore, in various forms blue- or red-shifted from the absorption band of bacteriorhodopsin [208]. The functional properties of such reconstituted bacteriorhodopsins have been studied with the aim of defining the site in which the retinal is found, and those features of the retinal which are necessary for light-absorption and transport. The results of these reconstitution studies are, in general, two kinds:

(a) interaction with the retinal analogue which results in a large bathochromic shift and produces a functional analogue of bacteriorhodopsin, as revealed by proton translocation or light-driven ATP synthesis in whole cells;

and (b) interaction which does not result in a large bathochromic shift but which competitively inhibits reconstitution with retinal.

Surprisingly, extensive changes in the β-ionone ring of the retinal are tolerated for the reconstitution [209,210]. However, any change in the length of the polyene chain results in inactive pigment [211,212]. Furthermore, short chain analogues are inhibitory to reconstitution with retinal, but only when present in the all-*trans* configuration [212], which suggests that an all-*trans* polyene chain from the C_7 to the C_{13} portion of retinal fits specifically into the retinal binding site. At least one double bond (between C_5 and C_6) and all of the methyl side chains are unnecessary for photocycling and proton transport [213]. Some substitutions on the β-ionone ring, such as 4-OH [211] and 3-Me [209], produce red-shifted species functional in proton transport, while others, such as 5,6-epoxy [211] and 4-keto [206] produce non-functional pigments. Chain substitutions near the aldehyde group, or bizarre isomeric forms, such as 9,13-*dicis* [214] fail to form a chromophore. Retinol [196,211] or retinoic acid [215] cannot produce a Schiff's base or a functional pigment.

Reduction of bacteriorhodopsin during illumination in the presence of borohydride produces a fluorescent species which absorbs at 360 nm [195,196]. The fine structure of the absorption band of this form is like that of retro-retinal, but the participation of such an isomer in the photocycle of bacteriorhodopsin [215,216] is unlikely. Rather, planarization of the chromophore by constraints of the retinal binding site might be responsible for the fine structure [110].

Because the chromophores in the bacteriorhodopsin trimer are in close proximity, favorable orientation of the retinal absorption vectors would produce electronic interactions within the trimer. The CD spectrum of bacteriorhodopsin is indeed different from that of vertebrate rhodopsin, which exists as a monomer. The observed splitting of the CD spectrum around 574 nm into a negative and a positive band is interpreted, accordingly, as exciton interaction among the three retinals [45–47,217]. Detergent treatment or disruption of the crystalline lattice by thermal motion in liposomes causes the disappearance of the CD splitting and the appearance of a single CD peak.

2.3. *Photochemical reactions*

Upon illumination bacteriorhodopsin undergoes a series of interconversions which are detectable by following changes in the position and amplitude of its absorption band. The photointermediates are designated by a letter (in analogy with photoproducts earlier described for visual rhodopsin), numbered with the wavelength of maximal absorption [34]. Time-resolved spectroscopy after flashes, and low temperature spectroscopy, show that absorption of light by light-adapted bacteriorhodopsin is followed by the appearance of a batho intermediate, K, whose absorption band was thought to be at 590 nm [21–23], but is more recently proposed to be at 610 nm. K-610 arises within picoseconds [218,219] at room temperature, and absorption of light converts it back [220] to BR-568, producing photo-stationary states [26,203, 221,222] between these two forms. In the dark K-610 is stable at 77 °K [20,23,26,203], but at higher temperatures it is thermally converted to L-550. The lifetime of K-610 at room temperature is about 1 μs. The existence of intermediates M-412 (rise time 50 μs, decay several ms), O-640 (rise time 1 ms, decay 5 ms), and N-530 (decay time 3 ms) has been proposed from time-resolved absorption changes at various wavelengths at ambient temperature, as well as from light-reactions at low temperatures. All of these intermediates arise by thermal decay from K-610 but are also photoactive, and will thus undergo photointerconversions [128,223,224]. While two-photon events can be avoided if the flash duration is short relative to the life-time of the intermediates, the physiologically relevant continuous illumination will cause photo-stationary states of uncertain composition. In spite of this complication, attempts have been made to devise simple kinetic schemes for the photocycle of bacteriorhodopsin. These assume a linear sequence of reactions, in the order BR-568, K-610, L-550, M-412, N-530 and O-640. Strictly speaking, linearity is at present only a simplifying assumption to aid kinetic analysis, rather than a conclusion supported by the data [225]. Branched pathways were suggested for the second half of the photocycle [24,226–230], as well as two spectroscopically and kinetically different [231–235] forms of M. Another possibility [236] is that the decay of L-550 proceeds via either M or O, depending on the pH.

The conditions used during the flash experiments, such as temperature, pH and ionic strength, affect the pathway of photochemical changes. At increasing temperatures between 0 and 40 °C the proportion of O-640 which accumulates will increase. At very low temperatures L-550 decays to BR without observable [222,237] intermediates. The pathway of BR recovery through the O-640 species predominates [26,236] at lower pH. When O-640 is not observable, e.g., at higher pH or low temperatures, the decay of M-412 corresponds to the rise of BR-568, obviating the need to interpose another intermediate into this part of the scheme. Thus, the place of N-530 in the kinetic scheme under these conditions is uncertain.

The 13-*cis* retinal-chromophore in dark-adapted bacteriorhodopsin exhibits a very different photocycle, whose predominant intermediate has an absorption maximum at 610 nm [199], and which contains no intermediate [202,238] analogous to M. The 610 nm intermediate will decay to either the 13-*cis* chromophore or the all-*trans* form, the latter pathway being responsible for the phenomenon of light-adaptation [199]. This pathway does not explain, however, why monomeric bacteriorhodopsin shows poor light-adaptation [168,239]. The chromophore in the 13-*cis* configuration is not associated with proton translocation [240]. Indeed, reconstitution of bacterio-opsin with 13-demethyl retinal, which traps the retinal moiety in the 13-*cis* configuration, results [241] in a non-transporting photocycle.

Effects of pH on the photochemical cycle of bacteriorhodopsin have special significance in view of the fact that this pigment is a proton pump. Thus, protonation and deprotonation of groups involved in the transfer of the protons across the protein spanning the membrane ought to influence those photochemical conversions which are directly coupled to the transport. The observed effect of protons on the photochemistry is complex. Although the absorption spectrum of bacteriorhodopsin is not altered between pH 4 and 10.5, increases in the rate of rise for M-412, and decreases for its rate of decay with pH, have been reported [29]. The overall magnitude of the flash-induced bleaching [242] is relatively pH independent, but separate kinetic components of the decay of M-412 (whose assignment is still an open question) are pH dependent [98]. The steady-state accumulation of M-412 in illuminated purple membrane preparation is reported to be pH independent [243], but in whole *H. halobium* cells was found to increase [85] with pH between 6 and 9. An added complexity is that, as mentioned above, the decay of M-412 at high pH proceeds through pathways which do not include O-640. Replacement of water with deuterium oxide affects the formation and decay of M-412 more than the other components of the photocycle [26, 244].

Valuable information about the photocycle is derived from low temperature spectroscopy, because thermal interconversions of the photointermediates have characteristic transition temperatures, and at certain chosen temperatures single intermediates or mixtures of only two intermediates will accumulate. Thus, the batho-intermediate (626 nm absorption peak at this temperature) is stable [222,245–247] at and below 150 °K, but is converted into the L form (543 nm absorption band) above this temperature [248,249]. The L intermediate is converted into the M form (418 nm absorption peak) upon warming to 220 °K, particularly at higher salt concentrations [48].

Dehydration of purple membrane films, deposited in an oriented fashion by drying down onto a glass surface [250], affect the thermal reactions of the bacteriorhodopsin photointermediates. While fully hydrated films show the same decay kinetics as purple membrane in suspension for the M and O intermediates, and for the 610 nm intermediate when dark-adapted samples are used, progressive dehydration slows down these reactions by as much as 3 orders of magnitude [251,252] and modifies the photocycle somewhat [253]. Low hydration will also prevent light-adaptation of the pigment [252]. Some membrane perturbants have effects on the photocycle which resemble those caused by removal of water. Thus, diethyl ether in saline solution [19], valinomycin plus beauvericin [254], gramicidin [255] and guanidine/HCl at alkaline pH [256] cause pronounced slowing of the decay of M-412, and therefore its significant accumulation during illumination.

The shifted absorption bands of the photointermediates obviously reflect changes at or near the retinal [257]. The nature of these changes is, for the most part, not clear, although at early steps in the photocycle the presence or absence of a fixed charge near the protonated Schiff's base of the retinal [258] has been proposed to play a part. More is known about the probable cause of the largest spectral shift, to the M-412 intermediate. Resonance Raman spectra [35,259,260] of BR-568, as well as ^{13}C-NMR data [261], have been interpreted as originating from a protonated

Schiff's base, but the spectra for the M-412 intermediate indicate that this proton is lost. The deprotonation of the Schiff's base is apparently after the K intermediate [262], and proposed to be during the L to M transition [209,263,264]. Reprotonation of the nitrogen is suggested to occur during the M-412 to O-640 conversion [265]. Part of the blue-shift in the formation of M-412 is, of course, explained by the fact that, in model retinal compounds, loss of the proton leads to a 440–380 nm shift [266], but other effects must also be present. Circumstantial evidence, which includes the finding of 13-*cis* retinal in M-like intermediates stabilized under somewhat denaturing conditions [198,267], favors the idea that the retinal is isomerized in the M intermediate, as do the more direct resonance Raman data [268,269]. In fact, the K and L intermediates seem already to contain the 13-*cis* isomer of retinal, as indicated by extraction of 13-*cis* retinal from the L intermediate [270] and spectroscopic data on the K and L intermediates [271–274]. The resonance Raman spectroscopy of bacteriorhodopsin photointermediates has been recently reviewed [275].

There is evidence that members of the trimers within the crystalline lattice of bacteriorhodopsin in purple membranes behave in a cooperative fashion. Hydroxylamine bleaching of the chromophore proceeds simultaneously within a trimer [276]. Regeneration of the pigment with retinal proceeds with non-linear binding kinetics [140], yielding a Hill coefficient of 3. The decay rate of the M-412 intermediate is more rapid from photostationary states containing greater M-412/bacteriorhodopsin ratios [277], and the kinetics agree well with a scheme containing three components showing allosteric interactions for the formation and decay of M-412. Addition of diethyl ether, which causes increased rotational mobility within the matrix, abolishes the observed cooperativity [277].

Quantum efficiencies for the conversion of BR to M have been reported to be about 0.25 at room temperature [48,278], and not much higher [48] at $-40\,^{\circ}$C. The apparent yield depends, of course, on the photostationary state between BR and K, as well as on any photoconversions involving the other intermediates of the photocycle. Such effects will be minimized with flashes of low intensity and sufficiently short duration, but thermal decay of the intermediates leading directly to BR, and other branching reactions are difficult to assess.

The chromophore of bacteriorhodopsin is weakly fluorescent, with an excitation maximum near 580 nm and a structured emission maximum near 670 and 720 nm (data from many sources summarized in Ref. 279). The lifetime of the emitting species is less than 100 ps at 77 [280] and 90 °K [281], and only a few ps at room temperature [280–282]. Spectroscopic studies of the fluorescence at 77 °K suggest [283–285] that the emitting species is excited bacteriorhodopsin, rather than a photointermediate of the photocycle, although the existence of a new photochemical species, 'pseudobacteriorhodopsin', to which the emission would be assigned, has also been suggested [286–288].

If bacteriorhodopsin is to be a light-driven proton pump, the photochemical reactions of the retinal chromophore will be accompanied by changes in residues not immediately adjacent to the retinal. The individual steps of the photochemical sequence in the pigment are associated with photopotentials arising from charge separation inside the protein [289], and externally applied electrical fields greatly influence the production of the photointermediates [290], or even give rise to spectral shifts in the dark suggestive of the photochemical process [291]. Charge movements in bacteriorhodopsin after absorption of light are detected on ns, μs and ms timescales [292]. Protonation and deprotonation of groups because of pK changes have been suggested [34] as a mechanism for proton migration. The pK changes, in turn, might be caused by redistributions of local charges and/or conformational changes. Unlike the alterations of the environment of the retinal, which absorbs in the visible region, and gives strong resonance Raman signals, these changes are not easily detected. The intrinsic fluorescence of bacterio-opsin might be considered a tool to detect conformational changes, as it originates primarily from tryptophan [293] and tyrosine residues. Quenching of this fluorescence during the photocycle has been observed, in both ether-saturated saline with slowed-down kinetics [19] and in untreated bacteriorhodopsin [27,294]. An unambiguous interpretation of the results is not possible, but it is likely that ionization of tyrosine residues is responsible for the fluorescence changes. Light-absorption by these residues will, in fact, result in efficient energy transfer to the retinal [295], and vice versa [175]. Changes in the protonation state of tyrosine(s) and tryptophan(s) during the photocycle are shown more convincingly by changes in UV absorption. Flash-induced absorbance changes in the UV region agree with steady-state difference spectra between illuminated and dark samples of bacteriorhodopsin [28], and show a decrease in absorbance at 275 nm, an increase near 296 nm [27,28,296], and another increase at 240 nm [297]. The decrease in absorption near 280 nm is detected also by difference spectroscopy of frozen samples [48], in which the L and M intermediates are accumulated. The absorption changes are interpreted as originating from the deprotonation of aromatic residues and conformational changes during the photocycle, because difference spectra between a mixture of tryptophan and tyrosine in ethanol (hydrophobic, protonated state) and in water at pH 12 (hydrophilic, deprotonated state) resemble the obtained light minus dark difference spectrum in bacteriorhodopsin [298]. An attempt at quantitation, based on such a model, identifies [298] the changes as deprotonation of two tryptophans and one tyrosine. In addition to these specific effects, illumination of bacteriorhodopsin, so as to accumulate L and M, causes a general decrease in UV absorption at all wavelengths, an effect which has been suggested to originate in a conformational change of the protein [296]. The rise kinetics put the UV absorption changes, as well as the intrinsic fluorescence changes [27,294], at or just before the L-550 to M-412 transition. The decay kinetics of the UV absorption changes do not agree with either M decay or BR recovery [28], a finding consistent with the UV fluorescence kinetics, which follow the O-640 to BR interconversion [27] rather than the M-412 decay. Other evidence which implicates tyrosines in the photocycle includes chemical modification studies [299]. Two tyrosines, specifically Tyr-26 (postulated to be near the cytoplasmic surface) and Tyr-64 (postulated to be near the extracellular surface) are particularly interesting because modification of these residues causes inactivation of proton transport by bacteriorhodopsin [30,59].

Other residues which might be involved in the proton translocation or in maintaining structure include arginine [55], but evidently not lysine, whose modifications do not remove functional properties in bacteriorhodopsin [300]. Segments of the polypeptide chain with residues 1–3, 68–72 and 231–248 are evidently also not essential, as they can be removed by limited proteolysis without loss of function [301]. Modification of COOH groups will alter the later events in the photocycle [150,302]. Tryptophan is implicated in the chromophore and the photocycle by results with *N*-bromosuccinimide modification [303]. It is possible that the p*K* of some of these residues is altered when an electrical potential is posed across the membranes, because such a potential slows down the photocycle kinetics [304] and proton pumping [305]. A 'back-pressure' effect of the entire proton motive force, which includes the pH difference across the membrane, on the photocycle is also evident [306]. A pulse of externally applied electric field causes proton uptake and then release from purple membrane in a manner similar to the events after a light flash, but in the opposite direction [307], an effect which also argues for a p*K* change.

Changes in residue interactions, protonation and deprotonation, etc., will result in changes in the IR absorption bands of specific vibrational modes. Measurements are on partially hydrated films of bacteriorhodopsin, which are transparent to the wavelength range which includes the rotational and stretching frequencies of such bonds of interest as CO, NH, COO^-, NH_3^+, etc. Some of the steady-state changes in the IR spectra are attributable to the retinal isomerization and deprotonation of the Schiff's base during the production of the M intermediate [308] and the K intermediate [309]. Others are consistent with the appearance of a protonated carboxylate group [262,308,310]. Kinetic IR measurements have also revealed a large number of absorption changes and band shifts in the IR region, some of which can be assigned to retinal isomerization [311], others seem to originate in the protein since their rise and decay times do not correlate with the absorption changes in the visible [311–313]. One of the IR bands is attributed to two COOH groups in different environments because it can be resolved kinetically into two components [314], one of which has the same rise-time as M-412.

Little specific information is available on conformational changes in the bacterio-opsin during the photochemical changes and proton translocation, but many results suggest at least some changes. After proton release, but before proton uptake, there is a large enthalpy deficit, which ought to be compensated for by an entropy decrease [315]. If so, at this point a large increase in molecular order would occur relative to the ground state. Effects of lipid viscosity [316] and aqueous phase viscosity [317] on the photocycle kinetics in bacteriorhodopsin reconstituted into liposomes and in purple membrane sheets, have been interpreted as influences on conformational changes in the protein. In solubilized bacteriorhodopsin certain groups become susceptible to chemical modification during illumination [318], indicating some change of structure. Crosslinking of bacteriorhodopsin inhibits the photocycle [319]. Dichroism measurements in bacteriorhodopsin embedded in Agar suggest, however, that little or no movement of the retinal chromophore (and of aromatic chromophores) takes place during the photocycle [320]. A similar conclusion was reached from resonance Raman measurements about the orientation of the retinal in the K intermediate [321], and from determinations of the anisotropy of the fluorescence of the borohydride-reduced chromophore [322].

2.4. Proton transport

The function of bacteriorhodopsin as a light-driven proton pump is well established from studies [14,70,83–85,323] of whole *H. halobium* cells, cell envelope vesicles prepared from the cells [78,324], and liposomes [17,18,135,191,325–327] as well as planar films [328–339] into which purple membrane was incorporated. In all of these cases light-dependent net translocation of protons across the membrane is observed, whose magnitude exceeds the number of bacteriorhodopsin molecules in the system by up to two orders of magnitude.

Many different methods [340] are available for the reconstitution of bacteriorhodopsin into liposomes: (a) sonication of purple membrane with dried, dispersed lipids [341]; (b) co-precipitation of lipids and bacteriorhodopsin from organic solvents, such as dimethylsulfoxide [340]; (c) dispersion of purple membrane sheets or monomeric bacteriorhodopsin and lipids in detergents, such as Triton X-100 [340], cholate [17,325], deoxycholate [170], or octyl-glucoside [327,342], followed by dialysis or removal of the detergent with Bio-Beads.

Demonstration of net proton transfer during illumination of the reconstituted systems depends on the asymmetric placement of bacteriorhodopsin into the membrane. Surprisingly, under most reconstitution conditions purple membrane has a marked tendency to become incorporated into liposomes with a biased orientation, in the sense reversed from its orientation in the original cytoplasmic membrane. The uniform orientation of the bacteriorhodopsin seems to originate from charge asymmetry in the protein [325], and it is not very dependent on the kind of lipid used in the reconstitution [343]. Thus, net light-induced pH gradients are in the acid-inside direction in the liposomes [17,18], unlike the alkaline-inside pH gradients which develop in cell envelope vesicles [78,324]. However, liposomes reconstituted with bacteriorhodopsin which translocate protons from inside to outside can be produced at low pH [344,345]. Enhancement of net proton movements in the presence of cationic ionophores, such as gramicidin or valinomycin, as well as their abolition in the presence of proton conductors, such as CCCP, FCCP or 1799, argue that the translocation of the protons is active and electrogenic [305,346]. Besides following pH during illumination, transport of protons into the liposomes or vesicles can be observed with ^{31}P-NMR of enclosed phosphate [347], or on a ms timescale by using modulated actinic light and pH dyes [348]. Unidirectional incorporation of purple membrane into planar films is possible either by taking advantage of statistical chance when very few patches are placed into the bilayer, or by fusing well-oriented liposomes with the films. The photocurrents originate from the bilayer portion of the lipid film in the septa [349]. Photocurrents during steady illumination in the presence of uncouplers or flash regimes have been measured [350–355], and correspond to net proton transport.

Dissociation of the purple membrane lattice, followed by incorporation of the bacteriorhodopsin into liposomes [327,356,357], yields a functional proton transport system. Thus, bacteriorhodopsin monomers will translocate protons upon illumination.

Purple membrane sheets (which do not enclose a compartment) show reversible net proton release above pH 5 and net proton uptake below during steady-state illumination [236]. The change in protonation is presumed to reflect the difference between the occupancy of proton-binding sites in bacteriorhodopsin and its photointermediates, of which M-412 is the one accumulated to the greatest extent. It is important to note, therefore, that as mentioned above, resonance Raman spectra indicate that the Schiff's base proton is lost from the M intermediate [35,259]. The kinetics of the protonation changes were resolved on a ms time-scale using pH indicator dyes, and these and the steady-state results indicate that the value of the protons/M intermediate varies with pH, from a net release of about 2 at acidic pH to net uptake at alkaline pH [21,23,358–361]. The release is observed to be in less than 1 ms. The uptake kinetics are slower but do not always agree with the recovery of BR-568. These results have suggested that the popular scheme of bacteriorhodopsin photocycle is coupled to proton release and uptake in such a way that the spectroscopically observable transitions correspond to protonated and deprotonated states, a simplification which assumes that no states arise in the protein which are not described by the spectroscopic changes in the chromophore.

Simplified schemes (such as shown in Ref. 20) might lead to the natural assumption that a single proton is translocated per photocycle. Measurements of the stoicheiometry of the pumping have shown, however, that this is not the case [37–40,361]. The number of protons detected with indicator dyes and with measurements of flash-induced volume changes is one or less at low ionic strength, but increases to a limiting value of 2 in 200 mM salt [37,39]. The ionic strength dependence of the proton release is confirmed by measurements of the surface potential produced upon the release of protons [362,363]. An apparent stoicheiometry higher than one in purple membrane sheets might reflect the release of Bohr protons (due to conformational changes), but such an interpretation is unlikely for whole cells and vesicles where the turn-over of bacteriorhodopsin is many-fold [38–40,364–366], because the binding of too many protons have to be invoked per bacteriorhodopsin molecule. Interestingly, measurement of flash-induced conductance changes in purple membrane suspension showed that, at pH 4, protons were taken up before being released, but at pH 8 an ion other than proton was released (e.g., a counterion) which was displaced by a proton [367]. At high ionic strength the quantum yield was 2–3.

Theoretical models for the functioning of bacteriorhodopsin must include: entry and leaving sites for protons on the two sides of the membrane, a proton conduction pathway, and the unidirectional translocation of protons across a potential barrier somewhere inside the protein so as to accomplish net transport against an electrochemical potential difference. Many proposed models postulate that the photoinduced displacement of the protonated Schiff's base relative to other protonated groups, e.g., one [268,295] or two [368] tyrosines, is sufficient to set off a vectorial movement of proton(s). Other groups suggested to be near the retinal nitrogen include lysine, on the basis of the observation that the apparent p*K* of the Schiff's base is lowered from above 12 to near 10 [369,370] or even much lower [371] during the initial steps of the photocycle. Movement of the Schiff's base nitrogen may be caused by light-induced isomerization around the C_{14}-C_{15} bond [372]. Conduction of protons to and from the Schiff's base could take place via an anhydrous path [31–33], i.e., a 'proton wire'. However, H/D exchange between the Schiff's base proton and bulk water in the dark is much too fast for conventional base-catalyzed proton migration [373], and a concerted exchange mechanism may have to be invoked.

3. Halorhodopsin

3.1. Spectroscopic and molecular properties

In the commonly used strains of *H. halobium* (e.g., R-1 or S-9), halorhodopsin is present in much smaller amounts than bacteriorhodopsin, and most spectroscopic measurements will show only the properties of the dominant pigment. This complication is avoided when strains negative for bacterio-opsin (e.g., ET-15 or L-33) are used [92,102,374,375], but strains which contain only the halorhodopsin (and lack the slowly cycling rhodopsin, see below) are not yet available. Early work on halorhodopsin did not recognize the existence of the third pigment, and some of the spectroscopic properties attributed [92,93,376] at that time to halorhodopsin refer, in fact, to a mixture of halorhodopsin and slow rhodopsin. It is the comparisons with strains containing only slow rhodopsin [102,375], as well as newly developed methods [94] for distinguishing the two pigments, which have allowed, at this time, to draw some reasonably firm conclusions about the spectroscopic properties of halorhodopsin.

The visible absorption band of halorhodopsin is red-shifted from that of bacteriorhodopsin. Earlier reports of difference spectroscopy of cell envelope vesicles [92,376], as well as action spectra for transport [86,87,89,377], ATP synthesis [374], and flash-induced absorbance changes [93], suggested that the absorption maximum is at 588 nm, but more recent results suggest [378] a maximum at 580 nm. Lowering the NaCl concentration from the physiological concentration of several molar to near zero results in a blue-shift of 10–15 nm, an effect which may be related to the well-known tendency of halobacterial proteins to denature at low ionic strengths [3], but in this case the effect is at least partially reversible [379,380]. The absorption maximum of the native pigment is maintained not by the sodium ions but specifically by chloride, although bromide or iodide are also found to be effective [380]. The chloride-dependent shift is not due to a change in isomeric composition of the retinal, but is caused presumably by a conformational change upon chloride binding [380]. A more dramatic spectral shift, to near 410 nm, is observed under a variety of conditions. In the absence of chloride but at high ionic strength to preserve the membrane structure, e.g., in 1.5 M sodium sulfate or 3 M potassium phosphate, the shift will occur increasingly at alkaline pH, such as 9, in the dark, and it is reversible with chloride [98] (apparent affinity constant 40 mM). In 4 M NaCl this transition takes place only near pH 11 when in the dark. During sustained illumination with yellow light in 4 M NaCl the shift occurs at much lower pH, e.g., near 7, although with increasing rate at increasing pH [98]. The photoconversion will take place in 0.4 M NaCl [379], but it is not observable in the absence of chloride [380] probably because the pigment is already bleached under these conditions. The nature of the change which gives rise to the large blue-shift is uncertain: the ubiquitous 410 nm species is apparently due partly to isomerization to the 13-*cis* form [379], and very likely also to deprotonation of the Schiff's base [98]. In purified halorhodopsin preparations the shift to 410 nm was shown to be accompanied by the appearance of a proton in the medium [96]. These results are rationalized in terms of a scheme [98], where halorhodopsin is in equilibrium with a chloride-bound form. At alkaline pH the chloride-bound form is stable up to pH 11, but can be photoconverted to the 410 nm species. Without chloride the conversion occurs in the dark at lower pH, in a reversible reaction. Chloride thus appears to raise the pK of a group, which is probably the Schiff's base. The significance of these characteristic changes for the functioning of halorhodopsin as a light dependent ion transport system remains to be seen.

Illumination causes cyclical absorption changes in halorhodopsin, as in bacteriorhodopsin, with an overall recovery time of 5–10 ms at high NaCl concentrations [93,98,381,382], and 1–2 ms in dilute salt [383] or in the absence of chloride but in 1.5 M sulfate (or 3 M phosphate) [98,99]. The slower photocycle is proposed to include intermediates designated as P-632, P-500 and P-380, arranged in linear sequence [93]. P-632 arises and decays too rapidly to be detected at room temperature at μs time resolution and is most conveniently observed [93,379] at 77 °K. P-500 is easily seen at room temperature [93] or at lower temperatures [380], such as −75 °C. Overlapping between the absorption band of this photointermediate and that of halorhodopsin is likely to have shifted the apparent absorption maximum in the flash-induced difference spectrum; we calculate [99] that P-500 is really P-520, with an extinction coefficient somewhat lower than that of halorhodopsin.

In contrast to the case of bacteriorhodopsin, little dependence on proton concentration between pH 5 and 9 is seen for the recovery of absorption changes at 570 nm after the flashes [99,382], a result predicted from steady-state measurements of light-dependent transport [383].

The photocycle in the absence of chloride proceeds [384] via intermediates P-630 and P-640 (or as we more recently found [99], P-660). Although this type of photocycle was earlier referred to as 'water cycle', it appears also at high ionic strength in the absence of chloride [99]. Addition of chloride or bromide under these conditions causes the reappearance of the photocycle via the P-520 intermediate and the disappearance of the photocycle via the P-660 intermediate, at halide concentrations consistent with the apparent affinity constant for the light-induced transport. At the same time, with increasing halide concentration the decay half-lives of both P-520 and P-660 increase from 1–2 to 5–15 ms. These findings [99] suggest that, similarly to halorhodopsin, the two principal photointermediates will also bind halide, and that the binding of halide slows down the decay kinetics.

Thus, it appears that the observed photochemical changes reflect the influence of the binding of halide to halorhodopsin, and thus they might reveal whether or not anion translocation is taking place in the system. One might expect this to be so for proton binding and transport in bacteriorhodopsin as well, but the greater complexity of that photocycle, and the fact that pH dependent groups must be involved in other effects (i.e., structural interactions), do not allow the unambiguous assignment of any part of the photochemistry to proton translocation pathways.

The molecular weight of halorhodopsin is estimated from SDS gel patterns of halo-opsin labeled with ^{3}H-labeled retinal [94] and of purified halorhodopsin [95,96], and it appears to be slightly lower than that of bacteriorhodopsin, i.e., about 25 000. Differences in the cleavage patterns from limited proteolysis, as well as lack of immunological crossreactivity, suggest that halo-opsin and bacterio-opsin are very different proteins, however [95].

3.2. Functional properties

Cell envelope vesicles prepared from bacterio-opsin negative *H. halobium* strains develop membrane potential (inside negative) and pH difference (inside acid) during illumination [88,89]. The proton influx is passive, because: (a) proton conductors, such as CCCP or 1799, will enhance its rate and not diminish its magnitude; (b) ionophores for alkali metal cations, which shortcircuit membrane potential, will abolish it; and (c) at steady state, established within a few minutes of illumination, the proton motive force is at or near zero [89] yet the membrane potential and pH difference (of opposing signs) persist. Thus, it is evident that these membranes contain a light-driven electrogenic pump for an ion other than protons. Vesicles prepared from a strain which contains both bacteriorhodopsin and halorhodopsin will exhibit more complex changes. Here the light-dependent uptake of protons is enhanced in the presence of uncouplers [89,90,385], and at acidic pH there is net proton extrusion. The results are consistent with the model of simultaneous (a) active proton extrusion by bacteriorhodopsin and (b) passive proton uptake in response to the membrane potential created by halorhodopsin. Transport by halorhodopsin requires the presence of retinal during growth of retinal-deficient cells [374], and extensive bleaching of halorhodopsin-containing membranes by illumination in the presence of hydroxylamine reversibly inactivates the system [92]. Action spectra identify the pigment responsible as one absorbing about 20 nm to the red from bacteriorhodopsin [86,87,89,377]. The identification of halorhodopsin as the pigment with the observed transport activity was facilitated by the availability of mutants which lack one or another of the retinal proteins [102,375,386]. Differences in the rate of bleaching in the presence of hydroxylamine between the pigment which cycles with a time constant of 5–10 ms (halorhodopsin) and another which cycles much slower (slow rhodopsin, discussed below) also allowed the assignment of halorhodopsin to the transport [98].

The ion transported by halorhodopsin under the conditions used must be either sodium or chloride, the former having to be extruded, the latter taken up in order to generate the membrane potential of the observed direction. Although the circulation of sodium ions is increased upon illumination [88,90], and a specific requirement for sodium for the light-induced pH changes was reported [88,90], net extrusion of sodium ions from the vesicles resulting in a concentration gradient for sodium could not be demonstrated [387]. In contrast, rapid chloride uptake (more than 10 $nmol \cdot min^{-1} \cdot mg^{-1}$ protein) into illuminated vesicles, which had been dialyzed against either 1.5 M sodium sulfate or 3 M potassium phosphate, can be shown [97]. This chloride uptake is enhanced by cationic ionophores, such as gramicidin or valinomycin, and results in significant volume increase, detected by light scattering [97] and with ESR probes [388]. The transport seems to generate a concentration gradient for chloride (inside > outside) [97]. The light-dependent creation of membrane potential, pH change and volume increase require the addition of chloride to the vesicle exterior only, with an apparent affinity constant of about 40 mM [97]. It appears therefore, that halorhodopsin behaves as an inward-directed chloride pump. In analogy with bacteriorhodopsin, conclusive proof of this will come from studies of transport by purified halorhodopsin reconstituted into liposomes, but the isolation of halorhodopsin with all of its functional properties intact, has not yet been reported.

The accurate measurement of net chloride transport is difficult, but indirect methods for quantitating net ion transport in this system are available. One of these takes advantage of the fact that initially the counterions to chloride movement are protons, particularly when their permeability is increased by adding large amounts of uncouplers. Initial rates of proton uptake, measured with a pH electrode in the medium, should be then equal to the rate of chloride uptake [94,205]. Another method is to follow the light dependent volume changes in the presence of gramicidin or valinomycin. Under such conditions the rate-limiting process in salt (and therefore water) accumulation is the chloride transport. When the volume changes are followed with light scattering the rates are obtained in relative units, but when included ESR probes are used [388], the volumes are estimated quantitatively. A third method is to follow the initial rate of increase in membrane potential (a parameter related to the net current developed by the pump) with the fluorescence of a potential-sensitive dye [97]. Using such methods it can be shown that halorhodopsin functions as a pump at a pH optimum near 6.8, about a unit above the optimum for bacteriorhodopsin [97]. Maximal pH difference across the vesicle membranes due to passive proton uptake and maximal rate of pH change were determined [383] with an ESR probe to be at pH 7.5.

Results using the above methods indicate that the apparent affinity constant for chloride is about 40 mM, a value not very different from the binding constant for chloride to the band 3 protein in erythrocyte membranes [389], i.e., 60–80 mM. That a distinct chloride binding site exists in halorhodopsin is indicated by the fact that inhibitors of eukaryotic chloride transport of the MK series inhibit chloride transport in the halobacterial system also, and with competitive kinetics [99].

4. Slowly cycling rhodopsin

A third retinal pigment, distinct from bacteriorhodopsin and halorhodopsin, is detectable in several *H. halobium* strains, but is likely to be present in small quantities in most commonly used strains. In strains which contain the opsin of this pigment alone, it can be detected by difference spectroscopy of the membranes with and without added retinal, in others only by the fact that its flash-induced absorbance decrease recovers very slowly, with a time constant of 0.8 ms, which is easily separated from the ms time-scale changes of halorhodopsin and bacteriorhodopsin [102,382]. Suitable strains for the study of 'slowly cycling rhodopsin' include those in which the other opsins contain a genetic defect (e.g., Flx37 or Flx3), and those in which the genetic defect is in retinal synthesis as well as in bacterio-opsin (e.g. L-07), because it appears that the biosynthesis of halo-opsin is controlled strictly by the presence of retinal, but that of slow opsin is not [390]. In vitro reconstitution of isolated membranes from the latter strains with retinal yields mostly slowly cycling pigment, with only a small extent of chloride transport activity [94]. The pigment so reconstituted absorbs near 590 nm, which should be taken as the absorption band of slow rhodopsin.

The photocycle of slow rhodopsin was studied with flash photolysis, as well as with continuous illumination since the kinetics are slow enough to allow the accumulation of a bleached intermediate. The pigment cycles between the 590 nm state and one absorbing at 375 nm, in a photoreversible way [102]. Thus, illumination with yellow light causes accumulation of the UV absorbing form, while illumination with blue or near UV light leads to a reversal of this change. This effect resembles the phototactic behavior of the cells [100,101,391], a light-dependent motility response suggested to involve a retinal protein [392,393]. For this reason, and because genetic lesions in the photochemical and behavioral properties seem to go together, it has been proposed that slow rhodopsin is, in fact, the phototactic pigment [102]. Indeed, reconstitution results with retinal analogues have already suggested that the photoactic pigment cannot be bacteriorhodopsin [394].

5. Light-driven ion transport in the halobacteria

Illumination of whole *H. halobium* cells under anaerobic conditions (to exclude contributions from respiration) produces increased proton motive force, as expected from the functioning of bacteriorhodopsin as a proton pump and of halorhodopsin as an electrogenic chloride pump. Changes in steady-state membrane potential and pH difference between inside and outside the cells, caused by the light, are easily determined with the commonly used probes of these bioenergetic parameters. Such measurements yield the result that the membrane potential increases by only 10–40 mV, and the pH difference by 0–1 pH units, depending on the initial pH [69,70,83,395,396]. These values are so low that in order for the cells to gain significant benefit from the illumination there must be a high 'resting' proton motive force in the dark, to which the light-dependent effects will add. Although their determination is fraught with technical difficulties (e.g., probe binding to membranes and cytoplasmic components), such resting potentials have been demonstrated in the cells, and are typically 100–110 mV (inside negative). Resting pH differences range from 2 at pH 5 to zero at pH 8.

The origin of the resting potentials cannot be respiration or continued metabolic activity, since reasonable precautions to exclude such causes have been taken. It is more likely that a large concentration gradient for K^+ (inside > outside), established under energized conditions and known to be of many hours duration [397–399], gives rise to it [395,396]. Diffusion potentials due to potassium have been detected in the halobacteria, although not the full extent predicted. However, the membrane is much more permeable to K^+ than to other ions [72,400]. Selective permeability of the cell membrane to K^+ is suggested also by the fact that K^+ uptake during illumination seems to be driven mostly by the electrical potential [401]. However, as discussed below the cell envelope vesicles prepared from the cells are found [402] to be very poorly permeable to K^+.

At physiological pH, which is near 7 for these cells, the principal effect of illumination is to increase the membrane potential [72]. This, in turn, will energize the uptake of K^+ and the extrusion of Na^+. In an experiment with starved cells which had accumulated Na^+ and lost K^+ the time-course and the extent of these transport processes were found to be nearly the same [72]: over a period of about one hour the cells gained 637 ± 7 mmol K^+ per kg cell water (up to a total of about 3000 $\text{mmol} \cdot \text{kg}^{-1}$ cell water) and lost 534 ± 5 mmol Na^+ per kg cell water. Taking into account gains and losses of other cations (protons, calcium, magnesium) during this period, a net anion gain must have occurred to the extent of 100–120 $\text{mmol} \cdot \text{kg}^{-1}$ cell water. Presumably, the anion is chloride. On the other hand, light-scattering studies of cell suspensions have suggested that under some conditions there is a small (ca. 1%) net loss of volume, and therefore of salt, during illumination [403]. The transport of K^+ is thought to occur by passive (although perhaps facilitated) diffusion [401]. Indeed, the concentration gradient for this ion achieved during energized conditions is 200–300 fold, or equivalent to about 140 mV membrane potential. The intracellular K^+ is largely free, rather than bound [404]. Other mechanisms of potassium transport in these cells have not been excluded.

The extrusion of Na^+, together with K^+ uptake, results in lowered intracellular sodium concentrations in energized cells relative to that in the medium. The concentration gradient will slowly collapse during starvation conditions, but becomes reestablished during illumination [72]. The mechanism of sodium transport is investigated most conveniently in cell envelope vesicles [78,405], prepared from the halobacteria. In addition to the cytoplasmic membrane, such vesicles contain essentially all of the bacteriorhodopsin in the cells, and the orientation of the membrane is conserved. Interference from halorhodopsin is minimized by the use of strains which contain very high amounts of bacteriorhodopsin (e.g., S-9), or strains which lack retinal but not bacterio-opsin (e.g., JW-5 or W-296). In the latter case halo-opsin biosynthesis is repressed [390], and in vitro reconstitution of the vesicles with small amounts of retinal yields a system with only bacteriorhodopsin activity [97]. Illumination of such vesicles in several molar NaCl causes the membrane potential to rise from zero to 100 mV or higher (inside negative), and a pH difference is created which depends on the initial pH: near pH 5 about 1.5, near pH 7 down to zero [324,402,406,407]. The pH changes which occur during illumination depend on the sodium content of the vesicles. Sodium inside the vesicles causes smaller pH changes to develop (and higher membrane potentials), thus vesicles loaded with mixtures of NaCl and KCl exhibit biphasic pH changes: lower pH changes initially, and higher pH changes once the vesicles are depleted of sodium ions [74]. Sodium efflux is measured by atomic absorption methods [77,79,402]. From such measurements it is also concluded that above pH 7 the pH gradient established by the illumination will be in the reverse direction, i.e., after transient proton extrusion there is net proton flux inward [74,97,388]. This characteristic pattern of light-induced proton movements depends critically on the magnitude of the proton motive force. When the latter is somewhat lowered, either by adding a limited amount of uncoupler, or by using vesicles with bacteriorhodopsin contents of less than 0.1 $nmol \cdot mg^{-1}$ protein, the light-dependent proton uptake is replaced by monophasic proton extrusion [97], as expected for transport by bacteriorhodopsin alone. These observations strongly suggest that the operation of a gated proton/sodium antiporter which functions in an electrogenic fashion, i.e., the proton/sodium ratio of the exchange is > 1. Experiments designed to address this question directly show that the rate of sodium extrusion is not proportional to the proton motive force, but increases [77] steeply above a threshold value of about -130 mV. Although such gating can be a property of specific structures designed to function in a nonlinear fashion, such as sodium and potassium channels in excitable membranes, it is observed also in transport systems where the rate-limiting step is the electrical potential-dependent translocation of a charge [408]. Estimation of the steady-state magnitude of the sodium gradient (by indirect means, using the sodium gradient-dependent transport of serine) allowed its comparison with the proton motive force at different light intensities [77], yielding a slope of 1.8, a value which suggests a stoicheiometry of 2 H^+/Na^+. It appears therefore, that the extrusion of a sodium ion is accompanied by the net inward movement of a positive charge.

When the only cation present is Na^+ the illumination results in volume decrease [97,388,402], a consequence of active sodium extrusion and passive chloride efflux due to the membrane potential created. When K^+ is also added, in order to create conditions similar to those which the cells face during growth, the volume decrease is less because potassium is taken up [402]. In envelope vesicles the permeability of K^+

is poor and significant transport occurs only at either high concentrations or in the presence of valinomycin [402], presumably because the mechanism for the uptake of this ion available to the cells is lost during the preparation of the vesicles.

Cell envelope vesicles prepared from *H. halobium* strains, which contain halorhodopsin but not bacteriorhodopsin (e.g., L-33), behave in a very different way from the vesicles described above. Energization of these membranes is by the transport of ions other than protons [88–90], i.e., chloride [97]. Upon illumination, the vesicles develop proton motive force only transiently, in the form of membrane potential (inside negative). This is followed by rapid passive proton influx, and a pH difference of opposite sign (inside acid) is created which cancels the net force on protons. After this time the net proton flux is near zero [89], and the principal counterion to chloride uptake is sodium [97]. Thus, the vesicles will swell during illumination [97,388]. Whole cells appear to behave in a similar way as far as external pH changes are concerned, but since membrane potentials have not been measured under these circumstances, unambiguous conclusions cannot be made. As expected for the passive uptake of protons, however, illumination of the cells will result in alkalinization of the medium external to the cells [86,87,375], and this is greatly increased in the presence of proton conductors, but abolished when agents which short-circuit membrane potential are added.

Similar to the behavior of cell envelope vesicles, the illumination of most halobacterial strains containing both bacteriorhodopsin and halorhodopsin will result in complex ionic movements across the cytoplasmic membrane. The pattern of pH changes depends on the previous history of the cells. Cells which have been recently metabolically active or illuminated will show rapid pH rise, followed within a minute by a lowering of pH [14,83,85]. The rise and fall in pH are separable by kinetic analysis of data at different light intensities [83], and by altering the initial pH of the cell suspensions: at pH lower than 6.5 the acidification response dominates, while above this value the alkalinization is dominant [85]. The two effects are separable also by previous treatment of the cells. Heating at 75 °C for 5 min abolishes the alkalinization, while bleaching in the presence of hydroxylamine abolishes the acidification [86,87]. All of these observations suggest, although do not conclusively prove, that the alkalinization reflects passive proton uptake in response to the increased membrane potential created by chloride uptake, and the acidification reflects active proton extrusion by bacteriorhodopsin. Results with triphenyl tin, a hydroxyl/chloride exchanger, suggest [409,410] that one of the functions of bacteriorhodopsin might be to remove the protons accumulated inside the cells in response to the photopotential created by halorhodopsin. This interpretation is somewhat compromised by the finding that proton conductors will abolish both alkalinization and acidification effects. Although this observation can be rationalized by arguments about the effect of internal pH on the transport, other explanations for the biphasic pH changes have been given, which attribute it to proton extrusion by bacteriorhodopsin and a gated proton leak pathway via the membrane ATPase and/or the sodium/proton antiporter [83,411].

References

1 Larsen, H. (1967) Adv. Microbiol. Physiol. 1, 97–132.
2 Kushner, D.J. (1978) In Microbial Life in Extreme Environments (Kushner, D.J., ed.) pp. 317–368, Academic Press, London.
3 Lanyi, J.K. (1974) Bacteriol. Rev. 38, 272–290.
4 Bayley, S.T. and Morton, R.A. (1978) CRC Crit. Rev. Microbiol. 6, 151–205.
5 Skulachev, V.P. (1978) FEBS Lett. 87, 171–179.
6 Lanyi, J.K. (1979) Biochim. Biophys. Acta 559, 377–397.
7 Lanyi, J.K. (1969) Arch. Biochem. Biophys. 128, 716–724.
8 Cheah, K.S. (1970) Biochim. Biophys. Acta 197, 84–86.
9 Hartmann, R., Sickinger, H.-D. and Oesterhelt, D. (1980) Proc. Natl. Acad. Sci. U.S.A. 77, 3821–3825.
10 Oesterhelt, D. and Krippahl, G. (1983) Ann. Microbiol. (Inst. Pasteur) 134B, 137–150.
11 Stoeckenius, W. and Rowen, R. (1967) J. Cell Biol. 34, 365–393.
12 Stoeckenius, W. and Kunau, W.H. (1968) J. Cell Biol. 38, 337–357.
13 Oesterhelt, D. and Stoeckenius, W. (1971) Nature New Biol. 233, 149–152.
14 Oesterhelt, D. and Stoeckenius, W. (1973) Proc. Natl. Acad. Sci. U.S.A. 70, 2853–2857.
15 Blaurock, A.E. and Stoeckenius, W. (1971) Nature New Biol. 233, 152–155.
16 Danon, A. and Stoeckenius, W. (1974) Proc. Natl. Acad. Sci. U.S.A. 71, 1234–1238.
17 Racker, E. and Stoeckenius, W. (1974) J. Biol. Chem. 249, 662–663.
18 Racker, E. and Hinkle, P.C. (1974) J. Membr. Biol. 17, 181–188.
19 Oesterhelt, D. and Hess, B. (1973) Eur. J. Biochem. 37, 316–326.
20 Stoeckenius, W. and Lozier, R.H. (1974) J. Supramol. Struct. 2, 769–774.
21 Dencher, N. and Wilms, M. (1975) Biophys. Struct. Mech. 1, 259–271.
22 Kung, M.C., Devault, D., Hess, B. and Oesterhelt, D. (1975) Biophys. J. 15, 907–911.
23 Lozier, R.H., Bogomolni, R.A. and Stoeckenius, W. (1975) Biophys. J. 15, 955–962.
24 Slifkin, M.A. and Caplan, S.R. (1975) Nature 253, 56–58.
25 Sherman, W.V. and Caplan, S.R. (1975) Nature 258, 766–768.
26 Lozier, R.H. and Niederberger, W. (1977) Fed. Proc. 36, 1805–1809.
27 Bogomolni, R.A., Stubbs, L. and Lanyi, J.K. (1978) Biochemistry 17, 1037–1041.
28 Hess, B. and Kuschmitz, D. (1979) FEBS Lett. 100, 334–340.
29 Scherrer, P., Packer, L. and Seltzer, S. (1981) Arch. Biochem. Biophys. 212, 589–601.
30 Lemke, H.-D. and Oesterhelt, D. (1981) Eur. J. Biochem. 115, 595–604.
31 Nagle, J.F. and Morowitz, H.J. (1978) Proc. Natl. Acad. Sci. U.S.A. 75, 298–302.
32 Nagle, J.F. and Mille, M. (1981) J. Chem. Phys. 74, 1367–1372.
33 Merz, H. and Zundel, G. (1981) Biochem. Biophys. Res. Commun. 101, 540–546.
34 Stoeckenius, W., Lozier, R.H. and Bogomolni, R.A. (1979) Biochim. Biophys. Acta 505, 215–278.
35 Lewis, A., Spoonhower, J., Bogomolni, R.A., Lozier, R.H. and Stoeckenius, W. (1974) Proc. Natl. Acad. Sci. U.S.A. 71, 4462–4466.
36 Marcus, M.A. and Lewis, A. (1977) Science 195, 1328–1330.
37 Ort, D.R. and Parson, W.W. (1979) Biophys. J. 25, 341–353.
38 Bogomolni, R.A., Baker, R.A., Lozier, R.H. and Stoeckenius, W. (1980) Biochemistry 19, 2152–2159.
39 Govindjee, R., Ebrey, T.G. and Crofts, A.R. (1980) Biophys. J. 30, 231–242.
40 Renard, M. and Delmelle, M. (1980) Biophys. J. 32, 993–1006.
41 Henderson, R. (1975) J. Mol. Biol. 93, 123–138.
42 Blaurock, A.E. and Wober, W. (1976) J. Mol. Biol. 106, 871–888.
43 Unwin, P.N.T. and Henderson, R. (1975) J. Mol. Biol. 94, 425–440.
44 Henderson, R. and Unwin, P.N.T. (1975) Nature 257, 28–32.
45 Heyn, M.P., Bauer, P.-J. and Dencher, N.A. (1975) Biochem. Biophys. Res. Commun. 67, 897–903.
46 Becher, B. and Cassim, J.Y. (1976) Biophys. J. 16, 1183–1200.
47 Becher, B. and Ebrey, T.G. (1976) Biochem. Biophys. Res. Commun. 69, 1–6.

48 Becher, B. and Ebrey, T.G. (1977) Biophys. J. 17, 185–191.
49 Ovchinnikov, Yu.A., Abdulaev, N.G., Feigina, M.Y., Kiselev, A.V. and Lobanov, N.A. (1979) FEBS Lett. 100, 219–224.
50 Khorana, H.G., Gerber, G.E., Herlihy, W.C., Gray, C.P., Anderegg, R.J., Nihei, K. and Biemann, K. (1979) Proc. Natl. Acad. Sci. U.S.A. 76, 5046–5050.
51 Dunn, R., McCoy, J., Simsek, M., Majumdar, A., Chang, S.H., RajBhandary, U. and Khorana, H.G. (1981) Proc. Natl. Acad. Sci. U.S.A. 78, 6744–6748.
52 Engelman, D.M., Henderson, R., McLachlen, A.D. and Wallace, B.A. (1980) Proc. Natl. Acad. Sci. U.S.A. 77, 2023–2027.
53 Agard, D. and Stroud, R.M. (1982) Biophys. J. 37, 589–602.
54 Gerber, G.E., Gray, C.P., Wildenauer, D. and Khorana, H.G. (1977) Proc. Natl. Acad. Sci. U.S.A. 74, 5426–5430.
55 Packer, L., Tristam, S., Herz, J.M., Russell, C. and Borders, C.L. (1979) FEBS Lett. 108, 243–248.
56 Harris, G., Renthal, R., Tuley, J. and Robinson, N. (1979) Biochem. Biophys. Res. Commun. 91, 926–931.
57 Dumont, M.E., Wiggins, J.W. and Hayward, S.B. (1981) Proc. Natl. Acad. Sci. U.S.A. 78, 2947–2951.
58 Katre, N., Wolber, P., Stoeckenius, W. and Stroud, R.M. (1981) Proc. Natl. Acad. Sci. U.S.A. 78, 4068–4072.
59 Lemke, H.-D., Bergmeyer, J., Straub, J. and Oesterhelt, D. (1982) J. Biol. Chem. 257, 9384–9388.
60 Bridgen, J. and Walker, I.D. (1976) Biochemistry 15, 792–798.
61 Ovchinnikov, Yu.A., Abdulaev, N.G., Testlin, V.I. and Zakis, V.I. (1980) Bioorg. Chem. USSR 6, 1427–1429.
62 Lemke, H.-D. and Oesterhelt, D. (1981) FEBS Lett. 128, 255–260.
63 Bayley, H., Huang, K.S., Radhakrishnan, R., Ross, A.H., Takagaki, Y. and Khorana, H.G. (1981) Proc. Natl. Acad. Sci. U.S.A. 78, 2225–2229.
64 Mullen, E., Johnson, A.H. and Akhtar, M. (1981) FEBS Lett. 130, 187–193.
65 Kriebel, A.N. and Albrecht, A.C. (1976) J. Chem. Phys. 65, 4576–4583.
66 Ebrey, T.G., Becher, B., Mao, B., Kilbridge, P. and Honig, B. (1977) J. Mol. Biol. 112, 377–397.
67 King, G.I., Stoeckenius, W., Crespi, H. and Schoenborn, B. (1979) J. Mol. Biol. 130, 395–404.
68 King, G.I., Mowery, P.C., Stoeckenius, W., Crespi, H.L. and Schoenborn, B.P. (1980) Proc. Natl. Acad. Sci. U.S.A. 77, 4726–4730.
69 Michel, H. and Oesterhelt, D. (1976) FEBS Lett. 65, 175–178.
70 Bakker, E.P., Rottenberg, H. and Caplan, S.R. (1976) Biochim. Biophys. Acta 440, 557–572.
71 Bogomolni, R.A. (1977) Fed. Proc. 36, 1833–1839.
72 Wagner, G., Hartmann, R. and Oesterhelt, D. (1978) Eur. J. Biochem. 89, 169–179.
73 Hubbard, J.S., Rinehart, C.A. and Baker, R.A. (1976) J. Bacteriol. 125, 181–190.
74 Lanyi, J.K. and MacDonald, R.E. (1976) Biochemistry 15, 4608–4614.
75 Eisenbach, M., Cooper, S., Garty, H., Johnstone, R.M., Rottenberg, H. and Caplan, S.R. (1977) Biochim. Biophys. Acta 465, 599–613.
76 Caplan, S.R., Eisenbach, M., Cooper, S., Garty, H., Klemperer, G. and Bakker, E.P. (1977) In Bioenergetics of Membranes (Packer, L., Papageorgiou, G. and Trebst, A., eds.) pp. 101–114, Elsevier/North Holland, Amsterdam.
77 Lanyi, J.K. and Silverman, M.P. (1979) J. Biol. Chem. 254, 4750–4755.
78 MacDonald, R.E. and Lanyi, J.K. (1975) Biochemistry 14, 2882–2889.
79 Lanyi, J.K., Yearwood-Drayton, V. and MacDonald, R.E. (1976) Biochemistry 15, 1595–1603.
80 Lanyi, J.K., Renthal, R. and MacDonald, R.E. (1976) Biochemistry 15, 1603–1610.
81 MacDonald, R.E., Greene, R.V. and Lanyi, J.K. (1977) Biochemistry 16, 3227–3235.
82 Oesterhelt, D. (1975) CIBA Fd. Symp. 31, 147–167.
83 Bogomolni, R.A., Baker, R.A., Lozier, R.H. and Stoeckenius, W. (1976) Biochim. Biophys. Acta 440, 68–88.
84 Danon, A. and Caplan, S.R. (1976) Biochim. Biophys. Acta 423, 133–140.
85 Wagner, G. and Hope, A.B. (1976) Austr. J. Plant Physiol. 3, 665–676.

86 Matsuno-Yagi, A. and Mukohata, Y. (1977) Biochem. Biophys. Res. Commun. 78, 237–243.
87 Matsuno-Yagi, A. and Mukohata, Y. (1980) Arch. Biochem. Biophys. 199, 297–303.
88 Lindley, E.V. and MacDonald, R.E. (1979) Biochem. Biophys. Res. Commun. 88, 491–499.
89 Greene, R.V. and Lanyi, J.K. (1979) J. Biol. Chem. 254, 10986–10994.
90 MacDonald, R.E., Greene, R.V., Clark, R.D. and Lindley, E.V. (1979) J. Biol. Chem. 254, 11831–11838.
91 Mukohata, Y., Matsuno-Yagi, A. and Kaji, Y. (1980) In Saline Environment (Morishita, H. and Masui, M., eds.) pp. 31–37, Business Center for Academic Soc., Japan.
92 Lanyi, J.K. and Weber, H.J. (1980) J. Biol. Chem. 255, 243–250.
93 Weber, H.J. and Bogomolni, R.A. (1981) Photochem. Photobiol. 33, 601–608.
94 Lanyi, J.K. and Oesterhelt, D. (1982) J. Biol. Chem. 257, 2674–2677.
95 Hegemann, P., Steiner, M. and Oesterhelt, D. (1982) EMBO J. 1, 1177–1183.
96 Steiner, M. and Oesterhelt, D. (1983) EMBO J. 2, 1379–1385.
97 Schobert, B. and Lanyi, J.K. (1982) J. Biol. Chem. 257, 10306–10313.
98 Lanyi, J.K. and Schobert, B. (1983) Biochemistry 22, 2763–2769.
99 Schobert, B., Lanyi, J.K. and Cragoe, E.J. Jr. (1983) J. Biol. Chem. 258, 15158–15164.
100 Hildebrand, E. and Dencher, N. (1975) Nature 257, 46–48.
101 Hildebrand, E. (1977) Biophys. Struct. Mech. 3, 69–77.
102 Bogomolni, R.A. and Spudich, J.L. (1982) Proc. Natl. Acad. Sci. U.S.A. 79, 6250–6254.
103 Oesterhelt, D. and Stoeckenius, W. (1974) Methods Enzymol. 31, 667–678.
104 Becher, B. and Cassim, J.Y. (1975) Prep. Biochem. 5, 161–178.
105 Blaurock, A.E., Stoeckenius, W., Oesterhelt, D. and Scherhopf, G.L. (1976) J. Cell Biol. 71, 1–22.
106 Kushwaha, S.C., Kates, M. and Stoeckenius, W. (1976) Biochim. Biophys. Acta 426, 703–710.
107 Kushwaha, S.C., Kates, M. and Martin, W.G. (1975) Can. J. Biochem. 53, 284–292.
108 Reynolds, J.A. and Stoeckenius, W. (1977) Proc. Natl. Acad. Sci. U.S.A. 74, 2803–2804.
109 Oesterhelt, D., Schuhmann, L. and Gruber, H. (1974) FEBS Lett. 44, 257–261.
110 Schreckenbach, T. and Oesterhelt, D. (1977) Fed. Proc. 36, 1810–1814.
111 Fisher, K.A. and Stoeckenius, W. (1977) Science 197, 72–74.
112 Fisher, K.A., Yanagimoto, K. and Stoeckenius, W. (1978) J. Cell Biol. 77, 611–621.
113 Neugebauer, D.-C. and Zingsheim, H.P. (1978) J. Mol. Biol. 123, 235–246.
114 Blaurock, A.E. and King, G.I. (1976) Science 196, 1101–1104.
115 Rothschild, K.J. and Clark, N.A. (1979) Biophys. J. 25, 473–487.
116 Cortijo, M., Alonso, A., Gomez-Fernandez, J.C. and Chapman, D. (1982) J. Mol. Biol. 157, 597–618.
117 Englander, J.J. and Englander, S.W. (1977) Nature 265, 658–659.
118 Konishi, T. and Packer, L. (1977) FEBS Lett. 80, 455–458.
119 Hayward, S.B. and Stroud, R.M. (1981) J. Mol. Biol. 151, 491–517.
120 Michel, H., Oesterhelt, D. and Henderson, R. (1980) Proc. Natl. Acad. Sci. U.S.A. 77, 338–342.
121 Michel, H. and Oesterhelt, D. (1980) Proc. Natl. Acad. Sci. U.S.A. 77, 1283–1285.
122 Henderson, R. (1980) J. Mol. Biol. 139, 99–109.
123 Heyn, M.P., Cherry, R.J. and Mueller, U. (1977) J. Mol. Biol. 117, 607–620.
124 Korenstein, R. and Hess, B. (1978) FEBS Lett. 89, 15–20.
125 Muccio, D.D. and Cassim, J.Y. (1979) J. Mol. Biol. 135, 595–605.
126 Keszthelyi, L. (1980) Biochim. Biophys. Acta 598, 429–436.
127 Keszthelyi, L. and Ormos, P. (1980) FEBS Lett. 109, 189–193.
128 Ormos, P., Dancshazy, Zs. and Keszthelyi, L. (1980) Biophys. J. 31, 207–214.
129 Huang, K.-S., Ramachandran, R., Bayley, H. and Khorana, H.G. (1982) J. Biol. Chem. 257, 13616–13623.
130 Zaccai, G. and Gilmore, D.J. (1979) J. Mol. Biol. 132, 181–191.
131 Razi Naqvi, K., Gonzales-Rodriguez, J., Cherry, R.J. and Chapman, D. (1973) Nature New Biol. 245, 249–251.
132 Cherry, R.J., Heyn, M.P. and Oesterhelt, D. (1977) FEBS Lett. 78, 25–30.
133 Hoffmann, W., Restall, C.J., Hyla, R. and Chapman, D. (1980) Biochim. Biophys. Acta 602, 531–538.

134 Cherry, R.J., Mueller, U. and Schneider, G. (1977) FEBS Lett. 80, 465–469.
135 Cherry, R.J., Mueller, U., Henderson, R. and Heyn, M.P. (1978) J. Mol. Biol. 121, 283–298.
136 Cherry, R.J., Mueller, U., Holenstein, C. and Heyn, M.P. (1980) Biochim. Biophys. Acta 596, 145–151.
137 Peters, R. and Cherry, R.J. (1982) Proc. Natl. Acad. Sci. U.S.A. 79, 4317–4321.
138 Heyn, M.P., Cherry, R.J. and Dencher, N.A. (1981) Biochemistry 20, 840–849.
139 Hiraki, K., Hanamaka, T., Mitsui, T. and Kito, Y. (1978) Biochim. Biophys. Acta 536, 318–322.
140 Rehorek, M. and Heyn, M.P. (1979) Biochemistry 18, 4977–4983.
141 Usukura, J., Yamada, E. and Mukohata, Y. (1981) Photochem. Photobiol. 33, 475–482.
142 Sigrist, H. and Zahler, P. (1980) FEBS Lett. 113, 307–311.
143 Chignell, C.F. and Chignell, D.A. (1975) Biochem. Biophys. Res. Commun. 62, 136–143.
144 Hoffmann, W., Clark, A.D., Turner, M., Wyard, S. and Chapman, D. (1980) Biochim. Biophys. Acta 598, 178–183.
145 Ekiel, I., Marsh, D., Smallbone, B.W., Kates, M. and Smith, I.C.P. (1981) Biochem. Biophys. Res. Commun. 100, 105–110.
146 Kinsey, R.A., Kintanar, A., Tsai, M.-D., Smith, R.L., Janes, N. and Oldfield, E. (1981) J. Biol. Chem. 256, 4146–4149.
147 Kinsey, R.A., Kintanar, A. and Oldfield, E. (1981) J. Biol. Chem. 256, 9028–9036.
148 Keefer, L.M. and Bradshaw, R.A. (1977) Fed. Proc. 36, 1799–1804.
149 Gerber, G.E., Anderegg, R.J., Herlihy, W.C., Gray, C.P., Biemann, K. and Khorana, H.G. (1979) Proc. Natl. Acad. Sci. U.S.A. 76, 227–231.
150 Ovchinnikov, Yu.A., Abdulaev, N.G., Dergachev, A.E., Drachev, L., Kaulen, A.D., Khitrina, L.V., Lazarova, Z.P. and Skulachev, V.P. (1982) Eur. J. Biochem. 127, 325–332.
151 Renthal, R., Harris, G. and Parrish, R. (1979) Biochim. Biophys. Acta 547, 258–269.
152 Campos-Cavieres, M., Moore, T.A. and Perham, R.N. (1979) Biochem. J. 179, 233–238.
153 Kimura, K., Mason, T.L. and Khorana, H.G. (1982) J. Biol. Chem. 257, 2859–2867.
154 Walker, J.E., Carne, A.F. and Schmitt, H.W. (1979) Nature 278, 653–654.
155 Ovchinnikov, Yu.A., Abdulaev, N.G., Feigina, M.Y., Kiselev, A.V. and Lobanov, N.A. (1977) FEBS Lett. 84, 1–4.
156 Wallace, B.A. and Henderson, R. (1982) Biophys. J. 39, 233–239.
157 Huang, K.-S., Bayley, H., Liao, M.-J., London, E. and Khorana, H.G. (1981) J. Biol. Chem. 256, 3802–3809.
158 Trewhella, J., Anderson, S., Fox, R., Gogol, E., Khan, S., Zaccai, G. and Engelman, D. (1983) Biophys. J. 42, 233–241.
159 Katre, N.V. and Stroud, R.M. (1981) FEBS Lett. 136, 170–174.
160 Wallace, B.A. and Henderson, R. (1982) Biophys. J. 39, 255–260.
161 Engelman, D.M. and Zaccai, G. (1980) Proc. Natl. Acad. Sci. U.S.A. 77, 5894–5898.
162 Rogan, P.K. and Zaccai, G. (1981) J. Mol. Biol. 132, 281–284.
163 Chang, S.H., Mujumdar, A., Dunn, R., Makabe, O., RajBhandary, U.L., Khorana, H.G., Ohtsuka, E., Tanaka, T., Taniyama, Y. and Ikehara, M. (1981) Proc. Natl. Acad. Sci. U.S.A. 78, 3398–3402.
164 Dellweg, H.-G. and Sumper, M. (1980) FEBS Lett. 116, 303–306.
165 Rothschild, K.J., Argarde, P.V., Earnest, T.N., Huang, K.-S., London, E., Liao, M.-J., Bayley, H., Khorana, H.G. and Herzfeld, J. (1982) J. Biol. Chem. 257, 8592–9595.
166 Argarde, P.V., Rothschild, K.J., Kawamoto, A.H., Herzfeld, J. and Herlihy, W.C. (1981) Proc. Natl. Acad. Sci. U.S.A. 1981, 1643–1646.
167 Dencher, N.A. and Heyn, M.P. (1978) FEBS Lett. 96, 322–326.
168 Casadio, R., Gutowitz, H., Mowery, P., Taylor, M. and Stoeckenius, W. (1980) Biochim. Biophys. Acta 590, 13–23.
169 Dencher, N.A. and Heyn, N. (1982) Methods Enzymol. 88, 5–10.
170 Huang, K.-S., Bayley, H. and Khorana, H.G. (1980) Proc. Natl. Acad. Sci. U.S.A. 77, 323–327.
171 Hartmann, R., Sickinger, H.-D. and Oesterhelt, D. (1977) FEBS Lett. 82, 1–6.
172 London, E. and Khorana, H.G. (1982) J. Biol. Chem. 257, 7003–7011.

173 Honig, B. and Ebrey, T.G. (1974) Annu. Rev. Biophys. Bioeng. 3, 151–177.
174 Honig, B., Greenberg, A., Dinur, U. and Ebrey, T.G. (1976) Biochemistry 15, 4593–4599.
175 Schreckenbach, T., Walckhoff, B. and Oesterhelt, D. (1978) Biochemistry 17, 5353–5359.
176 Fischer, U.C. and Oesterhelt, D. (1980) Biophys. J. 31, 139–146.
177 Nakanishi, K., Balogh-Nair, V., Arnaboldi, M., Tsujimoto, K. and Honig, B. (1980) J. Am. Chem. Soc. 102, 7945–7947.
178 Warshel, A. and Ottolenghi, M. (1979) Photochem. Photobiol. 30, 2981–2983.
179 Motto, M.G., Sheves, M., Tsujimoto, K., Balogh-Nair, V. and Nakanishi, K. (1980) J. Am. Chem. Soc. 102, 7947–7949.
180 Balogh-Nair, V., Carriker, J.D., Honig, B., Kamat, V., Motto, M.G., Nakanishi, K., Sen, R., Sheves, M., Tanis, M.A. and Tsujimoto, K. (1981) Photochem. Photobiol. 33, 483–488.
181 Crespi, H.L. and Ferraro, J.K. (1979) Biochem. Biophys. Res. Commun. 91, 575–582.
182 Fischer, U. and Oesterhelt, D. (1979) Biophys. J. 28, 211–230.
183 Mowery, P.C., Lozier, R.H., Chae, Q., Tseng, Y.W., Taylor, M. and Stoeckenius, W. (1979) Biochemistry 18, 4100–4107.
184 Moore, T.A., Edgerton, M.E., Parr, G., Greenwood, C. and Perham, R.N. (1978) Biochem. J. 171, 469–476.
185 Druckmann, S., Samuni, A. and Ottolenghi, M. (1979) Biophys. J. 26, 143–145.
186 Tsuji, K. and Rosenheck, K. (1979) FEBS Lett. 98, 368–372.
187 Maeda, A., Iwasa, T. and Yoshizawa, T. (1980) Biochemistry 19, 3825–3831.
188 Maeda, A., Iwasa, T. and Yoshizawa, T. (1981) Photochem. Photobiol. 33, 559–565.
189 Yoshihara, T., Suzuki, H. and Maeda, A. (1981) Photochem. Photobiol. 33, 501–510.
190 Maeda, A., Takeuchi, Y. and Yoshizawa, T. (1982) Biochemistry 21, 4479–4483.
191 Lind, Ch., Hoejeberg, B. and Khorana, H.G. (1981) J. Biol. Chem. 256, 8298–8305.
192 Bakker-Grunwald, T. and Hess, B. (1981) J. Membr. Biol. 60, 45–49.
193 Renthal, R. and Wallace, B. (1980) Biochim. Biophys. Acta 592, 621–625.
194 Oesterhelt, D., Meentzen, M. and Schuhmann, L. (1973) Eur. J. Biochem. 40, 453–463.
195 Peters, J., Peters, R. and Stoeckenius, W. (1976) FEBS Lett. 61, 128–134.
196 Schreckenbach, T., Walckhoff, B. and Oesterhelt, D. (1977) Eur. J. Biochem. 76, 499–511.
197 Jan, L.Y. (1974) Vision Res. 15, 1081–1086.
198 Pettei, M.J., Yudd, A.P., Nakanishi, K., Henselman, R. and Stoeckenius, W. (1977) Biochemistry 16, 1955–1959.
199 Sperling, W., Carl, P., Rafferty, Ch.N. and Dencher, N.A. (1977) Biophys. Struct. Mech. 3, 79–94.
200 Maeda, A., Iwasa, T. and Yoshizawa, T. (1977) J. Biochem. (Tokyo) 82, 1599–1604.
201 Sperling, W., Rafferty, C.N., Kohl, K.D. and Dencher, N.A. (1979) FEBS Lett. 97, 129–132.
202 Ohno, K., Takeuchi, Y. and Yoshida, M. (1977) Biochim. Biophys. Acta 462, 575–582.
203 Tokunaga, F., Iwasa, T. and Yoshizawa, T. (1976) FEBS Lett. 72, 33–38.
204 Dencher, N.A., Rafferty, Ch.N. and Sperling, W. (1976) Ber. Kernforsch. Juelich, No. 1374, pp. 1–42, Zentralbibl. d. Kernforschungsanlage, Juelich, F.R.G.
205 Oesterhelt, D. (1982) Methods Enzymol. 88, 10–17.
206 Sumper, M. and Herrmann, G. (1976) FEBS Lett. 71, 333–336.
207 Mukohata, Y., Sugiyama, Y., Kaji, Y., Usukara, J. and Yamada, E. (1981) Photochem. Photobiol. 33, 593–600.
208 Balogh-Nair, V. and Nakanishi, K. (1982) Methods Enzymol. 88, 496–506.
209 Marcus, M.A., Lewis, A., Racker, E. and Crespi, H. (1977) Biochem. Biophys. Res. Commun. 78, 669–675.
210 Towner, P., Gaertner, W., Walckhoff, B., Oesterhelt, D. and Hopf, H. (1980) FEBS Lett. 117, 363–367.
211 Oesterhelt, D. and Christoffel, V. (1976) Biochem. Soc. Trans. 4, 556–559.
212 Towner, P., Gaertner, W., Walckhoff, B., Oesterhelt, D. and Hopf, H. (1981) Eur. J. Biochem. 117, 353–359.
213 Mao, B., Govindjee, R., Ebrey, T.G., Arnaboldi, M., Balogh-Nair, V., Nakanishi, K. and Crouch, R. (1981) Biochemistry 20, 428–435.

214 Tokunaga, F., Govindjee, R., Ebrey, T.G. and Crouch, R. (1977) Biophys. J. 19, 191–198.
215 Fransen, M.R., Luyten, W.C.M.M., van Thuijl, J., Lugtenberg, J., Janssen, P.A.A., van Breugel, P.J.G.M. and Daemen, F.J.M. (1976) Nature 260, 726–727.
216 Sulkes, M., Lewis, A., Lemley, A.T. and Cookingham, R. (1976) Proc. Natl. Acad. Sci. U.S.A. 73, 4266–4270.
217 Bauer, P.-J., Dencher, N.A. and Heyn, M.P. (1976) Biophys. Struct. Mech. 2, 79–92.
218 Applebury, M.L., Peters, K.S. and Rentzepis, P.M. (1978) Biophys. J. 23, 375–382.
219 Ippen, E.P., Shank, C.V., Lewis, A. and Marcus, M.A. (1978) Science 200, 1279–1281.
220 Goldschmidt, C.R., Ottolenghi, M. and Kornestein, R. (1976) Biophys. J. 16, 839–843.
221 Hurley, J.B. and Ebrey, T.G. (1978) Biophys. J. 22, 49–66.
222 Iwasa, T., Tokunaga, F., Yoshizawa, T. and Ebrey, T.G. (1980) Photochem. Photobiol. 31, 83–85.
223 Ottolenghi, M. (1980) Adv. Photochem. 12, 97–200.
224 Dancshazy, Zs., Drachev, L.A., Ormos, P., Nagy, K. and Skulachev, V.P. (1978) FEBS Lett. 96, 59–63.
225 Nagle, J.F., Parodi, L.A. and Lozier, R.H. (1982) Biophys. J. 38, 161–174.
226 Sherman, W.V., Eicke, R.R., Stafford, S.R. and Wasacz, F.M. (1979) Photochem. Photobiol. 30, 727–729.
227 Eisenbach, M., Bakker, E., Korenstein, R. and Caplan, S.R. (1976) FEBS Lett. 71, 228–232.
228 Hurley, J.B., Becher, B. and Ebrey, T.G. (1978) Nature 272, 87–88.
229 Kalisky, O. and Ottolenghi, M. (1982) Photochem. Photobiol. 35, 109–115.
230 Sherman, W.V., Korenstein, R. and Caplan, S.R. (1976) Biochim. Biophys. Acta 430, 454–458.
231 Hess, B. and Kuschmitz, D. (1977) FEBS Lett. 74, 20–24.
232 Korenstein, R., Hess, B. and Kuschmitz, D. (1978) FEBS Lett. 93, 266–270.
233 Hoffmann, W., Grace-Miguel, M., Barnard, P. and Chapman, D. (1978) FEBS Lett. 95, 31–34.
234 Kriebel, A.N., Billbro, T. and Wild, U.P. (1979) Biochim. Biophys. Acta 546, 106–120.
235 Ohno, K., Takeuchi, Y. and Yoshida, M. (1981) Photochem. Photobiol. 33, 573–578.
236 Fischer, U. and Oesterhelt, D. (1976) In EMBO Workshop on Transd. Mech. of Photoreceptors, Juelich, p. 57.
237 Kalisky, O., Feilgelson, J. and Ottolenghi, M. (1981) Biochemistry 20, 205–209.
238 Ohno, T., Takeuchi, Y. and Yoshida, M. (1977) J. Biochem. (Tokyo) 82, 1177–1180.
239 Casadio, R. and Stoeckenius, W. (1980) Biochemistry 19, 3374–3381.
240 Fahr, A. and Bamberg, E. (1982) FEBS Lett. 140, 251–253.
241 Gaertner, W., Towner, P., Hopf, H. and Oesterhelt, D. (1983) Biochemistry 22, 2637–2644.
242 Ort, D.R. and Parson, W.W. (1978) J. Biol. Chem. 253, 6158–6164.
243 Renard, M. and Delmelle, M. (1981) FEBS Lett. 128, 245–248.
244 Korenstein, R., Sherman, W.V. and Caplan, S.R. (1976) Biophys. Struct. Mech. 2, 267–276.
245 Iwasa, T., Tokunaga, F. and Yoshizawa, T. (1979) FEBS Lett. 101, 121–124.
246 Iwasa, T., Tokunaga, F. and Yoshizawa, T. (1980) Biophys. Struct. Mech. 6, 253–270.
247 Iwasa, T., Tokunaga, F. and Yoshizawa, T. (1981) Photochem. Photobiol. 33, 539–545.
248 Mao, B. (1981) Photochem. Photobiol. 33, 407–411.
249 Tokunaga, F. and Iwasa, T. (1982) Methods Enzymol. 88, 163–167.
250 Korenstein, R. and Hess, B. (1982) Methods Enzymol. 88, 180–193.
251 Kornestein, R. and Hess, B. (1977) Nature 270, 184–186.
252 Kornestein, R. and Hess, B. (1977) FEBS Lett. 82, 7–11.
253 Lazarev, Y.A. and Terpugov, E.L. (1980) Biochim. Biophys. Acta 590, 324–338.
254 Brith-Lindner, M. and Avi-Dor, Y. (1979) FEBS Lett. 101, 113–115.
255 Tu, S.-I., Hutchinson, H. and Cavanaugh, J.R. (1982) Biochem. Biophys. Res. Commun. 106, 23–29.
256 Yoshida, M., Ohno, K. and Takeuchi, Y. (1980) J. Biochem. (Tokyo) 87, 491–495.
257 Ottolenghi, M. (1982) Methods Enzymol. 88, 470–491.
258 Honig, B., Ebrey, T., Callender, R., Dinur, U. and Ottolenghi, M. (1979) Proc. Natl. Acad. Sci. U.S.A. 76, 2503–2507.
259 Mendelsohn, R. (1976) Biochim. Biophys. Acta 427, 295–301.

260 Stockburger, M., Klusmann, W., Gattermann, H., Massig, G. and Peters, R. (1979) Biochemistry 18, 4886–4900.
261 Yamaguchi, A., Unemoto, T. and Ikegami, A. (1981) Photochem. Photobiol. 33, 511–516.
262 Bagley, K., Dollinger, G., Eisenstein, L., Singh, A.K. and Zimanyi, L. (1982) Proc. Natl. Acad. Sci. U.S.A. 79, 4972–4976.
263 Terner, J., Hsieh, C.L. and El-Sayed, M.A. (1979) Biophys. J. 26, 527–541.
264 Terner, J., Hsieh, C., Burns, R. and El-Sayed, M.A. (1979) Proc. Natl. Acad. Sci. U.S.A. 76, 3046–3050.
265 Terner, J., Hsieh, C.L., Burns, A.R. and El-Sayed, M.A. (1979) Biochemistry 18, 3629–3634.
266 Blatz, P., Mohler, J. and Navangul, H. (1972) Biochemistry 11, 848–855.
267 Tsuda, M., Glaccum, M., Nelson, B. and Ebrey, T.G. (1980) Nature 287, 351–353.
268 Aton, B., Doukas, A.G., Callender, R.H., Becher, B. and Ebrey, T.G. (1977) Biochemistry 16, 2995–2998.
269 Braiman, M. and Mathies, R. (1980) Biochemistry 19, 5421–5428.
270 Mowery, P.C. and Stoeckenius, W. (1981) Biochemistry 20, 2302–2306.
271 Kuschmitz, D. and Hess, B. (1982) FEBS Lett. 138, 137–140.
272 Braiman, M. and Mathies, R. (1982) Proc. Natl. Acad. Sci. U.S.A. 79, 403–407.
273 Hsieh, C.L., Nagumo, M., Nicol, M. and El-Sayed, M.A. (1981) J. Phys. Chem. 85, 2714–2717.
274 Pande, J., Callender, R.H. and Ebrey, T.G. (1981) Proc. Natl. Acad. Sci. U.S.A. 78, 7379–7382.
275 Lewis, A. (1982) Methods Enzymol. 88, 561–617.
276 Becher, B. and Cassim, J. (1977) Biophys. J. 19, 285–297.
277 Korenstein, R., Hess, B. and Marcus, M. (1979) FEBS Lett. 102, 155–161.
278 Goldschmidt, C.R., Kalisky, O., Rosenfeld, T. and Ottolenghi, M. (1977) Biophys. J. 17, 179–183.
279 Lewis, A. and Perreault, G.J. (1982) Methods Enzymol. 88, 217–229.
280 Shapiro, S.L., Campillo, A.J., Lewis, A., Perreault, G.J., Spoonhower, J.P., Clayton, R.K. and Stoeckenius, W. (1978) Biophys. J. 23, 383–393.
281 Alfano, R.R., Yu, W., Govindjee, R., Becher, B. and Ebrey, T.G. (1976) Biophys. J. 16, 541–545.
282 Hirsch, M.D., Marcus, M.A., Lewis, A., Mahr, H. and Frigo, N. (1976) Biophys. J. 16, 1399–1409.
283 Lewis, A., Spoonhower, J.P. and Perreault, G.J. (1976) Nature 260, 675–678.
284 Govindjee, R., Becher, B. and Ebrey, T.G. (1978) Biophys. J. 22, 67–77.
285 Sineshchekov, V.A. and Litvin, F.F. (1977) Biochim. Biophys. Acta 462, 450–466.
286 Gillbro, T., Kriebel, A.N. and Wild, U.P. (1977) FEBS Lett. 78, 57–60.
287 Gillbro, T. and Kriebel, A.N. (1977) FEBS Lett. 79, 29–32.
288 Gillbro, T. (1978) Biochim. Biophys. Acta 504, 175–186.
289 Hwang, S.-B., Kornebrot, J.I. and Stoeckenius, W. (1978) Biochim. Biophys. Acta 509, 300–317.
290 Lukashev, E.P., Vozary, E., Kononenko, A.A. and Rubin, A.B. (1980) Biochim. Biophys. Acta 592, 258–266.
291 Borisevitch, G.P., Lukashev, E.P., Kononenko, A.A. and Rubin, A.B. (1979) Biochim. Biophys. Acta 546, 171–174.
292 Drachev, L., Kaulen, L. and Skulachev, V.P. (1981) Eur. J. Biochem. 117, 461–470.
293 Sherman, W. (1981) Photochem. Photobiol. 33, 367–371.
294 Fukumoto, J., Hopewell, W., Karvaly, B. and El-Sayed, M. (1981) Proc. Natl. Acad. Sci. U.S.A. 78, 252–255.
295 Kalisky, O., Ottolenghi, M., Honig, B. and Kornestein, R. (1981) Biochemistry 20, 649–655.
296 Becher, B., Tokunaga, F. and Ebrey, T.G. (1978) Biochemistry 17, 2293–2300.
297 Rafferty, C.N. (1979) Photochem. Photobiol. 29, 109–120.
298 Kuschmitz, D. and Hess, B. (1982) Methods Enzymol. 88, 254–265.
299 Konishi, T. and Packer, L. (1978) FEBS Lett. 92, 1–4.
300 Ovchinnikov, Yu.A. (1982) FEBS Lett. 148, 179–191.
301 Abdulaev, N.G., Feigina, M.Y., Kiselev, A.V., Ovchinnikov, Yu.A., Drachev, L.A., Kaulen, A.D., Khitrina, L.V. and Skulachev, V.P. (1978) FEBS Lett. 90, 190–194.
302 Herz, J. and Packer, L. (1981) FEBS Lett. 131, 158–164.

303 Konishi, T. and Packer, L. (1977) FEBS Lett. 79, 369–373.
304 Quintanilha, A.T. (1980) FEBS Lett. 117, 8–12.
305 Hellingwerff, K.J., Arents, J.C., Scholte, B.J. and Westerhoff, H.V. (1979) Biochim. Biophys. Acta 547, 561–582.
306 Hellingwerff, K.J., Schuurmans, J.J. and Westerhoff, H.V. (1978) FEBS Lett. 92, 181–186.
307 Tsuji, K. and Neumann, E. (1981) FEBS Lett. 128, 265–268.
308 Rothschild, K.J., Zagaeski, M. and Cantore, W.A. (1981) Biochem. Biophys. Res. Commun. 103, 483–489.
309 Rothschild, K.J. and Marrero, H. (1982) Proc. Natl. Acad. Sci. U.S.A. 79, 4045–4049.
310 Merz, H. and Zundel, G. (1983) Chem. Phys. Lett. 95, 529–532.
311 Siebert, F. and Maentele, W. (1980) Biophys. Struct. Mech. 6, 147–164.
312 Siebert, F., Maentele, W. and Kreutz, W. (1981) Can. J. Spectrosc. 26, 119–125.
313 Maentele, W., Siebert, F. and Kreutz, W. (1982) Methods Enzymol. 88, 729–740.
314 Siebert, F., Maentele, W. and Kreutz, W. (1982) FEBS Lett. 141, 82–87.
315 Ort, D.R. and Parson, W.W. (1979) Biophys. J. 25, 355–364.
316 Sherman, W.V. and Caplan, S.R. (1978) Biochim. Biophys. Acta 502, 222–231.
317 Beece, D., Bowne, S.F., Czege, J., Eisenstein, L., Frauenfelder, H., Good, D., Marden, M.C., Marque, J., Ormos, P., Reinsch, L. and Yue, K.T. (1981) Photochem. Photobiol. 33, 517–522.
318 Renthal, R., Dawson, N. and Villareal, L. (1981) Biochem. Biophys. Res. Commun. 101, 653–657.
319 Konishi, T., Tristam, S. and Packer, L. (1979) Photochem. Photobiol. 29, 353–358.
320 Czege, J., Der, A., Zimanyi, L. and Keszthelyi, L. (1982) Proc. Natl. Acad. Sci. U.S.A. 79, 7273–7277.
321 El-Sayed, M.A., Karvaly, B. and Fukumoto, J.M. (1981) Proc. Natl. Acad. Sci. U.S.A. 78, 7512–7516.
322 Kouyama, T., Kimura, Y., Kinoshita, K. and Ikegami, A. (1981) FEBS Lett. 124, 100–104.
323 Bogomolni, R.A. and Stoeckenius, W. (1974) J. Supramol. Struct. 2, 775–780.
324 Renthal, R. and Lanyi, J.K. (1976) Biochemistry 15, 2136–2143.
325 Hwang, S.-B. and Stoeckenius, W. (1977) J. Membr. Biol. 33, 325–350.
326 Kayushin, L.P. and Skulachev, V.P. (1974) FEBS Lett. 39, 39–42.
327 Dencher, N.A. and Heyn, M.P. (1979) FEBS Lett. 108, 307–310.
328 Hwang, S.-B., Kornbrot, J.I. and Stoeckenius, W. (1977) J. Membr. Biol. 36, 137–158.
329 Hwang, S.-B., Kornbrot, J.I. and Stoeckenius, W. (1977) J. Membr. Biol. 36, 115–135.
330 Drachev, L.A., Frolov, V.N., Kaulen, A.D., Liberman, E.A., Ostroumov, S.A., Plakunova, V.A., Semenov, A.Yu. and Skulachev, V.P. (1976) J. Biol. Chem. 251, 7059–7065.
331 Boguslavsky, L.I., Kondrashin, A.A., Kozlov, I.A., Metelsky, S.T., Skulachev, V.P. and Volkov, A.G. (1975) FEBS Lett. 50, 223–226.
332 Karvaly, B. and Dancshazy, Zs. (1977) FEBS Lett. 76, 36–40.
333 Herrmann, T.R. and Rayfield, G.W. (1976) Biochim. Biophys. Acta 443, 623–628.
334 Shieh, P. and Packer, L. (1976) Biochem. Biophys. Res. Commun. 71, 603–609.
335 Herrmann, T.R. and Rayfield, G. (1978) Biophys. J. 21, 111–125.
336 Block, M.C. and van Dam, K. (1978) Biochim. Biophys. Acta 507, 48–61.
337 Block, M.C. and van Dam, K. (1979) Biochim. Biophys. Acta 550, 527–542.
338 Singh, K., Korenstein, R., Lebedeva, H. and Caplan, S.R. (1980) Biophys. J. 31, 393–402.
339 Bamberg, E., Dencher, N.A., Fahr, A. and Heyn, M.P. (1981) Proc. Natl. Acad. Sci. U.S.A. 78, 7502–7506.
340 van Dijk, P.W.M. and van Dam, K. (1982) Methods Enzymol. 88, 17–25.
341 Racker, E. (1973) Biochem. Biophys. Res. Commun. 55, 224–230.
342 Racker, E., Violand, B., Neal, S.O., Alonzo, M. and Telford, J. (1979) Arch. Biochem. Biophys. 198, 470–477.
343 Hellingwerff, K.J., Scholte, B.J. and van Dam, K. (1978) Biochim. Biophys. Acta 513, 66–77.
344 Happe,M. and Overath, P. (1976) Biochem. Biophys. Res. Commun. 72, 1504–1511.
345 Happe, M., Teather, R.M., Overath, P., Knobling, A. and Oesterhelt, D. (1977) Biochim. Biophys. Acta 465, 415–420.

346 Westerhoff, H.V., Scholte, B.J. and Hellingwerff, K.J. (1979) Biochim. Biophys. Acta 547, 544–560.
347 Block, M.C., Hellingwerff, K.J., Kaptein, R. and de Kruijff, B. (1978) Biochim. Biophys. Acta 514, 178–184.
348 Dewey, T.G. and Hammes, G.G. (1981) Proc. Natl. Acad. Sci. U.S.A. 78, 7422–7425.
349 Dancshazy, Zs., Ormos, P., Drachev, L.A. and Skulachev, V.P. (1978) Biophys. J. 24, 423–428.
350 Drachev, L.A., Kaulen, A.D. and Skulachev, V.P. (1978) FEBS Lett. 87, 161–167.
351 Drachev, L.A., Kaulen, A.D., Skulachev, V.P. and Voitsitsky, V.M. (1982) J. Membr. Biol. 65, 1–12.
352 Trissl, H.W. and Montal, M. (1977) Nature 266, 655–657.
353 Hong, F.T. and Montal, M. (1979) Biophys. J. 25, 465–472.
354 Fahr, A., Lauger, P. and Bamberg, E. (1981) J. Membr. Biol. 66, 51–62.
355 Rayfield, G. (1982) Biophys. J. 38, 79–84.
356 Heyn, M.P. and Dencher, N.A. (1982) Methods Enzymol. 88, 31–35.
357 Klausner, R.D., Berman, M., Blumenthal, R., Weinstein, J.N. and Caplan, S.R. (1982) Biochemistry 21, 3643–3650.
358 Lozier, R.H., Niederberger, W., Bogomolni, R.A., Hwang, S.-B. and Stoeckenius, W. (1976) Biochim. Biophys. Acta 440, 545–556.
359 Garty, H., Klemperer, G., Eisenbach, M. and Caplan, S.R. (1977) FEBS Lett. 81, 238–242.
360 Takeuchi, Y., Ohno, K., Yoshida, M. and Nagano, K. (1981) Photochem. Photobiol. 33, 587–592.
361 Kuschmitz, D. and Hess, B. (1981) Biochemistry 21, 5950–5957.
362 Carmeli, Ch., Quintanilha, A.T. and Packer, L. (1980) Proc. Natl. Acad. Sci. U.S.A. 77, 4707–4711.
363 Carmeli, Ch. and Gutman, M. (1982) FEBS Lett. 141, 88–92.
364 Arents, J.C., van Dekken, H., Hellingwerff, K.J. and Westerhoff, H.V. (1981) Biochemistry 20, 5114–5123.
365 Arents, J.C., Hellingwerff, K.J., van Dam, K. and Westerhoff, H.V. (1981) J. Membr. Biol. 60, 95–104.
366 Westerhoff, H.V., Arents, J.C. and Hellingwerff, K.J. (1981) Biochim. Biophys. Acta 637, 69–79.
367 Marinetti, T. and Mauzerall, D. (1983) Proc. Natl. Acad. Sci. U.S.A. 80, 178–180.
368 Lewis, A., Marcus, M.A., Ehrenberg, B. and Crespi, H.L. (1978) Proc. Natl. Acad. Sci. U.S.A. 75, 4642–4646.
369 Ehrenberg, B., Lewis, A., Porta, T.K., Nagle, J.F. and Stoeckenius, W. (1980) Proc. Natl. Acad. Sci. U.S.A. 77, 6571–6573.
370 Ehrenberg, B. and Lewis, A. (1978) Biochem. Biophys. Res. Commun. 82, 1154–1159.
371 Druckmann, S., Ottolenghi, M., Pande, A., Pande, J. and Callender, R.H. (1982) Biochemistry 21, 4953–4959.
372 Schulten, K. and Pavan, P. (1978) Nature 272, 85–86.
373 Doukas, A.G., Pande, A., Suzuki, T., Callender, R.H., Honig, B. and Ottolenghi, M. (1981) Biophys. J. 33, 275–280.
374 Wagner, G., Oesterhelt, D., Krippahl, G. and Lanyi, J.K. (1981) FEBS Lett. 131, 341–345.
375 Spudich, E.N. and Spudich, J.L. (1982) Proc. Natl. Acad. Sci. U.S.A. 79, 4308–4312.
376 Lanyi, J.K. (1982) Methods Enzymol. 88, 439–443.
377 Greene, R.V., MacDonald, R.E. and Perreault, G.J. (1980) J. Biol. Chem. 255, 3245–3247.
378 Spudich, J.L. and Bogomolni, R.A. (1983) Biophys. J. 43, 243–250.
379 Ogurusu, T., Maeda, A., Sasaki, N. and Yoshizawa, T. (1981) J. Biochem. (Tokyo) 90, 1267–1273.
380 Ogurusu, T., Maeda, A., Sasaki, N. and Yoshizawa, T. (1982) Biochim. Biophys. Acta 682, 446–451.
381 Tsuda, M., Hazemoto, N., Kondo, M., Kamo, N., Kobatake, Y. and Terayama, Y. (1982) Biochem. Biophys. Res. Commun. 108, 970–976.
382 Hazemoto, N., Kamo, N., Kobatake, Y., Tsuda, M. and Terayama, Y., (1984) Biophys. J. 45, 1073–1077.
383 Kamo, N., Takeuchi, M., Hazemoto, N. and Kobatake, Y. (1983) Arch. Biochem. Biophys. 221, 514–525.
384 Stoeckenius, W. and Bogomolni, R.A. (1982) Annu. Rev. Biochem. 52, 587–616.
385 Kanner, B.I. and Racker, E. (1975) Biochem. Biophys. Res. Commun. 64, 1054–1061.

386 Weber, H.J. and Bogomolni, R.A. (1982) Methods Enzymol. 88, 379–390.
387 Luisi, B.F., Lanyi, J.K. and Weber, H.J. (1980) FEBS Lett. 117, 354–358.
388 Mehlhorn, R.J., Schobert, B., Packer, L. and Lanyi, J.K. (1984) in preparation.
389 Knauf, P.A. (1979) Curr. Top. Membr. Transp. 12, 249–363.
390 Spudich, E.N., Bogomolni, R.A. and Spudich, J.L. (1983) Biochem. Biophys. Res. Commun. 112, 332–338.
391 Spudich, J.L. and Stoeckenius, W. (1979) Photobiochem. Photobiophys. 1, 43–53.
392 Dencher, N.A. and Hildebrand, E. (1979) Z. Naturforsch. 34c, 841–847.
393 Sperling, W. and Schimz, A. (1980) Biophys. Struct. Mech. 6, 165–169.
394 Schimz, A., Sperling, W., Hilderbrand, E. and Koehler-Hahn, D. (1982) Photochem. Photobiol. 36, 193–196.
395 Michel, H. and Oesterhelt, D. (1980) Biochemistry 19, 4607–4614.
396 Michel, H. and Oesterhelt, D. (1980) Biochemistry 19, 4615–4619.
397 Ginzburg, M., Ginzburg, B.Z. and Tosteson, D.C. (1971) J. Membr. Biol. 6, 259–268.
398 Gochnauer, M.B. and Kushner, D.J. (1971) Can. J. Microbiol. 17, 17–23.
399 Lanyi, J.K. and Hilliker, K. (1976) Biochim. Biophys. Acta 448, 181–184.
400 Wagner, G. and Oesterhelt, D. (1976) Ber. Deutsch. Bot. Ges. 89, 289–292.
401 Garty, H. and Caplan, S.R. (1977) Biochim. Biophys. Acta 459, 532–545.
402 Lanyi, J.K., Helgerson, S.L. and Silverman, M.P. (1979) Arch. Biochem. Biophys. 193, 329–339.
403 Wey, C., Ahl, P.L. and Cone, R.A. (1978) J. Cell Biol. 79, 657–662.
404 Lanyi, J.K. and Silverman, M.P. (1972) Can. J. Microbiol. 18, 993–995.
405 Lanyi, J.K. and MacDonald, R.E. (1979) Methods Enzymol. 56, 398–407.
406 Mehlhorn, R.J. and Probst, I. (1982) Methods Enzymol. 88, 335–344.
407 Kamo, N., Racanelli, T. and Packer, L. (1982) Methods Enzymol. 88, 356–360.
408 Gradmann, D., Hansen, U.-P. and Slayman, C.L. (1982) Curr. Top. Membr. Transp. 16, 257–276.
409 Mukohata, Y. and Kaji, Y. (1981) Arch. Biochem. Biophys. 206, 72–76.
410 Mukohata, Y. and Kaji, Y. (1981) Arch. Biochem. Biophys. 208, 615–617.
411 Hartmann, R. and Oesterhelt, D. (1977) Eur. J. Biochem. 77, 325–335.

Ernster (ed.) Bioenergetics
© 1984 Elsevier Science Publishers B.V.

CHAPTER 12

Biogenesis of energy-transducing systems

NATHAN NELSON [a] and HOWARD RIEZMAN [b]

[a] *Department of Biology, Technion-Israel Institute of Technology, Haifa, Israel and*
[b] *Swiss Institute for Experimental Cancer Research, CH-1066 Epalinges, Switzerland*

1. *Introduction*

The fundamental function of biological membranes is to separate components and to maintain different compositions of solutes in the separate spaces. Therefore, essentially every biological membrane functions in energy transduction. The maintenance of the different compositions in the two sides of the membrane is based on its functional asymmetry. The degree of asymmetry varies from uneven distribution of lipids in the bilayer up to absolute polarity of large protein complexes in the membrane. This asymmetry arises from the vectorial assembly of the individual protein complexes into the membranes in vivo where a high degree of specificity is maintained.

Research on the biogenesis of membranes involves a wide variety of subjects, some of which proliferated in the last 10 years through hundreds of published papers. Therefore, we elected to discuss in this chapter only the biogenesis of energy-transducing organelles of the eukaryotic cell. Mostly the biogenesis of mitochondria, chloroplasts and some secretory organelles will be mentioned. We will have to leave out exciting topics such as the biosynthesis and insertion of phospholipid into membranes, and we will not discuss the biogenesis of bacterial membranes which are, in fact, one of the most active energy-transducing membranes.

Among the unique features of the eukaryotic cell, the separate nucleus, the developed secretory pathway and the inclusion of semiautonomous organelles were very decisive for the development of its nature. Even though most of the genetic information is stored in the nucleus, chloroplasts and mitochondria retain control over synthesis of several of their proteins via their own unique DNA and RNA molecules and protein-synthesizing machinery. Some features of the protein-synthesizing machinery in chloroplasts and mitochondria closely resemble those of prokaryotes. This has led to a widely accepted hypothesis that the origins of chloroplasts and mitochondria were endosymbiosis between protoeukaryotic cells (with developed nuclei) and prokaryotes possessing a fully developed machinery for carrying out oxygenic photosynthesis and oxidative phosphorylation. Even though this suggestion can be traced back to the previous century, it was concisely formulated about a decade ago [1,2]. The mitochondrial genome codes for only about one dozen polypeptides, while the chloroplasts may synthesize in the order of fifty polypeptides [3–5]. Therefore the hypothetical symbiotic 'prokaryote' must have lost several hundreds of genes and transferred many of them to the nucleus of its host.

This apparent difficulty has led to an alternative hypothesis, that chloroplasts and mitochondria have evolved from episomes containing a limited amount of genes. The episomes were integrated with eukaryotic cells and organized the organelles through adaptation and selection over the course of evolution [6,7].

There are no known living intermediate states between prokaryotes (such as *Paracoccus denitrificans* [8,9]) and mitochondria. On the other hand, cyanelles can be considered as intermediates of red algae chloroplasts, and the recent discovery of the Prochloron might provide the missing link for the evolution of plant chloroplasts from chlorophyll *b*-containing prokaryotes [1,10–12].

Regardless of how mitochondria and chloroplasts evolved, most of their oligomeric membrane protein complexes are now synthesized partly by cytoplasmic and partly by organelle ribosomes. This complicates the biogenesis of the organelles, but also makes it an exciting subject to study [13–15].

2. Semiautonomous organelles

2.1. Organellar DNA

Chloroplasts and mitochondria are the only organelles which contain their own DNA. This implies that all of the enzymes necessary for DNA replication are present in the organelle. In addition, both mitochondria and chloroplasts contain all of the machinery for RNA synthesis and processing, and for protein synthesis.

Mitochondrial DNA has been characterized from various sources. The mitochondrial genome ranges in size from 16 569 basepairs in humans [16] to 75 kilobasepairs in yeast [17]. Plant mitochondria have a heterogeneous population of circular DNA ranging in size from 1.5 to 120 kilobasepairs and, in some instances, up to 2400 kilobasepairs [18]. The mammalian mitochondrial genome codes for all of its own ribosomal RNAs, tRNAs and a small portion of its polypeptides. In mammalian and yeast cells the genome codes for about a dozen polypeptides, most of which are membrane associated. The plant mitochondrial genome has a much higher coding capacity and indeed does synthesize a larger number of polypeptides [19].

The chloroplast genome is larger than the mitochondrial genome and is almost uniform in size of approximately 125–190 kilobasepairs [20,21]. The chloroplast genome codes for about fifty polypeptides, some of them quite hydrophilic [5]. The chloroplast-made polypeptides are found either in the stroma or in the thylakoid membrane.

There are a few striking differences between the mitochondrial and chloroplast genomes. The former has a slightly modified genetic code. In human mitochondrial DNA the UGA termination codon is read as a tryptophan, the AUA isoleucine codon appears to be read as a methionine, and the AGA or AGG arginine codon as a *STOP* [16]. Similar variations in the genetic code have been reported in fungi [22]. However, in plant mitochondria TGA codons have not been detected so far, and CGG may code for tryptophan instead of arginine [23]. In addition to having an altered genetic code, mitochondrial genomes from fungi and plants contain introns in several genes coding for rRNA or polypeptides [22,23]. In chloroplasts, genes coding for proteins seem to lack introns; however, several genes for both

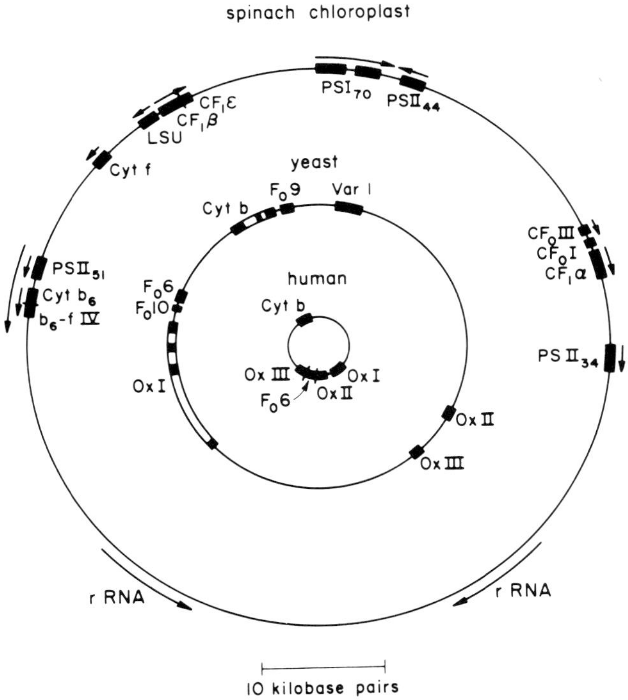

Fig. 12.1. Relative sizes of mitochondrial and chloroplast chromosomes and location of protein structural genes. The figure was constructed from published data [5,15,17,22,26–28]. The structural genes are marked by wide sections. Black areas code for proteins. White areas are introns. OxI, OxII and OxIII are subunits I, II and III of cytochrome *c* oxidase. Cyt *b*, cytochrome *b*. Fo_6 and Fo_9 are subunits 6 and 9 of the proton ATPase complex. In the chloroplast chromosome the arrows indicate the transcription direction and the size of the transcripts. $CF_1\alpha$, $CF_1\beta$, $CF_1\epsilon$ and CF_0III are subunits α, β, ϵ and III of the chloroplast proton ATPase complex [30]. $PSII_{51}$, $PSII_{44}$, and $PSII_{34}$ are subunits of photosystem II reaction center with the corresponding molecular weights of 51000, 44000 and 34000. PSI_{70} is subunit I of photosystem I reaction center. Cyt *f* is cytochrome *f*; cyt b_6 is cytochrome b_6; and b_6-fIV is subunit IV of cytochrome b_6-f complex.

mitochondrial ribosomal and tRNA do [24,25]. Fig. 12.1 compares the smallest mitochondrial genome (human) with the genomes of yeast mitochondria and spinach chloroplasts. The entire mitochondrial genomes of humans, cattle and mouse have been sequenced. On the other hand, only a few of the chloroplast protein genes have been mapped and sequenced so far [26]. Nevertheless, one can draw the following conclusion: each plastome gene exists as a single copy per chromosome. Unlike the rRNA genes, genes for thylakoid proteins are not arranged in operon structures, but are scattered over the entire plastid chromosome. Both complementary strands encode structural genes. Most of the genes are monocistronically transcribed. So far the genes for the β and ϵ subunits of CF_1 and for cytochrome b_6 and subunit IV of the cytochrome b_6/f complex were shown to have bicistronic mRNA. Untranslated regions can vary from 250 to more than 1000 basepairs, as was demonstrated for subunit III of the proton-ATPase and subunit I of photosystem I reaction center, respectively.

2.2. *Organellar protein synthesis*

Both chloroplasts and mitochondria can synthesize messenger RNA, process, and translate it. One difference is that mammalian mitochondria polyadenylate their messenger RNAs, whereas chloroplasts do not [27,28]. This synthesis can be performed in vitro using isolated organelles, in the absence of external macromolecules [13]. Like bacterial systems, both mitochondrial synthesis and chloroplast protein synthesis are inhibited by chloramphenicol but not by cycloheximide [29]. Many other common structural and mechanistic similarities have been found between organellar and bacterial ribosomes. Except for mammals, mitochondrial ribosomes and chloroplast ribosomes are approximately 70S compared with the 80S cytoplasmic ribosomes. These differences in organelle and cytoplasmic protein synthesis have been exploited heavily in biogenesis studies [3,4,29]. In vivo one can specifically pulse-label organellar proteins in the presence of cycloheximide and specifically inhibit their synthesis with chloramphenicol. This type of experimental design was used initially to identify the products of organellar protein synthesis. Among these polypeptides identified for mitochondria are subunits I, II and III of cytochrome *c* oxidase, cytochrome *b*, and 1–3 subunits of the ATPase complex and a ribosomal protein (in plants several other proteins are synthesized in the mitochondria), and in chloroplasts subunits α, β, ϵ, I and III of the proton ATPase complex [30], subunits I (cytochrome *f*), II (cytochrome b_6) and IV of the cytochrome b_6/f complex [5], subunit I of photosystem I reaction center [14], a 32 000 Da membrane protein [21], and the large subunit of RuBP carboxylase (see Fig. 12.1). The transcripts of the chloroplast genes are decoded to give the correct mature protein sizes and some of them are assembled into the membrane with an intact initiator methionine. So far two exceptions for this rule were observed, as the 32 kDa protein of photosystem II and cytochrome *f* are synthesized as larger precursors. The newly synthesized organellar proteins may undergo post-translational modifications, like heme insertion, but no glycosylation reactions seem to take place either in chloroplasts or mitochondria. This is in contrast with the proteins that are sent through the secretory pathway of eukaryotic cells.

While chloroplasts and mitochondria synthesize proteins, the majority of their proteins are encoded in the nucleus, synthesized on cytoplasmic ribosomes, and imported by the organelle [29]. For example, with only a few exceptions all of the proteins necessary for DNA, RNA and protein synthesis are imported from the cytoplasm. This raises several fundamental problems which are currently under intensive investigation:

(1) How are these specifically targeted to their proper location?
(2) What is the mechanism of import?
(3) How are the different genetic systems in the cell coordinately regulated?
(4) How are multisubunit complexes assembled?

3. *Import of proteins into chloroplasts, mitochondria and storage vesicles*

The eukaryotic cell contains a wide variety of membrane structures and organelles. A continuous process of biogenesis and degradation of these membranes takes place throughout the life cycle of the cell. How are proteins correctly directed to their target organelles, and how are they inserted into the membranes to assume their asymmetric orientation?

Classification of mechanisms might help to understand complicated processes, but quite often it might be misleading. There are many exceptions to each of the proposed mechanisms, and one can always criticize a proposed mechanism on the grounds of these exceptions. Therefore, playing the devil's advocate, we shall classify the process of protein biogenesis in membranes into three mechanisms: vectorial translation, vectorial processing and protein incorporation.

Vectorial translation [31,32]. Polypeptides are made on membrane-bound polysomes. Many of these proteins are synthesized with a 16–30 amino acid extension at the NH_2-terminus. This 'signal sequence' is hydrophobic in nature. Protein synthesis and translocation, into or across the membrane, are obligatorily linked. Therefore, the transmembrane movement is co-translational and it is coupled to the elongation of the polypeptide chain. Consequently, the completed polypeptide chain is never present in the compartment where it is synthesized. The polypeptides that do not yet cross the membrane are shorter than the mature protein. Addition of inhibitors of protein synthesis immediately arrest movement of the polypeptide across the membrane.

Vectorial processing [4,33,34]. The polypeptides are usually made on free polysomes. The transport of the polypeptide chain across the membrane is independent of protein synthesis. In most of the cases the protein is synthesized as a larger precursor, and the completed polypeptide chain is present and even might be accumulated in the compartment where it is synthesized. Addition of protein synthesis inhibitors will not prevent the transport of the completed chains across the membrane. During or immediately after transport across the membrane, chemical modification (usually proteolysis) takes place in order to convert the precursor into mature protein. In most cases, import is energy dependent.

Protein incorporation [35–37]. The polypeptides are made on free polysomes. The incorporation of the polypeptide into the membrane is independent of protein synthesis. The protein is synthesized as the mature size, and no chemical modification takes place during the insertion into the membrane. The completed polypeptide chain is present and might be accumulated in the compartment where it is synthesized. Inhibitors of protein synthesis should not prevent the transport of the completed chains into the membrane. In most cases, import is not energy dependent.

4. Vectorial translation — biogenesis of secretory vesicles and acetylcholine receptor

Neurotransmission is based on the secretion of neurotransmitters from secretory vesicles in the presynaptic membrane and the binding of the agonists by receptors on the postsynaptic membrane. The transmitters have to travel only about 20 nm across the synaptic cleft, whereas neurohormones may act on much more distant receptors. The biogenesis of the secretory vesicles and the receptors are intimately connected with the secretory pathway of the eukaryotic cells. In this system a series of membrane-bound structures mediate the transfer of exported proteins from their site of synthesis at the rough endoplasmic reticulum to their site of discharge at the plasma membrane. We will use the chromaffin granules (storage vesicles of the adrenal medulla) as an example for secretory vesicles, and the acetylcholine receptor for receptors of neurotransmitters.

4.1. Biogenesis of chromaffin granules

The function of the chromaffin granule is to store high concentrations of catecholamines and, upon stimulation of the chromaffin cell, to deliver the catecholamines into the extracellular space by exocytosis [38,39]. The uptake and storage of catecholamines is driven by a proton motive force which is formed by a membrane-bound proton-ATPase enzyme [39,40]. The uptake is facilitated by a special carrier for catecholamines [41]. The chromaffin granule membrane is also furnished with a vectorial electron transport chain that might function in the reduction of dehydroascorbate to ascorbate inside the granule [42]. Inside the granule there are high concentrations of soluble acidic proteins and a few enzymes that catalyze the interconversion of the catecholamines [38,39]. The secretion of the neurohormones from the chromaffin granules is controlled by the level of Ca^{2+} inside the cells and by the presence of a special protein, synexine [43,44]. The granules release their contents by exocytosis, and it is quite likely that part of the membrane-bound enzymes of the chromaffin granules can be recycled back to the newly formed organelles [45]. Therefore, during the biogenesis of chromaffin granules the complex energy transducing membrane must be properly assembled and segregated from other cellular membranes. This is in contrast with the biogenesis of chloroplast and mitochondria, whose components are directly incorporated into pre-existing membranes rather than being sorted out from other membranes.

Fig. 12.2 depicts a schematic pathway for the biogenesis of chromaffin granules, and for catecholamine uptake and release by exocytosis. The biogenesis of the granules starts in the rough endoplasmic reticulum. Most polypeptides destined for the chromaffin granules are probably synthesized on the rough endoplasmic reticulum via a vectorial translation process. Some are components of the granule membrane and others stay soluble in the cisternae. Several of these polypeptides are glycosylated [46]. The membranes then move to the Golgi apparatus where the specific proteins are probably sorted out to build up a chromaffin granule. It is interesting to note that the rate of the synthesis of the soluble proteins of the cisterna is about five-times greater than that of the chromaffin granule membrane. Therefore, it is likely that the membranes are recycled from the plasma membrane an average of five times before they are degraded by the lysosomes [45].

Once the chromaffin granule is formed, low molecular components such as catecholamines, ATP and Ca^{2+} are concentrated in the cisterna using the proton motive force provided by a proton ATPase [39,45]. Upon receiving stimulus, the granules fuse with the plasma membrane and the catecholamines are released from the cell. To prevent continual growth of the plasma membrane, portions of it must

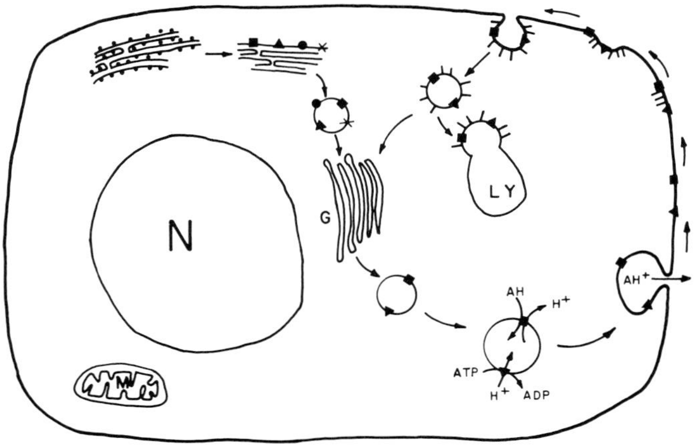

Fig. 12.2. Schematic representation of the biogenesis and function of chromaffin granules. The biogenesis of chromaffin granules starts in the rough endoplasmic reticulum. The ribosomes dissociate from the membranes and the enzymes are assembled. Vesicles containing enzymes of the chromaffin granule (▲ proton ATPase and ■ catecholamine carrier) as well as extrinsic enzymes (●x) are separated from the endoplasmic reticulum and are transferred to the Golgi apparatus (G). The Golgi apparatus sorts out the proteins destined to various locations in the cell. Catecholamines (AH) are accumulated and later on secreted by exocytosis. The enzymes of the chromaffin granules are sorted out from the cell membrane via coated vesicles and recycled through the Golgi apparatus, a post-Golgi compartment, or go to the lysosomes (LY) and are degraded. M, Mitochondrion; N, nucleus.

be removed by endocytosis. It is quite likely that some of the proteins may recycle through the Golgi apparatus or a post Golgi compartment to form new chromaffin granules [45].

4.2. Biogenesis of the acetylcholine receptor

Neurotransmission at the vertebrate neuromuscular junction is mediated by the release of acetylcholine from the nerve terminal and binding of the agonist to the acetylcholine receptor on the postsynaptic membrane. Electrophysiological experiments have shown that binding of the neurotransmitter to its receptor triggers the opening of large cation selective channels that subsequently close by a desensitization process [47]. The opening of the channel is suppressed by several pharmacological inhibitors including bungarotoxin and curare. Therefore, the functional acetylcholine receptor can be defined as the minimal structure that, upon binding of acetylcholine, brings about the opening of an ion channel across the membrane. An acetylcholine receptor from *Torpedo* electric organ has been isolated and consists of four different glycoprotein subunits. The monomeric receptor consists of two α subunits (M_r 38000) and one each of β (50000), γ (57000) and δ (64000) subunits [48,49].

Reconstitution and bilayer studies provided evidence that the four subunit receptor contains, not only the binding sites for acetylcholine but also the active cation channel, and the activated receptor by itself brings about the response of agonist-induced membrane permeability [50,51]. This is different from the β-adrenergic receptor in which the receptor must interact with another membrane protein in order to transmit the signal.

Immunological studies have revealed a high degree of homology among the various subunits of the acetylcholine receptor from a given animal, and receptors from different animals show strong immunological crossreactivity [52]. It was proposed that the antigenic similarities might reflect structural homologies among the various subunits of the receptor and with subunits of acetylcholine receptors from various sources. Analysis of amino acid sequences revealed strong homology among the various subunits of the *Torpedo* acetylcholine receptor [49]. Recently the primary structures of α, β and δ subunits of the *Torpedo* acetylcholine receptor were obtained from cDNA sequences [53,54]. The amino acid sequence homology among these three subunits was corroborated, and the information on specific domains in the subunits was further advanced. Bovine acetylcholine receptor was recently purified by affinity chromatography on toxin coupled to agarose [55]. Like the receptor from fish it is composed of four glycoprotein subunits. The subunits crossreacted with antibodies against the corresponding subunits of *Torpedo* acetylcholine receptor. It was found that immunization of rats with receptors from *Torpedo*, bovine and human muscles was very active in the induction of experimental autoimmune myasthenia gravis. The presented evidence suggests common functional structures in acetylcholine receptors from fish electric organ and mammalian muscle.

The complexity of each polypeptide is much more involved than imagination can envisage or a computer can predict. It is the biogenesis of each individual subunit that correctly assembles the functional units of the receptor in the membrane. Fig. 12.3 shows the initial steps in the biogenesis of acetylcholine receptor which is a typical vectorial-translation process. Signal sequences of 24, 17 and 21 amino acids were identified for the α, γ and δ subunits, respectively [54,56,57]. The synthesis of most secretory proteins is inhibited by 'signal recognition particle (SRP)' [58,59]. The 11S signal-recognition particle is composed of six subunits and a 7S RNA [60]. This soluble particle recognizes the NH_2-terminal extension of the protein as it emerges from the ribosome. Once the SRP is bound to the nascent chain-ribosome complex, translation is arrested. This arrest can only be relieved upon binding of the complex to a receptor on the rough endoplasmic reticulum. The receptor has been purified as a 72 000 Da rough endoplasmic reticulum membrane protein, largely exposed to the cytoplasm. It has been termed the 'docking protein' [61]. This mechanism ensures that the nascent polypeptide interacts exclusively with the rough endoplasmic reticulum and obligatorily couples translation and transport into or across the rough endoplasmic reticulum membrane (vectorial translation). When the SRP-ribosome complex is bound to the docking protein, the nascent chain is elongated across the membrane, removal of the signal peptide and glycosylation take place cotranslationally on the distal side of the membrane. This postulated mechanism has been derived from in vitro studies and has not yet been shown to function identically in vivo. Then the acetylcholine receptor is assembled from its various subunits and transported via the Golgi apparatus to secretory vesicles which fuse

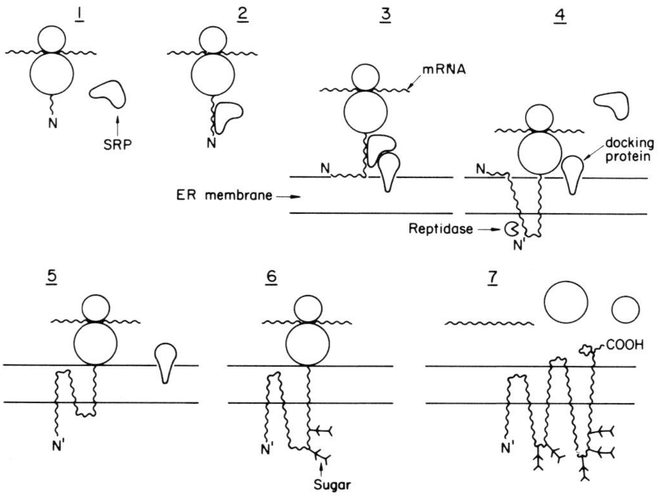

Fig. 12.3. Current model for vectorial translation depicted for one of the subunits of acetylcholine receptor. The various events are described in the text. ER, endoplasmic reticulum; SRP, signal recognition particle.

with the plasma membrane (Fig. 12.4). Due to the membrane fusion the glycosylated parts are now on the outer side of the plasma membrane. The final shape of the postsynaptic membrane is determined by association of the receptor with cytoskeletal proteins inside the cell.

The biogenesis of both acetylcholine receptor and chromaffin granules share several common properties. The specific polypeptides are synthesized and transported into the membrane by a vectorial translation process. The specific proteins are sorted out by the Golgi apparatus and eventually fuse with the plasma membrane via the secretory pathway. Yet the acetylcholine receptor functions on the plasma membrane, and therefore it should stay on this membrane for a long time (2–7 days). On the other hand, the function of chromaffin granules is to store neurotransmitters. Therefore they stay most of their lifetime inside the cell and their fusion with the plasma membrane is temporary. Soon after the secretion process, the constituents of the chromaffin granule membrane must be removed from the plasma membrane by endocytosis.

How is the final location of the acetylcholine receptor and secretory vesicles determined? Where is the information stored to determine this localization? There are two key locations in cells where sorting out of membrane proteins might occur. One of these is the Golgi apparatus and the second one involves the plasma membrane. After chromaffin granules have fused with the plasma membrane, their specific lipids and proteins are immediately and specifically removed from the latter membrane. The mechanism for this selective removal is unknown. In the Golgi apparatus, newly-synthesized membrane proteins are sorted out to various membrane vesicles. We propose a model for the sorting out of secretory vesicles from vesicles destined to deliver plasma membrane proteins.

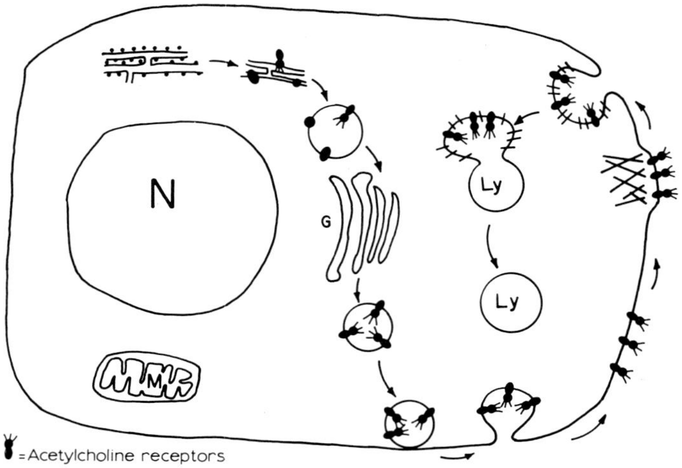

Fig. 12.4. Biogenesis of acetylcholine receptor. The pathway of acetylcholine receptor biogenesis resembles that of the chromaffin granules (Fig. 12.2) except that cytoskeleton elements (XY) function in formation of the receptor left on the cell membrane. The symbols are as in Fig. 12.2.

One of the key enzymes of the storage vesicles is a proton ATPase [39,62,63]. The assembly of this enzyme in the Golgi apparatus may initiate a local drop in the pH, creating a pH gradient within the Golgi apparatus [64]. This could be the signal and/or driving force to concentrate secretory protein in the nascent secretory vesicle. At the same time, protein destined for the plasma membrane may be repelled from such low local pH environment. When a certain concentration of secretory vesicle components is reached, these specialized patches of Golgi apparatus will bud off to form prosecretory vesicles.

An analogy for such a mechanism can be found in the yeast secretory pathway. Here the secretory pathway serves to secrete enzymes to the plasma membrane, the periplasmic space, and to the vacuoles. Sorting out of these two pathways occurs in the Golgi apparatus [65]. The vacuolar membrane also contains a proton ATPase that lowers the pH inside the vacuole. Such a proton ATPase could not be detected in secretory vesicles (R. Schekman, personal communication). Furthermore, it was shown that breaking down the pH gradient in vivo disturbed delivery of protein to the vacuole but not to the plasma membrane. Thus, the chromaffin granules may be analogous to the yeast vacuole. A special mechanism of Ca^{2+} and synexin-induced fusion has evolved in secretory cells which controls the secretion of components like neurotransmitters, etc. Since lysosomes also contain a H^+-ATPase, further sorting of lysosomal and secretory vesicle components may occur via a receptor-mediated process. Lysosomal glycoproteins contain a phosphorylated mannose residue which is recognized by a phosphomannosyl receptor [66,67].

5. Vectorial processing — import of proteins into chloroplasts and mitochondria

Most of the proteins of chloroplasts and mitochondria are synthesized in the cytoplasm and post-translationally imported into the organelle [29,34]. One can divide this process into four steps: (1) synthesis of the individual polypeptides as precursors on free ribosomes in the cytoplasm; (2) binding of the precursors to specific receptors on the organellar surface; (3) transmembrane movement; and (4) processing and sorting into the correct compartment.

5.1. Synthesis of cytoplasmic ribosomes

In higher eukaryotes messenger RNAs for chloroplast and mitochondrial proteins have been localized exclusively to free and not membrane-bound polysomes [4,68]. In yeast, the situation is not as clear; a portion of the polysomes for some of the proteins is found to be bound to mitochondria [69,70]. This may only reflect the speed at which the import process occurs in the different organisms. Most of the cytoplasmically-synthesized precursors have amino terminal extensions varying in size and nature. The size variation goes from no extension (cytochrome *c* and ATP-ADP translocator) to approximately 12000 Da (proteolipid of the proton

ATPase from *Neurospora crassa* mitochondria). Recently, the sequences of a few of the NH_2 terminal extensions of mitochondrial polypeptides have become available, and thus far there are no striking sequence homologies [71–73]. One common feature among most precursors is that the NH_2-terminal extension is highly basic in nature. This may play a role in recognizing the mitochondrial surface [74,75].

5.2. Binding of precursors to the organellar surface

In order to specifically select and take up proteins from the cytoplasm, mitochondria and chloroplasts must have specific receptors which recognize and concentrate the precursors at the organellar surface. The best evidence for the existence of receptors comes from work with mitochondria. There are probably at least three different receptors involved in protein uptake. One is a receptor for incorporation into the outer membrane (see Section 6). The second is a receptor specifically for precursors of proteins destined for internal mitochondrial compartments. This receptor binds precursors rapidly, reversibly and with high affinity. Binding is highly ligand specific; only precursors to mitochondrial proteins bind and not processed precursors nor mature forms. The binding is highly membrane specific; only outer membrane, and not inner, binds with high affinity. The receptor may be proteinaceous in character as it is protease sensitive [76]. The third receptor is the apocytochrome *c* receptor. Apocytochrome binds specifically to mitochondria although some binding may also occur to the rough endoplasmic reticulum [77,78]. The binding is of high affinity and is rapid and reversible. The import of in vitro synthesized apocytochrome *c* can be inhibited by adding a large excess of apocytochrome *c*. This large excess of apocytochrome *c* does not affect the import of any other precursor tested thus far [79], suggesting that cytochrome *c* is imported via a different receptor than most other mitochondrial proteins. Some initial observations indicate that similar receptors may function also on the chloroplast surface [80].

5.3. Transmembrane movement

There have been two main strategies to study the transport of proteins into mitochondria and chloroplasts:

(1) *In vitro pulse-chase experiments in the presence or absence of various inhibitors*; the products are analyzed by immunoprecipitation, gel electrophoresis and fluorography. Transport is usually measured as the amount of precursor converted to mature form. In some cases, after labelling cells are fractionated and transport quantified by measuring the accumulation of immunoprecipitable polypeptides in the organelle.

(2) *In vitro import*; precursors are synthesized in cell-free extracts and then incubated with chloroplasts or mitochondria. After incubation under various conditions organelles are reisolated and protease-treated. Import is quantified as the relative amount of protease-resistant polypeptide, and the reisolated organelle is compared to the total amount of the same polypeptide.

What have these two approaches told us about the transport step?

In vivo studies with *Chlamydomonas reinhardtii* were the first to show vectorial processing of a chloroplast enzyme [81]. After pulse labelling *Chlamydomonas* cells, a larger precursor form of the small subunit of the RuBP carboxylase was detected. During the chase this precursor was converted to the mature form, even in the presence of cycloheximide, giving the first indication for vectorial processing. This was confirmed and extended by in vitro studies using the plant enzyme in which the precursor was made in wheat germ extracts and post-translationally imported by isolated chloroplasts [82,83].

Similar observations were made for mitochondria. Pulse-labelled yeast spheroplasts were shown to contain precursor forms of three subunits of the proton ATPase. This precursor could be chased into the mature form [84,85]. An in vitro system was also developed to assay mitochondrial import. Precursors were synthesized in a reticulocyte lysate which was filtered to remove small molecules, then incubated with isolated mitochondria under various conditions. Mitochondria were collected by centrifugation and the supernatant and pellet analyzed by immunoprecipitation, gel electrophoresis and fluorography [86]. This import has been shown to reflect correctly the in vivo import process because precursors are correctly processed [87], refractory to high concentrations of GTP (see below), imported proteins are protected from protease digestion while the precursors are digested, the energy dependence is the same as in vivo (see below), preprocessed precursors are not imported, and polypeptides are imported into the correct suborganellar compartment. It will be noticed that the 'green' in vitro experiments use wheat germ extracts and the 'brown' in vitro experiments use reticulocyte lysate: this is due to the fact that reticulocyte lysate destroys chloroplast integrity, and many wheat germ extracts process mitochondrial precursors.

Import into chloroplasts and mitochondria is energy dependent. In vivo pulse labelling showed that precursors accumulate in yeast spheroplasts in which the mitochondria were de-energized by a combination of genetic lesions and specific inhibitors [33]. In vitro experiments with isolated chloroplasts also indicated that import into chloroplasts is energy dependent [88]. Import was stimulated by ATP or light and impaired by uncoupler and energy-transfer inhibitors in the light. It was concluded that molecular ATP is required for import into the chloroplast stroma and thylakoids. The situation is somewhat different in mitochondria. Here it was shown that a membrane potential across the inner membrane and not ATP itself is necessary for import [86,89]. The fact that rho^- yeast strains (strains containing no functional electron transport chain and proton ATPase) still transport proteins into their defective mitochondria, suggests that the membrane potential needed is very low. Therefore, it is unlikely that this membrane potential provides the energy for transmembrane movement. What is the function of the membrane potential? Since both binding [76] and processing [90] are energy independent, the membrane potential may be necessary for the transmembrane movement. One of the possible functions is to orient a special membranous transporter in a functional conformation. This conformation could also lead to functional associations with yet another necessary component of the transport mechanism.

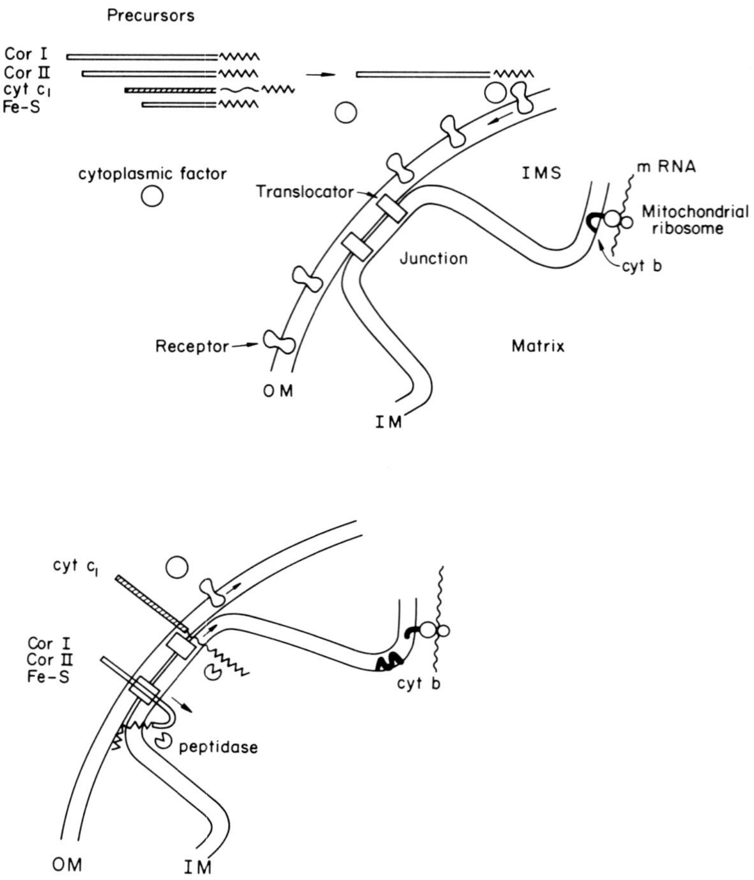

Fig. 12.5. Biogenesis and assembly of cytochrome b-c_1 complex in the inner mitochondrial membrane. Cytochrome b-c_1 complex contains at least five different subunits; COREI (corI), COREII (corII), nonheme iron protein (Fe-S), cytochrome c_1 (cyt c_1), and cytochrome b (cyt b). Cytochrome b is a mitochondrial gene product and is probably assembled into the inner membrane (IM) via vectorial translation by mitochondrial ribosomes. The other subunits are synthesized on cytoplasmic ribosomes as larger precursors. The precursors, perhaps in association with a 'cytoplasmic factor', are attached to receptors on the mitochondrial outer membrane (OM). The complex laterally diffuses to the junctions of the outer and inner membranes, and with the help of a hypothetical translocator the precursors are imported across the membrane. Pre-CorI, pre-CorII, and the pre-nonheme iron protein cross the two membranes, whereas cytochrome c_1 becomes anchored to the outer face of the inner membrane, facing the intermembrane space (IMS). Cytochrome b is assembled inside the inner membrane, and the nonheme iron protein and CorI and CorII are assembled into the matrix side of the inner membrane. The N-terminal extensions are removed by a soluble matrix protease. The N-terminal extension of cytochrome c_1 is removed in two steps; the first is catalyzed by the matrix protease and the second probably by a protease located on the outer face of the inner membrane.

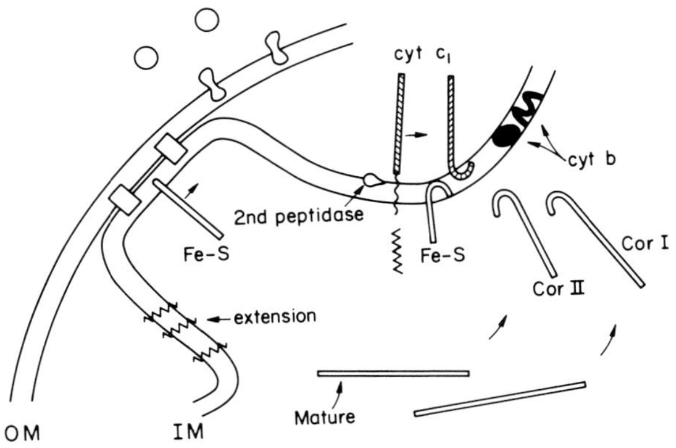

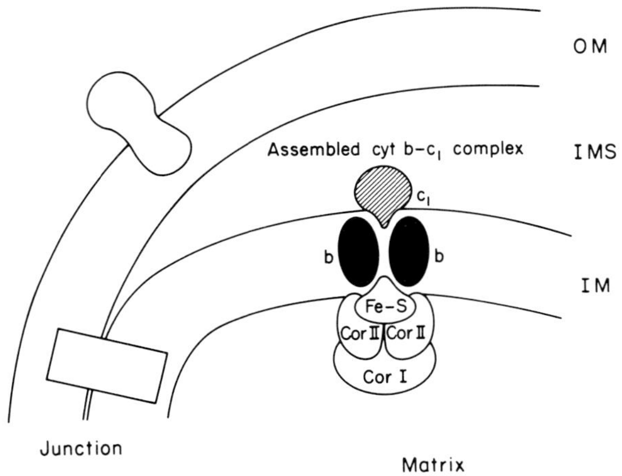

The difference between the energy requirement of chloroplasts and mitochondria for import is unclear. It may be the result of the difference in the nature of the membranes that are being crossed (a membrane with electron transport and oxidative phosphorylation activities in mitochondria and a membrane which mainly functions in solute translocation in chloroplasts). If this is the explanation, one may predict that plastocyanin, which is located in the cisternae of the thylakoids [91], would be imported in two steps; the first dependent on ATP and the final transthylakoid movement dependent on a membrane potential.

5.4. Processing of precursor and sorting into the correct compartment

In order to correctly process precursors to their mature form, a limited proteolysis reaction is necessary. Such a protease activity was first detected in chloroplasts [82] and was localized to a soluble compartment, most likely the stroma [83]. A similar protease activity was detected in mitochondria [90]. The mitochondrial processing protease is located in the matrix, is sensitive to metal chelators such as ortho-phenanthroline, EDTA and GTP, and stimulated by Zn^{2+} and Co^{2+} [90] or possibly Mn^{2+} [92]. This protease cleaves all mitochondrial precursor polypeptides thus far tested and no other proteins [90]. Processing is not energy dependent. The specificity and metal dependence of this protease have provided an important experimental advance. The protease has been used to preprocess precursors in order to show that the NH_2-terminal extension is neither necessary for binding [76] nor for import [35]. Since GTP inhibits the processing protease and is unable to cross the inner mitochondrial membrane, the addition of a large excess of GTP to the in vitro import assay will inhibit any processing protease which has leaked out of the mitochondria. This allows quantitative estimation of import efficiency by measuring processing alone (without protease treatment) [86].

The energy dependence of import into mitochondria has been exploited to accumulate large amounts of precursors in an uncoupler-poisoned living cell [93]. Precursor to the β-subunit of the proton ATPase has been purified from such cells by affinity chromatography on an antibody column, followed by chromatofocussing and isoelectric focussing. After renaturation, this precursor can be correctly processed by the matrix protease and can be imported into mitochondria, but only in the presence of a proteinaceous factor from the yeast cytoplasm [94]. A similar finding has been reported for import of precursors into rat liver mitochondria in which a factor is provided by the reticulocyte lysate [95,96].

In order to reach the correct subcellular compartment, precursors may follow different routes. Import into the outer membrane goes via 'protein incorporation' (see below). Import into the intermembrane space may go via two different pathways, one pathway exemplified by cytochrome *c* in which the precursor, apocytochrome *c*, binds to a specific receptor and is imported directly into the intermembrane space and processed by covalent heme attachment. This is an example of vectorial processing in which there is no NH_2-terminal extension nor proteolytic processing. Transport is not dependent on a membrane potential across the inner membrane but is dependent on heme attachment [77]. Cytochrome c_1 (on the outer face of the inner membrane), cytochrome b_2 and cytochrome *c* peroxidase (both soluble enzymes in the intermembrane space) go via a different route. All are processed by the matrix protease to intermediate forms [97] and then are subsequently processed by another protease (most likely located on the outer surface of the inner membrane). This two-step processing has been shown in vivo [97] and in vitro [74], and for cytochrome c_1 the second processing step is dependent on the presence of heme [74]. This example of vectorial processing is dependent on a membrane potential across the inner membrane. This energy dependence, along with the finding that the isolated matrix protease cleaves cytochrome b_2 and c_1 precursors to intermediate forms, shows that at least a portion of these precursors are imported into the mitochondrial matrix [98]. The intermediate forms of cytochrome *c* peroxidase and cytochrome *b* are embedded in mitochondrial inner membrane [97]. This proposed pathway is consistent with the amino acid sequence of the NH_2-terminal extension of cytochrome *c* peroxidase, where a long hydrophobic stretch flanked on both sides by basic sequences is present [73].

Import into the mitochondrial matrix and inner membrane may follow another route. Import is dependent on a membrane potential [86], probably goes via specific receptors on the outer membrane [76], and is usually accompanied by proteolytic processing of a precursor to the mature form. Fig. 12.5 summarizes the import pathways into the mitochondrial matrix, inner membrane and intermembrane space, exemplified by import of the various subunits of the cytochrome bc_1 complex.

6. *Protein incorporation*

One of the best-studied examples of protein incorporation is the mitochondrial outer membrane. In yeast and *Neurospora crassa* none of the outer membrane proteins studied thus far are made as larger precursors. All of these proteins are made on free ribosomes and are incorporated into the outer membrane post-translationally [35,69,99]. The import of the porin (a pore-forming protein) is time and temperature dependent but does not require energy [35,99]. The incorporation of porin in vitro was found to be membrane specific [35]. How is this membrane specificity determined, and what anchors the outer membrane protein to the outer membrane?

This problem has been approached using the 70 kDa mitochondrial outer membrane protein as an example. This protein is an integral outer membrane protein which has a large domain (at least 60 kDa) exposed to cytoplasm. This domain can be released intact from mitochondria or outer membrane by light protease treatment [75]. The gene for this protein has been cloned [100] and sequenced [101]. This gene has been destroyed by integrative transformation to create a mutant lacking the 70 kDa protein. While the precise function of this protein is unknown, the mutant has a definable phenotype which can be complemented by the intact gene carried on a plasmid. A truncated gene, coding for a 50 kDa protein, was found during the selection of the 70 kDa protein gene. This protein carries a large COOH-terminal deletion, yet it is properly targeted and probably incorporated into the outer membrane in vivo, demonstrating that the COOH terminus of this protein is not necessary for targeting to the outer membrane. This, in conjunction with the protease treatment described above and with the sequence, shows that the anchor of the 70 kDa protein is at its extreme NH_2-terminus and suggests that this same region of the molecule may act as an addressing signal. While the truncated protein is properly transported, it is unable to complement the 70 kDa mutation, indicating that the COOH terminus is necessary for function [100,101]. The construction of additional deletion mutations and expression of the modified proteins in yeast should give further insight into the mechanism of protein incorporation.

7. *Assembly of functional protein complexes*

Once proteins are transported into their proper location they must be assembled into well-defined protein complexes [102,103]. These protein complexes are characterized by a fixed subunit stoicheiometry, yet most of these complexes are composed of proteins encoded in both nuclear and organellar genomes. Moreover, the gene dosage for organellar genes is usually much higher than for the nuclear genes. This poses a challenging problem to understand how the protein complexes are correctly assembled, always maintaining proper stoicheiometry.

In vitro studies with isolated organelles indicate the possibility that newly synthesized subunits can be correctly assembled [104–107] into large complexes. However, none of these studies could eliminate the possibility that newly synthesized subunits merely exchanged places with pre-assembled ones. This leaves some doubt as to whether the in vitro assembly pathway is the same as in vivo. Also, these studies do not address the question of how correct subunit stoicheiometry is determined nor can they discover the order of assembly events.

One approach to study the control of subunit stoicheiometry is to artificially increase the amount of individual subunit by increasing the gene dosage. These experiments are possible because several mitochondrial and chloroplast genes have been cloned, and it is possible to introduce multicopy plasmids into yeast. The general finding from these types of experiments is that the overproduced subunit is properly imported into the correct compartment but cannot be assembled for the lack of its counterparts. Some of these overproduced, imported proteins are rapidly degraded and others seem to be stable [108,109]. In all cases tested thus far, the overproduction of one subunit does not influence the amounts of other subunits of the same complex. This makes it very unlikely that subunit stoicheiometry is determined by modulating protein import.

While the in vitro studies on assembly have provided relatively little information, in vivo data can give us some suggestions on possible pathway of assembly. For a long time it has been known that in rho^- yeast cells, where cytochrome *b* is not produced, cytochrome c_1 is still accumulated in the inner membrane. This agrees with the plasmid studies of overproduction. On the other hand the cytoplasmically synthesized subunits of cytochrome *c* oxidase accumulate in much lower quantities in the absence of subunits I, II and III, which are mitochondrial products. It is unlikely that this diminished accumulation is due to substantially reduced gene expression. This may indicate that certain subunits are stabilized by their counterparts.

8. *Regulation of membrane formation*

In order to understand the mechanism of membrane formation, one must first understand the environmental condition which influences membrane biogenesis. The two most obvious effectors of mitochondria and chloroplasts are O_2 and light, respectively. Under anaerobic conditions, mitochondria are de-differentiated into respiration-deficient organelles termed 'promitochondria' [110]. Upon introduction of O_2 a rapid development of the inner mitochondrial membrane occurs together with a rapid increase in respiration [111]. A similar developmental change occurs in chloroplast structure and function controlled by light [112]. Concomitant with the light-induced appearance of chlorophyll, a massive development of the thylakoid membrane occurs.

We know that chloroplasts and mitochondria are never synthesized de novo; their membranes grow by insertion of new elements into pre-existing membranes. How is the insertion of these new elements controlled? Are pre-existing elements inserted into the membrane or are newly synthesized elements assembled into the membrane? Is the insertion of these new elements controlled post-translationally or on the level of gene expression? Are all of the different complexes coordinately controlled and assembled at the same rate? Is assembly of a single protein complex the result of an ordered or a concerted mechanism?

Much of the control of the synthesis of mitochondrial proteins may be at the level of transcription. In many cases increased gene dosage gives a corresponding increase in gene expression [100,107,108]. Perhaps the clearest case of transcriptional control of protein synthesis is the effect of glucose-repression on the transcription of cytochrome *c*. In these studies the β-galactosidase gene was fused with those parts of the iso-1-cytochrome *c* gene which are upstream of the initiating ATG codon. As a result, expression of β-galactosidase was controlled by the cytochrome *c* promoter. It was suggested that there are two promoters used to synthesize cytochrome *c* mRNA. A weak promoter is utilized when yeast cells are grown on glucose as a carbon source and a stronger promoter is used when galactose is the carbon source [113]. The amounts of mRNAs for subunits of the cytochrome bc_1 complex are also greatly increased by growth on lactate compared with glucose. This has been exploited to clone several glucose-repressed nuclear-coded mitochondrial protein genes [100,107,108]. Until now, there was no good evidence for any post-transcriptional regulation of gene expression for nuclear-coded mitochondrial proteins. Perhaps the best candidate for an effector of membrane biogenesis that would act post-transcriptionally is heme. In a heme-deficient yeast mutant, cytochrome c_1 accumulates as its intermediate form, and most likely cytochrome *c* cannot be imported [77,98].

Are the different complexes of the mitochondria coordinately regulated? The answer is no, not all of them. It is well known that large amounts of the proton ATPase complex are present in promitochondria, whereas cytochrome oxidase is barely detectable [114]. Fig. 12.6 illustrates this for several mitochondrial polypeptides. Yeast cells were grown under anaerobic conditions and then shifted to an aerobic environment. At several time points, samples were taken and the relative amounts of different polypeptides quantified by electrotransfer and immunodecoration [115,116]. As can be seen, during an 8 h period of aerobic shift, the 29 kDa protein level was not changed, the β subunit of the proton ATPase complex was nearly doubled, whereas the level of cytochrome c_1 was increased over 10-fold. Not only are different complexes differentially regulated but also subunits within a single complex show different patterns of appearance. During the 8 h shift to an aerobic environment, subunits II, III and VII of cytochrome *c* oxidase appear earlier than subunits V and VI, and these appear earlier than subunit IV. This in vivo study may help to shed light on the regulation and pattern of assembly of the cytochrome *c* oxidase complex. This suggests that, at least for cytochrome *c* oxidase, assembly may be the result of a combination of concerted and ordered assembly; some subunits may only assemble together in a concerted fashion, whereas others are added to pre-existing subcomplexes.

Most of the studies on chloroplast biogenesis were performed with dark-grown seedlings or algae, which undergo greening on exposure to light. This system is somewhat artificial and may not be fully homologous to the chloroplast biogenesis under continuous light. Dark-grown seedlings contain protochlorophyllide, but chlorophyll is not present. Upon illumination the pre-existing protochlorophyllide undergoes phototransformation to chlorophyll *a*. After a lag phase a rapid chlorophyll accumulation takes place. If at any point the seedlings are returned to

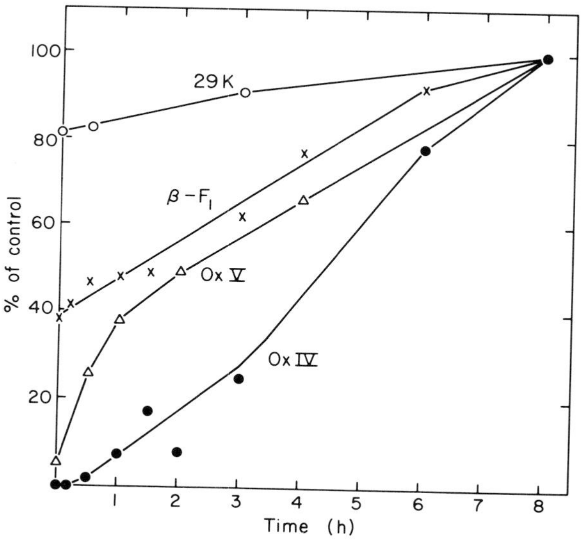

Fig. 12.6. The onset of synthesis of various mitochondrial polypeptides upon transferring anaerobically grown yeast cells to aerobic conditions. Yeast cells were grown overnight under anaerobic conditions. At time zero they were transferred to aerobic conditions, and at the indicated time periods samples of cells were removed and lysed in the presence of NaOH and mercaptoethanol. Samples containing about 50 μg of protein were electrophoresed in a sodium dodecyl sulfate-polyacrylamide gel. The proteins were electrotransferred to nitrocellulose sheets and decorated with specific antibodies and ^{125}I-labelled protein A. Paper pieces corresponding to the labelled protein spots were cut out from the immune blot and counted in a γ counter. The amount of counts obtained in the samples of 8 h aerobic conditions was taken as 100%. The antibodies used were directed against the following polypeptides: porin of the mitochondrial outer membrane (29 k); β subunit of the proton-ATPase (β-F_1); subunit IV of cytochrome *c* oxidase (OxIV) and subunit V of cytochrome *c* oxidase (OxV).

darkness, chlorophyll accumulation stops. If light is turned on again within a few hours, chlorophyll accumulation starts again. Concerning the light control of the chloroplast biogenesis, several questions can be raised:

Which photoreceptors are involved in light-induced chloroplast development, and how do they function?

What are the sequential events that lead to the synthesis of the various components necessary for photosynthesis?

How is the photosynthetic apparatus coordinately assembled?

Three main factors are involved in the regulation of the chloroplast biogenesis: direct light which triggers synthetic reactions within the organelle; the phytochrome controlling gene expression in the nucleus; and phytohormones (cytokinins) that may function in communicating signals between the nucleus and the chloroplast. It was observed that red light induced the formation of chlorophyll in etiolated plants, and far-red light suppressed the effect of red light [117]. The action spectrum for these effects was consistent with a phytochrome-controlled process [118]. The function of phytochrome is likely to be in the regulation of nuclear gene expression [119]. Recent studies with isolated nuclei of light- and dark-grown pea plants revealed that the light effect is mediated by an increase in transcription rather than by decrease in RNA degradation [120]. One of the systems which is under this phytochrome-mediated control is that of the apoproteins of the chlorophyll *a/b* light-harvesting complex [119]. It was shown that the light-induced synthesis, transport and assembly of the apoproteins will proceed in a barley mutant which lacks chlorophyll *b* [121]. Yet the polypeptides are not present in isolated mutant chloroplasts due to a rapid turnover in the absence of chlorophyll *b*. It was also reported that in the dark, when chlorophyll synthesis does not occur, the major polypeptide of the pea light-harvesting complex is rapidly turning over [122]. These observations show lack of communication between events that take place in the chloroplast with those that are proceeding in the nucleus. Algae may lack phytochrome but possess a blue-absorbing photoreceptor [123].

The formation of thylakoids in response to the illumination of etiolated leaves is not reversed by far-red light. Two 'photoreceptors' are functioning in this process within the chloroplast: (1) the protochlorophyllide of which its photoconversion into chlorophyllide may trigger the biogenesis of the thylakoid membrane; and (2) the developing photosynthetic activity which may provide some necessary substances as well as signals for further development [124]. Therefore it seems as if the delicate control of the chloroplast biogenesis is in the hands of the nucleus. A brief exposure to red light is sufficient to trigger the chloroplast biogenetic events that are coded in the nucleus. On the other hand, a continuous illumination is required to enhance the biogenetic events that take place in the chloroplasts. The coordinated assembly of the chloroplast looks more like a miracle than something that may be readily explained.

The transcriptions of several of the chloroplast genes are induced by light. The best studied example is the 32 kDa polypeptide of photosystem II [21]. It seems as if light induces the removal of a repressor or promotes the initiation of transcription by yet unknown mechanisms. There are indications that cytokinins may have similar effects on gene expression. Light has no effect on many other genes that code for chloroplast polypeptides. This phenomenon is reflected by the presence of several competent protein complexes in etioplasts, including ribosomes that are capable of protein synthesis. Therefore, tens of different polypeptides and several RNA species should be coordinately assembled in the proplastid prior to the light-induced events. Moreover, plastids of dark-grown leaves contain several components of the thylakoid membrane. Recently we looked for the onset of assembly of individual subunits into protein complexes following illumination of etiolated seedlings [125]. Fig. 12.7 depicts a quantitative determination of relative amounts of individual subunits during the greening of etiolated spinach seedlings. The following conclusions could be drawn. The protein complexes which are not involved in photochemical reactions were present in etiolated leaves and their amounts did not significantly change along a greening period of 24 h. On the other hand the synthesis of polypeptides of photosystem I reaction center, photosystem II reaction center, and the light harvesting complexes were induced by light. While the subunits of the proton-ATPase complex were assembled in a concerted fashion, a sequential synthesis and assembly of photosystem I reaction center was observed. Subunit II which is a cytoplasmic product was the first to be synthesized after exposing the plants to light. The synthesis of subunits III, IV and V followed the synthesis of subunit II in this order. Subunit I is the heart of the system, and it contains both the primary electron donor

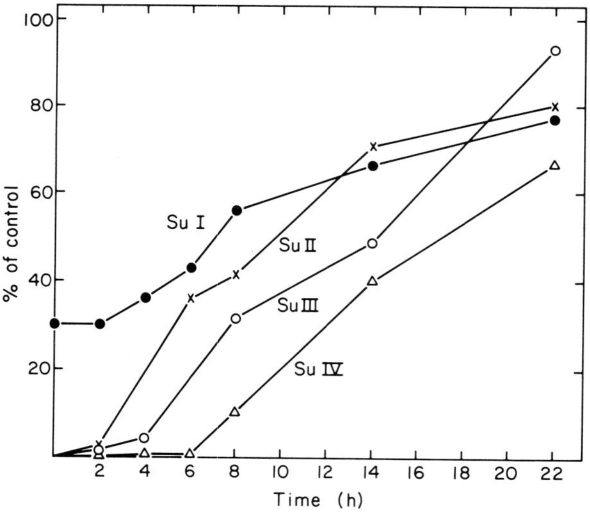

Fig. 12.7. The onset of synthesis of the various subunits of photosystem I reaction center after the illumination of etiolated spinach seedlings. Spinach seedlings were grown in the dark. At time zero they were transferred to light. Leaves were removed at the indicated time points and processed as was described in Fig. 12.6 for yeast cells. The antibodies were against individual subunits of photosystem I reaction center. The amount of each subunit in light-grown spinach plants was taken as 100% [125].

and acceptor as well as a primary light harvesting system. This subunit was present in etioplasts, probably assembled in the membrane. The first event after illumination may be the formation of chlorophyll *a* molecules and incorporation into subunit I. This will render the unit to be photochemically active. Since the cytochrome b_6/f

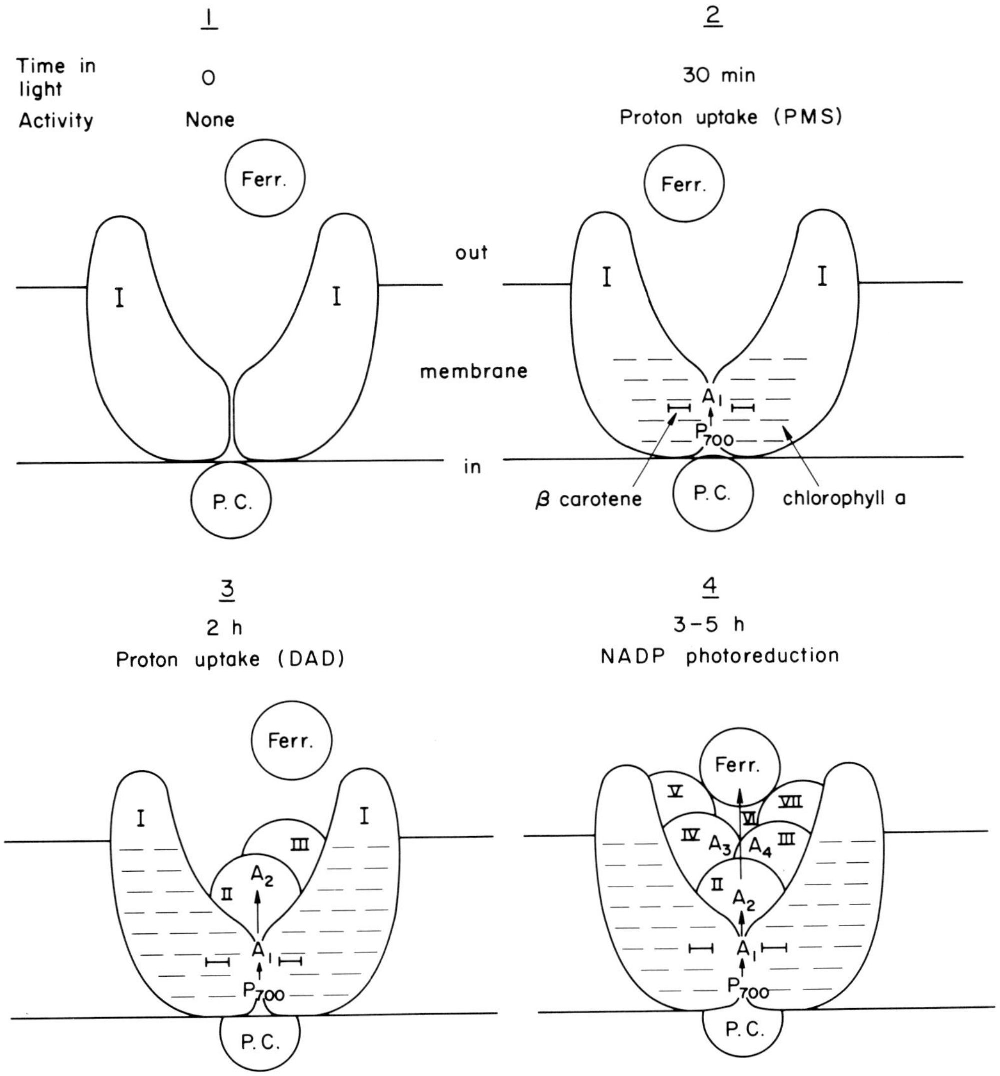

Fig. 12.8. Schematic representation of events occurring during biogenesis of photosystem I reaction center. The subunits are designated as I to VII, the abbreviations are: Ferr, ferredoxin; P.C., plastocyanin; A_1, A_2, A_3 and A_4, primary, secondary, tertiary and quaternary electron acceptors; PMS, phenazine methosulfate; DAD, diaminodurine.

complex, ferredoxin and plastocyanin are present in etioplasts, the system is competent in electron transport from reduced plastoquinone to oxygen. Shortly thereafter, perhaps with the appearance of subunit III, the system may function in cyclic photophosphorylation with diaminodurine (DAD). Later, when more subunits join the complex and photosystem II is being assembled, NADP photoreduction and water oxidation may take place. In contrast, the proton ATPase complex appears to be assembled in a concerted fashion, perhaps because partial assembly of the complex may cause a bioenergetic disaster due to a proton leak [106,126].

We have now uncovered only a few of the pieces from the puzzle of the chloroplast biogenesis. The following sequential events can be drawn.

(1) In etioplasts there are pre-existing functional protein complexes.

(2) Some nuclear and plastid genes are expressed in the dark. Most of the gene products will be correctly assembled in the chloroplast. Some of them will not be assembled and will be degraded.

(3) Brief exposure to light will induce mRNA formation in the nucleus, mediated by phytochrome action.

(4) Longer light reaction is required to convert the protochlorophyllide to chlorophyll *a*.

(5) At the same time several 'photogenes' will be induced in the chloroplast chromosome.

(6) Partial reactions take place in which cytochrome b_6/f complex and partially assembled photosystem I reaction center are involved.

(7) The various protein complexes are coordinately assembled and steady-state biogenesis takes over.

Fig. 12.8 depicts some steps in the light-induced assembly of the photosystem I reaction center. In this case, subunit I exists in etioplasts as an inactive unit. Light induces its initial activity within a few minutes, and it takes hours to synthesize and assemble the fully functional protein complex.

Genetic engineering and transformation of plant cells are likely to yield exciting results in the near future, providing that questions of biological interest will be asked and that biologists will not be transformed into engineers.

References

1 Margulis, L. (1970) Origin of Eukaryotic Cells, Yale University Press, New Haven and London.
2 Raven, P.H. (1970) Science 169, 641–646.
3 Schatz, G. and Mason, T.L. (1974) Annu. Rev. Biochem. 43, 51–87.
4 Chua, N.-H. and Schmidt, G.W. (1979) J. Cell Biol. 81, 461–483.
5 Alt, J., Westhoff, P., Sears, B.B., Nelson, N., Hurt, G., Hauska, G. and Hermann, R.G. (1983) EMBO J. 2, 979–986.
6 Raff, R.A. and Mahler, H.R. (1972) Science 177, 575–582.
7 Bogorad, L. (1975) Science 188, 891–898.
8 Ludwig, B. and Schatz, G. (1980) Proc. Natl. Acad. Sci. U.S.A. 77, 196–200.
9 Ludwig, B. (1980) Biochim. Biophys. Acta 594, 177–189.
10 Trench, R.K. (1979) Annu. Rev. Plant Physiol. 30, 485–531.

11 Trench, R.K. (1981) In On the Origins of Chloroplasts (Schiff, J.A., ed.) pp. 55–75, Elsevier/North Holland, Amsterdam.
12 Lewin, R.A. (1976) Nature 261, 697–698.
13 Ellis, R.J. (1977) Biochim. Biophys. Acta 463, 185–215.
14 Nechushtai, R. and Nelson, N. (1981) J. Biol. Chem. 256, 11624–11628.
15 Westhoff, R., Nelson, N., Bunemann, H. and Herrmann, R.G. (1981) Curr. Gen. 4, 103–120.
16 Anderson, S., Bankier, A.T., Barrell, B.G., deBruijn, M.H.L., Coulson, A.R., Drouin, J., Eperon, I.C., Nierlich, D.P., Roe, B.A., Sanger, F., Schreier, P.H., Smith, A.J.H., Staden, R. and Young, I.G. (1981) Nature 290, 457–465.
17 Macino, G., Scazzocchio, C., Waring, R.B., McPhail-Berks, M. and Davies, R.W. (1980) Nature 288, 404–406.
18 Dale, R.M.K. (1981) Proc. Natl. Acad. Sci. U.S.A. 78, 4453–4457.
19 Forde, B.G., Oliver, R.J.C. and Leaver, C.J. (1978) Proc. Natl. Acad. Sci. U.S.A. 75, 3841–3845.
20 Bedbrook, J.R. and Kalodner, R. (1979) Annu. Rev. Plant Physiol. 30, 593–620.
21 Edelman, M. (1981) In The Biochemistry of Plants, Vol. 6 (Marcus, A., ed.) pp. 249–301, Academic Press, New York.
22 Bonitz, S.G., Berlani, R., Coruzzi, G., Li, M., Macino, G., Nobrega, F.G., Nobrega, M.P., Thalenfeld, B.E. and Tzagaloff, A. (1980) Proc. Natl. Acad. Sci. U.S.A. 77, 3167–3170.
23 Fox, T.D. and Leaver, C.J. (1981) Cell 26, 315–323.
24 Weil, J.H. (1979) In Nucleic acids in plants (Hall, T.C. and Davies, J.W., eds.) Vol. I, pp. 143–192, CRC Press, Boca Raton.
25 Ojala, D., Montoya, J. and Attardi, G. (1981) Nature 290, 470–474.
26 Westhoff, P., Alt, J., Nelson, N., Bottomley, W., Bunemann, H. and Herrmann, R.G. (1983) J. Plant Mol. Biol. 2, 95–107.
27 Montoya, J., Ojala, D. and Attardi, G. (1981) Nature 290, 465–470.
28 Hermann, R.G., Westhoff, P., Alt, J., Winter, P., Tittgen, S., Bisanz, C., Sears, B.B., Nelson, N., Hurt, E., Hauska, G., Viebrock, A. and Sebald, W. (1982) In Structure and Function of Plant Genomes (Cifferi, O., ed.) pp. 143–154, Academic Press, New York.
29 Ellis, R.J. (1983) In Subcellular Biochemistry (Roodyn, D.B., ed.) pp. 237–261, Plenum, New York.
30 Nelson, N. (1981) Curr. Top. Bioenerg. 11, 1–33.
31 Blobel, G. and Dobberstein, B. (1975) J. Cell Biol. 67, 835–851.
32 Blobel, G. and Dobberstein, B. (1975) J. Cell Biol. 67, 852–862.
33 Nelson, N. and Schatz, G. (1979) Proc. Natl. Acad. Sci. U.S.A. 76, 4365–4369.
34 Schatz, G. (1979) FEBS Lett. 103, 201–211.
35 Gasser, S.M. and Schatz, G. (1983) J. Biol. Chem. 258, 3427–3430.
36 Wickner, W. (1979) Annu. Rev. Biochem. 48, 23–45.
37 Rachubinski, R.A., Verma, D.P.S. and Bergeron, J.J.M. (1980) J. Cell Biol. 84, 705–716.
38 Njus, D. and Radda, G.K. (1978) Biochim. Biophys. Acta 463, 219–244.
39 Njus, D., Knoth, J. and Zallakian, M. (1981) Curr. Top. Bioenerg. 11, 107–147.
40 Johnson, R.G., Carlson, N.J. and Scarpa, A. (1978) J. Biol. Chem. 253, 1512–1521.
41 Gabizon, R., Yetinson, T. and Schuldiner, S. (1982) J. Biol. Chem. 257, 15145–15150.
42 Njus, D., Knoth, J., Cook, C. and Kelley, P.M. (1983) J. Biol. Chem. 258, 27–30.
43 Creutz, C.E., Pazoles, C.J. and Pollard, H.B. (1979) J. Biol. Chem. 254, 553–558.
44 Hong, K., Duzgunes, N. and Papahadjopoulos, D. (1981) J. Biol. Chem. 256, 3641–3644.
45 Winkler, H. (1977) Neuroscience 2, 657–683.
46 Abbs, M.T. and Phillips, J.H. (1980) Biochim. Biophys. Acta 595, 200–221.
47 Anholt, R., Lindstrom, J. and Montal, M. (1984) In The Enzymes of Biological Membranes (Martonosi, A., ed.) Vol. III, pp. 335–401, Plenum Press, New York.
48 Lindstrom, J., Merlic, J. and Yoguswaran, O. (1979) Biochemistry 18, 4465–4470.
49 Raftery, M.A., Hunkapiller, M.W., Strader, C.D. and Hood, L.A. (1980) Science 208, 1454–1457.
50 Huganir, R.L., Schell, M.A. and Racker, E. (1979) FEBS Lett. 108, 155–160.
51 Nelson, N., Anholt, R., Lindstrom, J. and Montal, M. (1980) Proc. Natl. Acad. Sci. U.S.A. 77, 3057–3061.

52 Tsartos, S.J. and Lindstrom, J.M. (1980) Proc. Natl. Acad. Sci. U.S.A. 77, 755–759.
53 Noda, M., Takahashi, H., Tanabe, T., Toyosato, M., Furulani, Y., Hirose, T., Asai, M., Inayama, S., Miyata, T. and Numa, S. (1982) Nature 299, 793–797.
54 Noda, M., Takahashi, H., Tanabe, T., Toyosato, M., Kikyotani, S., Hirose, T., Asai, M., Takashima, H., Inayama, S., Miyata, T. and Numa, S. (1983) Nature 301, 251–255.
55 Einarson, B., Gullick, W., Conti-Tronconi, B., Ellisman, M. and Lindstrom, J. (1982) Biochemistry 21, 5295–5302.
56 Anderson, D.J. and Blobel, G. (1981) Proc. Natl. Acad. Sci. U.S.A. 78, 5598–5602.
57 Anderson, D.J., Walter, P. and Blobel, G. (1982) J. Cell Biol. 93, 501–506.
58 Walter, P. and Blobel, G. (1981) J. Cell Biol. 91, 551–556.
59 Walter, P. and Blobel, G. (1981) J. Cell Biol. 91, 557–561.
60 Walter, P., Ibrahimi, I. and Blobel, G. (1981) J. Cell Biol. 91, 545–550.
61 Meyer, D.I., Krause, E. and Dobberstein, B. (1982) Nature 297, 647–650.
62 Cidon, S. and Nelson, N. (1983) J. Biol. Chem. 258, 2892–2898.
63 Cidon, S., Ben-David, H. and Nelson, N. (1983) J. Biol. Chem. 258, 11684–11688.
64 Zhang, F. and Schnider, D.L. (1983) Biochem. Biophys. Res. Commun. 114, 620–625.
65 Stevens, T., Esmon, B. and Schekman, R. (1982) Cell 30, 439–448.
66 Fischer, H.D., Gonzalez-Noriega, A., Sly, W.S. and Morre, D.J. (1980) J. Biol. Chem. 255, 9608–9615.
67 Gonzalez-Noriega, A., Grubb, J.H., Talkad, V. and Sly, W.S. (1980) J. Cell Biol. 85, 839–852.
68 Raymond, Y. and Shore, G.C. (1979) J. Biol. Chem. 254, 9335–9338.
69 Suissa, M. and Schatz, G. (1982) J. Biol. Chem. 257, 13048–13055.
70 Butow, R.A. and Schatz, G. (1983) Cell 32, 316–318.
71 Viebrock, A., Perz, A. and Sebald, W. (1982) EMBO J. 1, 565–571.
72 Faye, G. and Simon, M. (1983) Cell 32, 77–87.
73 Kaput, J., Goltz, S. and Blobel, G. (1982) J. Biol. Chem. 257, 15054–15058.
74 Gasser, S.M., Ohashi, A., Daum, G., Bohni, P.C., Gibson, J., Reid, G.A., Yonetani, T. and Schatz, G. (1982) Proc. Natl. Acad. Sci. U.S.A. 79, 267–271.
75 Riezman, H., Hay, R., Gasser, S., Daum, G., Schneider, G., Witte, C. and Schatz, G. (1983) EMBO J. 2, 1105–1111.
76 Riezman, H., Hay, R., Witte, C., Nelson, N. and Schatz, G. (1983) EMBO J. 2, 1113–1118.
77 Hennig, B. and Neupert, W. (1981) Eur. J. Biochem. 121, 203–212.
78 Matsumura, S., Arpin, M., Hannum, C., Margoliash, E., Sabatini, D.D. and Morimoto, T. (1981) Proc. Natl. Acad. Sci. U.S.A. 78, 4368–4372.
79 Hennig, B., Koehler, H. and Neupert, W. (1983) Proc. Natl. Acad. Sci. U.S.A. 80, 4963–4967.
80 Pfisterer, J., Lachmann, P. and Kloppstech, K. (1982) Eur. J. Biochem. 126, 143–148.
81 Dobberstein, B., Blobel, G. and Chua, N.-H. (1977) Proc. Natl. Acad. Sci. U.S.A. 74, 1082–1085.
82 Highfield, P.E. and Ellis, R.J. (1978) Nature 271, 420–424.
83 Chua, N.-H. and Schmidt, G. (1978) Proc. Natl. Acad. Sci. U.S.A. 75, 6110–6114.
84 Maccecchini, M.L., Rudin, Y., Blobel, G. and Schatz, G. (1979) Proc. Natl. Acad. Sci. U.S.A. 76, 343–347.
85 Cote, C., Solioz, M. and Schatz, G. (1979) J. Biol. Chem. 254, 1437–1439.
86 Gasser, S.M., Daum, G. and Schatz, G. (1982) J. Biol. Chem. 257, 13034–13041.
87 Cerletti, N., Bohni, P.C. and Suda, K. (1983) J. Biol. Chem. 258, 4944–4949.
88 Grossman, A., Bartlett, S. and Chua, N.-H. (1980) Nature 285, 625–628.
89 Zimmermann, R. and Neupert, W. (1980) Eur. J. Biochem. 109, 217–229.
90 Böhni, P.C., Daum, G. and Schatz, G. (1983) J. Biol. Chem. 258, 4937–4943.
91 Hauska, G.A., McCarty, R.E., Berzborn, R.J. and Racker, E. (1971) J. Biol. Chem. 246, 3524–3531.
92 McAda, P.C. and Douglas, M.G. (1982) J. Biol. Chem. 257, 3177–3182.
93 Reid, G.A. and Schatz, G. (1982) J. Biol. Chem. 257, 13056–13061.
94 Ohta, S. and Schatz, G. (1984) EMBO J. 3, 651–658.
95 Argan, C., Lusty, C.J. and Shore, G.C. (1983) J. Biol. Chem. 258, 6667–6670.
96 Miura, S., Mori, M. and Tatibana, M. (1983) J. Biol. Chem. 258, 6671–6674.

97 Reid, G.A., Yonetani, T. and Schatz, G. (1982) J. Biol. Chem. 257, 13068–13074.
98 Ohashi, A., Gibson, J., Gregor, I. and Schatz, G. (1982) J. Biol. Chem. 257, 13042–13047.
99 Freitag, H., Janes, M. and Neupert, W. (1982) Eur. J. Biochem. 126, 197–202.
100 Riezman, H., Hase, T., van Loon, A.P.G.M., Grivell, L., Suda, K. and Schatz, G. (1983) EMBO J. 2, 2161–2168.
101 Hase, T., Riezman, H., Suda, K. and Schatz, G. (1983) EMBO J. 2, 2169–2172.
102 Bengis, C. and Nelson, N. (1975) J. Biol. Chem. 250, 2783–2788.
103 Nelson, N. and Cidon, S. (1984) J. Bioenerg. Biomembr. 16, 11–36.
104 Chua, N.-H. and Schmidt, G.W. (1978) Proc. Natl. Acad. Sci. U.S.A. 75, 6110–6114.
105 Smith, S.M. and Ellis, R.J. (1979) Nature 278, 662–664.
106 Nelson, N., Nelson, H. and Schatz, G. (1980) Proc. Natl. Acad. Sci. U.S.A. 77, 1361–1364.
107 Lewin, A.S. and Norman, D.K. (1983) J. Biol. Chem. 258, 6750–6755.
108 van Loon, A.P.G.M., de Groot, R.J., van Eyk, E., van der Horst, G.T.S. and Grivell, L.A. (1982) Gene 20, 323–337.
109 Emir, S.D., Schikman, R., Flessel, M.C. and Thorner, J. (1983) Proc. Natl. Acad. Sci. U.S.A. 80, 7080–7084.
110 Criddle, R.S. and Schatz, G. (1969) Biochemistry 8, 322–334.
111 Rouslin, W. and Schatz, G. (1969) Biochem. Biophys. Res. Commun. 37, 1002–1007.
112 Ohad, I., Siekevitz, P. and Palade, G.E. (1967) J. Cell Biol. 35, 521–584.
113 Guarente, L. and Mason, T.L. (1983) Cell 32, 1279–1286.
114 Woodrow, G. and Schatz, G. (1979) J. Biol. Chem. 254, 6088–6093.
115 Nelson, N. (1983) Methods Enzymol. 97, 510–523.
116 Riezman, H. (1984) Curr. Top. Bioenerg. 13, 257–280.
117 Price, L. and Klein, W.H. (1961) Plant Physiol. 36, 733–735.
118 Virgin, H.I. (1961) Physiol. Plant 14, 439–452.
119 Apel, K. (1979) Eur. J. Biochem. 97, 183–188.
120 Gallagher, T.F. and Ellis, R.J. (1982) EMBO J. 1, 1493–1498.
121 Bellemare, G., Bartlett, S.G. and Chua, N.-H. (1982) J. Biol. Chem. 257, 7762–7767.
122 Bennett, J. (1981) Eur. J. Biochem. 118, 61–70.
123 Egan, J.M., Dorsky, D. and Schiff, J.A. (1975) Plant Physiol. 56, 318–323.
124 Bradbeer, J.W. (1981) In The Biochemistry of Plants (Hatch, M.D. and Boardman, N.K., eds.) Vol. 8, pp. 423–472, Academic Press, New York.
125 Nechushtai, R. and Nelson, N. (1984) Plant Mol. Biol. in press.
126 Nelson, N. (1981) In Energy Coupling in Photosynthesis (Selman, B.R. and Selman-Reimer, S., eds.) pp. 261–268, Elsevier, Amsterdam.

Subject Index